Lecture Notes in Mathematics

Edited by A. Dold and B. Eckmann

1185

Group Theory, Beijing 1984

Proceedings of an International Symposium held in Beijing, Aug. 27–Sep. 8, 1984

Edited by Tuan Hsio-Fu

Springer-Verlag
Berlin Heidelberg New York Tokyo

Editor

TUAN Hsio-Fu
Department of Mathematics, Peking University
Beijing, The People's Republic of China

Mathematics Subject Classification (1980): 05-xx, 12F-xx, 14Kxx, 17Bxx, 20-xx

ISBN 3-540-16456-1 Springer-Verlag Berlin Heidelberg New York Tokyo
ISBN 0-387-16456-1 Springer-Verlag New York Heidelberg Berlin Tokyo

Printed in Germany

Printing and binding: Beltz Offsetdruck, Hemsbach/Bergstr.
2146/3140-543210

PREFACE

From August 27 to September 8, 1984 there was held in Peking University, Beijing an International Symposium on Group Theory.

As well said by Hermann Weyl: "Symmetry is a vast subject significant in art and nature. Whenever you have to do with a structure endowed entity, try to determine the group of those transformations which leave all structural relations undisturbed." This passage underlies that the group concept is one of the most fundamental and most important in modern mathematics and its applications. Group theory is indeed a vast branch of mathematics. The chief topics of the Symposium were finite groups and their connections with combinatorics, classical groups, algebraic groups and Lie groups (with emphasis on connections with algebra).

The Symposium was sponsored by our Ministry of Education. It was conducted by an Organization Committee consisting of seven Chinese professors of group theory:

- Tuan Hsio Fu (Duan Xuefu) (Peking University), Chairman
- Cao Xihua (East China Normal University), Vice Chairman
- Wan Zhexian (Institute of System Science, Academia Sinica), Vice Chairman
- Yen Zhida (Nankai University)
- Zhang Yuanda (Wuhan University)
- Zeng Kencheng (Graduate School, China University of Science and Technology)
- Wang Efang (Peking University)

Professor Hua Loo-Keng and Professor S.S.Chern kindly gave their advice and support to the Symposium.

The purpose of the Symposium was to provide a timely forum for the exchange of information and encouragement of international dialogue. For this purpose, we invited eight well-known mathematicians from abroad:

- John G. Thompson (University of Cambridge, U.K.)
- Michael Aschbacher (California Institute of Technology, U.S.A.)
- Charles W. Curtis (University of Oregon, U.S.A.)
- Donald Livingstone (University of Birmingham, U.K.)
- Alexander Hahn (University of Notre Dame, U.S.A.)
- E.R. Kolchin (Columbia University, U.S.A.)
- T.A. Springer (University of Utrecht, Netherlands)
- J.C. Jantzen (University of Bonn; University of Hamburg, FRG)

Each of them delivered a series of six one-hour lectures presenting an exposition and (or) providing an overview of recent advances on topics of current interest.

All together, there were fifty registered Chinese participants and twenty others. In general, registered participants and a few others gave a lecture (of half an hour to one hour) about their recent works either in a general meeting or in one of three sessions, emphasizing respectively on finite groups and combinatorics, classical groups, algebraic groups and Lie groups. In total, forty five papers were presented, including six joint works and two papers not read in person. On invitation, Professor Hua Loo-Keng gave an one-hour lecture entitled

Non-negative square matrices and planning economy.

In the Proceedings of the Symposium, contributions of seven of our eight foreign guests form Part A, while 9 papers of Chinese participants form Part B. In order to keep the Volume with a reasonable size, other papers of our Chinese participants are not included, and most of these have already appeared or will appear in other mathematical journals. However, a list of other Chinese participants with their affiliations and titles of their lectures will be given as an appendix for reference.

It is believed that the Symposium has achieved its purpose of international academic communication and would promote the research of group theory in China.

Thanks are due to Springer-Verlag for consenting to publish the Proceedings of the Symposium as a Volume in Lecture Notes in Mathematics.

Hsio-Fu Tuan (Duan Xuefu)
June 12, 1985

TABLE OF CONTENTS

FINITE SIMPLE GROUPS AND THEIR SUBGROUPS

Michael Aschbacher

California Institute of Technology

Pasadena, CA 91125, U.S.A.

INTRODUCTION

The material in this article corresponds roughly to the contents of six lectures given at the International Symposium on Group Theory at Peking University in September, 1984.

In essence the article describes the beginnings of a theory of permutation representations of finite groups based on the classification of the finite simple groups. Chapter 3 is devoted to an outline of the Classification, with emphasis on recent efforts to improve the proof of the Classification Theorem. Chapters 1, 2, and 6 discuss the finite simple groups themselves, a notion of geometry due to J. Tits, and a class of group theoretical techniques introduced by B. Fischer. Each of these topics plays a role in the theory of permutation representations under discussion. The heart of the theory is the study of the subgroup structure of the finite simple groups. Certain results on this structure are described in Chapters 4 and 5.

The author would like to thank the organizers of the Peking Symposium, particularly Professor Hsio-Fu Tuan, for all their efforts which made the conference and this article possible.

The author's work is partially supported by the National Science Foundation.

CHAPTER I

The Finite Simple Groups

I'm going to begin by advancing some basic principals of finite group theory which will serve has a foundation for these lectures.

One unifying notion is that of a group representation. I will use the term representation in a much broader sense than usual. Namely in these lectures a *representation* of a group G in a category C is a homomorphism of G into the automorphism group of some object of C. Among such representations the permutation representations, the linear representations, and the representations of groups on groups seem most fundamental.

As the groups we are considering are finite, their representations can be broken up into indecomposable or irreducible representations. Indecomposability can be defined precisely in terms of products and coproducts, but I do not wish to do so here. All we need to know is that each representation can be written in exactly one way as a finite sum of indecomposable representations which cannot themselves be decomposed. In the examples we'll consider it will be clear what the indecomposables are and what it means to write a representation as the sum of other representations.

There will also be a notion of irreducibility corresponding to the absence of G-invariant equivalence relations on the object. For example the simple groups act irreducibly on themselves in the category of groups and homomorphisms. Again we can associate to each representation a finite set of irreducible representations, but this time the set of irreducible constituents may not be uniquely determined and will not in general uniquely determine the representation. Still much information about the representation can be retrieved from its irreducible constituents.

Most questions about finite groups which interest me can be regarded as questions about representations of groups, where the term representation is used in the very general sense defined above. Moreover such questions are often best answered via a reduction to a question about irreducible permutation and linear representations of simple groups, followed by an appeal to the Classification Theorem and knowledge of the irreducible representations of the simple groups. This suggests to me that the most fundamental questions about finite groups are of the following three sorts:

1) Classification of the finite simple groups.

2) Description of the irreducible permutation representations and linear representations of the almost simple groups.

3) Generation of techniques to reduce questions about general finite groups to questions about irreducible representations of almost simple groups.

In addition there are the long term goals of better understanding finite groups,(Most particularly the simple groups) and of using finite group theory in other areas of mathematics.

I will be concentrating on several of the topics above which are closest to my own interests and expertise. In particular I will say something about the Classification, with emphasis on recent efforts to improve and simplify the proof of the Classification Theorem. I will also discuss work aimed at determining the irreducible permutation representations of the finite groups. I will devote a lecture to recent work on a class of geometries first introduced by J. Tits. This work is potentially of use in simplifying portions of the Classification and as a tool for describing the permutation representations of the simple groups. To set the stage, I'll spend the remainder of this lecture discussing the finite simple groups themselves, since they are the objects of interest in almost all that will follow.

The Classification Theorem says that every finite simple group is a member of one of the following classes of groups:

Groups of prime order

Groups of Lie Type

Alternating groups

26 sporadic simple groups

In some sense most of the finite simple groups are groups of Lie type, so I'm going to begin with those groups and devote most of my time to them. The groups of Lie type are the finite analogues of the semisimple Lie groups. The classical theory describing semisimple Lie groups generalizes to a large extent to the finite groups of Lie type, and is the principal tool available for investigating these groups. I will not attempt to describe this theory in any detail, leaving that to others speakers at this conference who are expert on the subject. I will however appeal to certain consequences of the theory.

There are at least five related descriptions of the groups of Lie type which I will discuss briefly, more or less in chronological order of appearance. The first of these describes only the so called classical groups. Let V be an n-dimensional vector space over a field F and f a bilinear form, hermitian symmetric sesquilinear form, or quadratic form on V. It will be assumed that either f is trivial or f is nondegenerate and satisfies one of several symmetry conditions. Denote by O(V,f) the isometry group of the space (V,f). For example if f is trivial then O(V,f) = GL(V) is the general linear group on V. Certain normal subgroups of O(V,f) and the images of these groups in the projective general linear group PGL(V) will be termed *classical groups*. Thus each classical group comes equipped with one or two interesting representations: the linear representation on V and the projective representation on the projective geometry of V. Of course in each case the classical group preserves the extra geometric structure defined by the form f. The classical groups are groups of Lie type although not all groups of Lie type are classical.

Assume next that L is a finite dimensional simple Lie algebra over the complex numbers. Associated to L is a root system Φ and a simple system π for Φ such that L possesses a so called *Chevalley basis* $(H_i, X_\alpha : i \in \pi, \alpha \in \Phi)$ with the property that for each $\alpha \in \Phi$ and each positive integer n, X_α is nilpotent and $X_\alpha^n/n!$ preserves the integral span of the Chevalley basis. Thus for any field F we can form the Z-module L_Z generated by the Chevalley basis and the F-algebra $L^F = F \otimes_Z L_Z$, and we see that for each $t \in F$ the map $U_\alpha(t) = \sum_n t^n X_\alpha^n/n!$ is a well defined automorphism of that algebra. Let $G = G(\Phi, F)$ be the group generated by the maps $U_\alpha(t)$, $t \in F$, $\alpha \in \Phi$. G is an (ordinary) *Chevalley group*. The maps $U_\alpha(t)$, t in $F^\#$, and their conjugates under G are called *root elements* and the groups $U_\alpha = \{U_\alpha(t) : t \text{ in } F\}$ and their conjugates are called *root groups*. The Lie algebra structure imposes certain relations on the root elements, the so called *Chevalley commutator relations* which show that the product U of the root groups of the positive roots is a subgroup of G. Moreover it can be shown that there exist subgroups B and N of G such that the following properties are satisfied:

(BN1) G is a group generated by the subgroups B and N, and $H = B \cap N \trianglelefteq N$.

(BN2) $W = N/H$ is generated by a set $S = \{s_1, \ldots, s_l\}$ of involutions.

(BN3) For each s in S and each w in W, $sBw \subseteq BwB \cup BswB$.

(BN4) For each s in S, $B^s \neq B$.

Indeed in our situation $U \trianglelefteq B$, H is a complement to U in B, and W is naturally represented as the group of automorphism of Φ with S the set of reflections through the roots in the simple system π. By convention if $w = nH \in W$ then $BwB = BnB$ and $B^w = B^n$; this is well defined as $H \leq B$.

If F is a finite field then the ordinary Chevalley group $G(\Phi,F)$ is a finite group of Lie type and is essentially simple. But there are other finite groups of Lie type called *twisted Chevalley groups*. The twisted groups are fixed points of suitable outer automorphisms of ordinary Chevalley groups. The twisted groups also have an associated root system, root groups, and subgroups B and N satisfying the properties (BNi), i = 1,2,3,4. The finite groups of Lie type are the ordinary and twisted Chevalley groups together with certain extensions of these groups.

Define a *Tits System* to be a quadruple (G,B,N,S) satisfying the properties (BNi), i = 1,2,3,4. G is also said to possess a BN-pair. Thus the finite groups of Lie type possess a BN-pair, and we have our third description of this class of groups. B,H,W and their conjugates are called *Borel subgroups*, *Cartan subgroups*, and *Weyl groups* of G, respectively. The order l of S is called the *Lie rank* of G. It is not difficult to show that if (G,B,N,S) is a Tits system then (W,S) is a Coxeter system and hence if the Weyl group W is finite there exists a root system Φ and a representation of W as the automorphism group of Φ with S the set of reflections through some simple system for Φ. Moreover we obtain the *Bruhat decomposition* of G as the union of the double cosets BwB, for w in W.

For each subset J of $I = \{1,\ldots,l\}$ define $S_J = \{s_i : i \in I\}$, $W_J = \langle S_J\rangle$, $P_J = BW_JB$, and let N_J be the preimage in N of W_J. It turns out P_J is a subgroup of G and indeed (P_J,B,N_J,S_J) is a Tits system. The subgroups $(P_J : J \subseteq I)$ and their conjugates are the *parabolic subgroups* of G. It is important to know that the map $J \mapsto P_J$ is a bijection of the power set of I with the set of overgroups of the Borel group B.

The parabolic subgroups of G define a geometry called the

building of G, and G is represented as a group of automorphisms of its building. I will wait till later for the discussion on geometries to define buildings. The representation of the groups of Lie type as groups of automorphisms of buildings supply our fourth description of these groups. Each building has a "rank" which turns out to be the Lie rank of the corresponding Tits system. In his fundamental paper [31], Tits classifies all buildings of rank at least 3 with finite Weyl groups and hence also all Tits systems of rank at least 3 with finite Weyl groups.

Our fifth and final description of the finite groups of Lie type involves algebraic groups. We have seen that each finite group G of Lie type is either an ordinary Chevalley group $G(\Phi,F)$ over some finite field F or the fixed points of some automorphism of such a group. Let $\bar{F}$ be the algebraic closure of F and $\bar{G} = G(\Phi,\bar{F})$ the corresponding Chevalley group over $\bar{F}$. Then $\bar{G}$ is an algebraic group and G can be realized as the fixed points of some endomorphism σ of $\bar{G}$. (σ is an endomorphism of algebraic groups.) This point of view is useful as it allows us to relate the linear representations and subgroup structure of G to that of $\bar{G}$, and $\bar{G}$ can be analyzed using algebraic group theory. In particular we will shortly need the following definition: A *maximal torus* of G is a subgroup of the form $C_T(\sigma)$ for some σ-invariant Cartan subgroup T of $\bar{G}$.

At this point it would be useful to take a brief look at some subgroups of groups of Lie type, particularly some local subgroups. Recall that if p is a prime then a *p-local subgroup* of a group G is the normalizer of some nontrivial p-subgroup of G. The finite simple groups are classified in terms of their local subgroups, so such subgroups are of great interest.

To discuss local subgroups intelligently we will also need the notion of the generalized Fitting subgroup of a finite group G. G is *quasisimple* if G is perfect (ie G is its own commutator group) and

$G/Z(G)$ is simple. A subgroup L of G is *subnormal* in G if there exists a series $L = G_0, \ldots, G_n = G$ of groups with G_i normal in G_{i+1} for each i. The *components* of G are its subnormal quasisimple subgroups. Write $E(G)$ for the subgroup of G generated by all components of G. It turns out $E(G)$ is the central product of the components of G and centralizes the Fitting subgroup $F(G)$ of G. (Recall $F(G)$ is the largest normal nilpotent subgroup of G and hence is the direct product of the groups $O_p(G)$, as p ranges over the prime divisors of the order of G, where $O_p(G)$ is the largest normal p-subgroup of G.) The *generalized Fitting subgroup* of G is the group $F^*(G) = F(G)E(G)$. $F^*(G)$ has the property that it contains its own centralizer in G. Hence, modulo the center of $F^*(G)$, G is faithfully represented as a group of automorphisms of the relatively uncomplicated group $F^*(G)$, so that we can analyze the structure of G in terms of its generalized Fitting subgroup.

When G is a finite group of Lie type there is a finite field associated to G. If $G = G(\Phi,F)$ is an ordinary Chevalley group the associated field is F; if G is the fixed points of an automorphism of $G(\Phi,F)$ the associated field is a certain subfield of F. One way to describe this field is as follows: let α be a long root in the root system for G; then $Z(U_\alpha)$ is isomorphic to the additive group of the associated field of G. In any case define the *characteristic* of G to be the characteristic of its associated field. If p is the characteristic of G then each root group is a p-group. Hence the product U of the positive root groups is also a p-group. It develops H is an abelian p'-group so U is Sylow in the product B, after which an easy argument using the Bruhat decomposition shows U is even a Sylow p-subgroup of G.

Let G be a finite group of Lie type and p the characteristic of G. The p-local structure of G is quite different from the r-local structure for primes r distinct from p. Recall that U is a Sylow p-subgroup of G, U is normal in B, and the parabolic subgroups P_J, $J \subseteq I$, are the overgroups of B in G. It is not difficult to see from these facts and the Bruhat decomposition that the maximal parabolics are even the maximal overgroups of U and that $B = N_G(U)$. A deeper result of Borel and Tits [12] says that every p-local is contained in a maximal parabolic.

A p-*superlocal* of a group X is a p-local Y such that $Y = N_G(O_p(Y))$. It is easy to see that if K is a p-local of X then there exists a p-superlocal Y such that $K \leq Y$ and $O_p(K) \leq O_p(Y)$. Hence the Borel-Tits Theorem can be restated: If G is a finite group of Lie type and characteristic p then the parabolic subgroups of G are the p-superlocals of G.

A group X is said to be of *characteristic p-type* if $F^*(Y) = O_p(Y)$ for each p-local Y of X. It is easy to see that if $F^*(X) = O_p(X)$ then X is of characteristic p-type. Moreover from the Chevalley commutator relations, $F^*(P) = O_p(P)$ for each proper parabolic P of a group G of Lie type and characteristic p. Hence each such group is of characteristic p-type.

The upshot of all this is that we have a good description of the p-locals of a group of Lie type and characteristic p in terms of its parabolic subgroups. Its a good exercise to see what all this means when G is a classical group in terms of its linear representation on the space (V,f). Here, with respect to a suitable basis for V, B is the set of upper triangular matrices, U the subgroup with 1's on the main diagonal, and H is (essentially) the subgroup of diagonal matrices. The parabolic subgroups are the stabilizers of the totally singular subspaces of V. From these observations it is easy to establish the Borel-Tits theorem for the classical groups.

Notice that the p-elements of G (ie those whose order is a power of p) are the *unipotent* elements of G; that is the elements which have all eigenvalues on V equal to 1. On the other hand the p'-elements of

G (those whose order is prime to p) are the *semisimple* elements of G: those which are diagonalizable over some extension of F. The same remarks hold for any group of Lie type and characteristic p in any linear representation over a field of characteristic p.

It turns out that if s is a semisimple element of a group G of Lie type in characteristic p then s is contained in some maximal torus T of G and $C_G(s)$ possesses a normal subgroup $X = TX_1 \ldots X_n$ such that the groups X_i are normal commuting subgroups of X and are groups of Lie type in characteristic p. Indeed with some degenerate exceptions the groups X_i are the components of $C_G(s)$. Hence G is almost never of characteristic r-type for primes r distinct from p.

Again it is useful to examine these concepts when G is a classical group from the point of view of its representation on (V,f). Now s is diagonal over some extension of F and to simplify matters assume s is actually diagonal on V. Then the groups X_i are essentially the classical groups on the eigenspaces of s on V.

We now have some idea of what the groups of Lie type and their local subgroups are like. Let us turn next to the remaining simple groups and compare their structure to that of the groups of Lie type.

First let G be the alternating group on a finite set Ω. The representation of G on Ω is an excellent tool for investigating G. For example it is easy to calculate centralizers from this representation. Let x be an element of G of prime order p. Then $C_G(x)$ has a subgroup of index at most 2 which is the direct product of the subgroups H and K of $C_G(x)$ fixing each point of Ω moved by x and fixed by x, respectively. Evidently H is an alternating group and $F^*(K) = O_p(K)$; in particular H is usually the unique component of $C_G(x)$. Thus this local has a structure which is a mixture of the structures of centralizers of unipotent and semisimple elements in groups of Lie type. In particular G usually has no natural characteristic. This, I believe, reflects that fact that the useful representations for the alternating groups are on combinatorial objects rather than algebraic objects.

The situation with the sporadic groups is even more extreme. We know of no nice representation for at least half the sporadic groups. The groups with no nice representation were discovered from a local group theoretic point of view. The sporadic groups which do possess useful representations are usually represented on combinatorial objects. For example the large Mathieu group M_{24} is the automorphism group of the Steiner system S(5,8,24) and the smaller Mathieu groups are stabilizers of interesting subobjects of this Steiner system. Several of the sporadic groups are represented on strongly regular graphs. For example this is true of the Fischer groups, where the graph is particularly useful as the vertices of the graph can be taken to be the set of 3-transpositions of the group. The large Conway group does have a beautiful algebraic representation on the Leech lattice. But even here the Leech lattice is defined combinatorially in terms of the Steiner system S(5,8,24).

The local structure of the sporadic groups is also mixed, although here in some cases the group may be thought of as possessing one or more characteristics. To introduce this new notion of characteristic we need some more definitions. Define a p-group P to *special* if $[P,P] = \Phi(P) = Z(P)$. P is *extra special* if P is special and Z(P) is cyclic. For example (with a few exceptions) if G is a group of Lie type and characteristic p and α is a long root then $F^*(C_G(U_\alpha(t)))$ is a special group whose center is the root group U_α, and if the associated field of G is of prime order this generalized Fitting group is extraspecial.

It turns out that in many of the sporadic groups G there exists a prime p and an element x of order p such that $F^*(C_G(x))$ is extraspecial and such that the set of overgroups of a Sylow p-group of G resemble the parabolics in a group of Lie type. We'll look at this more closely when we talk about geometries.

We've seen that most of the finite simple groups are of Lie type and we now have some idea of the nature of the structure of the local subgroups of the groups of Lie type. In particular we've seen that each group G of Lie type has a characteristic p and that the p-local structure of G is quite different from the local structure for other primes. Thus this characteristic can be retrieved from the local structure of the group. On the other hand we've seen that the alternating groups have no natural characteristic; there are similarities between the locals in the alternating groups and the groups of Lie type but there are also big differences. The sporadic groups fall somewhere in between. I've hinted that sporadic groups can often be assigned several characteristics, but that even when p is a characteristic in a sporadic group its p-locals are not as well behaved as those of a group of Lie type and characteristic p.

The Classification is based on an analysis of local subgroups with emphasis on characteristic as defined in terms of the generalized Fitting group of the various locals. In addition this characteristic is strongly reflected in the subgroup structure of the group and its associated geometries. Hence we will be returning to this concept frequently in later lectures.

CHAPTER II

Tits Geometries

In this chapter I'll be discussing a notion of geometry due to J. Tits. The theory appears to be of use both in the study of the subgroup structure of finite groups and in simplifying portions of the proof of the Classification Theorem. Unfortunately the discussion will begin with a long series of definitions; this seem unavoidable.

According to Tits, a *geometry* over an index set I is a triple $\Gamma = (\Gamma, \tau, *)$ where Γ is a set of objects, $\tau : \Gamma \to I$ is a type function, (ie for $u \in \Gamma$, $\tau(u)$ is the type of u) and * is a symmetric incidence relation. The only axiom is: If $u,v \in \Gamma$ are of the same type then $u*v$ if and only if $u = v$. The *rank* of Γ is the order of I. Γ will always be assumed to be of finite rank in this lecture.

A *flag* of Γ is a subset T of Γ such that each pair of members of T is incident. Together with our axiom this forces τ to be injective on T, and the image of T under τ is said to be the *type* of T. The order of T and the order of $I-\tau(T)$ are the *rank* and *corank* of T, respectively. A *morphism* of geometries is a map preserving type and incidence.

Here is a way to get geometries from groups. Let G be a group and $F = (G_i : i \in I)$ a family of subgroups of G. Define $\Gamma(G,F)$ to be the geometry whose set of objects of type i is the coset space G/G_i and with cosets incident if their intersection is nonempty. Evidently G is represented as a group of automorphisms of the geometry $\Gamma(G,F)$ via right multiplication. Observe that for $J \subseteq I$ and $x \in G$,

$$S_{J,x} = \{G_j x : j \in J\}$$

is a flag of Γ of type J. Write G_J for the subgroup $\bigcap_{j\in J} G_j$ and observe G_J is the stabilizer in G of the flag $S_J = S_{J,1}$.

The canonical example to keep in mind here is the case where G is a group of Lie type and F is the set of maximal parabolics over some fixed Borel group of G.

The *residue* of a flag T in a geometry Γ is

$$\Gamma_T = (\Gamma_T, \tau, *)$$

regarded as a geometry over $I-\tau(T)$, where Γ_T consists of the members of Γ-T incident with each member of T. I'll refer to the residues of flags of corank n as *rank* n *residues*. Often Γ is assumed to satisfy one or more of the following restrictions:

(α) The isomorphism type of the rank 2 residues is of some restricted type.

(β) Γ is *residually connected*. That is each residue of rank at least 2 is connected and rank 1 residues are nonempty. (Γ is said to be *connected* if the graph $(\Gamma,*)$ is connected.)

(γ) Γ admits a *flag transitive* group of automorphisms; that is a group transitive on flags of type J for each subset J of I. In that event the stabilizer of a flag of type I will be called a *Borel group* of the flag transitive group.

<u>Remarks</u> (1) In the G,F construction G is transitive on edges but in general is not flag transitive. However there are various general hypotheses which insure the flag transitivity of G on $\Gamma(G,F)$. See for example [9]. These hypotheses are invariably present in the examples we'll consider.

(2) $\Gamma = \Gamma(G,F)$ is connected if and only if F generates G. Hence setting $F_J = \{G_{J\cup\{i\}}: i \in J' = I-J\}$, it follows that if G is flag transitive on Γ then Γ is residually connected precisely when G_J is generated by F_J for each subset J of I.

(3) G is flag transitive on Γ if and only if the residue of $S_{J,x}$ is isomorphic to $\Gamma(G_J,F_J)$ for each $J \subseteq I$ and $x \in G$.

A *generalized* n-*gon* is a rank 2 geometry Γ such that the graph $(\Gamma,*)$ is of diameter n and contains no cycle of length less than $2n$. A geometry Γ is *thin*, *thick* if every flag of corank 1 is contained in two, more than two, flags of type I, respectively.

A *diagram* on I is a family $D = (D_J : J \subseteq I,\ |J| = 2)$ of isomorphism classes D_J of rank 2 geometries. A geometry Γ is said to *belong* to D if the residue of each flag of cotype J is in D_J for each subset J of I of order 2. Of course Γ belongs to a unique minimal diagram which will be called the *diagram of* Γ. The *Coxeter diagram* of a Coxeter matrix $M = (m_{ij})$ is the diagram D for which $D_{\{i,j\}}$ consists of the generalized m_{ij}-gons for each i,j from I. A Coxeter diagram is *spherical* if the corresponding Coxeter groups is finite.

Consider our canonical example $\Gamma = \Gamma(G,F)$ with G of Lie type and F the maximal parabolics over some fixed Borel. Then Γ is thick and residually connected, G is flag transitive on Γ, and Γ belongs to the Coxeter diagram of the Dynkin diagram of the root system of G.

Another useful notion is that of the *graph* of a diagram D. This is the graph on I obtained by joining i to j if $D_{\{i,j\}}$ contains a geometry which is *not* a generalized 2-gon. A geometry is the direct sum of geometries in a nontrivial manner if and only if the graph of its diagram is disconnected, so with little loss of generality we need only consider geometries with *connected diagrams*; that is with diagrams whose graph is connected. A *string diagram* is a diagram whose graph is a string under a suitable ordering.

A *building* is a geometry Γ possessing a set A of subgeometries (called *apartments*) such that:

(1) The members of A are thin residually connected geometries.

(2) Each pair of flags is contained in a member of A.

(3) If S and T are flags of Γ contained in Σ, Σ' in A, then there exists an isomorphism from Σ to Σ' which is the identity on S and T.

It turns out our canonical example is a building. The Borel group B is the stabilizer of a flag C of type I and we take the set of apartments to be the conjugates of Σ under G, where Σ is the set of images of objects in C under N, the second member of the BN-pair for G. Moreover Tits' fundamental paper [31] shows:

Theorem 1. Each thick building of rank at least 3 whose apartments are finite is a geometry $\Gamma(G,F)$ for some group G of Lie type (not necessarily finite) and some set F of parabolics.

A *covering* of a geometry Γ is a surjective morphism $\pi : \tilde{\Gamma} \to \Gamma$ such that for each $u \in \Gamma$, the map induced by π from the residue of u to the residue of $u\pi$ is an isomorphism. The covering is connected if $\tilde{\Gamma}$ is connected. A *universal cover* of Γ is a connected cover $\pi : \tilde{\Gamma} \to \Gamma$ such that whenever $\alpha : \Sigma \to \Gamma$ is a covering, $a \in \tilde{\Gamma}$, and $b \in \Sigma$ such that $a\pi = b\alpha$, then there exists a morphism $\beta : \tilde{\Gamma} \to \Sigma$ such that $a\beta = b$ and $\beta\alpha = \pi$. We'll want to know that each connected geometry possesses a universal cover and of course this cover is unique up to isomorphism. Further if $\pi : \tilde{\Gamma} \to \Gamma$ is a universal cover then there exists a group B of automorphisms of $\tilde{\Gamma}$ and a surjective homomorphism of B onto $\mathrm{Aut}(\Gamma)$ commuting with π whose kernel is regular on the fibres of π. In particular if Γ possesses a flag transitive group of automorphisms then so does $\tilde{\Gamma}$. One more definition: Γ is *simply connected* if Γ is connected and its own universal cover.

In [30] Tits proves:

Theorem 2. If Γ is a residually connected geometry with a Coxeter diagram and rank 3 residues of Γ of type C_3 and H_3 are buildings, then the universal cover of Γ is a building. If Γ itself is a building then Γ is simply connected.

Define a rank 2 geometry to be *classical* if it is the parabolic geometry of a finite group of Lie type; from the point of view of finite group theory the interesting rank 2 geometries with Coxeter diagram are classical. There are no classical geometries of type H_2 so for our purpose the following result (found in [8]) fills the gap in the Tits theorem:

Theorem 3. Let Γ be a residually connected geometry with Coxeter diagram of type C_3 whose rank 2 residues are classical or generalized 2-gons and which admits a flag transitive group G of automorphisms with finite Borel group. Then either Γ is a building or G is A_7 and Γ is the 7-geometry.

The 7-*geometry* is the geometry $\Gamma(G,F)$ where G is the alternating group A_7 and $F = \{G_1, G_2, G_3\}$ where G_1 is the stabilizer of one of the 7 points permuted by G, G_2 is the global stabilizer of 3 of these points, G_3 is isomorphic to $L_3(2)$, and the G_i's contain a common Sylow 2-subgroup of G.

Combining the three results above with Seitz' theorem classifying the flag transitive subgroups of the finite groups of Lie type [25] and a little extra work, we obtain the following result from [8]:

Theorem 4. If Γ is a residually connected geometry with a spherical connected Coxeter diagram whose rank 2 residues are classical or generalized 2-gons and G is a flag transitive group of automorphisms of Γ with a finite Borel group, then Γ is a building or the 7-geometry and G is a known group.

If the hypothesis that the diagram be spherical is removed then there are more examples.

We've seen one way to obtain a geometry $\Gamma(G,F)$ from a group G and a family F of subgroups of G. In our canonical example F was the set of maximal parabolics over some fixed Borel group in a group of Lie type. There is another point of view corresponding to the minimal parabolics. Here we start with a collection $X = (X_i : i \in I)$ of subgroups of G, for $J \subseteq I$ we let $X_J = \langle X_j : j \in J \rangle$, $J' = I-J$ the complement of J in I, and consider the geometry $\Delta(G,X) = \Gamma(G,F)$, where $F = (X_{i'} : i \in I)$. In our canonical example we would choose X to be the set of minimal parabolics over some fixed Borel group. This suggest the following definition: A family $X = (X_i : i \in I)$ is a *parabolic system of characteristic* p if

(1) G is generated by X but by no proper subfamily of X.

(2) The members of X contained a common Sylow p-group which is also Sylow in X_J for each subset J of I of order 2.

(3) $X_i = O^{p'}(X_i)$ and $X_i/O_p(X_i)$ is a group of Lie type of characteristic p and Lie rank 1 for each i in I.

(4) $X_J/O_p(X_J)$ is a group of Lie type of characteristic p and Lie rank 2 for each subset J of I of order 2.

In (4) it is understood that the central product of groups of Lie type is itself of Lie type with Lie rank equal to the sum of the ranks of the factors. As a matter of fact for optimal geometric results, in (4) we should also allow any group which is the product (not necessarily central) of groups of rank 1, since these correspond to generalized 2-gons, but the next result I'll state is established under the stronger hypothesis above. Modulo this problem and a little work, results on parabolic systems can be translated into results about geometries with a Coxeter diagram and vise-versa. The following result is due to Niles and appears in [23]:

Theorem 5. Let X be a parabolic system in G, T a common Sylow group of the members of X and assume:

(1) T contains no nontrivial normal subgroup of G.

(2) $X_J/O_p((X_J)$ is not of type $A_1(2)$, $A_1(3)$, ${}^2A_2(4)$, or ${}^2B_2(2)$ for any subset J of I of order 1, or of type $A_2(4)$ for any subset J of order 2.

Then G is a group of Lie type and characteristic p and T is a Sylow p-subgroup of G.

As mentioned above, modulo a few problems which don't seem so difficult, this pretty much determines the geometries with a Coxeter diagram, classical residues, and a flag transitive group with finite Borel group, except when the residues are over the fields of order 2,3, or 4. Unfortunately these small cases are the most difficult and include lots of extra examples.

I'll next state two results of Timmesfeld from [27] and [28]. I'll state the results in geometric terms; Timmesfeld also states corresponding results about parabolic systems.

Theorem 6. Let Γ be a residually connected geometry with a connected Coxeter diagram whose rank 2 residues are generalized 2-gons or classical of characteristic 2 and which admits a flag transitive group G with finite Borel group. Then

(1) If the rank 2 residues of Γ are all either generalized 2-gons or projective planes then either Γ has a spherical diagram or Γ is in case (a) or (b) below.

(2) If Γ is of rank 3 then either Γ has a string diagram or Γ is in case (c) below.

In case (a) Γ is of rank 5, its diagram is a cross, the stabilizer in G of an object indexed by a node not in the center of the diagram is isomorphic to $D_4(2)$, and the stabilizer of an object of the remaining type is isomorphic to $E_8/SL_2(3)^4$. An example is the group

$D_4(3)$ on the cosets of the maximal overgroups of a Sylow 2-group. (cf [10]) In case (b) the diagram of Γ is a square and the stabilizer of each object is A_7. $U_4(3)$ acts on such a geometry; again see [10]. In case (c) the diagram is a triangle and each stabilizer is of order 21.

Notice in the exceptional cases the geometry and the group are not determined, just the proper residues and their stabilizers. This is because the universal cover of the geometry is infinite and possesses infinitely many morphic images satisfying the conclusions of the theorem. This situation seems to arise only in degenerate cases when the stabilizer of some object acts faithfully on its residue. In most cases the generalized Fitting subgroup of the stabilizer is a p-group, where p is the characteristic of the rank 2 residues, and this generalized Fitting group is in the kernel of the action on the residue.

The arguments used by Niles and Timmesfeld are basically group theoretic as opposed to geometric. Timmesfeld makes heavy use of so called "pushing up" arguments which are part of the proof of the Classification Theorem.

It seems likely that all finite geometries with a Coxeter diagram, classical rank 2 residues, and a flag transitive group will be classified, modulo the type of ambiguities present in Timmesfeld's results. However this class of geometries is not large enough for purposes of studying the finite simple groups. Most of the sporadic groups act on one or more geometries and the rank 2 residues of these geometries are often not generalized n-gons. Indeed if the p-structure of a sporadic group G is rich enough (eg a high power of p divides the order of G, say at least p^4 to get interesting examples) then there exists a collection F of maximal overgroups of a Sylow p-subgroup T of G such that the geometry $\Gamma = \Gamma(G,F)$ is residual connected, G is flag transitive on Γ, the diagram of Γ is uncomplicated, (usually a string diagram) and the rank 2 residues are well behaved. Usually the rank 2 residues are generalized 2-gons or classical generalized n-gons but a few other geometries occur. The collection F is usually a set of representatives for the conjugacy classes of maximal overgroups of T. I'll refer to such a geometry as a p-*overgroup geometry*.

Lets consider the case where G is a sporadic group and p = 2 as an example. Lets further only look at the case when the generalized Fitting subgroup of the centralizer of some involution is extraspecial; that is G has 2 as a characteristic in the sense of the first lecture. Let Γ be the 2-overgroup geometry of G. Then Γ is of rank at least 3 precisely when G is M_{24}, Sz, J_4, Aut(M(22)), M(24), Co_2, Co_1, F_2, or F_1, and the nonclassical rank 2 residues which appear are the truncation of the geometries of $L_4(2)$, $Sp_6(2)$, and M_{24}, and the 2-overgroup geometry of M_{22}. A *truncation* is obtained by omitting one or more types of objects from the geometry.

Again there seems to be little doubt that we can add in the extra rank 2 residues involved in the overgroup geometries of the sporadic groups and still obtain a classification of the resulting collection of geometries. Such an approach seems artificial however. It would be much better to find some natural class of rank 2 geometries containing all the relevant residues and classify the geometries whose rank 2 residues fall in this natural class.

The geometry of a group G of Lie type is intimately connected to the Bruhat decomposition of G which in turn can be used to define the root subgroups of G and all the related structure. Thus the representation of G on its geometry is an extremely useful tool for studying the group. For example we saw in the first lecture how the p-local structure of G was related to the geometry. So far the representation of the sporadic groups on their geometries is not as useful. I have (essentially) determined all the overgroups of a Sylow p-group of each sporadic group G whose order is divisible by p^2 and I am in the process of proving an analogue of the Borel-Tits Theorem with respect to these overgroups; in particular I intend to determine all superlocals. The object is to develop a reasonably uniform theory which can describe simultaneously portions of the subgroup structure of all finite simple groups.

It would be nice to obtain some analogue of the Bruhat decomposition and root groups for the sporadic groups. Geometrically the Bruhat decomposition corresponds to the existence of apartments. So far no one has been able to find useful analogues of apartments for sporadic groups.

It is also worth recalling that Curtis' treatment in [11] of the linear representations of characteristic p for groups of Lie type and characteristic p uses only the Bruhat decomposition and root group structure of these groups so the existence of analogous structure in sporadic groups could also have implications for the study of their linear representations.

In a later lecture we will see that the geometric theory discussed in this lecture also has potential for simplifying the proof of the Classification Theorem.

CHAPTER III

The Classification Thereom

In this chapter I will be discussing the Classification Theorem for finite simple groups. Of necessity my remarks will be very general. Emphasis will be on recent efforts to simplify the original proof. For a more detailed outline see [1] or [6], and for an in depth treatment of the Classification see Gorenstein's series of books. [17,18]

The Classification Theorem states that every finite simple group is a group of prime order, a group of Lie type, an alternating group or one of the 26 sporadic simple groups. Denote this collection of groups by K and define a group G to be a *K-group* if every simple section of G is in K. (A *section* of G is a homomorphic image of a subgroup of G.) The Classification proceeds by induction on group order so a minimal counter example to the Classification has the property that each proper subgroup of G is a K-group. This fact together with our knowledge of the members of K will be used repeatedly.

The finite simple groups are classified in terms of local subgroups, more particularly in terms of the centralizers of elements of prime order, and most particularly in terms of centralizers of involutions. More precisely given a prime p define a *p-standard subgroup* of a group G to be a quasisimple subgroup L of G such that $C_G(L)$ has nontrivial cyclic Sylow p-subgroups and L is a component of the centralizer in G of each element of order p centralizing L. We seek to show first that each simple group contains a standard subgroup and second that the isomorphism type of a standard subgroup determines the isomorphism type of the simple group. Suppose we know this and indeed we know that if L is standard in G with L/Z(L) in K then G is in K. Let G be our minimal counterexample to the Classification and L a standard subgroup of G. Since proper subgroups of G are K-groups, L/Z(L) is in K so the Classification is complete.

This is how the Classification proceeds once our minimal counter example is sufficiently large with respect to one of several appropriate measures of size. Unfortunately certain small groups do not possess standard subgroups. For this and other reasons special arguments are required for small groups.

Its probably useful to look at an example. Let G be a group of Lie type and characteristic p. Recall if x is a p-element (or unipotent element) of G then $F^*(C_G(x))$ is a p-group, so in particular $C_G(x)$ has no components. Hence G possesses no p-standard subgroup. On the other hand if x is a p'-element (or semisimple element) then $C_G(x)$ has a normal subgroup $X = TX_1 \ldots X_n$, where T is abelian and the X_i are normal groups of Lie type and usually quasisimple. Thus the X_i are usually components of $C_G(x)$, and under suitable conditions X_i will be standard in G. Thus in a group G of Lie type there will usually exist r-standard subgroups for suitable primes r distinct from p, but never p-standard subgroups. Moreover standard subgroups centralize semisimple elements, and as a matter of fact we can usually choose our standard subgroup to centralize an element in a Cartan subgroup. For example take G to a general linear group and x diagonal. Recall the X_i correspond to eigenspaces of x, so if we choose x to have a hyperplane as an eigenspace then a moments thought shows the corresponding X_i is standard.

We see that if a simple group G is of characteristic p-type then it has no p-standard subgroup, and in particular groups of Lie type and characteristic 2 have no 2-standard subgroup. This is important because the prime 2 plays a special role in simple group theory. Many arguments work for the prime 2 but not for odd primes, and vise versa. Hence the Classification is divided into four cases: Large and small groups of characteristic 2-type and large and small groups which are not of characteristic 2-type. If G is large we show it has a p-standard subgroup; when G is of characteristic 2-type we must take p odd, otherwise we hope to take p to be 2. On the other hand if G is small it may have no standard subgroup so we must characterize G in terms of some other local subgroup.

The appropriate notions of size are related to p-rank and

connectedness. Let $\Delta = \Delta_n^p$ be the set of elementary abelian p-subgroups of G of p-rank at least n (ie groups which are the direct product of at least n copies of the cyclic group of order p. The *p-rank* of G is the maximum rank of its elementary abelian p-subgroups.) and define a graph on Δ by joining vertices if they commute elementwise. Evidently G is represented as a group of automorphisms of this graph by conjugation. G is *n-connected for the prime* p if this graph is connected. Its not hard to see that if G is not 1-connected for the prime p then the stabilizer H of a connected component of this graph is *strongly p-embedded* in G; that is H is a proper subgroup of G of order divisible by p which contains the normalizer of each of its nontrivial p-subgroups.

A group G which is not of characteristic 2-type is defined to be small if G is 2-disconnected for the prime 2. It is a deep theorem that G is simple and 2-disconnected for the prime 2 precisely when G is of 2-rank 2, a group of Lie type of Lie rank 1 and characteristic 2, or the smallest sporadic group J_1. This theorem uses special properties of the prime 2, most particularly the fact that any pair of involutions generate a dihedral group. The fact that there is no proof of an analogous result for odd primes short of an appeal to the Classification causes big problems and is one reason for the four part subdivision in the proof of the Classification.

A group G of characteristic 2-type is small if e(G) is at most 2, where e(G) is defined to be the maximum over odd primes p of the 2-local p-rank of G and the *2-local p-rank* of G is the maximum p-rank of a 2-local of G. In a group of Lie type and even characteristic e(G) is a good approximation of the Lie rank. There is also another small case: the so called *Uniqueness Case*. In the Uniqueness Case for each odd prime p such that G is of 2-local p-rank at least 3, G has a strongly p-embedded 2-local.

The most difficult problem in handling the large groups is to establish the following general property of finite groups:

B_p-Property. If G is a finite group and p is a prime such that $O_{p'}(G) = 1$ then for each element x of G of order p,

$$E(C_G(x)/O_{p'}(C_G(x))) = E(C_G(x))O_{p'}(C_G(x))/O_{p'}(C_G(x)).$$

As a matter of fact for odd primes this property is only known as a corollary to the Classification. Given the B_p-Property, one can (essentially) demonstrate the existence of a p-standard subgroup in a simple group which is 2-connected for the prime p but not of characteristic p-type, and go on from there to determine the isomorphism type of G from that of its standard subgroup. While this demonstration is long and technically difficult it is in some sense straight forward. Thus one can also think of the proof of the Classification as made up of the following steps:

(1) Determine the simple groups 2-disconnected for the prime 2.

(2) Determine the groups of characteristic 2-type with e(G) at most 2; handle the Uniqueness Case.

(3) Establish the B_p-Property.

(4) Establish the existence of p-standard subgroups in each simple group G such that either:

(i) p = 2 ,G is 2-connected for the prime 2, and G is not of characteristic 2 type, or

(ii) G is of characteristic 2 type, e(G) > 2, and p is a suitable odd prime.

(5) Solve the *standard form problems*; that is show that if G is a simple group with a standard subgroup L such that L/Z(L) is in K then G is in K.

The emphasis in today's lecture is upon on going work to simplify or revise the proof of the Classification. The two programs I will mention are more in the nature of revisions than simplifications. The first program is being undertaken by Foote, Gorenstein, Lyons, and Solomon. Its aim is to reorganize and merge the proofs of steps (3), (4), and (5). The second program is that of Delgado, Stellmacher, Stroth, Timmesfeld, and various associates. This program seeks an alternate treatment of step (2).

I'll first discuss the efforts to revise step (2). In step (2) G is a minimal counter example to the Classification and is of characteristic 2-type. Let $\mathcal{M}$ be the set of maximal 2-local subgroups of G and T a Sylow 2-subgroup of G. For each subset X of G write $\mathcal{M}(X)$ for the set of members of $\mathcal{M}$ containing X. Recall that either e(G) is at most 2 or G is in the Uniqueness Case; in the the latter case remember that there exist members of $\mathcal{M}$ which are strongly p-embedded in G for various odd primes p. When e(G) is at most 2 our object is to show $\mathcal{M}(T)$ resembles the set of maximal parabolics in some group of Lie type and characteristic 2, and then use this information to show G is in K. In the Uniqueness Case we need to derive a contradiction. In each case we have at our disposal *uniqueness theorems*; that is statements of the form: $\mathcal{M}(X)$ possesses a unique member for each X in a suitable set $\mathcal{X}$ of subgroups of G. We also have available an extension of the result describing the groups 2-disconnected for the prime 2 which says that $\mathcal{M}(T)$ contains at least two members M and N.

Observe that if R is a nontrivial subgroup of T such that $N_M(R)$ contains a member of $\mathcal{X}$ then $M = \mathcal{M}(N_G(R))$. In particular $N_N(R)$ is contained in M and contains no member of $\mathcal{X}$. These observations restrict the structure of M and N. When e(G) is at most 2 they force M and N to resemble parabolics in a group of Lie type. In the Uniqueness Case $\mathcal{X}$ is too large, containing each p-subgroup of M or N of rank at least 3; hence we are able to derive a contradiction.

The methods used to restrict the structure of M and N are called *pushing up arguments*. The general idea is to produce a collection C of normal subgroups of T such that for each small subgroup H of G containing T with $F^*(H) = O_2(H)$, either some member of C is normal in H or the structure of H is restricted. Then if H and I are small subgroups of M and N containing T we can hope to find a member of C normal in both H and I, and hence conclude either H or I contains no member of $\mathcal{X}$. in particular $\mathcal{X}$ cannot be too large.

One choice for C is $\{Z(T), J(T)\}$ where J(T) is the *Thompson subgroup* of T; that is J(T) is generated by the elementary abelian subgroups of T of maximal order. Another choice is

$$\{W_i(T,V),\ Z(W_i(T,V)) : i \geq 0\}$$

where V is some normal elementary abelian subgroup of M and $W_i(T,V)$ is the subgroup generated by all G-conjugates of subgroups of index at most 2^i in V which are contained in T. A typical small group H would satisfy $F^*(H/O_2(H))$ simple. If no member of C is normal in H we expect $F^*(H/O_2(H))$ to be a group of Lie type of characteristic 2 or an alternating group, we expect few chief factors in $O_2(H)$, and we expect the representation of $H/O_2(H)$ on those chief factors to be quite special.

The treatment of case (2) in the existing literature is quite complicated. The program of Stellmacher et al seeks to use geometric ideas to shorten this treatment or at least make it easier to understand. Namely take a collection F of small subgroups of N and M containing T such that the groups N_F and M_F, generated by the members of F in N and M, respectively, each contain a member of $\mathcal{X}$. Hence G is generated by F. Consider the geometry $\Gamma = \Gamma(G,F)$. Pushing up arguments restrict the structure of the members of F and the residues of Γ, hopefully to the point that G can be determined using one of our theorems classifying such geometries. From the other direction, pushing up arguments are used by Timmesfeld to prove his results on geometries which were discussed in the previous lecture.

It is not yet clear precisely which geometric classifications would be necessary to implement this program particularly for the Uniqueness Case. The program requires theorems on rank 2 geometries which are of a different flavor than the theorems on higher dimensional geometries which take as part of their hypothesis knowledge of rank 2 residues. However most of the necessary theorems of this sort may already exist due to the work of Goldschmidt, his students, and Chermak, Stellmacher, and Stroth.

Let us turn now to the program of Gorenstein et al to revise steps (3), (4), and (5). This program involves two major modifications to the organization of the original proof plus numerous smaller modifications. The first major change is an attempt to do much of (3), (4), and (5) simultaneously for all primes p rather than treating the prime 2 as special. The second modification involves a change of the definition of groups of even characteristic. This change is along the lines of the notion of characteristic for the sporadic groups advanced in the first lecture. The following definition is not quite the Gorenstein-Lyons definition but is in the same spirit: A group G is a *W-group* if $F^*(H) = O_2(H)$ for each 2-local H of G containing a Sylow 2-group of G and G is of *even characteristic* if G is a W-group and for each involution x of G, $O_{2'}(C_G(x)) = 1$ and each component of $C_G(x)$ is a W-group. For example most of the sporadic groups which are of even characteristic with respect to the definition of the introduction are also of even characteristic with respect to this definition.

One advantage of this weaker notion of even characteristic is that it makes it easy to do many standard form problems. Namely if G is not of even characteristic but satisfies the B_2-conjecture then there exists an involution z in G and a component L of C(z) such that either L is not a W-group or z is in the center of a Sylow 2-subgroup of G. Thus we can hope to do the standard form problems for W-groups under the assumption that the normalizer of the standard subgroup contains a Sylow 2-group of G. Often it is easy to obtain a contradiction with this extra hypothesis.

Of course a disadvantage of this approach is that makes the treatment of groups of even characteristic more difficult. It is hoped however that the advantages will out weigh the disadvantages.

In the existing treatment of steps (3), (4), and (5), the groups of even and odd characteristic are handled quite differently. Gorenstein and Lyons propose to minimize these differences in their revision and indeed to do as much of the proof as possible simultaneously for all primes. Where changes are made the revised treatment will tend to resemble the existing treatment of groups of even characteristic rather than that in odd characteristic.

In broad outline, the proof would go like this: If our minimal counter example G is of odd characteristic let $\sigma = \{2\}$ and let $\mathcal{E}$ be the set of elementary abelian 2-subgroups of maximal rank; otherwise let σ consist of the odd primes p such that the 2-local p-rank is maximal rank n and let $\mathcal{E}$ consist of those elementary abelian p-subgroups of rank n in 2-locals. Using signalizer functor methods argue that either G is of even characteristic and in the Uniqueness Case or there exist p in σ, a p-group E in $\mathcal{E}$, and elements x and y in $E^{\#}$ such that components X and Y of $C_G(x)$ and $C_G(y)$ are p-standard in G and $C_G(\langle x,y\rangle)$ is large in a suitable sense. This last condition restricts the possibilities for X and Y and makes it easier to do the corresponding standard form problems. Notice we do not attempt to prove the B_p-conjecture directly.

Gorenstein and Lyons do not intend to use the so called Classical Involution Theorem, the Root Involution Theorem, or the classification of groups with a large extraspecial subgroup in their revision. Each of these results played a prominent role in the original proof. These omissions make me uneasy as I regard each of these results as natural and beautiful, and as each is extremely useful outside the context of the Classification. (As we will see in later chapters.)

CHAPTER IV

Permutation Representations

In the remaining chapters I'll discuss an emerging theory of permutation representations based on the Classification Theorem.

If one follows the philosophy advanced in chapter I, a theory of permutation representations should concentrate on the indecomposable and irreducible permutation representations of almost simple groups. On the one hand it should include a body of techniques for reducing questions about arbitrary permutation representations of general finite groups to questions about indecomposable or irreducible representations of almost simple groups, and on the other it should include descriptions of this latter more specialized class of representations.

The indecomposable permutation representations are the transitive representations. It is well known that each transitive permutation representation of a group G is equivalent to a representation by right multiplication on the cosets of some subgroup of G. So the study of permutation representations of G is essentially the same as the study of the subgroup structure of G. The irreducible permutation representations are the primitive representations. Of course a representation is primitive precisely when the stabilizer of a point in the representation is a maximal subgroup of G, so the study of primitive representations of G corresponds to a study of the maximal subgroups of G. Further two transitive representations of G are equivalent precisely when stabilizers of points in the representations are conjugate in G. So to a large extent we are really concerned with conjugacy classes of maximal subgroups.

Incidentally for purposes of this talk a group G is *almost simple* if $F^*(G)$ is a nonabelian simple group.

To reduce questions about transitive representations to questions about primitive representations one needs information about the maximal overgroups of suitable subgroups of G. Moreover we can hope that in most interesting transitive representations the stabilizer of a point is reasonably large in the sense that it contains some easily recognizable subgroup. Thus the maximal overgroups of such subgroups are of particular interest. Indeed we may wish to construct our theory so as to ignore small maximal subgroups.

Thus we will be interested in describing the maximal overgroups of suitable subgroups. It is not yet clear what type of description is most useful but various criteria suggest themselves. We would like our description to be reasonably elegant, to have a nice proof, and to apply simultaneously to as many of the almost simple groups as possible. For example simple descriptions of certain portions of subgroup structure of large classes of almost simple groups were particularly useful in the final stages of the Classification, where it was necessary to establish numerous facts about K-groups.

One way to describe the subgroup structure of a group is in terms of a representation of the group on some object. If the representation is particularly nice we can hope to describe most subgroups as stabilizers of suitable structures on the object. The representation of the alternating group of degree n on a set of order n and the representation of a classical group on its associated space are examples of such representations. To exploit such representations it will be useful to have available a few elementary observations of a general nature.

Let G be a group and C a category. Representations $\pi_i : G \to Aut(X_i)$, $i = 1,2$, in C are *equivalent* if there exists an isomorphism α of X_1 with X_2 such that $\pi_2 = \pi_1\alpha^*$, where $\alpha^* : Aut(X_1) \to Aut(X_2)$ is defined by $\beta\alpha^* = \alpha^{-1}\beta\alpha$ for $\beta \in Aut(X_1)$. Representations $\pi_i : G_i \to Aut(X_i)$ are *quasiequivalent* if there exists an isomorphism γ of G_1 with G_2 such that the representations π_1 and $\gamma\pi_2$ of G_1 are equivalent. The following elementary observation is very useful:

Lemma 1. Let π and σ be faithful representations of a group G on an object X. Then π and σ are quasiequivalent if and only if $G\pi$ and $G\sigma$ are conjugate in Aut(X). Moreover the group of automorphisms induced on $G\pi$ in Aut(X) is isomorphic to the stabilizer in Aut(G) of the equivalence class of π.

Lets begin to look at some specific results. The first theorem I'm going to discuss describes primitive permutation representations of general finite groups in terms of primitive representations of almost simple groups. It was obtained in collaboration with L. Scott in [5]. There is significant intersection of this work with that of F. Gross and L. Kovaks in [19].

Let G be a finite group and $N = N_G$ be the set of maximal subgroups of G. Let $M \in N$ and denote by $ker_M(G)$ the kernel of the permutation representation of G on the cosets of M. With little loss of generality we can take $ker_M(G) = 1$ by passing to $G/ker_M(G)$. That is we assume M is in $N^* = N^*_G$, the set of N in N with $ker_N(G) = 1$. Let C^* denote the set of orbits of G on N^* via conjugation.

Theorem 1. Let $D = F^*(G)$ and $M \in N^*$. Then one of the following holds:

A) D is an elementary abelian p-group and M is a complement to D in

G. $C^* \cong H^1(M,D)$.

B) D is the direct product of the G-conjugates of LK, where L and K are isomorphic nonconjugate simple components of G, $G = MD$, $M = N_G(M \cap D)$, and $M \cap D$ is the direct product of the M-conjugates of a full diagonal subgroup of LK. $C^* \cong m.(C_{Out(L)}(Out_G(L)))$, where m is the number of conjugates U of K such that $N_G(U) = N_G(L)$ and there is a $N_G(L)$-invariant full diagonal subgroup of LK.

C) D is the direct product of the G-conjugates of some simple component L of G. N^* is the disjoint union of G-stable subsets N_i^*, $1 \leq i \leq 3$, where:

1) N_1^* consists of complements M to D with $Inn(L) \leq Aut_M(L)$.

2) N_2^* consists of those subgroups M of G such that $G = MD$, $M = N_G(M \cap D)$, and $M \cap D$ is the direct product of the M-conjugates of a full diagonal subgroup of $\prod_{X \in \Gamma} X$ for some $\Gamma^G \in P^*(G)$.

3) N_3^* consists of those subgroups M of G such that $G = MD$, $M = N_G(M \cap D)$, $Aut_M(L) \in N^*_{Aut_G(L)}$, and $M \cap D$ is the direct product of the M-conjugates of $M \cap L$. $C_3^* \cong C^*_{Aut_G(L)}$.

A *full diagonal subgroup* of a direct product $L_1 x \ldots x L_n$ is a subgroup X such that the projection maps from X to L_i are isomorphisms for each i. In Case C2, $P^*(G)$ denotes the set of G-invariant partitions Γ^G of L^G, such that $|\Gamma| > 1$ and Γ^G possesses no proper refinement satisfying these constraints. $H^1(M,D)$ is of course the 1-cohomology group of M on D. There are parameterizations of C_1^* and C_2^* given in [5], but I omit them here as they are rather technical.

In some sense Theorem 1 reduces the study of primitive permutation representations of finite groups to two difficult but well defined problems:

Problem 1: Determine C_G^* for G an almost simple group.

Problem 2: Determine $H^1(G,V)$ for G an almost simple group and V an irreducible GF(p)G-module.

This type of reduction is in the spirit of the philosophy of chapter I. Problem 1 arises out of Case C3, while Problem 2 arises out of Case A, given the following theorem from [5]:

Theorem 2. Let G be a finite group, K a field of prime characteristic p, and V a faithful KG-module with $H^1(G,V) \neq 0$. Then:

1) $F^*(G)$ is the direct product of the G-conjugates of a simple component L with $p \in \pi(L)$.

2) V is the direct sum of the G-conjugates of $U = [L,V]$.

3) $N_G(L)$ is irreducible on U with $C_G(L) = C_G(U)$.

4) $H^1(G,V) \cong H^1(Aut_G(L),U)$.

5) $\dim(H^1(G,V)) \leq \dim(H^1(L,E))$ for each nontrivial irreducible KL-submodule E of V.

In a while I'll use Theorem 1 to prove a result of O'Nan and Scott which is an important contribution to the description of the maximal subgroups of the alternating groups. But first some remarks to set the stage for this result and an analogous result on the classical groups.

There are a finite number of sporadic groups and a finite number of families of exceptional groups of Lie type. (A group of Lie type is *exceptional* if it is not classical.) Moreover modulo some unusual behavior over small fields, we can expect the subgroup structure of the groups in each family of exceptional groups to be more or less the same. Thus we can hope to enumerate the maximal subgroups of the sporadic groups and the exceptional groups of Lie type by ad hoc methods. Hence there is some reason to treat the alternating groups and classical groups from the point of view of their natural representations, even though such representations may not be available for all simple groups.

In any event these representations are a useful tool for describing the subgroup structure of an alternating group or classical group G. The idea is to describe subgroups of G as stabilizers of certain natural structures on the object X upon which G is represented. These stabilizers are candidates for maximal subgroups. Knowledge of the action of a subgroup H of G on a class of structures is equivalent to knowledge about the restriction to H of the permutation representation of G on the cosets of the stabilizer of a member of the class. In particular if we know the action of G on the structures we know the conjugacy classes of maximal subgroups which are stabilizers.

The problem of describing the maximal subgroups of G which stabilize no structure remains. However at least we do know lots of things about such subgroups. For example each such subgroup H acts irreducibly on X and indeed the fact that H stabilizes no structure imposes a variety of other useful constraints upon the representation of H on X. Perhaps most important it will turn out that H is almost simple. Then using Lemma 1 from early in this section, the study of the maximal subgroups of G is reduced to a study of the irreducible representations of almost simple groups in the appropriate category.

The O'Nan-Scott Theorem carries out this analysis for the alternating groups. I'll state the O'Nan-Scott Theorem in a moment after setting the stage with some definitions and preliminary lemmas. In the next chapter I'll state a result of my own which completes the analysis for the classical groups.

Let X be a set of finite order n and G the symmetric group on X.

Given integers m and k such that $m^k = n$ let Δ be a set of order m, Γ the set product of k copies of Δ, and R the wreath product $S_m wr S_k$ of the symmetric group on Δ with the symmetric group on $I = \{1,...,k\}$. An element of R is of the form $r = r_1 ... r_k\sigma$ where σ is a permutation of I, r_i is a permutation of Δ, and R is faithfully represented on Γ via

$$(x_1,...x_k)r = (x_{1\gamma}r_{1\gamma},...,x_{k\gamma}r_{k\gamma})$$

where $\gamma = \sigma^{-1}$.Let P be the set of bijections from Γ to X. Then R and G act on P by left and right multiplication, respectively, and these actions induce a transitive representation of G on the orbits $R\alpha$ of R on P. Call these orbits (m,k)-*product structures* on X, and observe:

Lemma 2. The stabilizer in G of an (m,k)-product structure is isomorphic to $S_m wr S_k$.

Lemma 3. Let H be a group and D a subgroup of H such that D is the direct product of some set I of subgroups transitively permuted by H. Assume M is a subgroup of H such that H = MD and $M \cap D$ is the direct product of the groups $M \cap L$, for L in I. Then H stabilizes an (m,k)-product structure in its representation on the cosets of M, where m is the index of $M \cap L$ for L in I and k is the order of I.

The proof of Lemma 3 is not difficult. For $L \in I$ let $M_L = M \cap L$, let Δ_L be the coset space L/M_L. Pick a copy J of L and let $\beta_L : L \to J$ be an isomorphism with $M_L\beta_L = N$ independent of L and let $\gamma_L : \Delta_L \to \Delta = J/N$ be the bijection induced by β_L. Let X be the coset space H/M and α the bijection of Γ with X defined by

$$\alpha : (Nx_L : L \in I) \mapsto M(\prod_{L \in I} x_L\beta_L).$$

Then a typical element of H is of the form $h = (\prod h_L)u$ with $h_L \in L$ and $u \in M$. Further $\alpha h = r\alpha$ where $r = (\prod r_L)\sigma$ is the element of R such that

σ is the permutation of I induced by the action of u by conjugation, $r_L = (\gamma_L w)^{-1}(h_L w)c\gamma_L$, $w = u^{-1}$, and $c : \Delta_L w \to \Delta_L$ is the map induced by conjugation by u. In particular H stabilizes the product structure $R\alpha$.

With these preliminaries complete we can state and prove the O'Nan-Scott Theorem.

Theorem 3.(O'Nan-Scott) Let G be the symmetric group on a set X of order n and H a subgroup of G. Then one of the following holds:

(1) H stabilizes some proper nonempty subset of X.

(2) H stabilizes some nontrivial partition of X.

(3) $n = m^k$ and H stabilizes some (m,k)-product structure on X.

(4) n is the power of some prime p, $F^*(H) = E$ is an elementary abelian p-group regular on X, and $N_G(E)$ is the split extension of E by GL(E).

(5) $F^*(H) = D$ is the direct product of $k > 1$ copies of some nonabelian simple group L, $n = |L|^{k-1}$, and $N_G(D) \cong Out(L)/(L wr S_k)$.

(6) $F^*(H) = L$ is a nonabelian simple group, H acts primitively on X, and $N_G(L)$ is the stabilizer in Aut(L) of the equivalence class of the representation of L on X.

Assume the hypothesis of the Theorem and suppose H does not satisfy (1) or (2). Then H is primitive on X and hence is described in Theorem 1. Let $D = F^*(H)$. If case A of Theorem 1 holds then D is an elementary abelian p-group regular on X. Let U be the split extension of D by GL(D). Then the representation of U on the cosets of a complement to D is faithful and extends the representation of D on X, so we may take $U \leq G$. Then as $F^*(N_G(D)) = D$ by Theorem 1, it follows that $U = N_G(D)$ so that (4) holds.

So case B or C of Theorem 1 holds. Indeed by Theorem 1 one of the following holds:

(a) There exists a set I of subgroups of D such that the hypothesis of Lemma 3 are satisfied with respect to the stabilizer M in H of a point of X.

(b) D is the direct product of k > 1 components permuted by H and $M \cap D$ is a full diagonal subgroup of D.

(c) D is simple.

Of course in case (a) Lemma 3 says that (3) holds. In case (b) let V be the normalizer in Aut(D) of $M \cap D$. Then $V = V_1 x V_2$ with $V_1 \cong S_k$, $V_2 \cong Aut(L)$ where L is a component of D, and $V_1 D \cong L wr S_k$. Moreover the representation of VD on the cosets of V is faithful and extends the representation of D on X, so as above $VD \approx N_G(D)$ and (5) holds. Finally in case (c) Lemma 1 says that (6) holds, completing the proof.

To my knowledge it is not known which of the subgroups in (1)-(5) of Theorem 3 are maximal in G, although I don't imagine it should be to difficult to find out. Modulo this problem, Theorem 3 and Lemma 1 reduce the determination of the maximal subgroups of S_n and A_n to the following problem:

Problem 3. Let L be a finite simple group, $\pi : L \to G$ a transitive representation on X, and extend π to the stabilizer H in Aut(L) of the equivalence class of π. Assume H is primitive on X. When is $H\pi$ maximal in G?

Assuming H preserves no product structure on X, Theorem 3 seems to reduce Problem 3 to the question of when there exists an almost simple group J extending the representation of H. Thus we are returned once again to the study of the subgroup structure of almost simple groups.

In the next chapter I'll discuss a result on the classical groups analogous to Theorem 3.

CHAPTER V

Subgroups of Simple Groups

This chapter continues to study permutation representations of finite groups from the point of view of the Classification. In the previous chapter we encountered a result describing the maximal subgroups of the alternating group of degree n in terms of its permutation representation of degree n. I'm going to begin this chapter with the following analogous result on the classical groups which appears in [4]:

Theorem 1. Let G be an almost simple group such that $F^*(G) = G_0$ is a classical group. If $G_0 \cong P\Omega_8^+(q)$ assume no member of G induces a triality automorphism on G_0. Let $H \leq G$ with $G = G_0H$ and V the natural module for the covering group of G_0. Then either H is contained in a member of the collection S_G of subgroups of G defined below, or the following hold:

1) $F^*(H)$ is a nonabelian simple group.
2) Let L be the covering group of $F^*(H)$ and F the field of definition of V. Then V is an absolutely irreducible FL-module.
3) The representation of L on V is defined over no proper subfield of F.
4) If $a \in Aut(F)$, V^* is the dual of V, and V is FL-isomorphic to V^{*a} then either:

   (i) a = 1 and G_0 is orthogonal or symplectic, or

   (ii) a is an involution and G_0 is unitary.

There exists a covering Ω of G_0 and a form f on V such that Ω is the derived group of the isometry group of (V,f). For example if $G_0 \cong L_n(q)$ then f is trivial. Let Γ be the group of nonsingular semilinear transformations of V preserving f up to a scalar multiple and a twist by a field automorphism. Usually $Aut(G_0) = P\Gamma$ is the image of Γ under the projective map P, so that G is the image of some subgroup $\tilde{G}$ of Γ under P. For purposes of exposition I'll assume G is such an image. Then the members of S_G are the images of the stabilizers in $\tilde{G}$ of certain natural structures on V described in the next paragraph. Again for purposes of exposition, the definitions will be somewhat simplified and hence not entirely accurate.

The first class of structures are the nondegenerate or totally singular subspaces of V. Second are the sets S of isometric subspaces such that V is either the orthogonal direct sum of the members of S, or S consists of a pair of complementary maximal totally singular subspaces. Third are vector space structures over some extension of F of prime degree. Fourth are subspaces U of V over a subfield K of prime degree in F such that $V = F \otimes_K U$. Fifth are certain tensor product structures on V. The sixth type of stabilizer is the collection of normalizers of r-subgroups of Γ of symplectic type acting irreducibly on V. Finally we have the stabilizers of certain forms on V.

In addition the isomorphism type of the stabilizers is determined as are the orbits of G on the structures.

The following problem is still open as far as I know:

Problem 1. Determine which members of S_G are maximal in G.

Problem 1 can certainly be solved with available techniques and the Classification, but the next problem seems much more difficult:

Problem 2. Let L be a quasisimple group and $\pi : L \to O(V,f)$ an absolutely irreducible FL-representation defined over no proper subfield of F. When is $N_{\tilde{G}}(L\pi)$ maximal in $\tilde{G}$?

Gary Seitz has made some progress on Problem 2. Seitz seems to be close to showing that if F and L are both of characteristic p then for almost all π as in Problem 2, $N_{\tilde{G}}(L\pi)$ is maximal in GkT. In addition he expects to provide a list of the cases in which maximality fails. His approach is to reduce to the corresponding problem for algebraic groups. That is he expects to classify the conjugacy classes of maximal subgroups of the classical semisimple algebraic groups in prime characteristic. If Seitz is successful his result will represent a big step toward a solution to problem 2.

We've seen that certain representations of the alternating groups and classical groups are useful for the study of the subgroup structure of these groups. It would be nice to have representations of the exceptional groups of Lie type and the sporadic groups which could be utilized in a similar manner. The representation of an exceptional group on its building serves this purpose to a limited extent. Recall from an earlier chapter that for various primes p, and sporadic groups G, G possesses p-overgroup geometries similar to the building of a group of Lie type. I would now like to discuss how the representation of the sporadic groups on their overgroup geometries can be used to derive results on subgroup structure.

Let G be a sporadic group and p a prime such that p^2 divides the order of G. In the first place the p-overgroup geometry supplies us with a representative from each conjugacy class of maximal subgroup over some fixed Sylow p-subgroup T of G and the diagram of the geometry tells us how this particular collection of maximal subgroups intersect. I'll refer to these intersections and their conjugates as *parabolics*.

Next the p-local structure of G can be related to the geometry. You may recall that the p-local structure of a group can be retrieved from its p-superlocals and that in the case of a group of Lie type the parabolics are the p-superlocals of G and each is of characteristic p-type. The situation for the sporadic groups is not so nice but not so bad either. In the first place our sporadic group G usually contains p-locals which are not of characteristic p-type, and as the parabolics usually are of characteristic p-type, the p-locals which are not of characteristic p-type are not contained in parabolics. On the other hand it is easy to determine and describe p-locals which are not of characteristic p-type just from knowledge of the normalizers of subgroups of prime order. So we are reasonably content if the p-locals of characteristic p-type can be retrieved from the parabolics, and fortunately this is usually (but not quite always) the case. That is to say almost always p-locals of characteristic p-type are contained in a parabolic and indeed if P is a p-superlocal of characteristic p-type then, with a couple of exceptions, either P is a parabolic or there exists a parabolic Q such that $O_p(Q) \leq O_p(P)$ and $P/O_p(Q)$ is a p-superlocal of $Q/O_p(Q)$ which is not of characteristic p-type. This seems to be the correct analogue of the Borel-Tits Theorem for sporadic groups.

Many of the sporadic groups possess a subgroup Q with the following properties:

(I) Q is an extraspecial p-group. Let $Z = Z(Q)$.

(II) $Q \trianglelefteq N_G(Z)$ and $C_G(Q) \leq Q$.

(III) $\mathcal{N}_G(Q,p') = 1$.

Under these hypotheses I'll say Q is a *large extraspecial p-subgroup of* G. As a matter of fact usually if a high power of the prime p divides the order of a sporadic group G then G contains a large extraspecial p-subgroup. Recall also that if G is of Lie type over the field of order p then almost always G has a large extraspecial p-subgroup. Finally certain groups of Lie type over GF(2) and GF(3) have large extraspecial p-subgroups for p = 3 and 2, respectively.

The classification of groups with a large extraspecial 2-subgroup played a big role in the Classification. The theory of these groups also plays a big role in my analysis of overgroup geometries of sporadic groups. Thus I'm going to spend some time on the subject now.

Let G be an almost simple group with a large extraspecial p-subgroup Q, let $Z = Z(Q)$ and $H = N_G(Q)$. The following elementary observations get us started.

Lemma 1. Q is weakly closed in $N_G(Q)$.

Lemma 2. If M is a subgroup of G containing Q then one of the following holds:

(1) $F^*(M) = O_p(M)$, $V = \langle Z^M \rangle$ is elementary abelian, and M acts irreducibly on V.

(2) $F^*(M)$ is a nonabelian simple group.

(3) $p = 2$, $Q \cong Q_8 * Q_8$, and M is of index at most 2 in $S_5 wr Z_2$.

These observations hold because Q is a large extraspecial subgroup of any of its overgroups in G.

Moreover it is not difficult using the Classification to determine all groups G containing a large extraspecial subgroup and such that $F^*(G)$ is simple. Hence we have a list of possible isomorphism types for the subgroups M of G of type (2), and using the fact that the order of M divides the order of G and several other only slightly less elementary observations, it is possible to determine up to isomorphism the subgroups occurring in (2). Later observations will allow us to determine the subgroups of type (2) up to conjugacy in G. Subgroups of type (3) are even easier to determine, leaving the subgroups of type (1).

The next observation is also easy to establish.

Lemma 3. Suppose $Z \neq Z^g \leq Q$ and let $X = \langle Q, Q^g \rangle$, $U = ZZ^g$, and $Y = N_Q(Z^g)N_{Q^g}(Z)$. Then $Y = C_X(U)$ and $X/Y \cong SL_2(p)$ acts naturally on U. In particular X is transitive on $U^\#$ and $Z \leq Q^g$.

One important step in the analysis of groups with a large extra special p-subgroup is to determine those groups for which Z is weakly closed in Q. These groups include the unitary and symplectic groups over the field of order p plus a few others. But in the generic case there exist conjugates of Z in Q to which Lemma 3 can be applied. If $Z \neq Z^g \leq Q$ write $Z+Z^g$ for the set of conjugates of Z contained in ZZ^g; thus $Z+Z^g$ is of order p+1 and $Aut_G(ZZ^g)$ is an extension of $L_2(p)$ by Lemma 3. Call $Z+Z^g$ the *line* through Z and Z^g.

Let $Q \leq M \leq G$, and define

$$L(M) = \{Z^g+Z^h : g,h \in M\}.$$

Then we have a rank 2 geometry S(M) with point set Z^M and line set L(M), where incidence is defined by inclusion.

Let T be a Sylow p-subgroup of G containing Q. By Lemma 1, Q is weakly closed in T. In particular $Q \trianglelefteq T$ so $Z \trianglelefteq T$ and then as Z is of prime order $Z \leq Z(T)$. But by condition (II) in the definition of Q, $Z(T) \leq Z$, so $Z = Z(T)$. Write I(T) for the set of conjugates Z^g of Z such that the line $Z+Z^g$ is T-invariant. In the generic case it turns out that T is transitive on I(T) and indeed all elements of I(T) are on the same line. Moreover H and the normalizer of the unique member of I(T) are maximal overgroups of T so S(G) is a truncation of the overgroup geometry of T. Indeed usually the diagram of the overgroup geometry is a string diagram with H the stabilizer of an end node and the stabilizer of a line stabilizes the unique node adjacent to that end node.

We wish to determine the maximal overgroups of T. In the generic case it turns out that for each overgroup M of T which is not contained in H we have $M = \langle H \cap M, Q^g \rangle$ where $Q^g \in I(T)$. Of course $H \cap M$ is an overgroup of T in H and we know the overgroups of T in H in an inductive context. From this information and earlier remarks it is possible to determine the maximal overgroups of T in G. Moreover this approach also tells us how the overgroups of T intersect and hence gives the structure of the overgroup geometry.

I'm going to close this chapter with an example which will hopefully illustrate some of the theory I've been discussing in the last few chapters and motivate the material in the next chapter. The example I have in mind is a recent theorem of Bill Kantor and an application of this result to the theory of projective planes. J. Saxl and M. Liebeck have also obtained the group theoretic result independently.

Theorem C of Kantor in [22] describes the almost simple primitive permutation groups of odd degree. Kantor then uses this result to prove Theorem A of [22] which says that if P is a finite projective plane of order n admitting a flag transitive group G of automorphisms then either P is desarguesian and G contains $L_3(n)$, or the number n^2+n+1 of points of P is prime and G is a Frobenius group.

First a few remarks about Theorem A. The flag transitivity of G on P forces G to be primitive on the set X of points of P. Of course X is of order n^2+n+1. In particular X is of odd order so G is primitive on X of odd degree, and from Theorem 1 in chapter 4 and the special nature of the representation it is not difficult to see that either X is of prime order and G is Frobenius or G is almost simple. Hopefully at this point it is not hard to believe that Theorem C can then be used to complete the proof of Theorem A.

This then is an example of how a problem in geometry can be reduced to a question about permutation representations of finite groups, and that question can be reduced in turn to a question about primitive representations of almost simple groups. Moreover the question about primitive groups to which we've reduced is of the type we've been discussing since the maximal subgroups corresponding to the representation are overgroups of Sylow 2-subgroups of our almost simple group. (Since the representation is of odd degree.)

But back to Theorem C. Let G be an almost simple group, M a maximal subgroup of odd index, and $L = F^*(G)$. If L is an alternating group on a set Y then it is easy to show that (with a couple of known exceptions) M is imprimitive on Y and hence known. If L is of Lie type and even characteristic then we know M is a maximal parabolic. So difficulties occur only when L is sporadic or of Lie type and odd characteristic, and if L is sporadic we now know the maximal overgroups of Sylow 2-groups.

So we're left with the case when L is of Lie type and odd characteristic. The treatment of this case depends upon two deep results useful both in the Classification and in the study of subgroup structure. The first of these results determines all groups with a so called *classical involution* and the second is really a collection of results which on the one hand classify groups generated by so called *root elements* and on the other determines the subgroups of groups of Lie type containing root elements. I will discuss these results in the final chapter.

CHAPTER VI

Fischer Theory

In this final chapter I'll be discussing several theorems which are useful both in the Classification and in describing the subgroup structure of finite groups. Many of these results are in the spirit of B. Fischer's paper classifying groups generated by 3-transpositions. So part of this chapter will be devoted to techniques originating in that paper.

Let G be a finite group and Ω a set of subgroups of G permuted by G via conjugation. B. Fischer introduced the idea of studying G under the hypothesis that the isomorphism type of $\langle X,Y\rangle$ is contained in some restricted class $\Lambda = \Lambda(\Omega)$ of groups for each X,Y in Ω. For example define Ω to be *locally conjugate* in G if for each X,Y in Ω either $[X,Y] = 1$ or X is conjugate to Y in $\langle X,Y\rangle$. Then the following observation is almost immediate:

Lemma 1. Let Ω be locally conjugate in G, assume $G = \langle\Omega\rangle$, let Ω_i, $1 \leq i \leq n$, be the orbits of G on Ω via conjugation, and let $G_i = \langle\Omega_i\rangle$. Then G is the central product of the groups G_i and G_i is transitive on Ω_i.

In some of the examples we will be considering in a moment Ω will be locally conjugate in G. Even if Ω is not locally conjugate it will have some similar property which will tend to force the conclusion of Lemma 1 to hold in most interesting situations. Hence we can usually assume G is generated by and transitive on Ω.

Here is a more concrete example. Let G be a group of Lie type of characteristic p and Ω the set of long root subgroups of G. Exclude the cases where G is $^2F_4(q)$, $^2B_2(q)$, or $^2G_2(q)$. Then the members of Ω are isomorphic to E_q and Λ consists of E_q2, $SL_2(q)$, and a Sylow p-group q^{1+2} of $L_3(q)$. (q^{1+2} is a certain special group of order q^3.) Indeed if G is symplectic or unitary then q^{1+2} can be excluded, from which it follows that Ω is locally conjugate in G.

Here is a related example. This time take Ω to be the set of subgroups of order p generated by long root elements. Of course if q = p this is the same example, but in general it is more complicated. In this case if p is odd then Λ consists of E_p2, p^{1+2}, and $SL_2(r)$, for r dividing q or r = 5 if p = 3 and q > 3. If p = 2 then Λ consists of dihedral groups of order 2m, m = 2 or 4, or m odd. Moreover it turns out that if <X,Y> is dihedral of order 8 then [X,Y] is also in Ω.

This suggests the following definition of Timmesfeld: Let G be a finite group and Ω a set of involutions of G. Then Ω is said to be a set of *root involutions* of G if G = <Ω> and for each choice of x and y in Ω, the order of xy is 2,4, or an odd integer, and if xy is of order 4 then [x,y] is also in Ω.

Thus long root involutions of groups of Lie type and characteristic 2 are root involutions in the sense of Timmesfeld. (Unless G is $^2F_4(2)$.) An important theorem of Timmesfeld [29] (building on earlier work of Fischer [16] and Aschbacher [3]) classifies all groups generated by root involutions. In addition to the groups of Lie type in characteristic 2 one gets the symmetric groups, orthogonal groups over fields of order 3 and 5, and the Fischer groups generated by 3-transpositions.

To my knowledge a proof of the exact analogue of Timmesfeld's Root Involution Theorem in odd characteristic does not exist, although various partial results certainly are available. See Stark [26] for example. On the other hand there does exist a result which for most purposes is even stronger than an odd characteristic version of the Root Involution Theorem. That result is the so called Classical

Involution Theorem.

Let L be a group of Lie type and odd characteristic p, let $U = U_\alpha$ be a long root subgroup and $V = U_{-\alpha}$. Then V is a conjugate of U and K = <U,V> is isomorphic to $SL_2(q)$. K is a so called *fundamental subgroup* of L. Let z be the involution in K. It turns out that K is subnormal in $C_L(z)$ and that fact has implications for the fusion of 2-elements of K suggesting the following definition. Let G be a finite group and Ω a G-invariant set of subgroups of G. Then Ω is an *omega set* of G if for each choice of distinct K and J in Ω the following hold:

Ω1) K has a unique involution z(K) and nonabelian Sylow 2-subgroups.

Ω2) Either $K = O^{2'}(K)$ or $K/O(K) \cong SL_2(3)$.

Ω3) $|K \cap J|_2 \le 2$ with $[K,J] \le O(K) \cap O(J)$ in case of equality.

Ω4) If v is a 2-element in J-Z(J) centralizing z(K) then v normalizes K.

The fundamental subgroups in a group G of Lie type and odd characteristic form an omega set of G. The main theorem of [2] (Which is sometimes called the Classical Involution Theorem) classifies all groups generated by an omega set. In addition to the groups of Lie type and odd characteristic examples include M_{11}, M_{12}, $D_4(2)$, $Sp_6(2)$, and various groups which are not simple such as the subgroup of index 2 in the wreath product of Z_2 with A_n.

Let's look at some applications of the Root Involution Theorem and the Classical Involution Theorem. In the first place both results are used in the Classification. In particular the Classical Involution Theorem played a major role in establishing the B_2-property and dealing with the 2-standard form problems for groups of Lie type and odd characteristic. Both results are important in the linear representation theory involved in pushing up arguments. For example

Thompson's Quadratic Pair Theorem can be derived from the Classical Involution Theorem and some hard work. For a discussion of quadratic pairs see [20]. But in these lectures I wish to emphasize the applications to the study of the subgroup structure of the simple groups.

Let G be a group of Lie type and characteristic p and Ω the set of long root elements of G. Suppose M is a maximal subgroup of G containing a long root element. If $p = 2$ then $\Omega \cap M$ is a set of root involutions of M, so the Root Involution Theorem describes the normal subgroup $\langle M \cap \Omega \rangle$ of M up to isomorphism. Similarly if p is odd and $X \in M \cap \Omega$ then either $X \leq O_p(M)$ or there exists $Y \in \Omega$ with $K_0 = \langle X,Y \rangle \cong SL_2(r)$. In the former case M is a maximal parabolic by the Borel-Tits Theorem, so we may assume the latter holds. Let z be the involution in K_0. Then $K_0 \leq K_1 \cong SL_2(q)$ with K_1 subnormal in $C_G(z)$. Let $K = K_1 \cap M$. Then $K \cong SL_2(s)$ and K is subnormal in $C_M(z)$ so K^M is an omega set of M. Hence the Classical Involution Theorem describes $\langle K^M \rangle$ up to isomorphism.

Thus in either case a large normal subgroup of M is determined up to isomorphism. With this information and some hard work one can go on and determine up to conjugacy all maximal subgroups of G containing long root elements. This has been done by Kantor [21] and Cooperstein [13].

Recall Theorem C of Kantor from the previous chapter. That result classified the almost simple primitive permutation groups of odd degree. Remember that the hardest case in this result involved determining the maximal overgroups M of a Sylow 2-subgroup T of a group G of Lie type and odd characteristic p. The Classical Involution Theorem is a key tool here too. Namely there exists a fundamental subgroup K_1 of G such that $T \cap K_1$ is Sylow in K_1. Hence setting $K = M \cap M$, again K^M is an omega set of M and the Classical Involution Theorem is applicable. As a matter of fact that result says that usually K contains a long root element of K_1, so we can often reduce to the classification of subgroups of G containing long root elements.

The type of analysis used to prove the Root Involution Theorem and Classical Involution Theorem is of interest; I'll refer to such analysis as *Fischer theory*. In Fischer theory we are given a group G and a G-invariant set Ω of subgroups. Usually we can assume Ω generates G and G is transitive on Ω. Define $\Gamma = \Gamma(\Omega)$ to be the graph with vertex set Ω and with X and Y adjacent in Γ when $[X,Y] = 1$. Evidently G is represented as a group of automorphisms of Γ via conjugation. This representation is the principal tool in Fischer theory.

Under suitable hypotheses (Such as restrictions on $\Lambda(\Omega)$) Γ possesses lots of combinatorial and geometric structure. In the case of the Classical Involution Theorem and the Root Involution Theorem this structure is quite useful in the study of the subgroup structure of groups of Lie type; see [2] or [14] for example. Right now however I'd like to concentrate instead on the representation in the case of the three sporadic Fischer groups.

So assume G is one of the groups M(22), M(23), or M(24) and let Ω be the set of root involutions of G. Then G is actually a set of *3-transpositions* of G; that is for each $x,y \in \Gamma$, the order of xy is at most 3. For $x \in \Omega$ denote by D_x and A_x the set of elements in Ω distinct from x which are adjacent to x and not adjacent to x in Γ, respectively. Fischer proves the following fundamental fact about 3-transposition groups:

Lemma 2. Let Ω be a conjugacy class of 3-transpositions of a group G and assume $G' = G''$ and $O_3(G)$ and $O_2(G)$ are in the center of G. Then G is of permutation rank 3 in its action on Ω. That is $C_G(x)$ is transitive on D_x and A_x for each $x \in \Omega$.

Here G' denotes the commutator group of G. Lemma 2 has analogues in much more general situations; see for example [7].

Lets return to the special case where G is a Fischer group. Here many more statements about the transitivity of G on suitable substructures of Γ can be established, to the point where there is no doubt that the representation of G on Γ is an excellent tool for studying the group. However as far as I know the Fischer groups have never been systematically studied from the point of view of this representation. Some rather specialized results of Enright are available in [15] but his approach is not the one I have in mind. It would seem to me to be that someone should undertake such a systematic study. For example it should be possible to describe explicitly all subgroups of G containing 3-transpositions.

Its also worth remembering that Fischer discovered the Baby Monster F_2 as a {3,4}-transposition group. That is F_2 is generated by a class Ω of involutions such that the order of xy is at most 4 for each $x,y \in \Omega$. If $[x,y]$ were in Ω whenever xy is of order 4, then Ω would be a set of root involutions; but unfortunately this is not the case. The group F_2 should also be studied systematically from the point of view of its representation on its {3,4}-transposition graph.

I'm going to close by mentioning a few problems about the subgroup structure of finite simple groups that seem interesting to me. First it would be useful to have a linear representation of each of the exceptional groups of Lie type sharing many of the properties of the representation of a classical group on its associated space. That is we would like the module to be small and we would want to know the action of the group on the various subspaces of the module. One would then presumably establish an analogue of Theorem 1 in Chapter 5. A possible choice for our module is the module of minimal dimension for the group. It may be the case however that no representation exists with suitably nice properties.

It would be useful to have a good description of the r-local subgroups and overgroups of Sylow r-subgroups in the groups of Lie type and characteristic p for r distinct from p. In the case of the classical groups this can presumably be done in terms of the representation on the associated space with Theorem 1 in Chapter 5 supplying a good start. If analogous representations for the exceptional groups exist the same could be done for them. Recall that the Lie theory supplies a description of centralizers of r-elements. Seitz has results describing the isomorphism type of subgroups of groups of Lie type containing maximal tori. [24] These results are potentially of use in describing locals and overgroups of Sylow groups. However Seitz' theorems are only stated when the characteristic and field associated to the group are sufficiently large.

REFERENCES

1. Aschbacher, M., The classification of the finite simple groups, Math. Intelligencer. 3 (1981), 59-65.

2. Aschbacher, M., A characterization of Chevalley groups over fields of odd order, Ann. Math., 106 (1977), 353-468.

3. Aschbacher, M., Finite groups generated by odd transpositions, I, II, III, IV, Math. Z., 127 (1972), 45-56, J.Alg. 26 (1973), 451-491.

4. Aschbacher, M., On the maximal subgroups of the finite classical groups, Invent. Math., 76 (1984), 469-514.

5. Aschbacher, M., and Scott, L., Maximal subgroups of finite groups, J. Alg. 92 (1985), 44-80.

6. Aschbacher, M., The Finite Simple Groups and their Classification, Yale University Press, New Haven, 1980.

7. Aschbacher, M., A homomorphism theorem for finite graphs, Proc. AMS, 54 (1976), 468-470.

8. Aschbacher, M., Finite geometries of type C_3 with flag transitive groups, Geom. Ded., 16 (1984), 195-200.

9. Aschbacher, M., Flag structures on Tits geometries, Geom. Ded., 14 (1983), 21-32.

10. Aschbacher, M., and Smith, S., Tits geometries over GF(2) defined by groups over GF(3), Comm. Alg., 11 (1983), 1675-1684.

11. Borel, A. et al, Seminar on Algebraic Groups and Related Finite Groups, Springer-Verlag, Berlin, 1970,.

12. Borel, A. and Tits,J., Elements unipotents et sousgroupes paraboliques de groupes reductifs, Invent. Math., 12 (1971), 97-104.

13. Cooperstein, B., Subgroups of exceptional groups of Lie type generated by long root elements, I; Odd characteritic, J. Alg., 70 (1981), 270-282.

14. Cooperstein, B., The geometry of root subgroups in exceptional groups, I, Geom. Ded., 8 (1979), 317-381.

15. Enright, G., Subgroups generated by transpositions in F_{22} and F_{23}, Comm. Alg., 6 (1978), 823-837.

16. Fischer, B., Finite groups generated by 3-transpositions, University of Warwick preprint.

17. Gorenstein, D., The Classification of Finite Simple Groups, I, Plenum Press, New York, 1983.

18. Gorenstein, D., Finite Simple Groups; An Introduction to their Classification, Plenum Press, New York, 1982,.

19. Gross,F., and Kovacs, L., Maximal subgroups in composite finite groups, to appear.

20. Ho, C., On the quadratic pairs, J. Alg., 43 (1976), 338-358.

21. Kantor, W., Subgroups of classical groups generated by long root elements, Trans. AMS, 248 (1979), 347-379.

22. Kantor, W., Primitive permutation groups of odd degree, and an application to finite projective planes, to appear.

23. Niles, R., Finite groups with parabolic type subgroups must have a BN-pair, J. Alg., 75 (1982), 484-494.

24. Seitz, G., The root subgroups for maximal tori in finite groups of Lie type, to appear.

25. Seitz, G., Flag transitive subgroups of Chevalley groups, Ann. Math., 97 (1974), 27-56.

26. Stark, B., Another look at Thompson's quadratic pairs, J. Alg. 45 (1977), 334-342.

27. Timmesfeld, F., Tits geometries and parabolic systems of rank 3, to appear.

28. Timmesfeld, F., Tits geometries and parabolic systems in finite groups, to appear.

29. Timmesfeld, F., Groups generated by root involutions, I, II, J. Alg., 33 (1975), 75-134.

30. Tits, J., A local approach to buildings, The Geometric Vein, 517-547, Springer, , 1981.

31. Tits, J., Buildings of Spherical Type and Finite (B,N)-Pairs, Springer-Verlag, Berlin, 1974.

TOPICS IN THE THEORY OF REPRESENTATIONS OF FINITE GROUPS

Charles W. Curtis
Department of Mathematics
University of Oregon
Eugene, OR 97403 U.S.A.

Introduction.

These lectures are devoted to two currently active parts of the representation theory of finite groups. The first is an introduction to Brauer's theory of modular representations and blocks, taking account of the new methods introduced by Green, and some other recent developments. The second is a survey of the representation theory of finite groups of Lie type. One test of new ideas in representation theory, emphasized frequently by Brauer, is the extent to which they provide significant additional information in the classical theory of complex valued characters. We shall pursue each topic far enough so that at least some of these new insights should be apparent. Whenever possible, proofs of the main results are sketched, in order to show how the main ideas are applied. Most of the main theorems appear in at least one of the books of Curtis-Reiner [2], Feit [3], and Serre [8], where the reader will find references to the original sources. We have also included comments on several unsolved problems and areas for further research, with references to the literature.

1. Induced Modules.

The connection between representations of a finite group G and representations of subgroups of G is a theme whose importance was recognized from the beginning, in the work of Frobenius. The focus has changed, however, from an emphasis on calculations with characters to the use of the functors $\otimes$ and Hom of homological algebra. This section contains a survey of some of the basic results in this subject. While they are now quite well understood, they continue to suggest new ideas, and are essential background for results to be discussed later in these lectures.

1A. Frobenius Reciprocity and the Mackey Theorems.

In these lectures, G denotes a finite group, and RG the group algebra of G over a commutative ring R. An R-representation of G is a homomorphism

$T: G \to GL(M)$

$$(\lambda,\mu) = |G|^{-1} \sum_{x \in G} \lambda(x)\overline{\mu(x)} .$$

By a basic set of simple (or irreducible) CG-modules, we mean a complete set of representatives of isomorphism classes of simple CG-modules. A basic set of irreducible characters $\{\zeta^1,\ldots,\zeta^s\}$ is given by the characters of a basic set of simple CG-modules. The orthogonality relations assert that $\{\zeta^1,\ldots,\zeta^s\}$ form an orthonormal basis of $cf(G)$ with respect to the scalar product (λ,μ). In particular, if $\mu \in cf(G)$, its "Fourier expansion" in terms of the irreducible characters is given by

$$\mu = \Sigma(\mu,\zeta^i)\zeta^i$$

If μ is a character, the coefficients (μ,ζ^i) give the multiplicities with which the simple modules $\{Z_i\}$ affording $\{\zeta^i\}$ occur in a direct decomposition of a module affording μ as a direct sum of simple modules.

An easy calculation establishes:

(1.4) Frobenius Reciprocity Formula. Let $\zeta \in cf(G)$, $\lambda \in cf(H)$. Then

$$(\zeta,\lambda^G)_G = (\zeta|_H,\lambda)_H .$$

This formula is surprisingly powerful, and in many cases can be used, together with the definition (1.3), to find all the characters of a given group (for example the dihedral groups). The preceding formula, combined with Schur's Lemma and the orthogonality relations, shows that if M is a left CG-module, and L a CH-module, then

$$\dim_C \operatorname{Hom}_{CG}(M,L^G) = \dim_c \operatorname{Hom}_{CH}(M_H,L) .$$

This result suggests the following general result, which can be applied in cases where the modules are not necessarily semisimple.

(1.5) Proposition. Let R be a commutative ring, $H \le G$, M a left RG-module, and L a left RH-module. Then there exist isomorphisms of R-modules:

$$\operatorname{Hom}_{RH}(L,M_H) \cong \operatorname{Hom}_{RG}(L^G,M)$$

and

$$\operatorname{Hom}_{RG}(M,L^G) \cong \operatorname{Hom}_{RH}(M_H,L) .$$

The first is proved using the adjointness theorem:

$$\operatorname{Hom}_A(L', \operatorname{Hom}_B(M',N')) \cong \operatorname{Hom}_B(M' \otimes_A L',N'),$$

for rings A and B, and L' a left A-module, M' a (B,A)-bimodule, and N' a left B-module. For the second, let $L^G = \Sigma_{g \in G/H}(g \otimes L)$.

from G into the group of invertible R-endomorphisms of a finitely generated (f.g.) R-module M. Representations provide a way to study properties of G in terms of linear algebra. Each R-representation T defines a left RG-module, with

$$(\sum_{x\in G} a_x x)\cdot m = \sum_{x\in G} a_x T(x)m, \text{ for } a_x \in R,\ m \in M.$$

The classification of equivalence classes of R-representations is the same problem as the classification of isomorphism classes of left RG-modules.

Let H be a subgroup of G (notation: $H \le G$). Each RG-module M defines an RH-module M_H (or $\mathrm{res}_H^G M$) by restriction of scalars. Conversely, each RH-module L defines an RG-module $L^G = \mathrm{ind}_H^G L$, called an <u>induced module</u>, and given by

$$L^G = RG \otimes_{RH} L.$$

Since RG is a free right RH-module with a basis given by a cross section of the left cosets $\{gH\}$ (notation: $g \in G/H$), we obtain

$$L^G = \bigoplus_{g \in G/H} g\otimes L \tag{1.1}$$

using the basic properties of tensor products. From (1.1) we derive at once the

(1.2) <u>Imprimitivity Theorem</u>. <u>Let</u> $H \le G$, <u>and let</u> L <u>be an</u> <u>RH-submodule of</u> M_H, <u>for some f.g.</u> <u>RG-module</u> M. <u>Suppose</u>

$$M = \oplus_{g \in G/H}\, gL.$$

<u>Then</u> $M \cong L^G$.

The formula (1.1) also allows the representation afforded by L^G to be computed explicitly. In particular, if the ring of scalars is a field of characteristic zero, and λ is the <u>character</u> (trace function: $\lambda(x) = \mathrm{Tr}(x,L)$) of H afforded by L, then the character of G afforded by L^G is given by

$$\lambda^G(x) = |H|^{-1} \Sigma_{g \in G}\, \dot{\lambda}(g^{-1}xg), \quad \text{for } x \in G, \tag{1.3}$$

where

$$\dot{\lambda}(g) = \begin{cases} 0 & \text{if } g \in H \\ \lambda(g) & \text{if } g \in H \end{cases}.$$

Characters are constant on conjugacy classes, and hence belong to the vector space cf(G) of class functions. The formula (1.3) defines a map from the space of class functions cf(H) to cf(G).

In case the ring of scalars is the field of complex numbers C, we can define a hermitian scalar product $(\lambda,\mu) = (\lambda,\mu)_G$ on the vector space cf(G), where

Then each $\varphi \in \mathrm{Hom}_{RG}(M, L^G)$ defines a unique $\theta \in \mathrm{Hom}_{RH}(M_H, L)$ such that

$$\varphi(m) = \Sigma_{g \in G/H}\, g \otimes \theta(g^{-1}m), \quad m \in M,$$

and the map $\varphi \to \theta$ is the required isomorphism.

Now let $H, K \leq G$, and let L be a left RH-module. For $a \in G$, ${}^aH = aHa^{-1}$, and ${}^aL$ denotes the conjugate module, which is the $R({}^aH)$-module whose underlying R-module is L, and the action by elements ${}^ah = aha^{-1} \in {}^aH$ is given by

$$({}^ah).\ell = h\ell, \quad \text{for } h \in H,\ \ell \in L.$$

We now have

(1.6) Mackey's Subgroup Theorem. *Let $H, K \leq G$, and let*

$$G = \bigcup K aH \quad \text{(disjoint union of double cosets).}$$

Let L be a left RH-module. Then

$$(L^G)_K = \oplus\, ({}^aL)_{{}^aH \cap K}{}^K,$$

where the sum is taken over a cross section of the double cosets $K\backslash G/H$.

For the proof, let $L^G = \oplus_{g \in G/H}\, g \otimes L$. Then we have

$$(L^G)_K = \oplus_D W(D), \quad \text{where} \quad W(D) = \oplus_{gH \subset D}(g \otimes L),$$

and the first sum is taken over the (K,H)-double cosets D.
Let

$$H = \bigcup s_j\,({}^aH \cap K) \quad \text{(disjoint union of cosets).}$$

Then

$$KaH = \bigcup s_j aH \quad \text{(disjoint)}$$

and

$$W(D) = \oplus\, s_j a \otimes L = \oplus\, s_j\,(a \otimes L).$$

Noting that $a \otimes L \cong {}^aL$ as $R({}^aH \cap K)$-modules, the result follows at once from the Imprimitivity Theorem (1.2).

As an application, we have

(1.7) Mackey's Interwining Number Theorem. *Let L_1 and L_2 be left RH_1 and RH_2-modules, for subgroups $H_1, H_2 \leq G$. Then*

$$\mathrm{Hom}_{RG}(L_1{}^G, L_2{}^G) \cong \oplus_a\, \mathrm{Hom}_{R(H_1 \cap {}^aH_2)}(L_1, {}^aL_2),$$

where the sum is taken over a cross section of $H_2\backslash G/H_1$.

The proof is a nice illustration of the use of the general version of Frobenius reciprocity, combined with the subgroup theorem. We have, using (1.5) and (1.6),

$$\begin{aligned}\mathrm{Hom}_{RG}(L_1{}^G, L_2{}^G) &\cong \mathrm{Hom}_{RH_1}(L_1, (L_2{}^G)_{H_1}) \\ &\cong \oplus_a \mathrm{Hom}_{RH_1}(L_1, ({}^aL_2)_{H_1 \cap {}^aH_2}{}^{H_1}) \\ &\cong \oplus_a \mathrm{Hom}_{R(H_1 \cap {}^aH_2)}(L_1, {}^aL_2)\end{aligned}$$

as required.

For characters of CG-modules, $(\mu,\mu) = 1 \Leftrightarrow \mu \in \mathrm{Irr}G$. Thus we obtain, for $\lambda \in \mathrm{Irr}\, H$,

$$\lambda^G \in \mathrm{Irr}\, G \Leftrightarrow (\lambda, {}^x\lambda)_{H \cap {}^xH} = 0 \quad \text{for all} \quad x \in G - H.$$

Another application of these ideas is:

(1.8) <u>Clifford's Theorem.</u> <u>Let</u> F <u>be a field, let</u> M <u>be a simple FG-module, and</u> L <u>a simple submodule of</u> M_H, <u>for</u> H <u>a normal subgroup of</u> G. <u>Then</u>

(i) M_H <u>is a semisimple FH-module</u>.

(ii) <u>The simple submodules of</u> M_H <u>are all conjugates of</u> L, <u>and any two conjugates appear with the same multiplicity</u> in M_H.

(iii) <u>Let</u> $\tilde{H} = \{g \in H : gL \cong L\}$, <u>and let</u> $\tilde{L} = \Sigma_{gL \cong L}\, gL$.

<u>Then</u>

$$M = \oplus_{x \in G/\tilde{H}}\, x\tilde{L}$$

<u>and</u> $M \cong \mathrm{ind}_{\tilde{H}}^G \tilde{L}$.

The last statement follows from the Imprimitivity Theorem. The subgroup $\tilde{H}$ is the stabilizer of the homogeneous component of M_H containing L. Part (iii) gives no new information unless $\tilde{H} \neq G$. A condition for this to occur is:

(1.9) <u>Blichfeldt's Imprimitivity Theorem.</u> <u>Let</u> F <u>be an algebraically closed field, let</u> A <u>be an abelian normal subgroup of</u> G <u>not contained in the center, and let</u> M <u>be a simple FG-module on which</u> G <u>acts faithfully</u>: $gm = m$ <u>for all</u> $m \in M \Rightarrow g = 1$. <u>Then</u> $M \cong \mathrm{ind}_H^G L$ <u>for a proper subgroup</u> H, <u>and an FH-module</u> L.

This yields a definitive result on representations of nilpotent groups (for other types of groups we must work harder!)

(1.10) <u>Corollary.</u> <u>Let</u> G <u>be a finite nilpotent group, and let</u> F <u>be an algebraically closed field. Then every simple FG-module</u> M <u>can be expressed as an induced module</u>: $M \cong \mathrm{ind}_H^G L$, <u>for some subgroup</u> $H \leq G$, <u>and a one-dimensional FH-module</u> L.

1B. The Brauer Induction Theorem.

In this subsection, the ring of scalars will be the field of complex numbers C. Although not every irreducible character of a typical group is induced from a proper subgroup, Brauer proved a remarkable result which does express every character as a $\mathbb{Z}$-linear combination of induced characters. In order to understand the result, we need to observe that sums and products of characters of CG-modules are characters of CG-modules (afforded by the direct sum and tensor product of the modules affording them). It follows that the set of $\mathbb{Z}$-linear combinations of characters of CG-modules is a ring of complex valued functions on G, which we denote by ch(G), and call the ring of virtual characters (or generalized characters). The ring ch(G) is a free $\mathbb{Z}$-module, with a basis consisting of the irreducible characters $\zeta^1,\dots,\zeta^s$.

A second observation is the following result, which follows at once from the definition of induced class functions:

(1.11) Lemma. Let $H \le G$, and let $\lambda \in ch(H)$, $\mu \in ch(G)$. Then
$\mu\cdot(\lambda^G) = (\mu|_H\lambda)^G$

As a consequence, if $\mathcal{H}$ is any family of subgroups $H \le G$, the set of $\mathbb{Z}$-linear combinations of virtual characters of the form λ^G, where $\lambda \in ch(H)$ for some $H \in \mathcal{H}$, is an ideal in the ring ch(G). Thus, in order to prove that every virtual character of G is a $\mathbb{Z}$-linear combination of induced characters of the form λ^G, for $\lambda \in ch(H)$ and $H \in \mathcal{H}$, it is sufficient to prove that 1_G can be expressed in this way, where 1_G is the character of the trivial representation of G. This idea is the starting point of the proofs of many induction theorems; we shall state here only the most important one.

We require one other concept. A subgroup $H \le G$ is called p-elementary for a prime p if $H = \langle x\rangle \times B$ (direct product), where $\langle x\rangle$ is a cyclic p'-group, (of order prime to p), and B is a p-group. An elementary subgroup is a p-elementary sub-group for some prime p.

(1.12) The Brauer Induction Theorem. Let $\zeta \in ch(G)$. Then ζ can be expressed as a $\mathbb{Z}$-linear combination:

$$\zeta = \Sigma a_i(\lambda_i)^G,\ a_i \in \mathbb{Z}$$

where the $\{\lambda_i\}$ are linear characters (afforded by one dimensional CH_i-modules) of some elementary subgroups $\{H_i\}$ of G.

Note that elementary groups are nilpotent, so every irreducible character of an elementary group is induced from a linear character of a subgroup, by (1.10). Moreover, subgroups of elementary groups are elementary. Thus it is sufficient to prove, by the remarks preceding the statement of the Theorem, that 1_G is a $\mathbb{Z}$-linear combination of induced characters from elementary subgroups.

We mention two applications of Brauer's Theorem. The first gives a simple test for checking whether class functions are virtual characters.

(1.13) Brauer's Criterion for Virtual Characters. Let μ be a class function on G. Then μ is a virtual character if and only if $\mu|_H \in ch(H)$, for all elementary subgroups $H \le G$.

One way is clear. Now let $\mu|_H \in ch(H)$ for all elementary subgroups H. By (1.12), we have

$$1_G = \Sigma a_i \lambda_i^G, \ a_i \in \mathbb{Z}, \lambda_i \in ch(H_i),$$

for elementary subgroups H_i. Then by (1.11),

$$\mu = \mu \cdot 1_G = \Sigma a_i \mu(\lambda_i^G) = \Sigma a_i (\mu_{H_i} \cdot \lambda_i)^G,$$

and the result follows, since $\mu|_{H_i} \cdot \lambda_i \in ch(H_i)$ for each subgroup H_i.

Another application, in a different direction, involves the concept of a splitting field, which we discuss here for subfields of C. A CG-module M is realizable in a subfield F if there exists an FG-module $M_0 \le M$ such that $M \cong C \otimes_F M_0$. In other words, M_0 contains a C-basis of M, so the matrices representing elements of G, taken with respect to a suitable basis, all have entries in F.

(1.14) Brauer's Splitting Field Theorem. Let ω be a primitive m-th root of 1 in C, where m is the exponent of G. Then any field F containing ω is a splitting field for G.

It is clear that any representation of the form L^G, where L is a one-dimensional representation of a subgroup $H \le G$, is realizable in F. The Brauer Induction Theorem asserts that an arbitrary character of a CG-module is a difference of two characters of CG-modules which are realizable in F. A fairly straight-forward inductive argument completes the proof.

2. Modular Representations and Blocks.

In this section, we survey Brauer's theory of modular representations, with emphasis on new methods introduced by Green and others. Some points of special interest are: (i) the use of the trace map and the Mackey Subgroup Theorem as a common foundation of the theory of vertices of indecomposable modules, and defect groups of blocks; (ii) Brauer's Second Main Theorem as an illustration of how modular representations can be applied to obtain new results in character theory.

2A. Introduction to Modular Representations.

Let G be a finite group and p a prime number which will remain fixed throughout the discussion. A p-modular system is a triple (K,R,k) consisting of a discrete valuation ring (d.v.r) R with quotient field K of characteristic zero, and residue field k of characteristic p, where $k = R/\mathfrak{p}$, and $\mathfrak{p}$ is the unique maximal ideal in R. We assume always that R is complete in the $\mathfrak{p}$-adic topology, and that K is sufficiently large relative to G, so that both K and k are splitting fields for G. For example, K can be taken to be the completion of the algebraic number field $Q(\sqrt[m]{1})$ of m-th roots of 1 (where m is the exponent of G), with respect to the valuation defined by a prime ideal containing p.

The theory of modular representations is the study of modules over KG, RG and kG, and the relations between them. Since the characters of KG-modules can be identified with the complex valued characters of G, we can expect, as Brauer did, to obtain new information about complex characters from a study of kG-modules.

We begin with some preliminary remarks. The rings RG and kG are not semisimple, but we have

$$RG/\text{rad } RG \cong kG/\text{rad } kG,$$

so the simple modules for RG and kG can be identified with one another.

Each f.g. RG-module (or kG-module) is a direct sum of indecomposable modules, so their classification becomes a main object to study. The hypotheses imply that if M is a f.g. indecomposable RG-module, then $E = \text{End}_{RG}M$ is a local ring, that is $E/\text{rad}E$ is a division ring. Using this fact, it is not difficult to prove:

(2.1) Krull-Schmidt-Azumaya Theorem (K - S - A Theorem). Every f.g. RG-module M is a direct sum of a finite number of indecomposable modules, and these are uniquely determined up to isomorphism. (The same result holds for kG-modules.)

In the case of RG-modules, we shall be mainly interested in RG-lattices; these are f.g. RG-modules, which are projective as R-modules. This implies that they are free R-modules, since R is a principal ideal domain (P.I.D.).

Characters of kG-modules do not determine the modules up to isomorphism, as is shown by the simplest examples. So we use instead

<u>Grothendieck groups</u> as a substitute for characters. The Grothendieck group $K_0(\mathcal{C})$ of a category $\mathcal{C}$ of modules is the quotient group $\mathcal{I}/\mathcal{R}$, where $\mathcal{I}$ is the free abelian group with a $\mathbb{Z}$-basis consisting of the isomorphism classes (M) of modules in $\mathcal{C}$, and $\mathcal{R}$ is the subgroup of $\mathcal{I}$ generated by all differences $(M) - (M') - (M'')$ arising from all short exact sequences

$$0 \to M' \to M \to M'' \to 0$$

in $\mathcal{C}$. The Grothendieck group $G_0(A)$ of a ring A is $K_0(\mathcal{C})$, where $\mathcal{C}$ is the category of all f.g. A-modules. It follows from the Jordan-Hölder theorems that $G_0(KG)$ and $G_0(kG)$ are free abelian groups with bases $\{[Z_1],\ldots,[Z_s]\}$ and $\{[F_1],\ldots,[F_r]\}$ respectively where the $\{Z_i\}$ and $\{F_j\}$ are basic sets of simple modules. (Evidently $G_0(KG)$ is a commutative ring, isomorphic to chG.)

Now let $\mathcal{P}(RG)$ and $\mathcal{P}(kG)$ denote the categories of finitely generated projective RG- and kG-modules, respectively, and let $K_0(RG)$ and $K_0(kG)$ be the Grothendieck groups of the categories $\mathcal{P}(RG)$ and $\mathcal{P}(kG)$. These are also free Z-modules, with bases $\{[P_1],\ldots,[P_r]\}$ and $\{[U_1],\ldots,[U_r]\}$ respectively, where the $\{P_i\}$ and $\{U_j\}$ are indecomposable projective modules.

These groups are related by the <u>Cartan-Brauer triangle</u>, which is a commutative diagram

$$\begin{array}{ccc} G_o(KG) & \xrightarrow{\ d\ } & G_o(kG) \\ {\scriptstyle e}\nwarrow & & \nearrow{\scriptstyle c} \\ & K_o(kG) & \end{array}$$

The maps c, d, e are defined as follows.

Let V be a f.g. KG-module. Since R is a P.I.D., V contains a full RG-lattice M, which is an RG-lattice $M \subseteq V$ such that $K \otimes_R M \cong V$; in other words, an R-basis of M is a K-basis of V. If M is an RG-lattice, then

$$\overline{M} = M/\mathfrak{p}M$$

is a kG-module, obtained by <u>reduction mod</u> $\mathfrak{p}$. We note that $\overline{M} \cong k \otimes_R M$, where k is viewed as a (k,R)-bimodule using the natural map $R \to k$.

(2.2) <u>Definition of the decomposition map</u> d. Let $[V] \in G_o(KG)$, for a KG-module V. Let M be a full RG-lattice in V, and define

$$d[V] = [\overline{M}] \text{ in } G_o(kG), \text{ where } \overline{M} = M/\mathfrak{p}M.$$

Examples show that there may exist full RG-lattices M_1 and M_2 in V such that $\overline{M}_1 \not\cong \overline{M}_2$. Nevertheless, we always have $[\overline{M}_1] = [\overline{M}_2]$ in $G_o(kG)$, that is, $\overline{M}_1$ and $\overline{M}_2$ have the same composition factors. This is shown as follows. Since $M_1 + M_2$ is also a full RG-lattice in

V, it is sufficent to prove that $[\overline{M}_1] = [\overline{M}_2]$ in $G_o(kG)$ if $M_1 \subset M_2$, and in fact for the special case where M_1 is maximal in M_2. In that case we have

$$\mathfrak{p}M_1 \subset \mathfrak{p}M_2 \subset M_1 \subset M_2$$

(using Nakayama's lemma: if $\mathfrak{p}M_2 \not\subseteq M_1$, then $M_1 + \mathfrak{p}M_2 = M_2$ and $M_1 = M_2$). Thus it is enough to prove that M_2/M_1 and $\mathfrak{p}M_2/\mathfrak{p}M_1$ have the same composition factors, but this is clear, since $\mathfrak{p} = \pi R$ for some element π, and multiplication by π defines a kG-isomorphism $M_2/M_1 \to \mathfrak{p}M_2/\mathfrak{p}M_1$.

In particular we set

$$d[Z_i] = \Sigma_{i=1}^{s} d_{ij}[F_j] \text{ and } D = (d_{ij}).$$

Then the map d is determined by the <u>decomposition matrix</u> D.

(2.3) <u>Definition of the Cartan map</u> c. Let $U \in \mathcal{P}(kG)$. Then U defines elements $[U]_{K_o}$ in $K_o(kG)$ and $[U]_{G_o}$ in $G_o(kG)$, and

$$c[U]_{K_o} = [U]_{G_o}.$$

The map c is defined by the <u>Cartan matrix</u> $C = (c_{ij})$, where

$$c[U_i] = \Sigma_j c_{ij}[F_j].$$

(2.4) <u>Definition of</u> e. The indecomposable projective RG-modules $P_i \in \mathcal{P}(RG)$ are principal ideals in RG generated by primitive idempotents, and the same is true for $\mathcal{P}(kG)$. Since idempotents can be lifted from kG to RG using the completeness of R, it follows that if $\{P_1,\ldots,P_r\}$ is a basic set of projective indecomposable modules in $\mathcal{P}(RG)$, then $\{\overline{P}_1,\ldots,\overline{P}_r\}$ is a basic set of projective indecomposable modules in $\mathcal{P}(kG)$. The map e: $K_o(kG) \to G_o(KG)$ is defined on a basis of $K_o(kG)$ as follows:

$$e[\overline{P}_i] = [K \otimes_R P_i] \in G_o(KG).$$

Following Serre, we define bilinear forms

$$i_K: G_o(KG) \times G_o(KG) \to Z \text{ and } i_k: K_o(kG) \times G_o(kG) \to Z.$$

as follows. For left KG-modules V, V' we set $i_K([V],[V']) = \dim_K \operatorname{Hom}_{KG}(V,V')$; then we have $i_K = (\nu,\nu')$ where ν,ν' are the characters of V and V', and (ν,ν') is the scalar product defined in §1. The bilinear map i_k is defined in the same way;

$$i_k([U],[M]) = \dim_k \operatorname{Hom}_{kG}(U,M),$$

and can be shown to be a bilinear map on the Grothendieck groups using the fact that $\operatorname{Hom}(U,\cdot)$ is an exact functor when U is projective.

(2.5) <u>Theorem</u>. (Brauer, Serre). <u>The decomposition map</u> $d: G_o(KG) \to G_o(kG)$ <u>and the map</u> $e: K_o(kG) \to K_o(KG)$ <u>are transposes of each other with respect to the forms</u> i_K and i_k:

$$i_k(u,dz) = i_K(eu,z) \quad \text{for} \quad u \in K_o(kG),\ z \in G_o(KG).$$

(2.6) <u>Corollary</u>. $C = {}^tDD$.

The preceding results can perhaps best be understood using certain K-valued class functions, called <u>Brauer characters</u>. Let m be the exponent of G, and write $m = p^t m'$, with $\text{G.C.D.}(m',p) = 1$. Then the natural map from $R \to k = R/\mathfrak{p}$ defines an isomorphism between the groups of m'-th roots of 1 in K and k. Let $G_{p'}$ denote the set of p'-elements of G(= p'regular elements = elements of order prime to p). Each element $x \in G$ has a unique expression $x = su$, with $s \in G_{p'}$ and u a p-element commuting with s. Let L be a left kG-module. For each $x \in G_{p'}$, the eigenvalues of x on L are m'-th roots of 1. Using the isomorphism given above, these can be lifted to m'-th roots of 1 in $R \subset K$. Thus we can define a class function λ on $G_{p'}$, such that $\lambda(x)$ is the sum of the preimages of the eigenvalues of x on L, for each $x \in G_{p'}$. Then

$$\overline{\lambda(x)} = \operatorname{Tr}(x,L)$$

where $\overline{}$ is the natural map $R \to k$. The function λ on $G_{p'}$ is called the <u>Brauer character</u> of L. Two kG-modules L and M have the same Brauer characters if and only if $[L] = [M]$ in $G_o(kG)$. The Brauer characters $\{\varphi^1,\ldots,\varphi^r\}$ of the simple kG-modules $\{F_1,\ldots,F_r\}$ form a basis for the class functions on $G_{p'}$, so the number of simple kG-modules is equal to the number of p'-conjugacy classes. The decomposition map $d: G_o(KG) \to G_o(kG)$ can be described in terms of Brauer characters as follows:

$$\zeta^i|_{G_{p'}} = \Sigma d_{ij}\cdot\varphi^j$$

where ζ^i is the irreducible K-character afforded by Z_i, $1 \le i \le s$. In particular, a knowledge of the values of the φ^j and the decomposition matrix D gives the values of the irreducible characters $\{\zeta^i\}$ on $G_{p'}$.

A further property of the Cartan-Brauer triangle involves the theory of Brauer characters.

(2.7) <u>Theorem</u> (The Brauer lift). <u>Let</u> λ <u>be the Brauer character of a kG-module</u> L, <u>and define a class function</u> $\hat{\lambda}$ <u>on</u> G <u>by setting</u> $\hat{\lambda}(x) = \lambda(s)$, <u>where</u> $x = su$ <u>is the decomposition described previously</u>. <u>Then</u> $\hat{\lambda}$ <u>is a virtual K-character of</u> G.

The proof is a nice application of Brauer's Criterion (§1B). The result has only to be checked for elementary groups, where it is almost immediate.

(2.8) Corollaries. (i) <u>The decomposition map</u> d: $G_o(KG) \to G_o(kG)$ <u>is surjective</u>.

(ii) <u>The map</u> e: $K_o(kG) \to G_o(KG)$ <u>is injective</u>.

(iii) <u>Let</u> $P, P' \in \mathcal{P}(RG)$, <u>and suppose</u> $K\otimes_R P \cong K\otimes_R P'$ <u>as KG-modules</u>. <u>Then</u> $P \cong P'$ (<u>as RG-modules</u>).

Part (iii) of (2.8) has an important application to the theory of integral representation. It gives a key step in the proof of Swan's Theorem: a projective RG-lattice M is locally free, where R is the ring of algebraic integers in a number field.

Another application of the Brauer Induction Theorem shows that detC is a power of p (where C is the Cartan matrix.) An important problem is to learn more about the matrices C and D, at least in special cases where there is a possibility to compute them (see Humphreys [5] for a survey of this problem and for references to the literature).

Another intriguing problem is to decompose the Brauer lift as a $\mathbb{Z}$-linear combination of irreducible characters (see Green [4] and Lusztig [6]).

2B. <u>Defect Groups of Primitive Idempotents</u>.

In this subsection, R denotes either a field of characteristic p or a d.v.r with maximal ideal $\mathfrak{p}$, such that $R/\mathfrak{p}$ is a field of characteristic p, and R is complete in the $\mathfrak{p}$-adic topology.

A G-<u>algebra</u> A is an R-algebra with a finite basis over R, on which G acts as a group of R-algebra automorphisms. Thus there is an operation $x\cdot a \in A$ such that $a \to x\cdot a$ is an R-linear map, and

$$x(ab) = (xa)(xb) \text{ and } x(ya) = (xy)a$$

for all $x,y \in G$ and $a,b \in A$.

Examples of G-algebras to keep in mind are:

(i) $A = \mathrm{End}_R M$, for a left RG-module M. The G-action is given by

$$xf = xfx^{-1}, \text{ for } x \in G, f \in A.$$

(ii) $A = RG$, with G-action by inner automorphisms:

$$xa = xax^{-1}, x \in G, a \in A.$$

Let A be a G-algebra, and let $H \leq G$. We set $A_H = \{a \in A\colon ha = a \text{ for all } h \in H\}$. For example, if $A = RG$ as in (ii), then $A_G = c(RG)$, the center of the R-algebra A.

The most important idea for the study of G-algebras is the *trace map* $T_{H/D}$ defined for each pair of subgroups $D \leq H$. The trace map is the R-endomorphism of A defined by

(2.9) $$T_{H/D}a = \sum_{h \in H/D} h \cdot a$$

where the sum is taken over a cross section of H/D. We set $A_{H/D} = T_{H/D}A_D$; then $A_{H/D} \subseteq A_H$ and is independent of the choice of the cross section. We have

$$T_{H/D}(ba) = bT_{H/D}a \quad \text{and} \quad T_{H/D}(ab) = T_{H/D}(a)b$$

for all $a \in A_D$ and $b \in A_H$. Thus $A_{H/D}$ is a two-sided ideal in A_H. We also have

$$x \cdot A_H = A_{{}^xH} \quad \text{and} \quad xT_{H/D}(a) = T_{{}^xH/{}^xD}(xa)$$

for all $x \in G$ and $a \in A_D$.

The crucial property of the trace map is the precise analogue of the Mackey Subgroup Theorem (1.6).

(2.10) *Proposition. Let* A *be a* G-*algebra,* L *a subgroup of* G, *and* D, H *subgroups of* L. *Then*

$$T_{L/H}a = \Sigma T_{D/D\cap {}^xH}(xa) \quad \text{for all} \quad a \in A_H$$

where the sum is taken over a cross section of the double cosets $D\backslash L/H$.

The proof uses exactly the same calculation as in the proof of (1.6).

(2.11) *Corollary. Let* D, H, L *be as in* (2.10) *and let* $a \in A_H$, $b \in A_D$. *Then*

$$T_{L/H}(a)T_{L/D}(b) = T_{L/D\cap {}^xH}((x \cdot a)b)$$

where the sum is taken over a cross section of $D\backslash L/H$ *as before. In particular, we have*

$$A_{L/H}A_{L/D} \subseteq \Sigma_x A_{L/D\cap {}^xH}.$$

(2.12) *Rosenberg's Lemma. Let* A *be an* R-*algebra with a finite basis. Let* e *be a primitive idempotent in* A, *and suppose that* $e \in I + I'$, *for two-sided ideals* I *and* I' *in* A. *Then either* $e \in I$ *or* $e \in I'$.

Since e is a primitive idempotent, Ae is an indecomposable A-module, and $\mathrm{End}_A Ae \cong (eAe)^{\circ}$. Because of the assumptions on R, eAe is a local ring, and $e \in eIe + eI'e$ since e is idempotent.

Since eAe is local, one of the ideals eIe or $eI'e$ coincides with eAe, and the result follows.

(2.13) Theorem (Green). Let A be a G-algebra, $H \le G$, and let e be a primitive idempotent in A_H. Then there exists a p-subgroup $D \le H$ such that $e \in A_{H/D}$ with the further property that $D \le_H D'$ for any other subgroup D' such that $e \in A_{H/D'}$. (Note: $D \le_H D'$ means $D \le {}^hD'$ for some $h \in H$).

Since $A_{H/H} = A_H$, there are subgroups D such that $e \in A_{H/D}$. Let D_o be a minimal one. Let D' be another subgroup such that $e \in A_{H/D'}$. Then $e = e^2 \in A_{H/D}A_{H/D'}$ and hence

$$e \in \Sigma A_{H/(D_o \cap {}^xD')}$$

by (2.11). Since $A_{H/D}$ is a two-sided ideal in A_H for all subgroups D, we can apply (2.12), to get $e \in A_{H/(D_o \cap {}^xD')}$ for some x. The result now follows from the minimality of D_o. The fact that D_o is a p-subgroup is proved as follows. Let S be a Sylow p-subgroup of H; then $|H:S|$ is prime to p, and hence is a unit in R, and it follows that $e = T_{H/S}(|H:S|^{-1}e)$ so $e \in A_{H/S}$. Then $D_o \le_H S$, completing the proof.

Definition. The p-subgroup D_o, and its H-conjugates, which are associated to e by (2.13), are called defect groups of the primitive idempotent e.

2C. Application to Vertices and Sources.

The motivation for the preceding discussion comes, at least in part, from some earlier results of Gaschutz. Let $H \leqslant G$, and let M be a left RG-module. Then M is called (G,H)-projective if every exact sequence of RG-modules

$$0 \to M' \to M'' \to M \to 0$$

which splits as a sequence of RH-modules, also splits as a sequence of RG-modules. Similarly, M is (G,H)-injective if the corresponding statements hold for exact sequences

$$0 \to M \to M' \to M'' \to 0$$

These ideas are basic for the classification of indecomposable RG-lattices.

(2.14) Proposition. Let $H \leqslant G$, and let M be a left RG-lattice. Then the following statements are equivalent.

(i) M is (G,H)-projective.

(ii) M is (G,H)-*injective*.

(iii) $M|L^G$, *for some RH-lattice* L (*where* $M|L^G$ *means that* M *is isomorphic to a direct summand of* L^G).

(iv) $id_M = T_{G/H}f$, *for some* $f \in A_H$, *where* A *is the G-algebra* $End_R M$ *defined in* §2B, *and* id_M *is the identity map on* M.

By Theorem 2.13, we obtain:

(2.15) *Theorem*. (Green) *Let* M *be an indecomposable RG-lattice. Then there exists a p-subgroup* $D \leq G$, *uniquely determined up to G-conjugacy, such that* M *is* (G,D)-*projective, and having the further property that if* M *is* (G,H)-*projective for any subgroup* $H \leq G$, *then* $D \leq_G H$.

Note, in particular, that if R is a field, then $D = 1$ (in (2.15)) if and only if M is projective.

The subgroup D and its conjugates are called *vertices* of M. By the K - S - A theorem (2.1), there exists an indecomposable RD-lattice L such that $M|L^G$; these are called *sources* of M, and are uniquely determined up to conjugacy in $N_G(D)$.

These ideas give a straight-forward proof of half of the following result.

(2.16) *Theorem*. (D. G. Higman) *The group algebra* RG *is of finite representation type if and only if the Sylow p-subgroups of* G *are cyclic*. (Finite representation type means that the number of isomorphism classes of indecomposable RG-lattices is finite).

The theory of vertices is applied to the classification of indecomposable RG-lattices through the use of the Green Correspondence, which can be described as follows. Let D be a p-subgroup of G, let $H = N_G(D)$, and define the following families of subgroups of G:

$\mathfrak{X} = \{{}^xD \cap D,\ x \in G - H\}$

$\mathfrak{Y} = \{{}^xD \cap H,\ x \in G - H\}$

$\mathfrak{A} = \{D' \leq D,\ D' \not\leq_G X \text{ for all subgroups } X \in \mathfrak{X}\}$.

An RG-lattice $M = O(\mathfrak{X})$ if M is a direct sum of modules $\{M_i\}$ which are (G,X_i)-projective, for subgroups X_i in $\mathfrak{X}$. The notation $M = O(\mathfrak{Y})$ is defined similarly.

(2.17) *Green Correspondence*. *Let* D *be a p-subgroup of* G, *and let* $H = N_G(D)$. *There is a bijection* $(M) \leftrightarrow (N)$ *between isomorphism classes of indecomposable RG-lattices* M *with vertices in* $\mathfrak{A}$ *to isomorphism classes of indecomposable RH-lattices* N *with vertices in* $\mathfrak{A}$. *The*

<u>modules</u> M <u>and</u> N <u>correspond if and only if either of the following conditions hold</u>:

$$M_H = N \oplus O(\mathfrak{Y}) \quad \text{or} \quad N^G = M \oplus O(\mathfrak{X}).$$

<u>Corresponding modules have the same vertex</u>.

For example, if D is a T.I.-set (trivial intersection set), so that ${}^xD \cap D = 1$ if $x \notin H$, then $\mathfrak{X} = \{1\}$, so the Green Correspondence sets up a bijection between indecomposable non-projective RG- and RH-lattices (with vertices contained in D).

2D. <u>Application to Blocks</u>.

Let (K,R,k) be a p-modular system. A <u>block idempotent</u> b is a primitive idempotent in $c(RG)$ (the center of RG). Each block idempotent generates an indecomposable two-sided ideal in RG called the <u>block ideal</u> associated with b. A p-<u>block</u> $\underline{\underline{B}}$ of G consists of the categories of KG-modules, RG-modules, and kG-modules on which a given block idempotent b acts as the identity map (where b acts on kG-modules through the homomorphism $RG \to kG$, and on KG-modules through the injection $RG \subset KG$. Thus the p-blocks distribute the G-modules and their characters into subsets.

The distribution of the simple KG-modules $\{Z_1,\ldots,Z_s\}$ and their characters $\{\zeta^1,\ldots,\zeta^s\}$ into blocks is described as follows. Let $\{e_1,\ldots,e_s\}$ be the primitive idempotents in KG corresponding to the characters $\{\zeta^i\}$; then

$$e_i = \zeta^i(1)|G|^{-1}\Sigma_{x\in G}\zeta^i(x^{-1})x, \quad \text{for } 1 \le i \le s.$$

Since $c(RG) \le c(KG)$ (both have bases consisting of the class sums), a given block idempotent b can be expressed uniquely as a sum of a subset of the $\{e_i\}$:

$$b = e_{i_1} + \ldots + e_{i_f},$$

and the corresponding simple KG-modules $Z_{i_1},\ldots,Z_{i_f}$ are precisely the simple KG-modules in the block $\underline{\underline{B}}$ defined by b.

A first step in the investigation of the p-blocks was taken by Brauer-Nesbitt, who defined the defect of a block $\underline{\underline{B}}$. Let $G = p^e g'$, when GCD $(g',p) = 1$, and let $z_i = \zeta^i(1) = \deg \zeta^i = \dim_K Z_i$, $1 \le i \le s$. Let v_p be the exponential valuation on Q, so $v_p(a)$ is the exponent to which p occurs in the factorization of a, $(v_p(|G|) = e)$. The <u>defect</u> d of a block $\underline{\underline{B}}$ is defined by the formula:

$$d = e - \min\{v_p(z_i) : Z_i \in \underline{\underline{B}}\}.$$

Since z_i divides $|G|$ for all i (from basic character theory), d is a non-negative integer. Alternatively, d is defined by the conditions:

$$p^{e-d} \mid z_i \quad \text{for all} \quad Z_i \in \underline{\underline{B}} \quad \text{and} \quad p^{e-d+1} \nmid z_i \quad \text{for some} \quad Z_i \in \underline{\underline{B}}.$$

Example. At one extreme, we have blocks of defect zero. A character ζ^i belongs to a block of defect zero if and only if $v_p(z_i) = e$. In this case $\underline{\underline{B}}$ contains only the one simple module Z_i, and a full RG-lattice P_i in Z_i is an indecomposable projective RG-module in B. In particular, if p does not divide G, then all blocks have defect zero, and block theory has little new information to contribute.

Using orthogonality relations for Brauer characters, Brauer succeeded in relating the coefficients of the block idempotent $b \in \underline{\underline{B}}$ to the defect of the block $\underline{\underline{B}}$. Let

$$b = \Sigma a_i C_i, \; a_i \in R,$$

where the $\{C_i\}$ are the basis elements of $c(RG)$ consisting of class sums. If $C_i = \Sigma_{x \in \mathfrak{C}_i} x$ for the conjugacy class $\mathfrak{C}_i$, the class defect δ_i of $\mathfrak{C}_i$ is defined by

$$\delta_i = v_p \, | C_G(x_i) | \quad \text{for} \quad x_i \in \mathfrak{C}_i.$$

(2.18) Theorem. Let $b = \Sigma a_i C_i$ be a block idempotent belonging to a block of defect d. Then we have

$$b = \sum_{\delta_i = d} a_i C_i = \sum_{\delta_j < d} b_j C_j \quad (\text{mod } pG)$$

where some coefficient $a_i \not\equiv 0 \pmod{p}$.

This somewhat murky picture clears up at once when we apply the results on defect groups of idempotents from §2B. Let A be the G-algebra RG. Then $A_G = c(RG)$, and a block idempotent b in a block $\underline{\underline{B}}$ is a primitive idempotent in A_G. By (2.13), $b \in A_{G/D}$ for a p-subgroup D, which is the unique minimal subgroup, up to conjugacy, with this property. The group D and its conjugates are called defect groups of the block $\underline{\underline{B}}$.

An important question is, which p-subgroups of G are defect groups of p-blocks? We shall prove one basic result on this topic, due to Green.

The group algebra RG can be viewed as an $R(G\times G)$-lattice, where the action of the direct product $G\times G$ is given by

$$(x,y)a = xay^{-1}, \quad x,y \in G, \; a \in RG.$$

A block ideal $B = bRG$ is clearly an indecomposable $R(G\times G)$-lattice, so we can ask: what is its vertex? Let $\Delta : G \to G\times G$ be the diagonal map, so $\Delta(g) = (g,g)$, $g \in G$. Using the results of §2B and 2C, the following result is not difficult to prove.

(2.20) Theorem. Let B be a block ideal in RG, belonging to a block with defect group D. Then the vertex of B, as an indecomposable $R(G\times G)$-lattice, is ΔD.

We now have:

(2.21) Theorem. (Green) Let B be a block ideal with defect group D, and let $D \le S$ for a Sylow p-subgroup S of G. Then there exists an element $z \in C_G(D)$ such that

$$D = S \cap {}^zS.$$

We shall sketch the proof. We begin with:

(2.22) Lemma. Let H be a p-group, and let M be a transitive permutation RH-lattice, that is, M has an R-basis X such that X is a transitive G-set. Then M is an indecomposable RH-lattice, with vertex $H_X = \mathrm{Stab}_H x$, for some $x \in X$.

This result is a special case of a general theorem due to Green, but the following proof, due to M. Cabanes, is irresistible to include here. We have $M \cong (1_{H_x})^H$ as RG-modules. It is sufficient to prove that $k \otimes_R M$ is an indecomposable kH-module, and we have $k \otimes_R M \cong (1_{H_x})^H$ as kH-modules. Since k has characteristic p, the p-group H always has nontrivial fixed points on a vector space over k. So it is enough to prove that $\dim_k \mathrm{inv}_H(1_{H_x})^H = 1$ (where $\mathrm{inv}_H(V)$ is the space of fixed points under H in a kH-module V). For this we apply the general version of Frobenius reciprocity (1.5) and obtain

$$\mathrm{inv}_H(1_{H_x})^G \cong \mathrm{Hom}_{kH}(1_{kH}, (1_{H_x})^H) \cong \mathrm{Hom}_{kH}(1_{kH_x}, 1_{kH_x}) \cong k,$$

completing the proof that M is indecomposable. The second statement follows easily from the results in §2C.

For the proof of the Theorem, we follow the approach of Alperin-Burry. Let $G = \cup Sx_iS$ (disjoint union of double cosets). Then

$$RG = \oplus_i R(Sx_iS),$$

where $R(Sx_iS)$ denotes the $R(S\times S)$-module with a finite basis consisting of the elements of Sx_iS. Then each submodule $R(Sx_iS)$ is a transitive permutation module for the p-group $S\times S$, and hence is indecomposable by the Lemma. Since $B \mid RG$ as $R(G\times G)$-modules, we have $B_{S\times S} \mid (RG)_{S\times S}$, and by the K-S-A Theorem, $B_{S\times S}$ is isomorphic to a direct sum of the modules $\{R(Sx_iS)\}$. Since B has vertex ΔD, by (2.20), it follows that one of the modules $R(Sx_iS)$ also has vertex ΔD. By the Lemma, $R(Sx_iS)$ has vertex

$$\mathrm{Stab}_{S\times S}x_i = {}^{(x_i,1)}\Delta(S\cap {}^{x_i^{-1}}S)$$

and comparing this with ΔD, the result follows.

The deeper results on blocks involve the relation between blocks of G and blocks of subgroups of G.

(2.23) Brauer's First Main Theorem. There is a bijection between the set of blocks of G with defect group D and the set of blocks of $N_G(D)$ with defect group D, for each p-subgroup $D \le G$. If B' is a block ideal of $N_G(D)$ with defect group D, the corresponding block ideal of RG is the unique block ideal B such that $B' | B_{N_G(D)\times N_G(D)}$

More generally, for any subgroup H such that

$$DC_G(D) \le H \le N_G(D),$$

and a block ideal B' of RH with defect group D, there is a unique block ideal B of RG such that $B' | B_{H\times H}$. This defines a map $\underline{\underline{B}}' \to \underline{\underline{B}} = (\underline{\underline{B}}')^G$ of blocks called the Brauer Correspondence (and can be described in several other ways).

We finish with an application to character theory, which plays a key role in most investigations of blocks of characters.

Let $\underline{\underline{B}}_1,\ldots,\underline{\underline{B}}_t$ be the p-blocks of G. Let $\xi \in \mathrm{cf}_K G$. Then

$$\xi = \Sigma_i \xi_{\underline{\underline{B}}_i}, \quad \text{where} \quad \xi_{\underline{\underline{B}}_i} = \Sigma_{\zeta\in \underline{\underline{B}}_i} (\xi,\zeta)\zeta, \quad \text{for } \zeta \in \mathrm{Irr}G.$$

(2.24) Brauer's Second Main Theorem. Let $x \in G$, and let $H = C_G(u)$, where u is the p-part of x. Let ζ be an irreducible K-character of G belonging to a p-block B of G. Then for each p-block B' of H, we have

$$(\zeta|_H)_{\underline{\underline{B}}'}(x) = 0 \quad \text{if} \quad (\underline{\underline{B}}')^G \ne \underline{\underline{B}},$$

and, moreover,

$$\zeta(x) = \Sigma_{(\underline{\underline{B}}')^G = \underline{\underline{B}}} (\zeta|_H)_{\underline{\underline{B}}'}(x)$$

(where $\underline{\underline{B}}' \to (\underline{\underline{B}}')^G$ is the Brauer Correspondence).

An elegant proof, using the theory of vertices, was given by Nagao.

3. Representations of Finite Groups of Lie Type.

In the remaining lectures, I shall discuss some aspects of the representation theory, over the field of complex numbers, of finite Chevalley groups. There are two main approaches to the subject. The first, due to Harish-Chandra, is based on the concept of cuspidal representations, and leads to interesting problems in the decomposition of induced representations. The second, due to Deligne-Lusztig, is based on the analysis of certain virtual representations, arising from the actions of the finite groups on algebraic varieties. There will be some discussion of both approaches, and relations between them. I shall also make some comments on connections and analogies with the theory of blocks (§2). For a fuller discussion of these topics, and references to the literature, see Curtis [1], Srinivasan [9] and Lusztig [7].

3A. Finite Groups of Lie Type.

In §3, G denotes a finite Chevalley group, or a twisted type of Chevalley group (as in Chevalley's Tohoku Journal paper, or Steinberg's Lectures on Chevalley Groups). These groups can also be realized as the finite groups $G = \underline{\underline{G}}(F_q)$ of F_q-rational points on a connected, reductive, affine algebraic group $\underline{\underline{G}}$ defined over the finite field F_q (as in Borel-Tits, IHES Pub. Math, no. 27 (1965)). The Chevalley groups G are the counterparts, in finite group theory, of semisimple Lie groups over the fields of real or complex numbers. They include all finite simple groups, with the exception of the alternating and sporadic simple groups.

Each Chevalley group G has a BN-pair (or Tits system), and is associated with a finite Coxeter group W, called the Weyl Group of G. In more detail, G contains two subgroups B and N such that $G = \langle B,N\rangle$ (that is, G is generated by $B \cup N$),

$$T = B \cap N \text{ is a normal subgroup of } N,$$

and

$$W = N/T = \langle S\rangle$$

where $S = \{s_1,\dots,s_n\}$ is a set of involutions, called the distinguished generators of W. Then W has a presentation as a Coxeter group:

$$W = \langle s_1,\dots,s_n : s_i^2 = (s_is_j)^{n_{ij}} = 1\rangle$$

The number $n = |S|$ is uniquely determined, and is called the rank of the BN-pair. The other defining properties of a BN-pair describe the double cosets $B\backslash G/B$ of G relative to the Borel subgroup B:

$$s_iBw \subseteq BwB \cup Bs_iwB \quad \text{and} \quad s_iBs_i \neq B,$$

for each $s_i \in S$ and $w \in W$. (Note that these formulas are unambiguous, even though $W = N/T$ is not a subgroup of G, because of the fact that $T \leq B$).

From the axioms for a BN-pair, it follows that G has the <u>Bruhat decomposition</u>

$$G = \bigcup_{w \in W} BwB,$$

with $w \to BwB$ a bijection from $W \to B\backslash G/B$.

The Chevalley groups occur in infinite families $\{G(q)\}$, all with a fixed Weyl group W, and parametrized by prime powers $\{q\}$. In fact, each group $G(q)$ is associated with a finite field F_q of characteristic p, which is called the <u>natural characteristic</u> of $G(q)$. The parametrization is such that for each generator $s_i \in S$,

$$q_i = |Bs_iB/B| = q^{c_i}$$

for a fixed set of positive integers $\{c_1,\ldots,c_n\}$ depending only on the family. For example, for untwisted Chevalley groups, the c_i are all equal to 1, while in the case of finite unitary groups, this is not the case.

The finite general linear groups $G(q) = GL_n(F_q)$, for fixed n, are an example of a family of Chevalley groups. For these groups, the Borel subgroup is the group of upper triangular matrices, N is the group of monomial matrices, $T = B \cap N$ is the group of diagonal matrices, and the Weyl group $W = N/T$ is isomorphic to the symmetric group S_n. Thus each group $G(q)$ in the family has a BN-pair of rank $n-1$.

The objective is to find the representations and characters of all the groups $G(q)$ in a family at once, and, to a surprising extent, this turns out to be possible.

The BN-pair in G is determined up to conjugacy. The subgroups P containing B, and their conjugates, are called the <u>parabolic subgroups</u> of G. The <u>standard parabolic subgroups</u>, containing a fixed Borel subgroup B, are in bijective correspondence with the parabolic subgroups of the Weyl group W. These are the subgroups $\{W_J, J \subseteq S\}$ generated by subsets J of the distinguished generators S. The parabolic subgroup P_J corresponding to W_J is given by $P_J = BW_JB$. For example, in $GL_n(F_q)$, the parabolic subgroups of $W \cong S_n$ have the form $S_{n_1} \times \cdots \times S_{n_r}$ for a partition $(n_1,\ldots,n_r)$ of n, and the corresponding parabolic subgroups of G have the form

$$P_J = \left\{ \begin{array}{c} \\ n_1 \\ n_2 \end{array} \begin{array}{c} \begin{array}{ccc} n_1 & n_2 & \end{array} \\ \left[\begin{array}{c|c|c} * & * & * \\ \hline 0 & * & * \\ \hline & & \end{array}\right] \end{array} \right\}, \quad J \subseteq S.$$

In the general case, each parabolic subgroup P_J has a <u>Levi decomposition</u>.

$P_J = L_J V_J$ (semidirect product)

where $V_J = O_p(P_J)$ (the maximal normal p-subgroup, where p is the natural characteristic). Then $P_J = N_G(V_J)$, and L_J (the Levi subgroup of P_J) is a member of a family of Chevalley groups having the Weyl group W_J. In the example above, we have

$L_J \cong GL_{n_1} \times \ldots \times GL_{n_r}$.

The subgroup V_J is called the unipotent radical of P_J

3B. Generalized Restriction and Induction.

Let $G = G(q)$ belong to a family of Chevalley groups with Weyl group W. Generalizing the Bruhat decomposition, there exists a bijection

$P_I \backslash G/P_J \leftrightarrow W_I \backslash W/W_J$,

for all $I, J \subseteq S$, with $G = \bigcup P_I w P_J$, and $\{w\}$ taken from a cross section of the double cosets $W_I\backslash W/W_J$. This is the information needed to apply the Interwining Number Theorem to induced representations from P_I and P_J (see (1.7)). The natural subgroups of G to consider, however, are not the parabolic subgroups P_J, but their Levi subgroups L_J, which are Chevalley groups of smaller BN-rank. A way out of this dilemma is to define operations of generalized induction and restriction, which suppress the influence of the unipotent radicals $\{V_J\}$ of the standard parabolic subgroups.

(3.1) Definition. Let ch(G) denote the ring of virtual characters of the group G (over the field of complex numbers.) For $J \subseteq S$, let L_J be a Levi subgroup of the parabolic subgroup P_J. We shall define two Z-homomorphisms

$T^G_{L_J}$: $ch(G) \to ch(L_J)$ (truncation)

and

$R^G_{L_J}$: $ch(L_J) \to ch(G)$ (generalized induction).

For each $\xi \in ch(G)$, we set

$$T^G_{L_J}\xi(x) = |V_J|^{-1} \sum_{u \in V_J} \xi(xu), \quad \text{for } x \in L_J.$$

For $\lambda \in ch(L_J)$, we set

$$R^G_{L_J}\lambda = \mathrm{ind}^G_{P_J}\tilde{\lambda} = \tilde{\lambda}^G,$$

where $\tilde{\lambda}$ is the pullback of λ to P_J, with V_J in the kernel of $\tilde{\lambda}$.

(3.2) <u>Proposition</u>. (i) (Transitivity) <u>If</u> $I \subseteq J \subseteq S$, <u>then</u>

$T_{L_I}^{G} = T_{L_I}^{L_J} \circ T_{L_J}^{G}$ <u>and similarly for</u> $R_{L_I}^{G}$.

(ii) (Reciprocity) <u>For</u> $\xi \in \mathrm{ch}\, G$ <u>and</u> $\lambda \in \mathrm{ch}(L_J)$, <u>we have</u>

$(R_{L_I}^{G}\lambda, \xi)_G = (\lambda, T_{L_J}^{G}\xi)_{L_J}$.

(iii) (Intertwining Number Theorem). <u>For</u> $\lambda \in \mathrm{ch}(L_I)$, $\mu \in \mathrm{ch}(L_J)$, $I, J \subseteq S$,

$(R_{L_I}^{G}\lambda, R_{L_J}^{G}\mu) = \Sigma_x \, (T_{L_K}^{L_I}\lambda\, ^x(T_{L_{K'}}^{L_J}\mu))$

<u>where</u> x <u>is taken from a cross section of</u> $W_I \backslash W / W_J$, <u>and</u> $K = I \cap {}^xJ$, $K' = {}^{x^{-1}}K$ <u>and</u> $L_K = {}^xL_{K'}$.

(3.3) <u>Definition</u>. A virtual character $\xi \in \mathrm{ch}(G)$ is <u>cuspidal</u> if $T_{L_J}^{G}\xi = 0$ for all proper subsets $J \subset S$.

(3.4) <u>Proposition</u>. <u>Let</u> $\xi \in \mathrm{Irr}\, G$. <u>Then the following statements are</u> equivalent:

(i) ξ <u>is cuspidal</u>.

(ii) $(\xi, R_{L_J}^{G}\lambda) = 0$ for all $\lambda \in \mathrm{ch}(L_J)$ and $J \subsetneqq S$.

In other words, the cuspidal irreducible characters are those that are missing from all the generalized-induced characters $R_{L_I}^{G}\lambda$ for $\lambda \in \mathrm{ch}\, L_I$ and $I \subsetneqq S$.

(3.5) <u>Proposition</u>. <u>Let</u> $\zeta \in \mathrm{Irr}\, G$. <u>Then either</u> ζ <u>is cuspidal or there exists a proper subset</u> $I \subset S$ <u>and a cuspidal character</u> $\lambda \in \mathrm{Irr}\, L_I$ <u>such that</u> $(\zeta, R_{L_I}^{G}\lambda) \neq 0$.

(3.6) <u>Theorem</u> (Harish-Chandra). <u>Let</u> $I, J \subseteq S$, $\varphi \in \mathrm{Irr}\, L_I$, $\psi \in \mathrm{Irr}\, L_J$, <u>and assume</u> φ <u>and</u> ψ <u>are cuspidal</u>. <u>Then</u>:

(i) $(R_{L_I}^{G}\varphi, R_{L_J}^{G}\psi) = 0$ <u>unless there exists</u> $x \in W$ <u>such that</u> ${}^xL_I = L_J$ and ${}^x\varphi = \psi$.

(ii) <u>If</u> $(R_{L_I}^{G}\varphi, R_{L_I}^{G}\varphi) \neq 0$, <u>then</u> $R_{L_I}^{G}\varphi$ $R_{L_J}^{G}\psi$ <u>and we have</u>

$(R_{L_I}^{G}\varphi, R_{L_I}^{G}\varphi) = \Sigma(\varphi, {}^t\varphi) = |\mathrm{Stab}_{N_W(L_I)/W_I}\varphi|$

where t is taken from a cross section of $N_W(L_I)/W_I$, and $N_W(L_I) = \{w \in W\colon {}^wL_I = L_J\}$.

From the viewpoint of (3.6), the two main problems are: (i) to find all irreducible cuspidal characters of G; and (ii) (the Decomposition Problem) to decompose the characters $R_{L_I}^{G}\lambda$, where λ is a cuspidal character of L_I.

Both of these problems are essentially solved for groups $G = \underline{\underline{G}}(F_q)$, for $\underline{\underline{G}}$ a reductive group with a connected center. The starting point is the construction by Deligne and Lusztig of a family of virtual characters $\{R_T^G\theta\}$, where T ranges over all maximal tori in G and $\theta \in \mathrm{Irr}\, T$. For the maximal torus $T = B \cap N$ defined in §3A, T is a Levi subgroup, and $R_T^G\theta$ is the operation of generalized irr-duction, defined in (3.1). For other maximal tori, the construction of the $\{R_T^G\theta\}$ lies much deeper. In the general case, the $R_T^G\theta$'s are virtual characters defined by certain cohomology groups of algebraic varieties on which the finite group G acts.

(3.7) <u>Theorem</u> (Deligne-Lusztig). <u>Let $G = \underline{\underline{G}}(F_q)$ and let T be a maximal torus, and $\theta \in \mathrm{Irr}\, T$. Then $R_T^G\theta \in \mathrm{Irr}\, G$ if θ is in general position (that is, θ is not stabilized by any nontrivial element of the rational Weyl group associated with T). Moreover $R_T^G\theta$ is irreducible and cuspidal if θ is in general position, and T is minisotropic (that it, T is contained in no proper parabolic subgroup of G).</u>

Every irreducible character of $G = \underline{\underline{G}}(F_q)$ appears with nonzero multiplicity in some virtual character $R_T^G\theta$. A major unsolved problem has been to decompose the virtual characters $R_T^G\theta$. Perhaps the most important result obtained in the theory so far is Lusztig's solution [7] of this problem for reductive groups $\underline{\underline{G}}$ with connected center. This includes the explicit determination of the values of all irreducible characters of G on semisimple elements.

Returning to the more elementary situation of generalized induction from standard Levi subgroups, Theorem (3.6) implies that we have a partition

$$\mathrm{Irr}\, G = \cup \mathcal{E}_{(I)}(G) \quad \text{(disjoint)}$$

where the sum is taken over equivalence classes (I) of subsets $I \subseteq S$. The equivalence relation sets $I \sim I'$ if and only if $L_I = {}^{x}L_{I'}$ for some $x \in W$, and $\mathcal{E}_{(I)}(G)$ consists of all irreducible characters ζ of G such that $(\zeta, R_{L_I}^G\lambda) \neq 0$ for some irreducible cuspidal character λ of L_I. (In order to understand the equivalence relation, one should note that while different standard parabolic subgroups P_I and P_J are never conjugate in G, it is possible for their Levi subgroups L_I and L_J to be conjugate.)

At one extreme, when $I = S$, the "series" of characters $\mathcal{E}_S(G)$ contains the cuspidal characters; at the other extreme, when I is empty, $\mathcal{E}_\emptyset(G)$ consists of all components of R_T^G, for the split torus $T = B \cap N$. By analogy with Lie group representation theory, $\mathcal{E}_S(G)$

is sometimes called the <u>discrete series</u>, while $\mathcal{E}_{\emptyset}(G)$ is called the <u>principal series</u>.

The series of characters $\mathcal{E}_{(I)}(G)$ behave, in some respects, like blocks of characters. For example, we have:

(3.8) <u>Theorem</u>. <u>Let</u> $\zeta \in \mathcal{E}_{(I)}{}^G$, <u>and let</u> $x \in G$ <u>be an element such that</u> $C_G(x) \leq L_J$, <u>for</u> $I, J \subseteq S$. <u>Then</u> $\zeta(x) = 0$ <u>unless there is at least one subset</u> $I' \subseteq J$ <u>such that</u> $I' \sim J$, <u>and if that occurs, then</u>

$$\zeta(x) = T_{L_J}^G \zeta(x) = \sum_{\substack{I' \subseteq J \\ I' \sim I}} \sum_{\lambda \in \mathcal{E}_{(I)}(L_J)} (\zeta, R_{L_J}^G \lambda)\, \lambda(x).$$

This theorem is somewhat similar to Brauer's Second Main Theorem (2.23). In a generalized form, due to Lusztig, it has been applied by Fong and Srinivasan [10] in their determination of r-blocks of the classical groups $G = \underline{\underline{G}}(F_q)$, where r is a prime different from the natural characteristic. We also remark that using Green's Theorem (2.21) on defect groups, it can be shown that an adjoint Chevalley group $G = \underline{\underline{G}}(F_q)$ has only two p-blocks for the natural characteristic p: the principal block (containing 1_G) and a block of defect zero containing the Steinberg character (see §3C).

3C. <u>The Decomposition Problem for</u> $R_T^G 1_T$.

We shall discuss here only the extreme case of the decomposition problem, namely the analysis of $R_T^G 1_T = (1_B)^G$. It turns out that the other cases of the decomposition problem can be treated by the same method.

We begin with some general remarks on the decomposition of permutation representations $(1_H)^G$, for an arbitrary finite group G. The representation $(1_H)^G$ is afforded by the left ideal CGe, where e is the idempotent defined by the formula $e = |H|^{-1} \sum_{h \in H} h$, and affords the trivial representation of H. The decomposition of CGe is accomplished through a study of its endomorphism algebra. We have

$$\mathrm{End}_{CG}\ CGe \cong (eCGe)^{o} \quad \text{(where } {}^{o} \text{ denotes the opposite ring).}$$

The subalgebra $eCGe$ of CG is called the <u>Hecke algebra</u> $\mathcal{H}$ of the permutation representation $(1_H)^G$; it is simply an explicit realization of the endomorphism algebra as a subalgebra of CG. We note that $\mathcal{H}$ is a semisimple algebra, with identity element e.

The induced representation $(1_H)^G$ is equivalent to the representation of G on the space of complex valued functions f on G satisfying $f(hx) = f(x)$ for all $h \in H$, with G-action given by $(xf)(y) = f(yx)$, for $x, y \in G$. It is not hard to show that the Hecke algebra $\mathcal{H}$ is isomorphic to the algebra of all bi-invariant functions f, satisfying $f(hxh') = f(x)$ for all $h, h' \in H$, multiplied by convolution. Thus the representation theory of $\mathcal{H}$ and its application to decompose $(1_H)^G$ is analogous to the theory of spherical functions on a Lie group.

By the preceding realization of $\mathcal{H}$, it follows that $\mathcal{H}$ has a standard basis $\{b_1,\ldots,b_m\}$ consisting of the normalized characteristic functions on the double cosets $H\backslash G/H$. If $G = D_i$, where $D_i = Hx_iH$, the corresponding basis element is

$$b_i = |H|^{-1}\Sigma_{y\in D_i}y.$$

Then we have

$$b_ib_j = \Sigma\mu_{ijk}b_k, \quad \text{with } \mu_{ijk} = |(D_i \cap x_kD_j^{-1})/H| \in Z.$$

In what follows, it is convenient to view the irreducible characters of G as functions on the group algebra CG, so $\zeta(\Sigma\alpha_x x) = \Sigma\alpha_x\zeta(x)$, for $\zeta \in \mathrm{Irr}\, G$.

(3.9) Proposition. Let $\mathcal{H}$ be the Hecke algebra of $(1_H)^G$.

(i) For each $\zeta \in \mathrm{Irr}\, G$, $(\zeta,(1_H)^G) > 0 \Leftrightarrow \zeta|_{\mathcal{H}} \neq 0$.

(ii) The map $\zeta \to \zeta|_{\mathcal{H}}$ is a bijection from the irreducible characters of G appearing with positive multiplicity in $(1_H)^G$ to the set of all irreducible characters of $\mathcal{H}$. We have

$$\deg \zeta|_{\mathcal{H}} = (\zeta,(1_H)^G).$$

(iii) We have $(1_H^G,1_G) = 1$. The corresponding character of $\mathcal{H}$ is called ind, (the index character) and satisfies

$$\mathrm{ind}\, b_i = |D_i/H|, \quad 1 \le i \le m.$$

(iv) (Orthogonality) Let $\varphi,\varphi' \in \mathrm{Irr}\,\mathcal{H}$. Then

$$\Sigma(\mathrm{ind}\, b_i)^{-1}\varphi(b_i)\varphi'(\hat{b}_i) = \begin{cases} 0 & \text{if } \varphi \neq \varphi' \\ \dfrac{\deg\varphi\,|G:H|}{\zeta(1)} & \text{if } \varphi = \varphi' = \zeta|_{\mathcal{H}}. \end{cases}$$

(where $\hat{b}_i$ is the standard basis element corresponding to D_i^{-1}.)

(v) (degree formula) If $\varphi \in \mathrm{Irr}\,\mathcal{H}$, and $\varphi = \zeta|_{\mathcal{H}}$, for $\zeta \in \mathrm{Irr}\, G$, then

$$\zeta(1) = \frac{|G:H|\,\deg\varphi}{\Sigma(\mathrm{ind}\, b_i)^{-1}\varphi(b_i)\varphi(\hat{b}_i)}$$

(vi) (Character formula) For $\varphi = \zeta|_{\mathcal{H}}$, and $t \in G$, we have

$$\zeta(t) = \frac{|C_G(t)|\Sigma(\mathrm{ind}\, b_i)^{-1}\varphi(b_i)|\,|\mathfrak{C}\cap D_i|}{|H|\Sigma(\mathrm{ind}\, b_i)^{-1}\varphi(b_i)\varphi(\hat{b}_i)},$$

where $\mathfrak{C}$ is the conjugacy class containing t.

Now let $G = G(q)$ be a member of a family of Chevalley groups with Weyl group W, and let $\mathcal{H}$ be the Hecke algebra of $R_T^G 1 = (1_B)^G$. By the Bruhat decomposition, the standard basis elements of $\mathcal{H}$ are indexed by the elements of W:

$$b_w = |B|^{-1}\Sigma_{x\in BwB}x, \quad \text{for} \quad w\in W.$$

We let $\ell(w)$ denote the length function on W, where $\ell(w)$ is the minimal length of a factorization of W as a product of the distinguished generators $S = \{s_1,\ldots,s_n\}$.

(3.10) Theorem (Iwahori-Matsumoto). The multiplication in $\mathcal{H}$ satisfies:

$$b_{s_i} b_w = \begin{cases} b_{s_i w} & \text{if } \ell(s_i w) > \ell(w) \\ q_i b_{s_i w} + (q_i - 1)b_w & \text{if } \ell(s_i w) < \ell(w) \end{cases}$$

for $s_i \in S$, $w\in W$, and $q_i = \operatorname{ind} b_{s_i}$, $1 \le i \le n$. Moreover, $\mathcal{H}$ has a presentation as a $\mathbb{C}$-algebra with generators $\{b_{s_1},\ldots,b_{s_n}\}$ and relations

$$b_{s_i}^2 = q_i 1 + (q_i - 1)b_{s_i}, \quad 1 \le i \le n,$$

and if $i \neq j$,

$$(b_{s_i} b_{s_j})^{k_{ij}} = (b_{s_j} b_{s_i})^{k_{ij}}$$

$$(b_{s_i} b_{s_j})^{k_{ij}} b_{s_i} = (b_{s_j} b_{s_i})^{k_{ij}} b_{s_j},$$

depending on whether $m_{ij} = 2k_{ij}$ or $2k_{ij} + 1$, where m_{ij} is the order of $s_i s_j$ in W.

(3.11) Corollary. There is a second one-dimensional representation of $\mathcal{H}$ (besides ind) called sgn, such that $\operatorname{sgn} b_{s_i} = -1$, for $1 \le i \le n$. The corresponding representation of G is called the Steinberg representation St_G, and we have $St_G(1) = |G|_p$ so St_G belongs to a block of defect zero.

We illustrate the use of Hecke algebras with a discussion of how the degrees of other components of $(1_B)^G$ are obtained.

Let u be an indeterminate over $\mathbb{Q}$, and let $R = \mathbb{Q}[u]$. The generic ring of the family of Chevalley groups $\{G(q)\}$ with Weyl group W, is the R-algebra with a basis $\{a_w\}_{w\in W}$ satisfying

$$a_{s_i} a_w = \begin{cases} a_{s_i w} & \text{if } \ell(s_i w) > \ell(w) \\ u^{c_i} a_{s_i w} + (u^{c_i} - 1)a_w & \text{if } \ell(s_i w) < \ell(w). \end{cases}$$

Then A has a presentation as in (3.10), and therefore there exist homomorphisms IND and SGN from $A \to R$ as in (3.9) and (3.11).

For each homomorphism $F: R \to F$, for a field F, there is a <u>specialized algebra</u> $F \otimes_R A = A_f$, with basis $\{1 \otimes a_w\}$ and structure constants obtained by applying f to the structure constants of A. In particular, for the specializations f_1 and f_q defined by $f_1(u) = 1$, $f_q(u) = q$ we have

$$A_{f_1} \cong \mathbb{C}W \quad \text{and} \quad A_{f_q} \cong \mathcal{H}$$

where $\mathcal{H}$ is the Hecke algebra of $(1_B)^G$, for $G = G(q)$.

(3.12) <u>Theorem</u>. <u>Let</u> K <u>be the quotient field of</u> R. <u>Let</u> A <u>be an R-algebra with a finite basis</u> $\{a_i\}$ <u>and let</u> $f: R \to F$ <u>be a homomorphism such that</u> A_f <u>is a separable F-algebra</u>.

(i) <u>Then</u> $K \otimes_R A$ <u>is separable, and has the same numerical invariants as</u> $F \otimes_R A$ (numerical invariants = dimensions of simple modules over an algebraically closed extension field).

(ii) <u>Letting</u> K^*, F^* <u>denote algebraic closures of</u> K, F and R^* <u>the integral closure of</u> R <u>in</u> K^*, <u>we have</u>

$$\mu(a_i) \in R^* \quad \underline{\text{for each}} \quad \mu \in \operatorname{Irr}(K^* \otimes_R A).$$

<u>Now let</u> $f^*: R^* \to F^*$ <u>extend</u> f. <u>For each</u> μ, <u>define an F-linear map</u> μ_{f^*} <u>by</u>

$$\mu_{f^*}(1 \otimes a_i) = f^*(\mu(a_i)) \quad \text{for each } i.$$

<u>Then</u> $\mu_{f^*} \in \operatorname{Irr}(F^* \otimes A_f)$, <u>and the map</u> $\mu \to \mu_{f^*}$ <u>is a bijection of irreducible characters</u>.

Part (i) is due to J. Tits. It is possible to prove it by an adaptation of the argument used to prove that $C = {}^tDD$, where C is the Cartan matrix and D the decomposition matrix (see §2A).

(3.13) <u>Corollary</u>. $\mathbb{C}W \cong \mathcal{H}(q)$, <u>for the Hecke algebra</u> $\mathcal{H}(q)$ <u>of</u> $(1_B)^G$, <u>for each group</u> $G = G(q)$ <u>in the family</u>.

(3.14) <u>Proposition</u>. <u>For each</u> $G(q)$, <u>there is a bijection</u>

$$\varphi \to \zeta_{\varphi,q}$$

<u>from</u> $\operatorname{Irr} W \to \{\zeta \in \operatorname{Irr} G(q): (\zeta, (1_{B(q)})^{G(q)}) > 0\}$, <u>such that for all</u> $J \subseteq S$,

$$(\varphi, (1_{W_J})^W) = (\zeta_{\varphi,q}, (1_{P_J(q)})^{G(q)}).$$

(3.15) <u>Theorem</u> (Benson-Curtis). <u>Let</u> (W,S) <u>be indecomposable. Then each character</u> $\varphi \in \operatorname{Irr} W$ <u>is uniquely determined by the multiplicities</u> $(\varphi, (1_{W_J})^W)$ <u>with the following exceptions:</u>

<u>the characters of degree</u> 2 <u>if</u> $|W| = 12$ <u>or</u> 16;
<u>the two characters of degree</u> 512 <u>if</u> W <u>is of type</u> E_7;
<u>the four characters of degree</u> 4096 <u>if</u> W <u>is of type</u> E_8

(3.16) <u>Corollary</u>. (i) <u>The bijection</u> $\varphi \to \zeta_{\varphi,q}$ <u>is independent of the extensions</u> f_1^*, f_q^*, <u>with the exceptions noted in</u> (3.15).

(ii) <u>The characters of</u> $K^* \otimes A$ <u>are rational, in the sense that</u> $\mu(a_w) \in \mathbb{Q}[u]$ <u>for all</u> $\mu \in \mathrm{Irr}(K^* \otimes_R A)$ (<u>with exceptions as in (3.15)</u>).

(iii) <u>The characters</u> $\zeta_{\varphi,q}$ <u>are rational valued on</u> G(q), <u>with the exceptions noted.</u>

(3.17) <u>Definition</u>. Let $\varphi \in \mathrm{Irr}\, W$, and let μ be the corresponding irreducible character of $K^* \otimes A$. The <u>generic degree</u> associated with φ is the element of K^* defined by

$$d_\varphi = \frac{P(u) \deg \varphi}{\Sigma(\mathrm{IND}\, a_w)^{-1} \varphi(a_w)\varphi(\hat{a}_w)},$$

where

$$P(u) = \Sigma_w \mathrm{IND}\,(a_w) (= \Sigma_w u^{\ell(w)} \text{ if all } c_i = 1).$$

(3.18) <u>Proposition</u>. <u>For each group</u> G(q), <u>let</u> $f_q(u) = q$. <u>Then</u>

$$f_q^*(d_\varphi) = \deg \zeta_{\varphi,q}$$

<u>for all</u> $\varphi \in \mathrm{Irr}\, W$.

(3.19) <u>Theorem</u>. (Benson-Curtis) <u>Let</u> $\{G(q)\}$ <u>be a family of Chevalley groups such that</u> $\{q\}$ <u>contains almost all primes. Then</u> $d_\varphi \in \mathbb{Q}[u]$ <u>for all</u> $\varphi \in \mathrm{Irr}\, W$.

(3.20) <u>Corollary</u>. <u>Let</u> h(u) <u>be the group order polynomial (such that</u> $h(q) = |G(q)|$ <u>for all</u> q <u>for a family of universal Chevalley groups</u> $\{G(q)\}$. Then $d_\varphi | h(u)$ in $\mathbb{Q}[u]$.

The preceding results are used in the proof of:

(3.21) <u>Theorem</u> (Curtis, Kantor, Seitz). <u>The only 2-transitive permutation representations of the finite Chevalley groups are the known ones</u>.

Here is the idea of the proof. Let θ be the permutation character of a 2-transitive permutation representation of $G = G(q)$, where p is the natural characteristic of G. Then $\theta = 1 + \zeta$, for $\zeta \in \mathrm{Irr}\, G$. If $(\zeta, (1_B)^G) \neq 0$ then $(\theta, (1_B)^G) = 1$, and we have $G = LB$, where $\theta = (1_L)^G$. The possibilities for this situation to occur were determined previously by Seitz. So we may assume $(\zeta, (1_B)^G) > 0$, and that the BN-rank of G is ≥ 2. By a group theoretical argument, the theorem holds if $p \nmid \theta(1)$. So we may assume that p divides $\theta(1)$, and hence $p \nmid \zeta(1)$. The proof is completed by an analysis of the contribution of the prime p to the generic degree associated with ζ.

References

1. C. W. Curtis, Representations of finite groups of Lie type, Bull. Amer. Math. Soc. (N.S.) 1 (1979), 721-757.
2. C. W. Curtis and I. Reiner, Methods of representation theory, I, Wiley-Interscience, New York, 1981; II, to appear.
3. W. Feit, The representation theory of finite groups, North-Holland, Amsterdam, 1982.
4. J. Green, The characters of the finite general linear groups, Trans. Amer. Math. Soc. 80(1955), 402-447.
5. J. Humphreys, Cartan invariants, Bull. London Math. Soc. 17(1985), 1-14.
6. G. Lusztig, On the discrete series of GL_n over a finite field, Princeton University Press, Princeton, 1974.
7. G. Lusztig, Characters of reductive groups over a finite field, Princeton University Press, Princeton, 1984.
8. J.-P. Serre, Representation theory of finite groups, Springer-Verlag, New York, 1982.
9. B. Srinivasan, Representations of finite Chevalley groups, Springer Lecture Notes 764, Springer-Verlag, Berlin, 1979.
10. P. Fong and B. Srinivasan, The blocks of finite general linear and unitary groups, Invent. math. 69, 109-153(1982).

ALGEBRAIC K-THEORY, MORITA THEORY, AND THE CLASSICAL GROUPS*

Alexander J. Hahn
University of Notre Dame
Notre Dame, IN 46556/USA

For a ring R (always associative with 1), let $GL_n(R)$ be the group of invertible n x n matrices with coefficients in R. For $r \in R$ and $1 \leq i,j \leq n$ with $i \neq j$, let $e_{ij}(r)$ be the matrix with 1 along the diagonal, r in the (i,j) position and 0 elsewhere, and let $E_n(R)$ be the subgroup of $GL_n(R)$ generated by all the $e_{ij}(r)$.

If R is a division ring there are the following three classical theorems:

(A) $E_n(R)$ <u>is a normal subgroup of</u> $GL_n(R)$ <u>and if</u> $n \geq 2$,

$$GL_n(R)/E_n(R) \approx R^*/DR^*,$$

where R^* is the group of non-zero elements of R, and DR^* is its commutator subgroup.

(B) $E_n(R)/\mathrm{Cen}\,(E_n(R))$ <u>is a simple group</u> (with two exceptions when n = 2).

(C) If R' <u>is another division ring</u>, <u>then</u> $E_n(R) \approx E_m(R')$ <u>if and only if</u> n = m <u>and</u> R <u>or</u> R^{op} <u>is isomorphic to</u> R'. Here R^{op} is the opposite ring of R. (There are exceptions when n or m is 2).

There are analogous theorems for the other classical groups over division rings.

An obvious questions that arises is this: Do the theorems (A)-(C) and their analogues for the other classical groups remain valid for more general rings R? If not, what are the correct generalizations? These questions and related ones have been the object of intensive research, particularly over the past 15 years. In reference to (A) and (B) the most important factor has been the introduction and study of the groups K_1 and K_2 and their unitary versions. As regards (C) the key contribution is the theory of "full" groups. In these lectures I will try to sketch both developments.

It is a chief strategy of algebraic K-theory to go first to certain infinite dimensional or "stable" linear groups and then to go back down to finite dimensions by "stablity" theorems. This is especially evident in Sections I-III below. "Full" groups are geometric abstractions which allow for the simultaneous treatment of the isomorphism problem for large classes of groups. See Sections VI and XIII below. Morita theory is a tool which facilitates the transfer from more difficult to more easily manageable rings, e.g. from group rings of a finite group to division rings. It thus helps to establish a connection between K-theory and algebraic topology. See Section XIII. When combined with the theory of full groups it yields important additional information about the isomorphisms of the classical groups. See VII and XIII. These lectures are only intended to serve as an introduction to these topics; they are neither definitive nor complete. The 2-dimensional cases are almost always exceptional and are largely ignored.

References for the thirteen sections below are as follows. Many more could have been added, but there are many basic sources through which others can be traced. In this regard refer in particular to the bibliographies in references [22], [29], [65], [74] and [76].

SECTION I: [1], [6], [8], [14], [26], [35], [50], [64]

SECTION II: [6], [8], [17], [26], [54], [56], [60], [61], [62], [65], [75]

SECTION III: [26], [27], [29], [35], [50], [52], [53]

SECTION IV: [15], [29], [30], [33], [34], [35], [50], [51], [52], [66]

SECTION V: [4], [5], [6], [10], [24], [35], [43], [50]

SECTION VI: [14], [22], [25], [37], [38], [40], [41], [47], [63], [73], [74]

SECTION VII: [6], [19], [22], [25], [38], [57]

SECTION VIII: [2], [3], [7], [9], [14], [39], [41], [42], [55], [58], [60], [64], [68], [70], [71]

SECTION IX: [2], [3], [7], [9], [32], [39], [55], [59], [60], [64]

SECTION X: [2], [3], [7], [9], [32], [55], [59], [60], [64], [69], [71]

SECTION XI: [3], [4], [7], [9], [10], [13], [30], [33], [43], [49], [52], [53]

SECTION XII: [2], [7], [13], [28], [33], [36], [59], [60]

SECTION XIII: [3], [11], [12], [14], [16], [18], [20], [21], [22], [23], [25], [31], [39], [40], [41], [44], [45], [46], [67], [69]

Important general references are [1], [3], [6], [7], [14], [15], [22], [24], [35], [37], [39], [48], [50], [53], [57], [58], [61], [74], [76], and in Chinese [72]. See the Springer Lecture Notes 108, 341-3, 551, 854, 966-7 and 1046, the volume "Current Trends in Algebraic Topology" Canadian Math. Soc. Conf. Proc. 3, and the issue of the Journal of Pure and Applied Algebra, 34 (1984), for the explosion in the field of algebraic K-theory.

I. THE STABLE LINEAR GROUPS AND K_1

Let R be a ring. Consider $GL_n(R) \subseteq GL_{n+1}(R)$ by identifying $g \in GL_n(R)$ with $\begin{pmatrix} g & 0 \\ 0 & 1 \end{pmatrix} \in GL_{n+1}(R)$, form the chain

$$GL_1(R) \subseteq GL_2(R) \subseteq GL_3(R) \subseteq \ldots ,$$

and observe that

$$E_1(R) \subseteq E_2(R) \subseteq E_2(R) \subseteq \ldots .$$

Define the <u>stable linear groups</u> GL(R) and E(R) by

$$GL(R) = \bigcup_{n \geq 1} GL_n(R) \text{ and } E(R) = \bigcup_{n \geq 1} E_n(R).$$

<u>Notation</u>: For a group G, DG will be its commutator subgroup. For a ring R, R* will be its groups of units. Normal subgroup is denoted $\triangleright$

<u>Theorem 1</u> (<u>Whitehead's Lemma</u>). DGL(R) = E(R) = DE(R). <u>So</u> $GL(R) \triangleright E(R)$ <u>and</u> $K_1(R) = GL(R)/E(R)$ <u>is an Abelian group</u>.

It is easy to see that a ring homomorphism $f : R \to R'$ induces a group homomorphism

$$K_1(f) : K_1(R) \to K_1(R')$$

and that K_1 is a functor from rings to Abelian groups.

Now let $\mathfrak{a} \subseteq R$ be a (two-sided) ideal of R. Consider the homomorphism

$$GL_n(R) \to GL_n(R/\mathfrak{a})$$

induced by $R \to R/\mathfrak{a}$, and let $GL_n(R,\mathfrak{a})$ be its kernel. Observe that $GL_n(R,\mathfrak{a}) = \{X \in GL_n(R) \mid X - I \in Mat_n(\mathfrak{a})\}$, where $Mat_n(\mathfrak{a})$ is the set of n x n matrices with coefficients in $\mathfrak{a}$. Let $E_n(R,\mathfrak{a})$ be the <u>normal</u> subgroup of $E_n(R)$ generated by the elements $e_{ij}(r)$ with $r \in \mathfrak{a}$. Then

$$E_n(R,\mathfrak{a}) \subseteq GL_n(R,\mathfrak{a}).$$

Define,

$$GL(R,\mathfrak{a}) = \bigcup_{n\geq 1} GL_n(R,\mathfrak{a}) \text{ and } E(R,\mathfrak{a}) = \bigcup_{n\geq 1} E_n(R,\mathfrak{a}).$$

The groups $GL_n(R,\mathfrak{a})$ and $E_n(R,\mathfrak{a})$ are <u>congruence groups</u> and $GL(R,\mathfrak{a})$ and $E(R,\mathfrak{a})$ are their "stable" versions.

<u>Theorem 2 (Bass)</u>. (1) <u>Any subgroup</u> H <u>satisfying</u> $E(R,\mathfrak{a}) \subseteq H \subseteq GL(R,\mathfrak{a})$ <u>is normal in</u> GL(R).

(2) <u>If</u> $H \subseteq GL(R)$ <u>is normalized by</u> E(R), <u>then there is a unique ideal</u> $\mathfrak{a}$ <u>in</u> R <u>such that</u>

$$E(R,\mathfrak{a}) \subseteq H \subseteq GL(R,\mathfrak{a}).$$

It is easy to see that in particular, E(R) is a simple group if and only if R is a simple ring.

Define the group $K_1(R,\mathfrak{a})$ by $K_1(R,\mathfrak{a}) = GL(R,\mathfrak{a})/E(R,\mathfrak{a})$. This can be shown to be an Abelian group. Note that $K_1(R,R) = K_1(R)$, and that $K_1(R,0) = 1$. In view of the theorem above, the determination of these groups classifies all the normal subgroups of GL(R). Refer also to Section V below.

Denote the ring of integers by Z and let Z_m be the quotient ring Z/mZ.

<u>Example 3</u>. For R a division ring, $K_1(R) \simeq R^*/DR^*$. This is a consequence of the Dieudonné determinant. The usual determinant gives that $K_1(Z) \simeq Z_2$; $K_1(Z,mZ)$ will be computed later.

II. STABILITY AND THE NORMAL SUBGROUPS OF $GL_n(R)$

Consider the free left R-module R^n. Then $x = (x_1,\ldots,x_n) \in R^n$ is <u>unimodular</u> if

$$r_1x_1 + \ldots + r_nx_n = 1$$

for some $r_1,\ldots,r_n$ in R.

<u>The condition</u> $(S)_d$: <u>For all</u> $n > d$ <u>and for any unimodular</u> $(x_0,x_1,\ldots,x_n)$ <u>in</u> R^{n+1}, <u>there exist</u> $s_1,\ldots,s_n$ <u>in</u> R <u>such that</u> $(x'_1,\ldots,x'_n)$ <u>with</u> $x'_i = x_i + s_ix_0$, <u>is unimodular in</u> R^n. Observe that $(S)_d \Rightarrow (S)_{d+1}$.

Example 4. For R a division ring, a local ring, or a semisimple Artinian ring, $(S)_0$ holds. For R Dedekind, eg. $R = Z$, $(S)_1$ holds. If F is a field and $R = F[t_1,\ldots,t_d]$ is a polynomial ring in d indeterminates, then $(S)_d$ holds. In fact, more generally, if S is a commutative, Noetherian ring of Krull dimension d, and R is an S-algebra which is finitely generated over S as S-module, then $(S)_d$ holds for R.

This "stable range" condition is related to the action of $E_n(R)$ on unimodular vectors:

$(S)_d$ implies that $E_{n+1}(R)$ is transitive on the unimodular elements of R^{n+1}, for all $n \geq d + 1$.

Now let $\mathfrak{a} \subseteq R$ be any ideal and consider the set $GL_n(R,\mathfrak{a})/E_n(R,\mathfrak{a})$ of cosets. Define the function

$$s_n : GL_n(R,\mathfrak{a})/E_n(R,\mathfrak{a}) \longrightarrow K_1(R,\mathfrak{a}).$$

by $s_n(gE_n(R,\mathfrak{a}))) = gE(R,\mathfrak{a})$.

Theorem 5 (Bass, Vaserstein). Suppose $(S)_d$ holds. Then for $n \geq d+2$, $E_n(R,\mathfrak{a}) \triangleleft GL_n(R)$ and s_n is an isomorphism. So

$$GL_n(R,\mathfrak{a})/E_n(R,\mathfrak{a}) \simeq K_1(R,\mathfrak{a}).$$

Returning to the study of the normal subgroups of $GL_n(R)$, consider for an ideal $\mathfrak{a}$, the composite

$$GL_n(R) \longrightarrow GL_n(R/\mathfrak{a}) \longrightarrow GL_n(R/\mathfrak{a})/C,$$

where $C = \{ \begin{pmatrix} s & 0 \\ 0 & s \end{pmatrix} \mid s \in (\text{Cen } R/\mathfrak{a})^*\}$, and denote the kernel by $GL_n'(R,\mathfrak{a})$. Observe that

$$E_n(R,\mathfrak{a}) \subseteq GL_n(R,\mathfrak{a}) \subseteq GL_n'(R,\mathfrak{a}).$$

Theorem 6 (Bass). Assume $(S)_d$ and that $n \geq \max(3, d+2)$. Then a subgroup H of $GL_n(R)$ is normalized by $E_n(R)$ if and only if there is a (unique) ideal $\mathfrak{a}$ of R such that

$$E_n(R,\mathfrak{a}) \subseteq H \subseteq GL_n'(R,\mathfrak{a}).$$

Corollary. Suppose R is a division ring and $n \geq 3$. Then $E_n(R)/\text{Cen } E_n(R)$ is simple.

Suppose $(S)_d$ holds for R and that $n \geq \max(3, d+2)$. In reference to the quotient $GL_n'(R,\mathfrak{a})/E_n(R,\mathfrak{a})$ observe that $GL_n'(R,\mathfrak{a})/GL_n(R,\mathfrak{a})$ is isomorphic to a subgroup of $(\text{Cen } R/\mathfrak{a})^*$. So the classification of the subgroups of $GL_n(R)$ normalized by $E_n(R)$ reduces in essence to the analysis of the groups $K_1(R,\mathfrak{a})$.

Wilson has proved that the conclusion of Theorem 6 holds for any commutative ring (without any stability assumptions) and $n \geq 4$, and Golubchik lowered this bound to $n \geq 3$. More recently, Vaserstein has established a theorem which contains the theorems of Bass, Wilson, and Golubchik all as special cases.

III. THE STEINBERG GROUPS AND K_2.

Let R be a ring and consider $E_n(R)$. The following relations hold for the generators $e_{ij}(r)$:

(S1) $e_{ij}(r)\, e_{ij}(s) = e_{ij}(r+s)$

(S2) $[e_{ij}(r), e_{jk}(s)] = e_{ik}(rs)$, i, j, k distinct

(S3) $[e_{ij}(r), e_{hk}(s)] = I$, if $j \neq h$, $i \neq k$.

Here $[g,h] = ghg^{-1}h^{-1}$. Since $E_n(R) \subseteq E(R)$, the same relations hold in $E(R)$ where i, j can now be any two positive integers with $i \neq j$ (not bounded by n).

Define the <u>Steinberg group</u> $St(R)$ to be the group generated by all symbols $x_{ij}(r)$, where $i \neq j$ and $r \in R$, subject to the <u>Steinberg relations</u>, i.e. those obtained by replacing e by x in (S1)-(S3).

The map $\phi : St(R) \longrightarrow E(R)$ given by $\phi(x_{ij}(r)) = e_{ij}(r)$ is a surjective homomorphism. Define $K_2(R) = \ker \phi$. Observe that $K_2(R) = 1$ if and only if ϕ is an isomorphism. So $K_2(R)$ is a measure of the additional relations which the e_{ij} satisfy.

It turns out that $St(R)$ is the universal central extension of $E(R)$, $K_2(R) = \text{Cen } St(R)$, and K_2 is a functor from rings to Abelian groups: If $f : R \longrightarrow R'$ is a ring homomorphism, $x_{ij}(r) \to x_{ij}(fr)$ determines a homomorphism $St(R) \longrightarrow St(R')$ which induces $K_2(f) : K_2(R) \to K_2(R')$.

In a completely analogous way define $St_n(R)$, and let $K_2(n,R)$ be the kernel of the canonical map $St_n(R) \longrightarrow E_n(R)$. For commutative R and

$n \geq 5$, $St_n(R)$ is the universal central extension of $E_n(R)$. Since the diagram

$$\begin{array}{ccc} St(R) & \longrightarrow & E(R) \\ \uparrow & & \uparrow \\ St_n(R) & \longrightarrow & E_n(R) \end{array}$$

commutes, the restriction of the map on the left gives a homomorphism $K_2(n,R) \longrightarrow K_2(R)$.

There is a "stability" theorem due to Dennis, Van der Kallen, Suslin, Tulenbaev and Vaserstein.

Theorem 7. Suppose $(S)_d$ *holds. Then for* $n \geq d+3$, *the homomorphism* $K_2(n,R) \longrightarrow K_2(R)$ *is an isomorphism.*

If F is a finite field, it is not hard to show that $K_2(F) = 1$. Therefore by Theorem 7 and Example 4:

Corollary 8. Suppose F *is a finite field and* $n \geq 3$. *Then*

$$SL_n(F) \approx St_n(F).$$

So $SL_n(F)$ *is generated by the elementary matrices, subject only to relations* (S1) - (S3).

So if R is a finite field all relations among the elementary matrices are a consequence of (S1)-(S3). This is rarely the case in general. For example

$$\left[\begin{pmatrix} 1 & 1 \\ & 1 \end{pmatrix} \begin{pmatrix} 1 & \\ -1 & 1 \end{pmatrix} \begin{pmatrix} 1 & 1 \\ & 1 \end{pmatrix}\right]^4 = \begin{pmatrix} 0 & 1 \\ -1 & 0 \end{pmatrix}^4 = 1 ,$$

and this relation is never a consequence of the others in any $SL_n(Z)$ with $n \geq 3$.

IV. THE THEOREMS OF MATSUMOTO, MERKURJEV-SUSLIN, AND SYLVESTER

We continue with our investigation of $K_2(R)$. Suppose R is a commutative ring, and consider $St(R)$. For $u \in R^*$ and $i \neq j$, put $w_{ij}(u) = x_{ij}(u)\, x_{ji}(-u^{-1})\, x_{ij}(u)$ and $h_{ij}(u) = w_{ij}(u)\, w_{ij}(-1)$. Of course,

$h_{ij}(u) \in St(R)$. An easy computation shows that under $\phi : St(R) \longrightarrow E(R)$,

$$\phi(h_{12}(u)) = \begin{pmatrix} u & 0 & 0 \\ 0 & u^{-1} & 0 \\ 0 & 0 & 1 \end{pmatrix} \quad \text{and} \quad \phi(h_{13}(v)) = \begin{pmatrix} v & 0 & 0 \\ 0 & 1 & 0 \\ 0 & 0 & v^{-1} \end{pmatrix}$$

Define $\{u,v\} = [h_{12}(u), h_{13}(v)]$, and observe that

$$\{u,v\} \in \ker \phi = K_2(R).$$

It turns out that $\{u,v\}$ is independent of the pairs of indices (1,2) and (1,3), i.e. repeating with $h_{ij}(u)$ and $h_{ik}(v)$, i, j, k distinct, yields the same element in $K_2(R)$. The element $\{u,v\}$ is called a _symbol_, it is related to the classical Lengendre and Hilbert symbols of number theory. Routine computations show:

Proposition 9. _For_ u, u_1, u_2, v, v_1, _and_ v_2 _in_ R^*,

$$\{u_1,v\}\{u_2,v\} = \{u_1u_2,v\}$$
$$\{u,v_1\}\{u,v_2\} = \{u,v_1v_2\}$$
$$\{u,1-u\} = 1, \text{ if } 1-u \in R^*, \text{ and}$$
$$\{u,-u\} = 1.$$

Consider in particular, the element $\{-1,-1\}$ in $K_2(R)$. Since $\{-1,-1\} \cdot \{-1,-1\} = \{1,-1\} = 1$, it has order ≤ 2 in $K_2(R)$.

Theorem 10 (_Sylvester_). $K_2(Z) \approx Z_2$. _More precisely_, $K_2(Z)$ _consists of_ 1 _and_ $\{-1,-1\}$ (_which are distinct_).

Corollary 11. _For_ $n \geq 3$, $SL_n(Z)$ _is generated by the_ $n(n-1)$ _elements_ $e_{ij}(1)$ _which are subject only to the relations_.

$$[e_{ij}(1), e_{jk}(1)] = e_{ik}(1), \text{ if } i, j, k \text{ distinct}$$

$[e_{ij}(1), e_{k\ell}(1)] = 1$, _if_ $j \neq k$, $i \neq \ell$, _and_ $(e_{12}(1)\, e_{21}^{-1}(1)\, e_{12}(1))^4 = 1$.

For later reference we need:

<u>Theorem 12</u> (<u>Sylvester</u>). <u>For</u> m $\geq$ 0,

$$K_2(Z_m) = \begin{cases} 1, \{-1,-1\} \text{ (distinct), if } m = 0 \pmod 4 \\ 1, \text{ otherwise} \end{cases}$$

<u>In particular</u>, <u>if</u> $f : Z \longrightarrow Z_m$ <u>is the natural map, then</u>

$$K_2(f) : K_2(Z) \longrightarrow K_2(Z_m)$$

<u>is surjective</u>.

If F is any field, the relations of Proposition 9 are sufficient to characterize $K_2(F)$.

<u>Theorem 13</u> (<u>Matsumoto</u>). $K_2(F) \approx F^* \otimes_Z F^* / \langle \{r \otimes (1-r) \mid r \neq 0,1\} \rangle$. <u>This isomorphism is given by</u> $r \otimes s \longrightarrow \{r,s\}$. <u>In other words, the elements</u> $\{r,s\}$ <u>along with the first three relations of Proposition</u> 9 <u>provide a presentation of</u> $K_2(F)$ <u>as Abelian group</u>.

A very important theorem is that of Merkurjev-Suslin which relates $K_2(F)$ to the Brauer group Br(F) of F. Suppose F contains m-distinct roots of X^m-1. Let

$$Br_m(F) = \{z \in Br(F) \mid z^m = 1\} \text{ and let } K_2(F)^m = \{y^m \mid y \in k_2(F)\}.$$

<u>Theorem 14</u> (<u>Merkurjev-Suslin</u>). $K_2(F)/K_2(F)^m \approx Br_m(F)$.

This is a very deep result whose proof involves both étale cohomology and spectral sequences for K_2 of Brauer-Severi varieties. The special case m = 2 is already a remarkable result. In this case F is any field of characteristic not 2 and the isomorphism above is given by

$$\{r,s\} + K_2(F)^2 \xrightarrow{\alpha_F} [(\tfrac{r,s}{F})],$$

where $(\frac{r,s}{F})$ is the quaternion algebra defined by the generators e, f and equations $e^2 = r$, $f^2 = s$ and $ef = -fe$.

The surjectivity of α_F says that every central simple F-algebra A which is anti-isomorphic to itself is similar to a tensor product of quaternion algebras. This settles an old question in the theory of algebras. The injectivity of α_F answers an important question in quadratic forms: Let W(F) be the Witt ring of F and let I(F) be its ideal of even dimensional forms. The Clifford invariant induces a homomorphism

$$I(F)^2 \longrightarrow Br_2(F) .$$

The fact that α_F is injective answers the question as to whether $I(F)^3$ is the kernel of this map in the affirmative. The surjectivity of α_F implies that this map is surjective also.

Note, finally, that it follows from Proposition 9 that the "squaring" map

$$K_2(F) \longrightarrow K_2(F)^2$$

has $\{F^*,-1\}$ in its kernel. Another theorem of Merkurjev-Suslin says that, for a field of characteristic not 2, the kernel is equal to $\{F^*,-1\}$

V. A K_1-K_2 EXACT SEQUENCE AND THE CONGRUENCE SUBGROUP PROBLEM

This section considers an exact sequence of Milnor relating K_1 and K_2 and describes its connection with the congruence subgroup problem.

Theorem 15 (Milnor). *Suppose R and R' are rings and $f : R \longrightarrow R'$ is a surjective homomorphism with kernel $\mathfrak{a}$. Then there is an exact sequence*

$$K_2(R) \xrightarrow{K_2(f)} K_2(R') \longrightarrow K_1(R,\mathfrak{a}) \longrightarrow K_1(R) \xrightarrow{K_2(f)} K_1(R').$$

Notice, therefore, that information about $K_2(R)$ and $K_1(R)$ gives information about $K_1(R,\mathfrak{a})$. See Section II. We pursue this in a special case.

Suppose R is commutative, let $\mathfrak{a}$ be an ideal of R and define

$$SL_n(R,\mathfrak{a}) = SL_n(R) \cap GL_n(R,\mathfrak{a})$$

where $SL_n(R)$ is the subgroup of $GL_n(R)$ consisting of the elements of determinant 1. Set

$$SL(R,\mathfrak{a}) = \bigcup_{n\geq 1} SL_n(R,\mathfrak{a}).$$

By Theorem 2, $E(R,\mathfrak{a}) \triangleleft SL(R,\mathfrak{a})$. Put $SK_1(R,\mathfrak{a}) = SL(R,\mathfrak{a})/E(R,\mathfrak{a})$. Denote $SK_1(R,R)$ by $SK_1(R)$. Using the determinant it is easy to see that

$$K_1(R,\mathfrak{a}) \approx SK_1(R,\mathfrak{a}) \times GL_1(R,\mathfrak{a}),$$

where $GL_1(R,\mathfrak{a}) = \{r \in R^* | r - 1 \in \mathfrak{a}\}$. See Section I.

Example 16. Since $SL_n(Z) = E_n(Z)$ for $n \geq 3$, $SK_1(Z) = 1$.

Corollary 17. *If R and R' are commutative rings and $f : R \longrightarrow R'$ is a surjective homomorphism with kernel $\mathfrak{a}$, there is an exact sequence*

$$K_2(R) \longrightarrow K_2(R') \longrightarrow SK_1(R,\mathfrak{a}) \longrightarrow SK_1(R) \longrightarrow SK_1(R'),$$

obtained by restricting the sequence of Theorem 15.

Corollary 18. $SK_1(Z,mZ) = 1$, *and* $K_1(Z,mZ) \approx \begin{cases} Z_2, & \text{if } m = 1,2 \\ 1, & \text{if } m \geq 3. \end{cases}$

For a quick proof of the last corollary, consider $f : Z \longrightarrow Z_m$, and apply Theorem 12, Example 16 and Corollary 17. This corollary in turn implies the solution of the congruence subgroup problem for $SL_n(Z)$, $n \geq 3$.

Theorem 19 (Bass, Milnor, Serre, Mennicke). *Let $n \geq 3$. If G is a subgroup of* $SL_n(Z)$ *of finite index, then*

$$G \supseteq SL_n(Z,mZ),$$

for some $m \neq 0$.

This theorem has recently been extended to much more general number theoretic classical and algebraic groups by Bak-Rehmann and Prasad-Ragunathan.

VI. ISOMORPHISM OF FULL GROUPS

We turn our attention to the study of the isomorphisms between the linear groups.

Let R and R' be rings and consider the following isomorphisms from $GL_r(R)$ onto $GL_n(R')$.

(i) Let $\alpha : R \longrightarrow R'$ be a ring isomorphism and define

$$GL_n(R) \longrightarrow GL_n(R')$$

by applying α to a matrix entries.

(ii) Let $\alpha : R \longrightarrow R'$ be an anti-isomorphism and define

$$GL_n(R) \longrightarrow GL_n(R')$$

by applying α to the matrix entries of $(X^{-1})^t$, the transpose of $X^{-1} \in GL_n(R)$.

Any isomorphism $GL_n(R) \longrightarrow GL_n(R')$ which is obtained by composing an isomorphism of type (i) or (ii) with an inner automorphism of $GL_n(R')$ is _standard_.

To handle larges classes of rings and groups simultaneously, O'Meara introduced the concept of a full group:

Let D be a division ring and consider $GL_n(D)$. A subgroup G of $GL_n(D)$ is _full_ if for every line L and hyperplane H in D^n with $L \subseteq H$, there is a v in D^n and a δ in the dual space of D^n, with $Dv = L$ and $\ker \delta = H$, such that T defined by

$$Tx = x + \delta(x)v$$

is in G. Any such T is a _transvection_.

Example 20. $SL_n(D)$ is full in $GL_n(D)$. More importantly, if R is a domain with division ring of quotients D, and if $\mathfrak{a}$ is an ideal of R, then $GL_n(R)$, $SL_n(R) = GL_n(R) \cap SL_n(D)$, $GL_n(R,\mathfrak{a})$ and $SL_n(R,\mathfrak{a}) = GL_n(R,\mathfrak{a}) \cap SL_n(D)$ are all full in $GL_n(D)$. In particular, $SL_n(Z)$ is full in $GL_n(Q)$, Q the rational numbers.

Theorem 21 (_O'Meara_). _Let_ D _and_ D' _be division rings and let_ G _and_ G' _be full subgroups of_ $GL_n(D)$ _and_ $GL_m(D')$ _respectively._

Assume $n, m \geq 3$ _and let_ $\phi : G \longrightarrow G'$ _be an isomorphism. Then_ $n = m$, _and there is a standard isomorphism_ $\beta : GL_n(D) \longrightarrow GL_n(D')$, _and a homomorphism_ $\chi : G \longrightarrow (\text{Cen } D')^*$, _such that_

$$\phi(g) = \chi(g) \cdot \beta(g), \text{ all } g \in G.$$

In particular, D _or_ D^{op} (_the opposite of_ D) _is isomorphic to_ D'.

This theorem in combination with Example 20 provides a wide ranging isomorphism theory for the linear groups. It was originally proved by O'Meara under the additional assumption that $n, m \geq 5$ if D or D' is not commutative. The remaining case was handled by Sosnovskii. Vaserstein has given a number of characterizations of full groups. One of these is as follows: Let R be a subring of D which does not necessarily contain 1 and call R full, if for any $d \in D$ there are nonzero r and s in R such that rd and ds are in R.

Theorem 22 (Vaserstein). Suppose $n \geq 3$. Then $G \subseteq GL_n(D)$ is full if and only if $G \supseteq E_n(R)$ for a full subring R of D.

VII. AN APPLICATION OF MORITA THEORY

We continue the discussion begun in the previous section, and let R and R' be rings with division rings of quotients. Suppose $n, m \geq 3$ and that

$$\phi : GL_n(R) \longrightarrow GL_m(R')$$

is an isomorphism. By Example 20, we know that Theorem 21 applies to ϕ (and in particular $n = m$). Additional questions remain, however. Can β be taken as standard "over R" and not just "over D"? Does the existence of ϕ imply that R or R^{op} is isomorphic to R'? While the answer to both questions is no in general, Morita theory provides some insight.

Two rings R and R' are Morita equivalent if there is an equivalence $E : \mathcal{M}_R \longrightarrow \mathcal{M}_{R'}$. Here $\mathcal{M}_R$ is the category of right R-modules and an equivalence is an additive functor with "inverse."

Example 23. Two division rings or two commutative rings are Morita equivalent if and only if they are isomorphic. Any ring R is equivalent, for any k, to the matrix ring $Mat_k(R)$. If S is a Dedekind domain with quotient field F, then any two maximal S-orders R and R' in a finite dimensional division algebra over F are equivalent.

It is easy to see that if $E : \mathcal{M}_R \longrightarrow \mathcal{M}_{R'}$ is an equivalence, and if M in $\mathcal{M}_R$ is any module, then

$$\text{(a)} \quad GL(M) \longrightarrow GL(E(M))$$

defined by $g \longrightarrow E(g)$ is an isomorphism. If M in $\mathcal{M}_R$ is reflexive, let $GL(M) \longrightarrow GL(M^*)$, where M^* is the dual of M, be the contragredient isomorphism. For an equivalence $E : \mathcal{M}_{R^{op}} \longrightarrow \mathcal{M}_{R'}$, there is the isomorphism

$$(b)\ GL(M) \longrightarrow GL(EM^*)$$

obtained by composition with the contragredient. Denote the isomorphism of (a) or (b) by E in either case. We now have the following theorem.

Theorem 24 (Hahn). Suppose R and R' are rings with division rings of quotients and assume that $n, m \geq 3$. Suppose $G \subseteq GL_n(R)$ is a subgroup containing $DGL_n(R)$ and that

$$\phi : G \longrightarrow GL_m(R')$$

is a monomorphism such that $G \supseteq DGL_m(R')$. Then there is an equivalence $E : \mathcal{M}_R \longrightarrow \mathcal{M}_{R'}$ with $E(R^n) = (R')^m$, or $E : \mathcal{M}_R\text{op} \longrightarrow \mathcal{M}_{R'}$ with $E((R^n)^*) = (R')^m$, and a homomorphism $\chi : G \longrightarrow (\text{Cen } R')^*$, such that

$$\phi(g) = \chi(g) \cdot E(g), \text{ all } g \in G.$$

In particular, R or R^{op} is Morita equivalent to R'.

This description of ϕ establishes the connection between R and R' and does not necessitate their rings of quotients. Morita theory provides a description of the equivalences $\mathcal{M}_R \longrightarrow \mathcal{M}_{R'}$. They arise as follows: A collection $(R, R', {}_RP_{R'}, {}_{R'}Q_R, \mu, \tau)$ where R, R' are rings; P and Q are bimodules as indicated; and

$$\mu : P \otimes_{R'} Q \longrightarrow R \text{ and } \tau : Q \otimes_R P \longrightarrow R'$$

are bimodule isomorphisms, satisfying

$$\mu(p \otimes q)\, p' = p\, \tau(q \otimes p') \text{ and } \tau(q \otimes p) q' = q\mu(p \otimes q').$$

for all p, p' in P and q, q' in Q, is a Morita context or set of equivalence data. It can be shown that $\_\_\otimes_R P$ defines an equivalence $\mathcal{M}_R \to \mathcal{M}_{R'}$ with inverse $\_\_\otimes_{R'} Q$, and that any equivalence $\mathcal{M}_R \longrightarrow \mathcal{M}_{R'}$ arises essentially in this way. The modules P and Q arising in a Morita context are progenerators. In particular, they are finitely generated and projective. If such modules are free then one gets more precise information from the theorem above. For example:

Corollary 25. Let $n \geq 3$ and let ϕ be an automorphism of $SL_n(Z)$. Then either ϕ is the restriction of an inner automorphism of $GL_n(Z)$ or the composite of such an automorphism with "transpose-inverse."

By the fact that it establishes equivalences between categories of modules, Morita theory is often an important tool with which module theoretic questions over more difficult rings, e.g. group rings, can be transformed to easier settings, e.g. division rings. See also Section XIII in this connection.

VIII. FORM RINGS AND UNITARY GROUPS

We have completed our brief survey of results for the linear classical groups. The study of the other classical groups has traditionally been partitioned as follows:

(i) Symplectic groups (of alternating forms)
(ii) Unitary groups (of hermitian forms), and corresponding to quadratic forms,
(iii) Orthogonal groups in characteristic not 2,
(iv) Non-defective orthogonal groups in characteristic 2, and
(v) Defective orthogonal groups in characteristic 2.

The introduction of form rings and generalized quadratic forms makes it possible to treat all these cases simultaneously and uniformly.

Suppose R is a ring equipped with an anti-automorphism J and a unit ε, such that $\varepsilon^J = \varepsilon^{-1}$ and J^2 is conjugation by ε. Let $R^{-\varepsilon} = \{r \in R \mid r^J\varepsilon = -r\}$ and $R_{-\varepsilon} = \{r^J\varepsilon - r \mid r \in R\}$, and observe that $R_{-\varepsilon} \subseteq R^{-\varepsilon}$. A <u>form parameter</u> Λ for R is a subgroup of $(R,+)$ satisfying $R_{-\varepsilon} \subseteq \Lambda \subseteq R^{-\varepsilon}$ and $r^J\Lambda r \subseteq \Lambda$, for all $r \in R$. A <u>form ring</u> is a pair (R,Λ), where R is a ring equipped with J and ε, and Λ is a form parameter. If there is an $s \in \mathrm{Cen}\, R$ such that $s^J + s \in R^*$, then $R^{-\varepsilon} = R_{-\varepsilon}$ and this is the only form parameter. This is the case if $2 \in R^*$.

Let (R,Λ) be a form ring and M a right R-module. A <u>sesquilinear form</u> $f : M \times M \longrightarrow R$ is a biadditive map satisfying $f(xr,ys) = r^J f(x,y)s$ for all x, y in M and r,s in R. For such an f consider

$$h : M \times M \longrightarrow R \text{ and } q : M \longrightarrow R/\Lambda,$$

respectively defined by $h(x,y) = f(x,y) + f(y,x)^J\varepsilon$, and $q(x) = f(x,x) + \Lambda$, for all x and y in M. The pair (h,q) is a (R,Λ)-<u>quadratic form on</u> M, and M equipped with (h,q) is a quadratic module over (R,Λ). A quadratic module M is <u>non-singular</u> if the map $M \longrightarrow M^*$ induced by h is bijective.

The unitary group $U(M)$ of an (R,Λ)-quadratic form (h,q) on M is defined to be the set of g in $GL(M)$ preserving both h and q. Taking $J = 1$ and $\varepsilon = -1$; or $J^2 = 1$ $(J \neq 1)$ and $\varepsilon = -1$, with $\Lambda = R^{-\varepsilon}$ in either case, specializes $U(M)$ to the symplectic group or the classical unitary group. Taking $J = 1$, $\varepsilon = 1$, with $\Lambda \neq R^{-\varepsilon}$, gives the defective orthogonal groups if $\Lambda \neq 0$, and the other orthogonal groups if $\Lambda = 0$.

Let M with (h,q) be a quadratic module. A vector u in M is <u>isotropic</u> if $q(u) = 0$, and two vectors u and v in M are <u>orthogonal</u> if $h(u,v) = 0$. Let u and v be orthogonal vectors with u isotropic. For any $r \in q(v)$, define the <u>Eichler transformation</u> $g_{u,v,r}$ in $U(M)$ by the equation

$$g_{u,v,r}(x) = x + uh(v\varepsilon,x) - vh(u,x) - u\varepsilon^{-1}r\, h(u,x), \text{ for all } x \in M.$$

Denote by $EU(M)$ the subgroup of $U(M)$ generated by all Eichler transformations. The following theorem, essentially due to Dieudonné, is the fundamental theorem in the theory of the classical groups:

<u>Theorem 26 (Dieudonné). Suppose</u> R <u>is a division ring and that</u> M <u>is finite dimensional, non-singular and contains isotropic vectors. If</u> $\Lambda = 0$ <u>(classical orthogonal case) and</u> $\dim M = 4$, <u>assume that any two orthogonal isotropic vectors are collinear. Then</u>

$$EU(M)/Cen(EU(M))$$

<u>is a simple group (with seven exceptions in all of which</u> R <u>is the Galois field</u> F_2 <u>or</u> F_3 <u>and</u> $\dim M \leq 4$).

It should be remarked that this theorem contains all simplicity theorems for the classical groups and that it has a completely uniform proof.

IX. THE HYPERBOLIC UNITARY GROUPS

Let (R,Λ) be a form ring with underlying J and ε. Suppose M is a quadratic module over (R,Λ) which is hyperbolic of rank $2n$, i.e., suppose M has a basis $\mathfrak{X} = \{x_1,\ldots,x_{2n}\}$ of isotropic vectors, such that

$$(h(x_i,x_j)) = \left(\begin{array}{cc|ccc} 0 & 1 & & & 0 \\ \varepsilon & 0 & & & \\ \hline & & \ddots & & \\ & & & & \\ 0 & & & 0 & 1 \\ & & & \varepsilon & 0 \end{array}\right)$$

Such a basis is called a <u>hyperbolic basis</u>. Note that hyperbolic modules are non-singular.

For $i \neq j$ and $h(x_i,x_j) = 0$, and $r \in R$, define

$$g_{i,j}(r) = g_{x_i r, x_j, 0}.$$

and for $r \in \Lambda$, define

$$g_{i,i}(r) = g_{x_i \varepsilon, 0, -r}$$

Let $EU_{\mathfrak{X}}(M)$ be the subgroup of $U(M)$ generated by all the $g_{ij}(r)$. If R is a division ring, then $EU_{\mathfrak{X}}(M)$ is the group $EU(M)$ of Section VIII.

For a matrix $X = (x_{ij}) \in \mathrm{Mat}_n(R)$, define $X^J = (x_{ij}^J)^t$, where t = transpose, and consider the set

$$\Lambda_n = \{X \in \mathrm{Mat}_n(R) \mid X^J = -X\varepsilon^{-1} \text{ and } x_{ii} \in \Lambda, \text{ all } i\}.$$

Define the hyperbolic unitary group $U_{2n}(R,\Lambda)$ to be the group

$$\left\{\begin{pmatrix} A & B \\ C & D \end{pmatrix} \in GL_{2n}(R) \;\middle|\; A^J D + C^J \varepsilon B = I,\ A^J C \in \Lambda_n \text{ and } B^J D \in \Lambda\right\}$$

Notice that the matrix

$$\left(\begin{array}{c|c} e_{ij}(r) & \\ \hline & e_{ji}(-r^J) \end{array}\right)$$

is in $U_{2n}(R,\Lambda)$ for any elementary matrix $e_{ij}(r) \in E_n(R)$. Let $\gamma_{ij}(r) \in \mathrm{Mat}_n(R)$ be the matrix with (i,j) entry r, (j,i) entry $-r^J\varepsilon$ and 0 elsewhere, and assume that $r \in \Lambda$ if $i = j$. Then the matrices

$$\left(\begin{array}{c|c} 0 & 0 \\ \hline \gamma_{ij}(r) & I \end{array}\right) \quad \text{and} \quad \left(\begin{array}{c|c} 0 & \gamma_{ij}(r)\varepsilon^{-1} \\ \hline I & 0 \end{array}\right)$$

are in $U_{2n}(R,\Lambda)$ for any $\gamma_{ij}(r)$. Denote the above matrices respectively by $E_{2i-1,2j}(r)$, $E_{2i,2j}(r)$, and $E_{2i-1,2j-1}(r)$, and let $EU_{2n}(R,\Lambda)$ be the subgroup of $U_{2n}(R,\Lambda)$ generated by all the $E_{k,\ell}(r)$. The group $EU_{2n}(R,\Lambda)$ is the <u>elementary unitary group</u>.

Consider the base $\{x_1,x_3,\ldots,x_2,\ldots,x_{2n}\}$ of M. Taking matrices in this base gives

$$U(M) \simeq U_{2n}(R,\Lambda) \text{ and } EU_{\mathfrak{X}}(M) \simeq EU_{2n}(R,\Lambda),$$

with $g_{2i-1,2j}(r) \to E_{2i-1,2j}(r)$, $g_{2i-1,2j-1}(r) \to E_{2i-1,2j-1}(r)$ and $g_{2i,2j}(r) \to E_{2i,2j}(r)$.

Now let $\mathfrak{a}$ be a two-sided ideal of R such that $\mathfrak{a}^J = \mathfrak{a}$. Let $\Lambda_{\mathfrak{a}}$ be the additive subgroup of $(R,+)$ generated by

$$\{a - a^J\varepsilon \mid a \in \mathfrak{a}\} \quad \cup \quad \{a^J\lambda a \mid a \in \mathfrak{a},\ \lambda \in \Lambda\}.$$

Let Γ be a subgroup of $(R,+)$ satisfying

$$\Lambda_{\mathfrak{a}} \subseteq \Gamma \subseteq \Lambda \cap \mathfrak{a}, \text{ and } a^J\Gamma a \subset \Gamma, \text{ for all } a \in \mathfrak{a}\ .$$

The pair $(\mathfrak{a},\Gamma)$ is an ideal of (R,Λ). Note that $(\mathfrak{a},\Lambda_{\mathfrak{a}})$ and $(\mathfrak{a},\Lambda \cap \mathfrak{a})$ are ideals of (R,Λ).

Define the groups $GL(M;\mathfrak{a})$ and $U(M;\mathfrak{a},\Gamma)$ respectively by

$$\{g \in GL(M) \mid (g - 1)M \subseteq M\mathfrak{a}\}, \text{ and}$$

$$\{g \in U(M) \cap GL(M;\mathfrak{a}) \mid f(gx,gx) - f(x,x) \in \Gamma, \text{ for all } x \in M\},$$

where f is any sesquilinear form that induces the quadratic form on M. The group $GL(M;\mathfrak{a})$ is the "non-matrix" analogue of the group $GL_{2n}(R,\mathfrak{a})$ of Section I. One can prove that the group $U(M;\mathfrak{a},\Gamma)$ is a normal subgroup of $U(M)$, which is independent of the choice of f.

Define the group $EU_{\mathfrak{X}}(M;\mathfrak{a},\Gamma)$ to be the <u>normal</u> subgroup of $EU_{\mathfrak{X}}(M)$ generated by

$$\{g_{i,j}(r) \mid r \in \mathfrak{a}, \text{ with } r \in \Gamma \text{ if } i = j\}.$$

Finally, denote by $U_{2n}(\mathfrak{a},\Gamma)$ and $EU_{2n}(\mathfrak{a},\Gamma)$, the respective images of $U_{\mathfrak{X}}(M;\mathfrak{a},\Gamma)$ and $EU_{\mathfrak{X}}(M;\mathfrak{a},\Gamma)$ under the isomorphism $U(M) \simeq U_{2n}(R,\Lambda)$ above. We have

$$EU_{2n}(\mathfrak{a},\Gamma) \subseteq U_{2n}(\mathfrak{a},\Gamma).$$

X. THE UNITARY K_1-GROUPS

Again let (R,Λ) be a form ring. Consider $U_{2n}(R,\Lambda) \subseteq U_{2(n+1)}(R,\Lambda)$ by identifying the matrix

$$\begin{pmatrix} A & B \\ C & D \end{pmatrix} \quad \text{with} \quad \left(\begin{array}{cc|cc} A & 0 & B & 0 \\ 0 & 1 & 0 & 0 \\ \hline C & 0 & D & 0 \\ 0 & 0 & 0 & 1 \end{array}\right).$$

So $U_2(R,\Lambda) \subseteq U_4(R,\Lambda) \subseteq U_6(R,\Lambda) \subseteq \ldots$. It follows that under this identification,

$$EU_2(R,\Lambda) \subseteq EU_4(R,\Lambda) \subseteq \ldots .$$

Define the stable unitary groups $U(R,\Lambda)$ and $EU(R,\Lambda)$ by

$$U(R,\Lambda) = \bigcup_{n\geq 1} U_{2n}(R,\Lambda) \text{ and } EU(R,\Lambda) = \bigcup_{n\geq 1} EU_{2n}(R,\Lambda).$$

The analogue of Theorem 1 of Section I is Vaserstein's

<u>Theorem 27</u> (<u>Unitary Whitehead Lemma</u>). $DU(R,\Lambda) = EU(R,\Lambda) = DEU(R,\Lambda)$.

Define $KU_1(R,\Lambda)$ to be the Abelian group $U(R,\Lambda)/EU(R,\Lambda)$. Analogous to the linear case, KU_1 is a functor from form rings to Abelian groups.

Now let $(\mathfrak{a},\Gamma)$ be an ideal of (R,Λ). Define the groups $U(\mathfrak{a},\Gamma)$ and $EU(\mathfrak{a},\Gamma)$, respectively, by

$$U(\mathfrak{a},\Gamma) = \bigcup_{n\geq 1} U_{2n}(\mathfrak{a},\Gamma) \quad \text{and} \quad EU(\mathfrak{a},\Gamma) = \bigcup_{n\geq 1} EU_{2n}(\mathfrak{a},\Gamma).$$

Clearly $EU(\mathfrak{a},\Gamma) \subseteq U(\mathfrak{a},\Gamma)$. The analogue of Theorem 2 is

Theorem 28 (Bass). (1) Any subgroup H satisfying $E(\mathfrak{a},\Gamma) \subseteq H \subseteq U(\mathfrak{a},\Gamma)$ is normal in $U(R,\Lambda)$.

(2) If $H \subseteq U(R,\Lambda)$ is normalized by $EU(R,\Lambda)$ then there is a unique ideal $(\mathfrak{a},\Lambda)$ of (R,Λ) such that

$$EU(\mathfrak{a},\Gamma) \subseteq H \subseteq U(\mathfrak{a},\Gamma).$$

It follows that $EU(R,\Lambda)$ is a simple group if and only if R is J-simple, i.e., has no J-invariant ideals other than 0 and R.

Define for any ideal $(\mathfrak{a},\Gamma)$ the group $KU_1(\mathfrak{a},\Gamma)$ by

$$KU_1(\mathfrak{a},\Gamma) = U(\mathfrak{a},\Gamma)/EU(\mathfrak{a},\Gamma).$$

This is an Abelian group. As in the linear case the determination of these groups classifies all normal subgroups of $U(\Lambda,R)$.

Example 29. If R is a division ring, then

$$KU_1(R,\Lambda) \simeq Z_2 \times R^*/(R^*)^2,$$

if $\Lambda = 0$ (so that $\varepsilon = 1$, $J = id_R$, and R is commutative); and

$$KU_1(R,\Lambda) \simeq R^*/\langle(\lambda^{-1}\Lambda)\rangle DR^*$$

if $\Lambda \neq 0$, where λ is any nonzero element in Λ and $\langle(\lambda^{-1}\Lambda)^{\bullet}\rangle$ is the subgroup of R^* generated by the nonzerc element of the form $\lambda^{-1}\lambda_1$, $\lambda_1 \in \Lambda$. This is in essence due to G.E. Wall.

It follows from work of C.T.C. Wall that for $R = Z$ and $\Lambda = 0$, $U_{2n}(R,\Lambda)/EU_{2n}(R,\Lambda) \simeq Z_2 \times Z_2$ for $n \geq 1$. Bass has shown that $KU_1(R,\Lambda) \simeq Z_2 \times Z_2$ in this case. If $R = Z$ and $\Lambda = Z$ (so that $J = id_R$ and $\varepsilon = -1$), then $EU_{2n}(R,\Lambda) = U_{2n}(R,\Lambda)$ for $n \geq 1$, and in particular $KU_1(R,\Lambda) = 1$. Compare with Theorem 32 below.

XI. THE UNITARY STEINBERG GROUP AND $KU_2(R,\Lambda)$

Let M be a hyperbolic quadratic module over (R,Λ) with hyperbolic base $\mathfrak{X} = \{x_1,\ldots,x_{2n}\}$. Refer to Section IX. Let

$$S_{2n} = \{(i,j) \mid h(x_i,x_j) = 0\}.$$

Observe that $(i,j) \in S_{2n}$ if and only if $1 \leq i,j \leq 2n$ and $j \neq i+1$, for i odd, and $j \neq i-1$ for i even. Note that $g_{ij}(r)$ is defined for any $(i,j) \in S_{2n}$ and $r \in R$, with $r \in R$, with $r \in \Lambda$ if $i = j$.

The following seven relations are readily checked for the generators $g_{ij}(r)$ of $EJ_{\mathfrak{X}}(M)$. Since $g_{i\ell}(t)$ is defined, $(i,\ell) \in S_{2n}$ in (4)-(7).

(1) $g_{ij}(r) = g_{ji}(-r^J\varepsilon)$

(2) $g_{ij}(r)\, g_{ij}(s) = g_{ij}(r+s)$

(3) $[g_{ij}(r), g_{k\ell}(s)] = 1$, if (i,k), (i,ℓ), (j,k) (j,ℓ) are in S_{2n}.

(4) $[g_{ij}(r), g_{k\ell}(s)] = g_{i\ell}(rs)$, if $(j,k) \notin S_{2n}$, j is even, and i,j,k,ℓ are distinct.

(5) $[g_{ij}(r), g_{ki}(s)] = g_{ii}(rs-(rs)^J\varepsilon)$, if $(j,k) \notin S_{2n}$, j is even, and i,k are distinct.

(6) $[g_{ii}(r), g_{k\ell}(s)] = g_{i\ell}(rs)g_{\ell\ell}(-s^Jrs)$, if $(i,k) \notin S_{2n}$, i is even, and i,k,ℓ are distinct.

(7) $[g_{ii}(r), g_{k\ell}(s)] = g_{i\ell}(-r^Js)\, g_{\ell\ell}(-(r^Js)^Js)$, if $(i,k) \notin S_{2n}$, i is odd, and i,k,ℓ are distinct.

Let $(i,j) \notin S_{2n}$, and let $r \in R$ with $r \in \Lambda$ if $i = j$. The unitary Steinberg group $StU_{2n}(R,\Lambda)$ is by definition the group generated by all symbols $x_{ij}(r)$, with (i,j) and r as above, subject to the seven relations obtained by replacing g_{ij} by x_{ij}.

By Section IX, there is a canonical homomorphism

$$\phi_{2n} : StU_{2n}(R,\Lambda) \longrightarrow EU_{2n}(R,\Lambda)$$

given by $\phi_{2n}(x_{ij}(r)) = E_{ij}(r)$. Define $KU_2(2n,R,\Lambda)$ to be its kernel.

Let S be the set of all ordered pairs (i,j) of positive integers such that $j \neq i+1$ if i is odd, and $j \neq i-1$ if i is even. Replacing

S_{2n} by S in the definition above gives the "stable" unitary Steinberg group $StU(R,\Lambda)$.

The equation that defines ϕ_{2n} also defines a homomorphism

$$\phi : StU(R,\Lambda) \longrightarrow EU(R,\Lambda).$$

Define $KU_2(R,\Lambda)$ to be the kernel of ϕ.

<u>Theorem 30</u> (<u>Steinberg</u>, <u>Sharpe</u>, <u>Bak</u>). $StU(R,\Lambda)$ <u>is the universal central extension of</u> $EU(R,\Lambda)$ <u>and</u>

$$KU_2(R,\Lambda) = \text{Cen } StU(R,\Lambda) .$$

In particular, KU_2 is a functor from form rings to Abelian groups. The above presentation for the unitary Steinberg group is patterned after Steinberg's original approach and is much simpler than that found in some of the literature.

<u>Example 31</u>. It follows from the work of Steinberg, Matsumoto, and Deodhar, that if R is a finite field with char $\neq 2$, then $KU_2(R,\Lambda) = 1$ if $\Lambda \neq 0$, and $KU_2(R,\Lambda) \simeq Z_2$ if $\Lambda = 0$. Similar results hold for $KU_2(2n,R,\Lambda)$. Masumoto and Bak have proved analogues of Theorem 13 in the hermitian case.

Sharpe has an important exact sequence which relates $KU_2(R,\Lambda)$ with $K_2(R)$. If R is a field F of characteristic not 2, then Suslin studied the sequence of Sharpe in the symplectic case using symbols. This work completes certain exact sequences arising from Section IV to the following exact diagram:

$$\begin{array}{ccccccccc}
 & & & & 1 & & 1 & & \\
 & & & & \downarrow & & \downarrow & & \\
 & & 1 & \longrightarrow & K_2(F)/\{F^*,-1\} & \xrightarrow{\text{squ}} & K_2(F)^2 & \longrightarrow & 1 \\
 & & \downarrow & & \downarrow{\scriptstyle F} & & \downarrow & & \\
1 & \longrightarrow & \text{Ker } H & \longrightarrow & KSp_2(F) & \xrightarrow{H} & K_2(F) & \longrightarrow & 1 \\
 & & \downarrow & & \downarrow & & \downarrow & & \\
1 & \longrightarrow & I^3(F) & \longrightarrow & I^2(F) & \longrightarrow & Br_2(F) & \longrightarrow & 1 \\
 & & \downarrow & & \downarrow & & \downarrow & & \\
 & & 1 & & 1 & & 1 & &
\end{array}$$

Here squ is the squaring map, and F and H are induced by the forgetful and hyperbolic maps respectively.

Finally, Bass has proved the analogue of Theorem 15 for the unitary K-groups and, as already remarked, Bak, Rehmann, Prasad and Ragunathan have extended the congruence subgroup theorem to the unitary groups in number theoretic situations.

XII. UNITARY STABILITY

Let (R,Λ) be a form ring with $J^2 = id_R$. So $\varepsilon \in$ Cen R. Suppose that S is a commutative Noetherian ring with Krull dimension d, and that R is an S-algebra which is finitely generated as module over S. For example, for $d = 1$ we can take $R = S$ to be a Dedekind domain.

For (R,Λ) with R as above, there is the Stability Theorem of Vaserstein.

Theorem 32. *Let $(\mathfrak{a},\Gamma)$ be an ideal of (R,Λ) and suppose that $n \geq d + 2$. Then*

$$EU_{2n}(\mathfrak{a},\Gamma) \lhd U_{2n}(R,\Lambda) \text{ and } U_{2n}(\mathfrak{a},\Gamma)/EU_{2n}(\mathfrak{a},\Gamma) \simeq KU_1(\mathfrak{a},\Gamma).$$

Compare this with Section II, particularly Theorem 5. Recall from Example 29, that if $R = Z$ and $\Lambda = Z$ (so $J = id$ and $\varepsilon = -1$), then for $n \geq 2$, $EU_{2n}(R,\Lambda) = U_{2n}(R,\Lambda)$.

Retain the hypotheses of the theorem above. Form $U_{2n}(R,\Gamma)/EU_{2n}(\mathfrak{a},\Gamma)$ and let G be the centralizer of $EU_{2n}(R,\Lambda)/EU_{2n}(\mathfrak{a},\Gamma)$. Define the group $U'_{2n}(\mathfrak{a},\Gamma)$ by the equation $G = U'_{2n}(\mathfrak{a},\Gamma)/EU_{2n}(\mathfrak{a},\Gamma)$.

Consider the quotient ring $R/\mathfrak{a}$ and the subgroup $\Lambda/\mathfrak{a} = (\Lambda+\mathfrak{a})/\mathfrak{a}$ of $(R/\Lambda,+)$. The form ring structure of (R,Λ) makes $(R/\mathfrak{a},\Lambda/\mathfrak{a})$ into a form ring in a natural way. The natural map $R \longrightarrow R/\mathfrak{a}$ induces a homomorphism $U_{2n}(R,\Lambda) \longrightarrow U_{2n}(R/\mathfrak{a},\Lambda/\mathfrak{a})$. In the special case $\Gamma = \Lambda \cap \mathfrak{a}$, $U'_{2n}(\mathfrak{a},\Gamma)$ is the inverse image of the center of $U_{2n}(R/\mathfrak{a},\Lambda/\mathfrak{a})$.

The analogue of Theorem 6 of Section II is the following theorem of Bak. Keep the assumptions of Theorem 32 on (R,Λ).

Theorem 33. *Suppose $n \geq d + 3$. Then a subgroup H of $U_{2n}(R,\Lambda)$ is normalized by $EU_{2n}(R,\Lambda)$ if and only if there is a (unique) ideal $(\mathfrak{a},\Gamma)$ of (R,Λ), such that*

$$EU_{2n}(\mathfrak{a},\Gamma) \subseteq H \subseteq U'_{2n}(\mathfrak{a},\Gamma).$$

It follows as in the linear case that the determination of the normal subgroups of $U_{2n}(R,\Lambda)$ reduces in large part to the study of groups $KU_1(\mathfrak{a},\Gamma)$.

Now let (R,Λ) be arbitrary. Assume only that $J^2 = id_R$, and hence that $\varepsilon \in \text{Cen}\, R$. The analogue of $(S)_d$, see Section II, is the unitary stable range condition

$(US)_d$: *Suppose* $(S)_d$ *holds and in addition that for all* $n > d$ *and all unimodular* $(a,b) \in R^{2n}$, *there are* u, v *in* R^n *such that* $u(v^J) \in \Lambda$ *and* $au^t + bv^t = 1$, *where* t *denotes transpose*.

It can be shown that if R is local, or more generally semi-local, then $(US)_o$ holds. The (R,Λ) preceeding Theorem 32 satisfies $(US)_d$. It is likely that both Theorems 32 and 33 hold more generally for rings satisfying $(US)_d$. In fact Vaserstein has announced a substantial generalization of Theorem 33, to rings which include rings that are finitely generated as modules over their centers, and also to non-hyperbolic situations.

Kolster (unpublished) and later Mustafa-Zade, have proved the "surjective stability" for KU_2:

Theorem 34. *Suppose* (R,Λ) *satisfies* $(US)_d$. *Then if* $n \geq d + 3$, *the natural map*

$$KU_2(2n,R,\Lambda) \longrightarrow KU_2(R,\Lambda)$$

is surjective.

If $\Lambda = R^{-\varepsilon} \neq 0$, and R is a division ring, then Bak (unpublished) has proved the "injective stability" analogue of the theorem above. This result is probably true in the generality of Theorem 34. If $\Lambda = 0$ on the other hand, then, at least for a finite field R, the kernel of the map is non-trivial, i.e., Z_2. Is it Z_2 in more general situations?

XIII. APPLICATIONS OF HERMITIAN MORITA THEORY

Suppose R is a ring with division ring of quotients D, and let D be equipped with a form parameter Λ. Let V be a finite dimensional quadratic module over (D,Λ), see Section VIII, and consider the unitary group

U(V). One can define full groups analogous to those in Section VI. There is an analogue of Theorem 21 but it is not quite complete. It is not known if a full group for the case $\Lambda \neq 0$ can in general be isomorphic to a full group for the case $\Lambda = 0$. In some cases full groups have been characterized analogously to Theorem 22.

There is also an analogue of Morita Theory. To simplify things we drop the reference to the form parameter.

Let (R,J,ε) be a ring equipped with an anti-automorphism J and an associated unit ε, and let (R',J',ε') be another such triple. Let

$$(R,\ S,\ P \xrightarrow{\theta} Q,\ \mu,\ \tau)$$

be a Morita context, see Section VII, equipped with an additive bijection θ that satisfies

$$\theta(rps) = s^{J'}\theta(p)r^{J}, \text{ and } \mu(\varepsilon p \otimes \theta(p_1)) = \mu(p_1 \otimes \theta(p\varepsilon'))^{J},$$

for all $r \in R$ and $s \in S$, and p and p_1 in P. Now let $H_{(R,J,\varepsilon)}$ be the category of right R-modules equipped with (J,ε)-hermitian forms. Tensoring with P defines a category equivalence $E : H_{(R,J,\varepsilon)} \longrightarrow H_{(R',J',\varepsilon')}$, and if $E(M) = N$, there is an induced isomorphism

$$E : U(M) \longrightarrow U(N) ,$$

where here U(M) is the isometry group of M and U(N) is that of N. In hyperbolic situations over rings with division rings of quotients one has a converse.

Theorem 35. (Hahn-Li). *Let (R,J,ε) and (R',J',ε') be given and assume that R and R' have division rings of quotients. Suppose M and N are free hyperbolic modules over R and R' respectively, of ranks ≥ 6. If*

$$\phi : U(M) \longrightarrow U(N)$$

is an isomorphism, then there is an equivalence $E : H_{(R,J,\varepsilon)} \longrightarrow H_{(R',J',\varepsilon')}$ (isomorphic to the E constructed above) with $E(M) = N$, and a homomorphism $\chi : U(M) \longrightarrow (\mathrm{Cen}\ R')^{*}$, such that

$$\phi(g) = \chi(g) \cdot E(g), \text{ all } g \in U(M).$$

Unitary K-theory, with hermitian Morita theory as an important tool, has seen significant application in algebraic topology via the "L-groups":

Let (R,Λ) be any form ring, and consider the groups $K_1(R)$ and $KU_1(R,\Lambda)$ of Sections I and X. Define a homomorphism $\phi : K_1(R) \longrightarrow KU_1(R,\Lambda)$ by

$$\phi(gE(R)) = \left(\begin{array}{c|c} g & 0 \\ \hline 0 & (g^{-1})^J \end{array}\right) EU(R),$$

for $g \in GL_n(R)$. Denote by $L(R,\Lambda)$ the cokernel $KU_1(R,\Lambda)/\phi K_1(R)$ of this map.

Theorem 36 (Connolly, G.E. Wall). Let R be a division ring. If $\Lambda \neq 0$, then $L(R,\Lambda) = 0$, and if $\Lambda = 0$, then $L(R,\Lambda) = Z_2$.

A similar fact is true for $R = Z$.

There are a multiplicity of related "L-groups". Using Hermitian Morita theory, localizations, exact sequences, etc., one can then show that these are trivial, or at least the resident topological invariants are, in many cases where π is a finite group and R its group ring over Z. This in turn leads to insight into important topological questions.

I do not want to miss the opportunity to thank the organizers of the Conference, Professor Tuan Hsio-Fu of Peking University and Professor Wan Zhe-xian of the Academia Sinica (and their many assistants) for their wonderful hospitality throughout an exciting visit.

BIBLIOGRAPHY

[1] E. Artin, Geometric Algebra, Wiley Interscience, New York, 1957.

[2] A. Bak, On modules with quadratic forms, pp. 55-66, in Lecture Notes in Mathematics 108, Springer Verlag, Berlin 1969.

[3] A. Bak, K-Theory of Forms, Annals of Mathematics Studies 98, Princeton University Press, 1981.

[4] A. Bak, Le probleme des sous-groupes de congruence et le probleme metaplectique pour les groupes classiques de rang > 1, C.R. Acad. Sc. Paris, 292, 307-310 (1981).

[5] A. Bak and U. Rehmann, The congruence subgroup and metaplectic problems for $SL_{n\geq 2}$ of division algebras, J. Algebra, 78(1982), 475-547.

[6] H. Bass, Algebraic K-Theory, Benjamin, New York, 1968.

[7] H. Bass, Unitary Algebraic K-Theory, pp. 57-265, in Lecture Notes in Mathematics 343, Springer Verlag, Berlin 1973.

[8] H. Bass, Introduction to some Methods of Algebraic K-Theory, American Mathematical Soc., Providence, R.I., 1974.

[9] H. Bass, Clifford algebras and spinor norms over a commutative ring, Amer. J. Math. 96 (1974), 156-206.

[10] H. Bass, J. Milnor, and J.P. Serre, Solution of the Congruence Subgroup Problem for SL_n and Sp_{2n}. Publ. Math. IHES 33(1967), 59-137.

[11] D. Callan, The isomorphisms of unitary groups over non-commutative domains, J. Algebra 52(1978), 475-503.

[12] F. Connolly, Linking numbers and surgery, Topology 12, 1973, 389-409.

[13] V.V. Deodhar, On central extensions of rational points of algebraic groups, Amer. J. Math. 100 (1978), 303-386.

[14] J. Dieudonne, La Geometrie des Groupes Classiques, 3rd ed., Springer Verlag, Berlin-New York, 1971.

[15] P. Draxl, Skew Fields, Cambridge University Press, 1982.

[16] A. Fröhlich and E.M. McEvett, Forms over rings with involution, J. Algebra, 12 (1969), 79-104.

[17] I. Golubchik, On the general linear group over an associative ring, Uspekhi Mat. Nauk, 28:3 (1973), 179-180 (Russian).

[18] A. Hahn, Isomorphism theory for orthogonal groups over arbitrary integral domains, J. Algebra, 51 (1978), 233-287.

[19] A. Hahn, Category equivalences and linear groups over rings, J. Algebra, 77 (1982), 505-543.

[20] A. Hahn, A hermitian Morita theorem for algebras with anti-structure, J. Algebra, 93 (1985), 215-235.

[21] A. Hahn and Z.X. Li, Hermitian Morita theory and hyperbolic unitary groups, to appear in J. Algebra.

[22] A. Hahn, D. James and B. Weisfeiler, Homomorphisms of algebraic and classical groups: a survey, in Canadian Mathematical Society Conference Proceedings, Volume 4 (1984), 249-296.

[23] I. Hambleton, L. Taylor and B. Williams, An introduction to maps between surgery obstruction groups, pp. 49-127, in Lecture Notes in Mathematics 1051, Springer Verlag 1982.

[24] J. Humphreys, Arithmetic Groups, Lecture Notes in Mathematics 789, Springer Verlag, Berlin, 1980.

[25] D. James, W. Waterhouse and B. Weisfeiler, Abstract homomorphisms of algebraic groups: problems and bibliography, Comm. Algebra, 9 (1981), 95-114.

[26] W. van der Kallen, Generators and relations in Algebraic K-Theory, pp. 305-210, in Proceedings of the International Conference of Mathematicians, Helsinki, 1978.

[27] W. van der Kallen, Stability for K_2 in Dedekind rings of arithmetic type, pp. 217-248, in Lecture Notes in Mathematics 854, Springer-Verlag, Berlin, 1980.

[28] M. Kolster, Surjective stability for unitary K-groups, preprint, 1975.

[29] M. Kolster, General symbols and presentation of elementary linear groups, J. für reine u. angew. Math. 353 (1984), 132-164.

[30] T.Y. Lam, The Algebraic Theory of Quadratic Forms, 2nd ed., Benjamin, New York, 1980.

[31] K. Leung, The isomorphism theory of projective pseudo-orthogonal groups, J. Algebra, 61 (1979), 367-387.

[32] J. Mennicke, Zur Theorie der Siegelschen Modulgruppe, Math. Ann. 159 (1965), 115-129.

[33] H. Matsumoto, Sur les sousgroupes arithmetiques des groupes semisimple deployes, Ann. Sci. Ecole Norm. Sup. (4) 2(1969), 1-62.

[34] A. Merkurjev and A. Suslin, K-Cohomologies of Severi-Brauer varieties and norm residue homomorphism, Izv. Akad. Nauk SSSR, 46 (1982), 1011-1046.

[35] J. Milnor, Introduction to algebraic K-Theory, Annals of Mathematical Studies 72, Princeton University Press, 1971.

[36] N.M. Mustafa-Zade, On epimorphic stability of a unitary K_2-functor, Russian Math. Surveys (1980), 99-100.

[37] O.T. O'Meara, Lectures on Linear groups, Amer. Math. Society, Providence, R.I., 1974.

[38] O.T. O'Meara, A general isomorphism theory for linear groups, J. Algebra, 44 (1977), 93-142.

[39] O.T. O'Meara, Symplectic groups, Math. Surveys, Amer. Math. Soc. Providence, R.I., 1978.

[40] O.T. O'Meara, A survey of the isomorphism theory of the classical groups, pp. 225-242, in "Ring theory and Algebra III", Dekker, New York, 1980.

[41] W. Pender, Automorphisms and Isomorphisms of the indefinite modular classical groups, Ph.D. Thesis, Sydney University (1972).

[42] W. Pender, Classical groups over division rings of characteristic 2, Bull. Aust. Math. Soc. 7 (1972), 191-226.

[43] G. Prasad and M.S. Ragunathan, On the congruence subgroup problem: determination of the "metaplectic kernel", Invent. math. 71, (1983), 21-42.

[44] H.G. Quebbemann, W. Scharlau, and M. Schulte, Quaratic and hermitian forms in additive and abelian categories, J. Algebra, 59 (1979), 264-289.

[45] A. Ranicki, The algebraic theory of surgery I, Foundations, Proc. London Math. Soc. (3) 40 (1980), 87-192.

[46] A. Ranicki, Exact Sequences in the Algebraic Theory of Surgery, Princeton Mathematical Notes, Princeton University Press, 1981.

[47] H.S. Ren and Z.X. Wan, Automorphisms of $PSL_2^{\pm}$ (K) over any skew field K, Acta. Math. Sinica, 25, (1982), 484-492.

[48] J.P. Serre, Trees, Springer-Verlag, Berlin, New York, 1980.

[49] R. Sharpe, On the structure of the unitary Steinberg group, Ann. Math. 96 (1972), 444-479.

[50] J. Silvester, Introduction to Algebraic K-Theory, Chapman and Hall, London, 1981.

[51] G. Soule, K_2 et le groupe de Brauer [d'apres A.S. Merkurjev et A.A. Suslin]. Seminare Bourbaki, 1982/83, No. 601 (1982).

[52] R. Steinberg, Generateurs, relations et revetements de groups algebriques, Colloque de Bruxelles, 1962, 113-127.

[53] R. Steinberg, Lecture Notes on Chevalley Groups, Yale University, 1967.

[54] A. Suslin, On the structure of the special linear group over polynomial rings, Math. USSR Izvestija, Vol. II (1972), No. 2, 221-328.

[55] A. Suslin and V. Kopeiko, Quadratic modules and the orthogonal group over polynomial rings, Zap. Naucn. Sem. Leningrad. Otdel. Math. Inst. Steklov. (LOMI) 71 (1977), 216-250.

[56] A. Suslin, Reciprocity laws and the stable rank of polynomial rings, Math. USSR Izvestija, Vol 15(1980), No. 3, 589-623.

[57] R. Swan, K-Theory of Finite Groups and Orders, Lecture Notes in Mathematics 149, Springer-Verlag, Berlin, 1970.

[58] J. Tits, Buildings of spherical type and finite BN-pairs, Lecture Notes in Mathematics 383, Springer-Verlag, Berlin, 1974.

[59] L. Vaserstein, Stabilization of unitary and orthogonal groups over a ring with involution, Math. USSR Sbornik, Vol. 10 (1970), 307-326.

[60] L. Vaserstein, The stabilization for classical groups over rings, Math. USSR Sbornik 22, (1974), 271-303.

[61] L. Vaserstein, Foundations of algebraic K-theory, Russian Math. Surveys, 31:4 (1976), 89-156.

[62] L. Vaserstein, On the normal subgroups of GL_n over a ring, pp. 456-465, in Lecture Notes in Mathematics 854, Springer-Verlag, Berlin, 1981.

[63] L. Vaserstein, On full subgroups in the sense of O'Meara, J. Algebra, 75 (1982), 437-44.

[64] L. Vaserstein, Classical groups over rings, in Canadian Mathematical Society Conference Proceedings, Volume 4 (1984).

[65] L. Vaserstein and A. Suslin, Serre's problem on projective modules over polynomial rings and algebraic K-theory, Math. USSR Izvestija, Vol. 10 (1976), No. 5, 937-1001.

[66] A. Wadsworth, Merkurjev's elementary proof of Merkurjev's theorem, Boulder Conference in Algebraic K-theory, to appear.

[67] C.T.C. Wall, Surgery on Compact Manifolds, Academic Press, 1970.

[68] C.T.C. Wall, On the axiomatic foundation of the theory of Hermitian forms, Proc. Camb. Phil. Soc., 67 (1970), 243-250.

[69] C.T.C. Wall, Foundations of algebraic L-Theory, pp. 266-300, in Lecture Notes in Mathematics 343, Springer Verlag, Berlin, 1973.

[70] C.T.C. Wall, On the classification of Hermitian Forms III, semisimple rings, Invent. Math., 18 (1972), 119-141.

[71] G.E. Wall, The Structure of a unitary factor group, Publ. Math., IHES, No. 1, (1959), 7-23.

[72] Z.X. Wan, The Classical Groups, Shanghai University Press, 1981.

[73] Z.X. Wan and J.G. Yang, Automorphisms of the projective quaternion unimodular group in dimension 2, Chinese Annals of Math., 3(1982), 395-402.

[74] B. Weisfeiler, Abstract homomorphisms of big subgroups of algebraic groups, pp. 135-181, in Topics in the theory of Algebraic Groups, Notre Dame Mathematical Lectures, No. 10, University of Notre Dame Press, 1982.

[75] J.S. Wilson, The normal and subnormal structure of general linear groups, Proc. Camb. Phil. Soc. 71(1972), 163-177.

[76] Zalesky, Linear groups, Russian Math. Surveys, 36, No. 5, (1981), 63-128.

MODULAR REPRESENTATIONS OF REDUCTIVE GROUPS

J.C. Jantzen

Mathematisches Institut
der Universität
Wegelerstraße 10
D-5300 Bonn 1
(until 31 March 1985)

Mathematisches Seminar
der Universität
Bundesstr. 55
D-2000 Hamburg 13
(after 1 April 1985)

Federal Republic of Germany

Introduction

The representation theory of a connected reductive algebraic group in characteristic zero is rather simple: There is a nice classification of all irreducible representations, there is a nice character formula (due to H. Weyl) for these simple modules and any representation is completely reducible.

In prime characteristic the situation is much more complicated. We still have the same classification of simple modules, but we do not know a character formula. So, to find such a formula may be regarded as the first important problem in this theory. Furthermore, not all modules are semi-simple (except for the case of a torus). A description of all indecomposable modules, which is the next thing we might ask for, seems to be impossible. In order to get at least some information we should like to try to compute extension groups (e.g. of simple modules) and the submodule structure of "important" modules (e.g. of injective hulls of simple modules). There has been done considerable work on these problems, there are as well computations for small groups as well as general theorems which however do not yet tell the whole story. But these results have at least led to a conjecture in [38] on a character formula.

These notes follow more or less (with more details and more indications of proof added) my six lectures in Beijing, each section corresponding to one lecture. Within that time it was impossible to give a survey of the whole theory as it stands now. So I have concentrated mostly on those results related to the character formula and which can be proved mainly using one method (the cohomology theory of line bundles on the flag variety). I had not time enough to discuss other, equally important aspects of the theory (computation of cohomology groups, relations with the representation theory of its p-Lie algebra and of finite Chevalley groups).

The first two sections contain general results about representations of an arbitrary algebraic group G. The point of view of algebraic groups is the same as in the books by Humphreys [22] and Springer [43] to which I refer for all terms not explained here. (The audience of my talks and the readers of these notes are supposed to be familiar with the content of these books.) Besides the definition of a representation and its elementary properties, section 1 contains a discussion of induced representations. Using the induction we can construct injective resolutions in the category of G-modules and use them to define derived functors. This is done in section 2 where also some spectral sequences relating different derived functors are constructed.

The remaining four sections deal with the representation theory of a connected reductive algebraic group G in characteristic $p \neq 0$. At first the simple modules are described. One chooses a maximal torus T in G and associates to each character λ of T a line bundle $\mathcal{L}(\lambda)$ on the flag variety X of G. The cohomology groups $H^i(X,\mathcal{L}(\lambda))$ are G-modules in a natural way. One has $H^0(X,\mathcal{L}(\lambda)) \neq 0$ if and only if λ is "dominant" and for such λ there is a unique simple submodule in $H^0(X,\mathcal{L}(\lambda))$. Each simple G-module can be embedded into exactly one of these $H^0(X,\mathcal{L}(\lambda))$. An important theorem (due to G. Kempf) says that $H^i(X,\mathcal{L}(\lambda)) = 0$ for all $i > 0$ and all λ "dominant". These and related results are described in section 3.

In characteristic 0 the Borel-Bott-Weil theorem determines $H^i(X,\mathcal{L}(\lambda))$ for all λ and all i. This theorem does not generalize to prime characteristic. It still holds (at least on the level of characters) for SL_2 (or, more generally, for G of semi-simple rank 1) and for "small" λ. These things are discussed in section 4.

There is an affine Weyl group acting on the set of characters on T, and there are some special fundamental domains, called alcoves. The linkage principle (section 5) says that some $H^i(X,\mathcal{L}(\lambda))$ and $H^j(X,\mathcal{L}(\mu))$ can have a common composition factor only if λ,μ are conjugate under the affine Weyl group. The translation principle (section 6) says how to get from a knowledge of all composition factors of some $H^i(X,\mathcal{L}(\lambda))$ with λ in the "interior" of some alcove the same information for all $H^i(X,\mathcal{L}(\mu))$ with μ in the same alcove. Without these two principles the conjecture about the character formula could not have been formulated.

We assume throughout these notes that K is an algebraically closed field. Terms like vector space, variety, algebraic group without reference to a specific ground field, always refer to K as a ground field. We denote by $K[X]$ for any variety X the algebra of regular functions on X. More notation will be introduced at the beginning of section 3 which will be used from then on until the end of these notes.

The last three sections are preceded by a short list of references to the papers from which the results are taken. The results in the first three sections are usually generalizations of results known for finite groups or Lie groups so I did not think it necessary always to trace the original proofs. (There are some exceptions where you can find the references within the text.)

1. Representations and Induction

Throughout this section let G be an algebraic group.

1.1 A representation of G (considered as an abstract group) on a vector space V is simply a group homomorphism $\rho : G \to GL(V)$. If V is finite dimensional, then $GL(V)$ carries a natural structure as an algebraic group. In this case we call ρ a representation of G as an algebraic group (or V a rational G-module), if ρ is a homomorphism of algebraic groups. An equivalent condition is that the map $G\times V \to V$, $(g,v)\mapsto \rho(g)v$ is a morphism of varieties. If $v_1,\ldots,v_n$ is a basis of V and if ρ is an (abstract) representation, then there are functions $m_{ij}: G \to K$ with $\rho(g)v_j = \sum_{i=1}^{n} m_{ij}(g)v_i$. (The m_{ij} are called the matrix coefficients of ρ with respect to this basis.) Now ρ is a representation of G as an algebraic group, if and only if all m_{ij} are regular functions on G, i.e. belong to $K[G]$. If $V' \subset V$ is a G-stable subspace, then we get a representation of G on V' by restricting each $\rho(g)$ to V'. Using the condition with matrix coefficients one concludes: If V is a rational G-module, then so is V'.

We may now drop the assumption that V be finite dimensional.

We call in general ρ a representation of G as an algebraic group (or V a rational G-module), if V is the sum of its finite dimensional G-stable subspaces V' and if each such V' is a rational G-module in the old sense. (By the remark above, this gives the old definition for $\dim V < \infty$.)

To demand that V is the sum of its finite dimensional G-stable subspaces is equivalent to the following: Each finite dimensional subspace of V is contained in a finite dimensional G-stable subspace. Usually we say for this property that V is a <u>locally finite</u> G-module.

<u>1.2</u> Let us mention that the usual operations on representations preserve the class of rational G-modules. The following statements are usually proved by reducing to the finite dimensional case and looking at matrix coefficients.

(1) If V is a rational G-module and $V' \subset V$ is a G-stable subspace, then V' and V/V' are rational G-modules.

(2) A direct sum of rational G-modules is rational.

(3) If V_1, V_2 are rational G-modules, then so is $V_1 \otimes V_2$.

(4) If V is a rational G-module, then so are all $S^r V$ and $\Lambda^r V$.

(5) If V is a rational G-module with $\dim V < \infty$, then so is V^*.

Combining (3) and (5) we see that $\mathrm{Hom}_K(V_1, V_2) \cong V_1^* \otimes V_2$ is a rational G-module, if V_1, V_2 are so with $\dim V_1 < \infty$.

From now on we shall write "representation" instead of "representation as an algebraic group" and "G-module" instead of "rational G-module". For the operation of G on a G-module V we shall simply write $(g,v) \mapsto gv$, if no confusion is possible.

<u>1.3</u> The sum of all simple submodules of any module is called the <u>socle</u> of this module. We denote the socle of a G-module V by $\mathrm{soc}_G V$. We have

(1) If $V \neq 0$, then $\mathrm{soc}_G V \neq 0$.

This is obvious for $\dim V < \infty$ and follows in general using the local finiteness of V. This property also implies that each simple G-module is finite dimensional.

Obviously the subspace of <u>G-fixed points</u>

(2) $V^G = \{v \in V \mid gv = v \quad \text{for all} \quad g \in G\}$

is always contained in the socle of V. It is the sum of all simple submodules isomorphic to the <u>trivial module</u> K (where each $g \in G$ operates as the identity).

Standard results about solvable groups yield (cf. [22], 17.5/6)

(3) <u>If</u> G <u>is unipotent, then</u> K <u>is the only simple</u> G-<u>module. Hence</u> $\mathrm{soc}_G V = V^G$ <u>for all</u> G-<u>modules</u> V <u>in this case</u>.

(4) <u>If</u> G <u>is connected and solvable, then each simple</u> G-<u>module has dimension</u> 1.

In general G-modules of dimension one correspond to the characters of G, i.e. to the homomorphisms (of algebraic groups) from G into the multiplicative group $K^\times$. These characters form a commutative group which we usually denote by $X(G)$. The group law in $X(G)$ is usually written additively. For each $\lambda \in X(G)$ and each G-module V we call

(5) $V_\lambda = \{v \in V \mid gv = \lambda(g)v \quad \text{for all} \quad g \in G\}$

the λ-weight space of V. The sum of all V_λ is direct and contained in the socle of V.

We know (e.g. from [43] 2.5.2)

(6) <u>If</u> G <u>is diagonalizable, then</u> $V = \bigoplus_{\lambda \in X(G)} V_\lambda$ <u>for each</u> G-<u>module</u> V.

If $\dim(V_\lambda) < \infty$ for all λ (and G diagonalizable) then we call

(7) $\mathrm{ch}(V) = \sum_{\lambda \in X(G)} \dim(V_\lambda)e(\lambda)$

the formal character of V. (Here $e(\lambda)$ is the canonical basis of the group ring $\mathbb{Z}[X(G)]$, so we have $e(\lambda)e(\lambda') = e(\lambda+\lambda')$ for all λ,λ'.)

We have obviously $\mathrm{ch}(V) = \mathrm{ch}(V') + \mathrm{ch}(V/V')$ for each submodule V' of V and $\mathrm{ch}(V \otimes E) = \mathrm{ch}(V)\mathrm{ch}(E)$ for each finite dimensional G-module E.

<u>1.4</u> Operations of G on varieties produce many important representations of G. Let X be an affine variety on which G acts from the right. This action is given by a morphism $\alpha: X\times G \to X$, $(x,g) \mapsto xg$. Then G operates on $K[X]$ through $(gF)(x) = F(xg)$ for all $F \in K[X]$, $g \in G$ and $x \in X$. The comorphism $\alpha^*: K[X] \to K[X\times G] \cong K[X] \otimes K[G]$ maps F to some element of the form

$\sum_{i=1}^{r} F_i \otimes f_i$ with $F_i \in K[X]$, $f_i \in K[G]$. Then an easy computation shows

$$gF = \sum_{i=1}^{r} f_i(g)F_i \in \sum_{i=1}^{r} KF_i.$$

Therefore $K[X]$ is locally finite under the operation of G. If we choose the F_i from a basis of $K[X]$ then the f_i above are matrix coefficients. They belong to $K[G]$ which easily implies that $K[X]$ is a rational G-module.

We can work similarly with an operation of G on X from the left. We just have to define gF through $(gF)(x) = F(g^{-1}x)$.

As an example let us take $X = G$ and let G act on itself by right or left multiplication. This yields two representations of G on $K[G]$ which we call the <u>right</u> (ρ_r) and <u>left</u> (ρ_ℓ) <u>regular representation</u>. So we have for all $f \in K[G]$ and $g, g' \in G$

(1) $(\rho_r(g)f)(g') = f(g'g)$,

(2) $(\rho_\ell(g)f)(g') = f(g^{-1}g')$.

<u>1.5</u> If V is a finite dimensional vector space, then it has a natural structure as an affine variety. So for any variety X the set $\mathrm{Mor}(X,V)$ of all regular maps from X to V is well defined, it has an obvious structure as a vector space. If we associate to each $f \otimes v$ with $f \in K[X]$ and $v \in V$ the map $x \mapsto f(x)v$, then we get an isomorphism $K[X] \otimes V \xrightarrow{\sim} \mathrm{Mor}(X,V)$ of vector spaces.

For an arbitrary V we define $\mathrm{Mor}(X,V)$ as the union of all $\mathrm{Mor}(X,V_1)$ over all finite dimensional subspaces V_1 of V. Then again $K[X] \otimes V \simeq \mathrm{Mor}(X,V)$ under the "same" map as above.

There is a natural structure as a G-module on $\mathrm{Mor}(G,V)$. For each $g,g' \in G$ and $F \in \mathrm{Mor}(G,V)$ set $(gF)(g') = F(g^{-1}g')$. If we identify $K[G] \otimes V$ and $\mathrm{Mor}(G,V)$ as above, then we get on $K[G] \otimes V$ the tensor product of the left regular representation with the trivial representation on V. (This implies that $\mathrm{Mor}(G,V)$ is a rational G-module.)

<u>1.6</u> Let H be closed subgroup of G (and keep this assumption until the end of section 1).

For each H-module M we set

(1) $\operatorname{ind}_H^G(M) = \{F \in \operatorname{Mor}(G,M) \mid F(gh) = h^{-1}F(g) \quad \text{for all} \quad g \in G,\ h \in H\}$.

This is obviously a G-submodule of $\operatorname{Mor}(G,M)$ for the representation defined in 1.5. We call $\operatorname{ind}_H^G M$ the <u>induced module</u> (by M).

We can describe $\operatorname{ind}_H^G(M)$ also as follows. Let H operate on $\operatorname{Mor}(G,M)$ through $(h*F)(g) = h(F(gh))$ for all $h \in H$, $g \in G$, $F \in \operatorname{Mor}(G,M)$. If we take the identification $\operatorname{Mor}(G,M) \simeq K[G] \otimes M$, then this is the tensor product of the right regular representation on $K[G]$ (restricted to H) with the given representation on M. (So $\operatorname{Mor}(G,M)$ is a rational H-module.) Now

(2) $\operatorname{ind}_H^G M = \operatorname{Mor}(G,M)^H \simeq (K[G] \otimes M)^H$.

If we take the trivial H-module K, then obviously

(3) $\operatorname{ind}_H^G K = K[G]^H \simeq K[G/H]$,

especially

(4) $\operatorname{ind}_{\{1\}}^G K \simeq K[G]$

(with respect to ρ_ℓ).

<u>1.7</u> Let us mention some properties. The first two are rather obvious:

(1) ind_H^G <u>is a left exact functor from</u> {H-<u>modules</u>} <u>to</u> {G-<u>modules</u>}.

(2) ind_H^G <u>commutes with direct sumes</u>.

For any H-module M the map

$$\varepsilon_M\colon \operatorname{ind}_H^G M \to M, \quad F \mapsto F(1)$$

is called the <u>evaluation map</u>. It is easily checked to be a homomorphism of H-modules. One can now formulate the <u>universal property of induction</u> which is also called <u>Frobenius reciprocity</u>:

(3) <u>For each</u> G-<u>module</u> V <u>and each</u> H-<u>module</u> M <u>the map</u> $\varphi \mapsto \varepsilon_M \circ \varphi$ <u>is an isomorphism</u>

$$\operatorname{Hom}_G(V, \operatorname{ind}_H^G M) \xrightarrow{\sim} \operatorname{Hom}_H(V,M).$$

The inverse map sends any ψ to $\tilde{\psi}$ with $\tilde{\psi}(v)(g) = \psi(g^{-1}v)$ for all

$g \in G$ and $v \in V$. If we denote by res_H^G the forgetful functor from {G-modules} to {H-modules}, then we can express (3) also as follows:

(4) ind_H^G is right adjoint to res_H^G.

Another important property is the transitivity of induction: If $H' \subset H$ is a closed subgroup, then we have for each H'-module M a canonical isomorphism of G-modules

(5) $\mathrm{ind}_H^G\, \mathrm{ind}_{H'}^H M \xrightarrow{\sim} \mathrm{ind}_{H'}^G M$.

One associates to any F on the left hand side $\bar{F} \in \mathrm{Mor}(G,M)$ with $\bar{F}(g) = F(g)(1)$, and the inverse map sends any $F' \in \mathrm{ind}_{H'}^G M$ to $\tilde{F}'$ with $\tilde{F}'(g)(h) = F'(gh)$ for all $g \in G$, $h \in H$.

Finally we have the tensor identity. For each G-module V and each H-module M we have a canonical isomorphism of G-modules

(6) $\mathrm{ind}_H^G(M \otimes V) \xrightarrow{\sim} (\mathrm{ind}_H^G M) \otimes V$

We can obviously embed both sides into $\mathrm{Mor}(G, M \otimes V)$. Then the automorphisms α, β of the vector space $\mathrm{Mor}(G, M \otimes V)$ with $(\alpha F)(g) = (\mathrm{id}_M \otimes g)F(g)$ and $(\beta F)(g) = (\mathrm{id}_M \otimes g^{-1})F(g)$ interchange these two subspaces and are compatible with the action of G.

2. Injective Modules and Derived Functors

Throughout this section let G be an algebraic group and H a closed subgroup.

2.1 The forgetful functor res_H^G is obviously exact. Therefore the adjoint (1.7(4)) functor ind_H^G maps injective objects in the category {H-modules} to injective objects in the category {G-modules}. We call such injective objects simply injective H-modules resp. injective G-modules.

In the case of $H = \{1\}$ any H-module M is injective, hence any $\mathrm{ind}_{\{1\}}^G M$ is an injective G-module. For example $K[G]$ under ρ_ℓ is injective by 1.6(4). More generally we get for any G-module V that

$$V \otimes K[G] \cong V \otimes \mathrm{ind}_{\{1\}}^G K \cong \mathrm{ind}_{\{1\}}^G (V)$$

is injective. As $v \mapsto v \otimes 1$ is a homomorphism $V \to V \otimes K[G]$ of G-modules we get

(1) Any G-module can be embedded into an injective G-module.

Of course, we can then continue and construct injective resolutions.

For any G-module V let us denote by V_{tr} the vector space V considered as a trivial G-module. As V and V_{tr} are isomorphic when considered as {1}-modules, we get from the tensor identity an isomorphism

$$V \otimes K[G] \simeq V_{tr} \otimes K[G]$$

of G-modules. If V is an injective G-module, then the embedding $V \to V \otimes K[G] \simeq V_{tr} \otimes K[G]$ has to split. Using this one gets easily:

(2) A G-module is injective if and only if it is isomorphic to a direct summand of a G-module of the form $M \otimes K[G]$ where M is a vector space considered as a trivial G-module.

This implies (using again the tensor identity)

(3) If V is an injective G-module and V' an arbitrary G-module, then $V \otimes V'$ is injective.

2.2 We can use the injective resolutions in the category {G-modules} to construct (right) derived functors of left exact functors from {G-modules} to other abelian categories.

For example, the fixed point functor $?^G$ from {G-modules} to {vector spaces} which associates to any V the subspace V^G is left exact. Its (right) derived functors are denoted by $H^n(G,?)$ and the $H^n(G,V)$ are called the Hochschild cohomology groups of V.

For any G-module M the functor $V \mapsto \mathrm{Hom}_G(M,V)$ from {G-modules} to {vector spaces} is left exact. Its derived functors are denoted by $\mathrm{Ext}^n_G(M,?)$. As usually one can describe the Ext-groups $\mathrm{Ext}^n_G(M,V)$ as sets of equivalence classes of certain finite exact sequences. For $M = K$ we have a canonical isomorphism $\mathrm{Hom}_G(K,V) \xrightarrow{\sim} V^G$ where $\varphi \mapsto \varphi(1)$. This yields isomorphisms $\mathrm{Ext}^n_G(K,V) \simeq H^n(G,V)$.

Finally also ind_H^G: {H-modules} $\to$ {G-modules} admits derived functors which we denote by $R^n\mathrm{ind}_H^G$.

2.3 We are going to describe some relations between these derived functors. This will be done using Grothendieck's spectral sequence: Consider abelian categories $\underline{\underline{C}}_1$, $\underline{\underline{C}}_2$, $\underline{\underline{C}}_3$ such that each object admits an injective resolution in each of these categories. Let $\Gamma_1: \underline{\underline{C}}_1 \to \underline{\underline{C}}_2$ and $\Gamma_2: \underline{\underline{C}}_2 \to \underline{\underline{C}}_3$ be left exact functors. Suppose Γ_1 maps injective objects to objects acyclic for Γ_2 (i.e. to objects M such that the higher derived functors $R^i\Gamma_2$ with $i > 0$ vanish on M). Then there is for each object V in $\underline{\underline{C}}_1$ a spectral sequence (depending canonically on V) with

(1) $E_2^{i,j} = (R^i\Gamma_2)(R^j\Gamma_1)V \Longrightarrow R^{i+j}(\Gamma_2 \circ \Gamma_1)V.$

One may consult [37] for more details and a proof. (This is not contained in the first edition of [37].)

2.4 As the induction functor maps injective modules to injective modules (2.1), we can certainly apply 2.3 to the case where $\Gamma_1 = \text{ind}$.

If $H' \subset H$ is a closed subgroup, then we can interprete the transitivity of induction (1.7(5)) as an isomorphism of functors

$$\text{ind}_H^G \circ \text{ind}_{H'}^H \simeq \text{ind}_{H'}^G.$$

Hence we get for each H'-module M a spectral sequence with

(1) $E_2^{i,j} = (R^i\text{ind}_H^G)(R^j\text{ind}_{H'}^H)M \Longrightarrow (R^{i+j}\text{ind}_{H'}^G)M$

If V is a G-module, then we can interprete Frobenius reciprocity (1.7(3)) as an isomorphism of functors

$$\text{Hom}_G(V,?) \circ \text{ind}_H^G \simeq \text{Hom}_H(V,?),$$

hence get for each H-module a spectral sequence with

(2) $E_2^{i,j} = \text{Ext}_G^i(V,(R^j\text{ind}_H^G)M) \Longrightarrow \text{Ext}_H^{i+j}(V,M).$

2.5 Suppose in this subsection that H is normal in G. Fix a finite dimensional G-module E and a G/H-module M. We can regard $V \mapsto \text{Hom}_H(E,V) \simeq (E^* \otimes V)^H$ as a left exact functor from {G-modules} to {G/H-modules}. It maps any $V_{tr} \otimes K[G]$ to

$$(E^* \otimes V_{tr} \otimes K[G])^H \simeq E^*_{tr} \otimes V_{tr} \otimes K[G]^H \simeq E_{tr} \otimes V_{tr} \otimes K[G/H],$$

hence injective G-modules to injective (G/H)-modules by 2.1(2). If we combine this functor with the (exact) forgetful functor from {G/H-modules} to {vector spaces}, we see that its n-th derived functor maps any G-module V to a G/H-module which is isomorphic to $\mathrm{Ext}_H^n(E,V)$. We shall denote also these derived functors by $\mathrm{Ext}_H^n(E,?)$.

We have an isomorphism of functors

(1) $\mathrm{Hom}_{G/H}(M,?)\circ\mathrm{Hom}_H(E,?) \cong \mathrm{Hom}_G(M\otimes E,?)$,

mapping any $\varphi \in \mathrm{Hom}_{G/H}(M,\mathrm{Hom}_H(E,V))$ to the map $m\otimes e \mapsto \varphi(m)(e)$ for all $m\in M$, $e\in E$. So we get a spectral sequence for each G-module V

(2) $E_2^{i,j} = \mathrm{Ext}^i_{G/H}(M,\mathrm{Ext}^j_H(E,V)) \Longrightarrow \mathrm{Ext}_G^{i+j}(M\otimes E,V)$.

In the special case $M = E = K$ we get the Hochschild-Serre spectral sequence

(3) $H^i(G/H,H^j(H,V)) \Longrightarrow H^{i+j}(G,V)$.

<u>2.6</u> Let V be a G-module. We can interprete the tensor identity as an isomorphism of functors

(1) $(?\otimes V)\circ\mathrm{ind}_H^G \cong \mathrm{ind}_H^G\circ(?\otimes V)$.

As tensoring with V is exact and maps injective modules to injective modules by 2.1(3) we get two spectral sequences converging to the same abutment, both degenerating, hence for each H-module M isomorphisms of G-modules

(2) $R^i\mathrm{ind}_H^G(M\otimes V) \cong (R^i\mathrm{ind}_H^G M)\otimes V$

for each $i \in \mathbf{N}$. This is called the <u>generalized tensor identity</u>.

<u>2.7</u> By the definition of induction we have an isomorphism of functors

(1) $?^H\circ(?\otimes K[G]) \cong F\circ\mathrm{ind}_H^G$

where F: {G-modules} → {vector spaces} is the forgetful functor. As in 2.6 we get for each H-module and each $i \in \mathbf{N}$ an isomorphism of vector spaces

(2) $R^i\mathrm{ind}_H^G M \cong H^i(H,M\otimes K[G])$.

<u>2.8</u> The groups $R^i\mathrm{ind}_H^G M$ can also be interpreted as cohomology

groups of a certain sheaf $\mathcal{L}(M)$ on G/H. Let $\pi: G \to G/H$ be the canonical projection. Then we set

(1) $\mathcal{L}(M)(U) = \{f \in \mathrm{Mor}(\pi^{-1}U, M) \mid f(xh) = h^{-1}f(x)$

for all $x \in \pi^{-1}U$, $h \in H\}$.

Obviously

(2) $H^0(G/H, \mathcal{L}(M)) = \mathcal{L}(M)(G/H) = \mathrm{ind}_H^G M$.

One can then show (without too much difficulty) that $\mathcal{L}(M)$ is a quasi-coherent sheaf of $\mathcal{O}_{G/H}$-modules where $\mathcal{O}_{G/H}$ is the structural sheaf on G/H. If $\dim M < \infty$, then $\mathcal{L}(M)$ is coherent. Furthermore $\mathcal{L}$ is an exact functor from {H-modules} to {$\mathcal{O}_{G/H}$-modules}. One has $\mathcal{L}(K[H]) = \pi_*\mathcal{O}_G$ where $K[H]$ is taken under ρ_ℓ or ρ_r. As π is affine, one gets $H^n(G/H, \mathcal{L}(K[H])) = 0$ for all $n > 0$, hence by 2.1(2) also $H^n(G/H, \mathcal{L}(M)) = 0$ for all $n > 0$ and all injective M. We may interprete (2) as an isomorphism of functors $H^0(G/H, ?) \circ \mathcal{L} \simeq \mathrm{ind}_H^G$ and have just seen that $\mathcal{L}$ maps injective H-modules to acyclic objects for $H^0(G/H, ?)$. Hence we get for the derived functors (for all H-modules M and $i \in \mathbb{N}$)

(3) $H^i(G/H, \mathcal{L}(M)) \simeq R^i\mathrm{ind}_H^G M$.

Standard results about sheaf cohomology (cf. [20], ch. III) imply:

(4) $R^i\mathrm{ind}_H^G = 0$ if $i > \dim G/H$.

(5) <u>If</u> G/H <u>is affine, then</u> $R^i\mathrm{ind}_H^G = 0$ <u>for all</u> $i > 0$.

(6) <u>If</u> G/H <u>is projective, and</u> $\dim M < \infty$, <u>then</u> $\dim R^i\mathrm{ind}_H^G M < \infty$ <u>for all</u> $i \in \mathbb{N}$.

If H is normal in G, then G/H is affine, hence $R^i\mathrm{ind}_H^G = 0$ for all $i > 0$ by (5), i.e. ind_H^G is exact.

<u>2.9</u> Let $N \subset H$ be a closed subgroup which is normal in G. Each H/N-module M is an H-module in a natural way which we denote for the moment by i^*M. Similarly any G/N-module V yields a G-module j^*V. Obviously i^* and j^* are exact functors. Elementary considerations show that we have an isomorphism of functors

(1) $j^* \circ \mathrm{ind}_{H/N}^{G/N} \simeq \mathrm{ind}_H^G \circ i^*$.

We claim: If M is an injective H/N-module, then i^*M is acyclic for ind_H^G. Using 2.1(2) it is enough to look at $M = K[H/N]$, hence $i^*M = K[H]^N = \mathrm{ind}_N^H K$. As N is normal in H and G, we know by 2.8 that ind_N^G and ind_N^H are exact. On the other hand we have by 2.4(1) a spectral sequence with

$$(R^i\mathrm{ind}_H^G)(R^j\mathrm{ind}_N^H)K \Longrightarrow R^{i+j}\mathrm{ind}_N^G K.$$

The exactness of ind_N^H makes the sequence degenerate to isomorphisms $R^i\mathrm{ind}_H^G(\mathrm{ind}_N^H K) \cong R^i\mathrm{ind}_N^G K$, the exactness of ind_H^G yields the vanishing of $R^i\mathrm{ind}_H^G(\mathrm{ind}_N^H K)$ for $i > 0$, i.e. the acyclicity of $i^*K[G/N]$.

Now (1) yields (by the same type of arguments as in 2.6) isomorphisms for all i

(2) $j^* \circ R^i\mathrm{ind}_{H/N}^{G/N} \cong R^i\mathrm{ind}_H^G \circ i^*$.

3. Simple Modules for Reductive Groups

The notations introduced in 3.1 will be used throughout until the end of these notes. (The objects discussed here should be known from [22] or [43].)

3.1 Let G be a connected reductive algebraic group over K, let $B \subset G$ be a Borel subgroup and $T \subset B$ a maximal torus. Let $R \subset X(T)$ be the root system, let R^+ be the positive system in R, such that B corresponds to the negative roots and let $S \subset R^+$ be the corresponding set of simple roots. Let $\leq$ be the order relation on $X(T)$ such that $\lambda \leq \mu$ if and only if $\mu-\lambda \in \sum_{\alpha\in S} \mathbb{N}\alpha$. For each $\alpha \in R$ denote by U_α the corresponding root subgroup. Then $U = \langle U_\alpha \mid \alpha < 0\rangle$ is the unipotent radical of $B = TU$ and $U^+ = \langle U_\alpha \mid \alpha > 0\rangle$ is the unipotent radical of the opposite Borel subgroup $B^+ = TU^+$ corresponding to the positive roots. Denote the Weyl group by $W \cong N_G(T)/T$ and denote by $s_\alpha \in W$ for each $\alpha \in R$ the corresponding reflection. There is a dual root $\alpha^\vee \in \mathrm{Hom}_{\mathbb{Z}}(X(T),\mathbb{Z})$ with $s_\alpha(\lambda) = \lambda - \langle\lambda,\alpha^\vee\rangle\alpha$ for all $\lambda \in X(T)$. The Weyl group W is generated by the s_α with $\alpha \in S$; we denote by $\ell(w)$ for $w \in W$ the smallest m such that $w = s_{\beta_1}s_{\beta_2}\dots s_{\beta_m}$ for suitable $\beta_i \in S$. Denote by w_0 the unique element in W with $w_0(R^+) = -R^+$. For each subset $I \subset S$ set $W_I = \langle s_\alpha \mid \alpha \in I\rangle \subset W$ and $P_I = BW_IB$. The elements in

$$X(T)_+ = \{\lambda \in X(T) \mid \langle\lambda,\alpha^\vee\rangle \geq 0 \quad \text{for all } \alpha \in R^+\}$$
$$= \{\lambda \in X(T) \mid \langle\lambda,\alpha^\vee\rangle \geq 0 \quad \text{for all } \alpha \in S\}$$

are called the dominant weights of T. Choose $\omega_\alpha \in X(T)\otimes_{\mathbb{Z}}\mathbb{Q}$ for all $\alpha \in S$ such that $\langle\omega_\alpha,\beta^\vee\rangle = \delta_{\alpha\beta}$ (the Kronecker delta) for all $\alpha,\beta \in S$. If G is semisimple, then the ω_α are uniquely determined and are called the fundamental weights. One has always

$$X(T)_+ = X(T) \cap (\sum_{\alpha\in S} \mathbb{N}\,\omega_\alpha + X(T)_0 \otimes_{\mathbb{Z}} \mathbb{Q})$$

where

$$X(T)_0 = \{\lambda \in X(T) \mid \langle\lambda,\alpha^\vee\rangle = 0 \quad \text{for all } \alpha \in S\}.$$

In the semisimple case one has $X(T)_0 = 0$. Set $\rho = \frac{1}{2} \sum_{\alpha\in R^+} \alpha \in X(T)\otimes_{\mathbb{Z}}\mathbb{Q}$.

Then obviously $2\rho \in X(T)$. One can show $\langle\rho,\alpha^\vee\rangle = 1$ for all $\alpha \in S$, hence $\rho - \sum_{\alpha\in S}\omega_\alpha \in X(T)_0 \otimes_{\mathbb{Z}} \mathbb{Q}$. It follows $s_\alpha\rho-\rho \in \mathbb{Z}R$ for all $\alpha \in S$, hence $w\rho-\rho \in \mathbb{Z}R \subset X(T)$ for all $w \in W$. We can therefore define a new operation (the dot action) of W on $X(T)$ through $w\cdot\lambda = w(\lambda+\rho)-\rho$ for all $w \in W$ and $\lambda \in X(T)$.

<u>3.2</u> If V is a G-module, then we denote by V_λ for each $\lambda \in X(T)$ the λ-weight space of V considered as a T-module, i.e. we apply the definition 1.3(5) to T instead of G. As T is diagonalizable, we have $V = \bigoplus_{\lambda\in X(T)} V_\lambda$ by 1.3(6). We apply also 1.3(7) to T, so $\mathrm{ch}(V)$ is an element in $\mathbb{Z}[X(T)]$ if it is defined. For all $w \in W$ and $\lambda \in X(T)$ one has $\dot{w}V_\lambda = V_{w\lambda}$ if $\dot{w} \in N_G(T)$ is a representative of w. Therefore $\dim V_\lambda = \dim V_{w(\lambda)}$, so $\mathrm{ch}(V)$ is a W-invariant element in $\mathbb{Z}[X(T)]$ under the action given by $we(\lambda) = e(w\lambda)$:

(1) $\mathrm{ch}(V) \in \mathbb{Z}[X(T)]^W$.

As U and U^+ are unipotent subgroups of G normalized by T we get from 1.3(1),(3):

(2) <u>For any G-module $V \neq 0$ the subspaces V^U and V^{U^+} are non-zero T-submodules.</u>

<u>3.3</u> Any $\lambda \in X(T)$ defines via $B/U \cong T$ and $B^+/U^+ \cong T$ a one-

dimensional B-module and B^+-module. We use both of the times the notation K_λ. (It will be clear which case we are dealing with.)

If V is a G-module, $V \neq 0$, then there is (by 3.2(2)) some $\lambda \in X(T)$ such that K_λ can be embedded into V as a B-submodule. Dualizing we get for any finite dimensional V that there is some $\lambda \in X(T)$ with $\mathrm{Hom}_B(V,K_\mu) \neq 0$. Using Frobenius reciprocity (1.7(3)) we get:

(1) For each simple G-module V there is some $\mu \in X(T)$ such that V is isomorphic to a submodule of $\mathrm{ind}_B^G K_\mu$.

The properties of the Bruhat decomposition imply that $U^+B = U^+TU$ and $BU^+ = UTU^+$ are dense and open subsets of G. So any regular function on G is uniquely determined by its restriction to U^+B. By definition

$$\mathrm{ind}_B^G K_\mu = \{f \in K[G] \mid f(gtu) = \mu(t)^{-1} f(g) \quad \text{for all } g \in G,\ t \in T,\ u \in U\}.$$

The operation of G is given by ρ_ℓ. Hence any $f \in (\mathrm{ind}_B^G K_\mu)^{U^+}$ satisfies $f(u_1 tu) = \mu(t)^{-1}$ for all $u_1 \in U^+$, $t \in T$, $u \in U$. Therefore f is determined by $f(1)$, hence $\dim(\mathrm{ind}_B^G K_\mu)^{U^+} \leq 1$. Furthermore if this space is non-zero, then T operates on it through μ. (Consider the restriction of f to U^+B !)

Any simple submodule of $\mathrm{ind}_B^G K_\mu$ has to contribute at least a subspace of dimension 1 to $(\mathrm{ind}_B^G K_\mu)^{U^+}$ by 3.2(1). So the upper estimate from above implies:

(2) If $\mathrm{ind}_B^G K_\mu \neq 0$, then it has a simple socle.

Let us denote the unique simple submodule of $\mathrm{ind}_B^G K_\mu$ by $L(\mu)$ (for those $\mu \in X(T)$ for which $\mathrm{ind}_B^G K_\mu \neq 0$). The calculation above implies:

(3) The space $L(\mu)^{U^+}$ has dimension 1 and T operates on it through μ.

Combining (1) and (3) we see:

(4) Each simple G-module is isomorphic to exactly one $L(\mu)$.

It remains to determine for which μ we have an $L(\mu)$. Let me refer to [22], 31.4 for a proof of:

(5) <u>We have</u> $\mathrm{ind}_B^G K_\mu \neq 0$ <u>if and only if</u> $\mu \in X(T)_+$.

So (4), (5) say: The simple G-modules are parametrized by the dominant weights of T.

<u>3.4</u> We have observed in 3.3 that U^+B is open and dense in G. It is one of the properties of the Bruhat decomposition that the multiplication $(g_1, g_2) \mapsto g_1 g_2$ induces an isomorphism of varieties $U^+ \times B \to U^+B$. So the canonical map $\pi: G \to G/B$ maps U^+B into an open and dense subset of G/B such that $u \mapsto \pi(u)$ is an isomorphism $U^+ \tilde{\to} \pi(U^+B)$ of varieties.

This implies that π is locally trivial. This means that G/B has a covering by open subsets Y such that there is a section $\sigma_Y: Y \to G$, i.e. a morphism with $\pi \circ \sigma_Y = \mathrm{id}_Y$. (Just take all $\pi(gU^+B)$ with $g \in G$.) For any such Y the map $(y,b) \mapsto \sigma_Y(y)b$ is an isomorphism $Y \times B \to \pi^{-1}(Y)$. For any B-module M we get then (cf. the definition in 2.8)

$$\mathcal{L}(M)(Y) \simeq K[Y] \otimes M = \mathcal{O}_{G/B}(Y) \otimes M.$$

This shows that each $\mathcal{L}(M)$ is a locally free $\mathcal{O}_{G/B}$-module, or (in other words) a vector bundle of rank equal to dim M. We get especially that each $\mathcal{L}(K_\lambda)$ is a line bundle on G/B.

The tangent space at $1B \in G/B$ can be identified with Lie G/Lie B, hence $\Lambda^n(\mathrm{Lie}\ G/\mathrm{Lie}\ B)^*$ with $K_{-2\rho}$ as a B-module where $n = \dim G/B = \dim U^+ = |R^+|$. Using a homogeneity argument it is easy to show that $\mathcal{L}(K_{-2\rho})$ is the canonical sheaf (cf. [20], p. 180) on G/B. Furthermore one can show that the Serre duality (cf. [20], III.7) for the cohomology groups $H^i(G/B, \mathcal{L}(M)) = (R^i \mathrm{ind}_B^G)(M)$ is compatible with the G-action, i.e. we have for each finite dimensional B-module M and each $i \in \mathbb{N}$, $i \leq n = \dim G/B$ an isomorphism of G-modules

(1) $(R^i \mathrm{ind}_B^G M)^* \simeq R^{n-i} \mathrm{ind}_B^G (M^* \otimes K_{-2\rho})$.

<u>3.5</u> For all $\lambda \in X(T)$ with $\langle \lambda, \alpha^\vee \rangle > 0$ for all $\alpha \in S$ the line bundle $\mathcal{L}(K_\lambda)$ on G/B is ample (even very ample). One may consult [20], II.7 and III.5 for the definition and properties of ampleness, and [24] or [34] for more details about line bundles on G/B. Now Kodaira's vanishing theorem [20], III.7.15) together with the

description of the canonical sheaf in 3.4 implies for all $\lambda \in X(T)_+$ that $H^i(G/B, \mathcal{L}(K_\lambda)) = 0$ for all $i > 0$ provided char $K = 0$.

The same result can also be proved in arbitrary characteristic, i.e. we have always (using 2.8(3))

(1) <u>If</u> $\lambda \in X(T)_+$ <u>and</u> $i > 0$, <u>then</u> $R^i \mathrm{ind}_B^G(K_\lambda) = 0$.

This result is due to Kempf [33] (if char $K \neq 0$) and is hence called <u>Kempf's vanishing theorem</u>. There are now proofs of this theorem available which are considerably simpler that Kempf's original one. These new proofs were discovered independently by Andersen [4] and Haboush [19] and use the action of the Frobenius morphism on these cohomology groups. The same method has been used very recently to prove a similar vanishing theorem also for the restrictions of these line bundles to all Schubert varieties $\overline{BwB/B} \subset G/B$ with $w \in W$ and then to prove that these Schubert varieties are normal, cf. [39], [10], [41], [42] for more details.

<u>3.6</u> For each $\lambda \in X(T)$ set

(1) $\chi(\lambda) = (\sum_{w \in W} \det(w) e(w \cdot \lambda)) / (\sum_{w \in W} \det(w) e(w \cdot 0))$.

This is to start with an element in the quotient field of $\mathbb{Z}[X(T)]$ as the $w \cdot 0 = w\rho - \rho$ with $w \in W$ are pairwise different. One has obviously

(2) $\chi(w \cdot \lambda) = \det(w) \chi(\lambda)$

for all $\lambda \in X(T)$ and $w \in W$. If $\langle \lambda + \rho, \alpha^\vee \rangle = 0$ for some $\alpha \in R$, then $s_\alpha \cdot \lambda = \lambda$ and $\det(s_\alpha) = -1$, hence $\chi(\lambda) = 0$. If $\langle \lambda + \rho, \alpha^\vee \rangle \neq 0$ for all $\alpha \in R$, then there is a (unique) $w \in W$ with $w \cdot \lambda \in X(T)_+$. For $\lambda \in X(T)_+$ we may now appeal to Weyl's character formula which implies $\chi(\lambda) = \mathrm{ch}\, L(\lambda)$ if char $K = 0$. This proves for any λ using (2) and for any K using a group over $\mathbb{Z}$ with the same root data as G that $\chi(\lambda) \in \mathbb{Z}[X(T)]$.

In fact, even more is true. One has for each $\lambda \in X(T)$:

(3) $\chi(\lambda) = \sum_{i \geq 0} (-1)^i \,\mathrm{ch}\, R^i \mathrm{ind}_B^G K_\lambda$.

If char $K = 0$, then each G-module is semi-simple, hence 3.3(2) implies $L(\lambda) = \mathrm{ind}_B^G K_\lambda$ for each $\lambda \in X(T)_+$. Therefore (3) is just Weyl's character formula for $\lambda \in X(T)_+$ and follows in general from the Borel-Bott-Weil theorem (cf. 4.4(3)) if char $K = 0$.

For char $K = p \neq 0$ one can deduce the result from those over using a suitable group scheme over $\mathbb{Z}$, cf. the discussion in [9]. There is however also a direct argument using complete information in the case of semi-simple rank 1 together with the fact that the formula holds for $\lambda = 0$, see [16] for the details.

More generally, one can prove character formulas for the cohomology groups of the restrictions of these line bundles to Schubert varieties. This is done in [10].

4. The Semi-Simple Rank 1 Case (and Applications)

References: [1],[2],[14],[7]

4.1 Let us suppose in this subsection that $G = SL_2(K)$. We may assume that T consists of all diagonal matrices (in G) and B of all lower diagonal matrices. If we denote the canonical basis of K^2 by e_1, e_2, then

$$B = \{g \in G \mid ge_2 \in Ke_2\}$$

and

$$U = \{g \in G \mid ge_2 = e_2\}.$$

We have $X(T) = \mathbb{Z}\rho$ and $X(T)_+ = \mathbb{N}\rho$. For each $n \in \mathbb{N}$ the induced module

$$\mathrm{ind}_B^G(K_{n\rho}) = \{f \in K[G] \mid f(gtu) = \rho(t)^{-n} f(g) \quad \text{for all } g \in G,\ t \in T,\ u \in U\}$$

is contained in

$$\{f \in K[G] \mid f(gu) = f(g) \quad \text{for all } g \in G,\ u \in U\} = K[G/U].$$

The map $g \mapsto ge_2$ induces a G-equivariant isomorphism of varieties $G/H \to Ge_2 = K^2 \setminus 0$ as the tangent map is obviously surjective. As $te_2 = \rho(t)^{-1} e_2$ for all $t \in T$ we get an isomorphism

$$\mathrm{ind}_B^G(K_{n\rho}) = \{f \in K[K^2 \setminus 0] \mid f(av) = a^n v \quad \text{for all } a \in K^\times,\ v \in K^2 \setminus 0\}.$$

As K^2 is a smooth, hence normal variety each regular function on $K^2\setminus 0$ can be extended (uniquely, of course) to K^2. Thus we get

$$\operatorname{ind}_B^G(K_{n\rho}) = \{f \in K[K^2] \mid f(av) = a^n f(v) \quad \text{for all} \quad v \in K^2, a \in K\},$$

i.e.

(1) $\operatorname{ind}_B^G(K_{n\rho}) \cong S^n((K^2)^*)$.

This formula can also be proved using the identification $G/B \cong \mathbb{P}^1$ and known results on the cohomology of line bundles on the projective line.

Note that we get all $R^i\operatorname{ind}_B^G(K_{r\rho})$ with $i \in \mathbb{Z}$ and $r \in \mathbb{N}$ from (1) together with 2.8(4) (by which only $i = 0,1$ have to be considered) and with the Serre duality 3.4(1) according to which $R^1\operatorname{ind}_B^G(K_{r\rho})$ is dual to $\operatorname{ind}_B^G(K_{-(r+2)\rho})$. As $\operatorname{ind}_B^G(K_{r\rho}) = 0$ for $r \leq -1$ by (the easy part of) 3.3(5), we get $R^1\operatorname{ind}_B^G(K_{r\rho}) = 0$ for $r \geq -1$ and

$$R^1\operatorname{ind}_B^G(K_{r\rho}) \cong (S^{-(r+2)}(K^2)^*)^* \quad \text{for} \quad r \leq -2.$$

Note that $s_\alpha \cdot (r\rho) = -(r+2)\rho$ in this case (where $\{\alpha\} = S$).

<u>4.2</u> One can generalize the results of 4.1 easily to all cases where G has semi-simple rank one by applying 2.9(2) to the surjections $\mathcal{D}G \times Z(G)^0 \to G$ and $SL_2(K) \to \mathcal{D}G$ and using an obvious formula for direct products. (Here $\mathcal{D}G$ is the derived group and $Z(G)^0$ is the connected component of 1 of the centre of G.)

Let us look at a more general situation. Take a simple root $\alpha \in S$ and look at the corresponding parabolic subgroup $P(\alpha) = P_{\{\alpha\}} \supset B$. Then $P(\alpha)/R_u(P(\alpha))$ is a connected reductive group of semi-simple rank one. (Here $R_u(?)$ denotes the unipotent radical.) As $R_u(P(\alpha)) \subset B$ operates trivially on each K_λ we can use 2.9(2) together with the remarks above to compute all $R^i\operatorname{ind}_B^{P(\alpha)}(K_\lambda)$. Again, we have to look only at $i = 0,1$. Let us at first record the vanishing results:

(1) <u>If</u> $\langle\lambda,\alpha^\vee\rangle = -1$, <u>then</u> $R^\bullet\operatorname{ind}_B^{P(\alpha)}(K_\lambda) = 0$.

(2) <u>If</u> $\langle\lambda,\alpha^\vee\rangle \geq 0$, <u>then</u> $R^1\operatorname{ind}_B^{P(\alpha)}(K_\lambda) = 0$.

(3) <u>If</u> $\langle\lambda,\alpha^\vee\rangle \leq -2$, <u>then</u> $\operatorname{ind}_B^{P(\alpha)}K_\lambda = 0$.

We want to describe the operation of $P(\alpha)$ on the non-vanishing

$R^i\mathrm{ind}_B^{P(\alpha)}(K_\lambda)$ explicitly. We use in 4.1 the basis of $S^n((K^2)^*)$ consisting of homogeneous polynomials and its dual basis. As $R_u(P(\alpha))$ operates trivially on any $R^i\mathrm{ind}_B^{P(\alpha)}(K_\lambda)$ we have to look only at the action of the standard Levi factor $\langle T,U_\alpha,U_{-\alpha}\rangle$. Choose root homomorphisms $x_\alpha: K \to U_\alpha$ and $x_{-\alpha}: K \to U_{-\alpha}$ as in [43], 11.2.1. (They come from an epimorphism $SL_2(K) \to \langle U_\alpha,U_{-\alpha}\rangle = \mathcal{D}\langle T,U_\alpha,U_{-\alpha}\rangle$.) Then we get from 4.1:

(4) <u>If</u> $\langle\lambda,\alpha^\vee\rangle = n \geq 0$, <u>then</u> $\mathrm{ind}_B^{P(\alpha)}(K_\lambda)$ <u>has a basis</u> $v_0,v_1,\dots,v_n$ <u>such that for all</u> i $(0 \leq i \leq n)$:

a) $tv_i = (\lambda-i\alpha)(t)v_i$ <u>for all</u> $t \in T$,

b) $x_\alpha(a)v_i = \sum_{j=0}^{i} \binom{i}{j} a^{i-j} v_j$ <u>for all</u> $a \in K$,

c) $x_{-\alpha}(a)v_i = \sum_{j=i}^{n} \binom{n-i}{n-j} a^{j-i} v_j$ <u>for all</u> $a \in K$.

(5) <u>If</u> $\langle\lambda,\alpha^\vee\rangle = n \geq 0$, <u>then</u> $R^1\mathrm{ind}_B^{P(\alpha)}(K_{s_\alpha\cdot\lambda})$ <u>has a basis</u> $v'_0,v'_1,\dots,v'_n$ <u>such that for all</u> i $(0 \leq i \leq n)$

a) $tv'_i = (\lambda-i\alpha)(t)v'_i$ <u>for all</u> $t \in T$,

b) $x_\alpha(a)v'_i = \sum_{j=0}^{i} \binom{n-j}{n-i} a^{i-j} v'_j$ <u>for all</u> $a \in K$,

c) $x_{-\alpha}(a)v'_i = \sum_{j=i}^{n} \binom{j}{i} a^{j-i} v'_j$ <u>for all</u> $a \in K$.

<u>4.3</u> We see that for each $\lambda \in X(T)$ at most one $R^i\mathrm{ind}_B^{P(\alpha)}(K_\lambda)$ is different from 0. Therefore the spectral sequence from 2.4(1)

$$(R^j\mathrm{ind}_{P(\alpha)}^G)(R^i\mathrm{ind}_B^{P(\alpha)})(K_\lambda) \Longrightarrow (R^{i+j}\mathrm{ind}_B^G)(K_\lambda)$$

degenerates.

(1) <u>If</u> $\lambda \in X(T)$ <u>and</u> $\alpha \in S$ <u>with</u> $\langle\lambda,\alpha^\vee\rangle = -1$, <u>then</u> $R^\bullet\mathrm{ind}_B^G(K_\lambda) = 0$.

(2) <u>If</u> $\lambda \in X(T)$ <u>and</u> $\alpha \in S$ <u>with</u> $\langle\lambda,\alpha^\vee\rangle \geq 0$, <u>then for all</u> i:

$$R^i\mathrm{ind}_B^G(K_\lambda) \cong R^i\mathrm{ind}_{P(\alpha)}^G(\mathrm{ind}_B^{P(\alpha)}K_\lambda)$$

<u>and</u>

$$R^i\mathrm{ind}_B^G(K_{s_\alpha\cdot\lambda}) \cong R^{i-1}\mathrm{ind}_{P(\alpha)}^G(R^1\mathrm{ind}_B^{P(\alpha)}K_\lambda).$$

In the situation of 4.2(4),(5) the map $v'_i \mapsto \binom{n}{i}v_i$ is a homomorphism of $P(\alpha)$-modules from $R^1\mathrm{ind}_B^{P(\alpha)}(K_{s_\alpha\cdot\lambda})$ to $\mathrm{ind}_B^{P(\alpha)}(K_\lambda)$. (In case $G = P(\alpha)$ its image is equal to $L(\lambda)$.) If all $\binom{n}{i}$ with $0 \le i \le n$ are non-zero in K, then it is an isomorphism. This is always the case if $\mathrm{char}\, K = 0$. If $\mathrm{char}\, K = p \neq 0$, then it is satisfied for $n < p$ (and more generally for $n = ap^r-1$ with $a,r \in \mathbb{N}$ and $0 < a < p$).

Set

$$p = \begin{cases} \mathrm{char}(K) & \text{if } \mathrm{char}(K) \neq 0, \\ \infty & \text{if } \mathrm{char}(K) = 0. \end{cases}$$

Then the discussion above and (2) imply

(3) <u>If</u> $\lambda \in X(T)$ <u>and</u> $\alpha \in S$ <u>with</u> $0 \le \langle\lambda+\rho,\alpha^\vee\rangle \le p$, <u>then</u> $R^i\mathrm{ind}_B^G K_\lambda \cong R^{i+1}\mathrm{ind}_B^G K_{s_\alpha\cdot\lambda}$ <u>for all</u> $i \in \mathbb{N}$.

We use here and below the convention $R^i\mathrm{ind}_B^G = 0$ for $i < 0$.

<u>4.4</u> Set

(1) $\bar{C}_{\mathbb{Z}} = \{\lambda \in X(T) \mid 0 \le \langle\lambda+\rho,\beta^\vee\rangle \le p \quad \text{for all } \alpha \in R^+\}$

where p is as in 4.3. For all $w \in W$ and $\alpha \in S$ with $w^{-1}\alpha > 0$ we have

$$0 \le \langle w\cdot\lambda+\rho,\alpha^\vee\rangle = \langle w(\lambda+\rho),\alpha^\vee\rangle = \langle\lambda+\rho,(w^{-1}\alpha)^\vee\rangle \le p$$

for all $\lambda \in \bar{C}_{\mathbb{Z}}$. Therefore 4.3 implies

$$R^i\mathrm{ind}_B^G(K_{w\cdot\lambda}) \cong R^{i+1}\mathrm{ind}_B^G(K_{s_\alpha w\cdot\lambda})$$

for all $i \in \mathbb{Z}$. Using induction on $\ell(w)$ and starting with Kempf's vanishing theorem resp. with 4.3(1) we get

(2) <u>If</u> $\lambda \in \bar{C}_{\mathbb{Z}}$ <u>with</u> $\lambda \notin X(T)_+$, <u>then</u> $R^\bullet\mathrm{ind}_B^G(K_{w\cdot\lambda}) = 0$ <u>for all</u> $w \in W$.

(3) <u>If</u> $\lambda \in \bar{C}_{\mathbb{Z}} \cap X(T)_+$, <u>then for all</u> $w \in W$:

$$R^i\mathrm{ind}_B^G(K_{w\cdot\lambda}) \cong \begin{cases} \mathrm{ind}_B^G(K_\lambda) & \underline{\text{for}}\ i = \ell(w), \\ 0 & \underline{\text{otherwise}}. \end{cases}$$

If $\operatorname{char} K = 0$, then each $\mu \in X(T)$ is of the form $w \cdot \lambda$ with $w \in W$ and $\lambda \in X(T)$. So in this case we have determined all $R^i \operatorname{ind}_B^G K_\mu$. The result is known as the Borel-Bott-Weil theorem. The approach above is due to Demazure.

If $\operatorname{char} K \neq 0$, then (2),(3) do not generalize to all $\lambda \in X(T)$ with $\langle\lambda+\rho,\alpha^\vee\rangle \geq 0$ for all $\alpha \in S$. If R has a component of rank at least 2, then there are $\mu \in X(T)$ with $R^i \operatorname{ind}_B^G K_\mu \neq 0$ for two different values of i. This was first observed by D. Mumford for $SL_3(K)$ and $p = 2$. Then W.L. Griffiths determined for every $SL_3(K)$ all (μ,i) with $R^i \operatorname{ind}_B^G K_\mu \neq 0$. Later on, this was done for all rank 2 cases by H.H. Andersen, cf. [2],[6].

In the case of semi-simple rank 1 we get from 4.2 that (2),(3) generalize to all λ (with $\langle\lambda+\rho,\alpha^\vee\rangle \geq 0$ for all $\alpha \in S$) on the level of characters, i.e. we have $\operatorname{ch} R^{\ell(w)} \operatorname{ind}_B^G(K_{w\cdot\lambda}) = \operatorname{ch} \operatorname{ind}_B^G(K_\lambda)$ in (3) whereas the modules themselves may be non-isomorphic. A similar result holds for arbitrary G for those λ for which $\operatorname{ind}_B^G(K_\lambda)$ has generic decomposition behaviour in the sense of [27]. This is proved in [13].

<u>4.5</u> In order to illustrate the differences between characteristic 0 and $\neq 0$ let us look at $R^1 \operatorname{ind}_B^G K_\mu$ and let us assume $\operatorname{char} K = p \neq 0$.

We need some general remarks about ind_P^G for any parabolic $P \supset B$. Let $U(\bar{P})$ be the product of all root subgroups U_α not contained in P and let $H \supset T$ be the standard Levi factor of P. Then $\bar{P} = HU(\bar{P})$ is a parabolic subgroup of G with $\bar{P} \supset B^+$, with unipotent radical $U(\bar{P})$ and Levi factor H. For any $\bar{P}$-module M the subspace $M^{U(\bar{P})}$ is an H-module.

The subset $U(\bar{P})P$ is dense in G. Therefore any $f \in \operatorname{ind}_P^G M$ (for some P-module M) is determined by $f|_{U(\bar{P})}$ and the evaluation map $\varepsilon_M\colon \operatorname{ind}_P^G M \to M$, $f \mapsto f(1)$ is injective on $(\operatorname{ind}_P^G M)^{U(\bar{P})}$.

One can show for any $\lambda \in X(T)_+$ that $L(\lambda)^{U(\bar{P})}$ is the simple H-module corresponding to the dominant weight λ. Therefore $(\operatorname{soc}_G V)^{U(\bar{P})} \subset \operatorname{soc}_H V$ for any G-module V. If M is a P-module with $\operatorname{soc}_H M$ simple corresponding to λ, then the injectivity of ε_M on $(\operatorname{ind}_P M)^{U(\bar{P})}$ implies that $\operatorname{soc}_G(\operatorname{ind}_P^G M) \simeq L(\lambda)$ or $\operatorname{ind}_P^G M = 0$. In the first case necessarily $\lambda \in X(T)_+$.

Suppose now that $R^1 \operatorname{ind}_B^G K_\mu \neq 0$ for some $\mu \in X(T)$. By Kempf's vanishing theorem and by 4.3(1) there has to be $\alpha \in S$ with $\langle\mu+\rho,\alpha^\vee\rangle$

< 0, hence by 4.3(2) with

$$R^1 \mathrm{ind}_B^G K_\mu \cong \mathrm{ind}_{P(\alpha)}^B (R^1 \mathrm{ind}_B^{P(\alpha)} K_\mu)$$

If $\langle s_\alpha \cdot \mu, \alpha^\vee \rangle = ap^r - 1$ with $a, r \in \mathbb{N}$ and $0 < a < p$, then $R^1 \mathrm{ind}_B^G K_\mu \cong \mathrm{ind}_B^G K_{s_\alpha \cdot \mu}$ by the proof of 4.3(3), so in this case $R^1 \mathrm{ind}_B^G K_\mu \neq 0$ if and only if $s_\alpha \cdot \mu \in X(T)_+$ and, if so, then $L(s_\alpha \cdot \mu) = \mathrm{soc}_G R^1 \mathrm{ind}_B^G K_\mu$.

Suppose now that $\langle s_\alpha \cdot \mu, \alpha^\vee \rangle$ is not of this form and write $\langle s_\alpha \cdot \mu, \alpha^\vee \rangle = \sum_{i=0}^{m} a_i p^i$ with $0 \leq a_i \leq p$ for all i and $a_m \neq 0$. Using the explicit description of $R^1 \mathrm{ind}_B^{P(\alpha)} K_\mu$ in 4.2(5) and some elementary arguments one shows that $R^1 \mathrm{ind}_B^{P(\alpha)} K_\mu$ has a simple socle corresponding to the weight $\mu + a_m p^m \alpha$. So: If $R^1 \mathrm{ind}_B^G K_\mu \neq 0$, then $\mu + a_m p^m \alpha \in X(T)_+$ and $L(\mu + a_m p^m \alpha)$ is the socle of this G-module. On the other hand one can also prove the converse: If $\mu + a_m p^m \alpha \in X(T)_+$, then $\mathrm{ind}_{P(\alpha)}^G (R^1 \mathrm{ind}_B^{P(\alpha)} K_\mu) \neq 0$. See [2] for the details.

It is more convenient to describe the result the other way round:

(1) <u>Let</u> $\lambda \in X(T)_+$. <u>Denote by</u> $a(r,\alpha)$ <u>for each</u> $\alpha \in S$ <u>and</u> $r \in \mathbb{N}$ <u>the unique integer with</u>

$$(a(r,\alpha)-1)p^r < \langle \lambda+\rho, \alpha^\vee \rangle \leq a(r,\alpha)p^r.$$

<u>Then</u> $L(\lambda)$ <u>is the socle of all</u> $R^1 \mathrm{ind}_B^G K_\mu$ <u>with</u> μ <u>of the form</u> $\lambda - a(r,\alpha)p^r \alpha$ <u>for some</u> $\alpha \in S$ <u>and</u> $r \in \mathbb{N}$, <u>whereas</u> $\mathrm{Hom}_G(L(\lambda), R^1 \mathrm{ind}_B^G K_\mu) = 0$ <u>for all other</u> μ.

For example $L(0) = K$ is the socle of each $R^1 \mathrm{ind}_B^G K_{-p^r \alpha}$ with $\alpha \in S$ and $r \in \mathbb{N}$. If $r > 0$ and if $\alpha \in S$ does not belong to a component of type A_1, then $s_\alpha \cdot (-p^r \alpha) = (p^r - 1)\alpha \notin X(T)_+$, but $R^1 \mathrm{ind}_B^G K_{-p^r \alpha} \neq 0$. In this way we get many μ with $R^1 \mathrm{ind}_B^G K_\mu \neq 0$, but $\mu \notin \bigcup_{\alpha \in S} s_\alpha \cdot X(T)_+$ which does not happen over $\mathbb{C}$.

There are other methods to get non-vanishing $R^i \mathrm{ind}_B^G (K_\mu)$, cf. e.g. [4].

<u>4.6</u> Using the spectral sequence (from 2.4(2))

$$\mathrm{Ext}_G^i (V, R^j \mathrm{ind}_B^G K_\lambda) \Longrightarrow \mathrm{Ext}_B^{i+j} (V, K_\lambda)$$

for each G-module V and each $\lambda \in X(T)$ results about the $R^j \mathrm{ind}_B^G K_\lambda$

give information about B-cohomology. For example, Kempf's vanishing theorem implies for all $\lambda \in X(T)_+$ that

(1) $\operatorname{Ext}^i_G(V, \operatorname{ind}_B^G K_\lambda) \cong \operatorname{Ext}^i_B(V, K_\lambda)$

for all i and V.

In general, the spectral sequence leads to a five term exact sequence

$$0 \to \operatorname{Ext}^1_G(V, \operatorname{ind}_B^G K_\lambda) \to \operatorname{Ext}^1_B(V, K_\lambda) \to$$

$$\to \operatorname{Hom}_G(V, R^1\operatorname{ind}_B^G K_\lambda) \to \operatorname{Ext}^2_G(V, \operatorname{ind}_B^G K_\lambda) \to \operatorname{Ext}^2_B(V, K_\lambda)$$

This implies for all $\lambda \notin X(T)_+$:

(2) $\operatorname{Ext}^1_B(V, K_\lambda) \cong \operatorname{Hom}(V, R^1\operatorname{ind}_B^G K_\lambda)$.

So the results in 4.5 show

(3) $H^1(B, K_\lambda) \cong \begin{cases} K & \text{if } \lambda = -p^r\alpha \text{ for some } \alpha \in S,\ r \in \mathbb{N} \\ 0 & \text{otherwise,} \end{cases}$

as soon as we prove $H^1(B, K_\lambda) = 0$ for all $\lambda \in X(T)_+$. But $\lambda \in X(T)_+$ implies $\lambda \nless 0$ and therefore each exact sequence $0 \to K_\lambda \to M \to K \to 0$ of B-modules splits.

For $H^2(B, K_\lambda)$ so far only partial results are known, cf. [40] and [7].

5. The Linkage Principle

References: [3], [26], [5], [35], [14], [21], [30], [18], [23], [15]

Let us assume from now on that $\operatorname{char} K = p \neq 0$. Furthermore we shall assume $\rho \in X(T)$. Usually the final results hold also without this assumption and one can prove them by going over to a suitable covering group for some time.

5.1 Let $P \supset B$, $P \neq B$ be a parabolic subgroup of G and let M be a P-module. We have $R^\cdot\operatorname{ind}_B^P K_{-\rho} = 0$ as $\langle -\rho, \alpha^\vee \rangle = -1$ for all $\alpha \in S$, cf. 4.3(1). So the generalized tensor identity 2.6(2) implies $R^\cdot\operatorname{ind}_B^P(M \otimes K_{-\rho}) = 0$, hence by 2.4(2)

(1) $R^{\cdot}\mathrm{ind}_B^G(M \otimes K_{-\rho}) = 0$.

The generalized tensor identity implies also

$$R^i\mathrm{ind}_B^P(M) = R^i\mathrm{ind}_B^P(M \otimes K) \cong M \otimes R^i\mathrm{ind}_B^P(K),$$

and Kempf's vanishing theorem applied to $K = K_0$ and to $P/R_u(P)$ yields $R^i\mathrm{ind}_B^P K = 0$ for $i > 0$ and $\mathrm{ind}_B^P K = K$. So we get from 2.4(2) for all i

(2) $R^i\mathrm{ind}_B^G M \cong R^i\mathrm{ind}_P^G M$.

5.2 Let us use for all $\lambda \in X(T)$ and $i \in \mathbb{N}$ the abbreviation

$$H^i(\lambda) = R^i\mathrm{ind}_B^G(K_\lambda).$$

Consider $\lambda \in X(T)$ and $\alpha \in S$ with $n = \langle\lambda,\alpha^\vee\rangle \geq 0$. Set $H_\alpha^0(\lambda) = \mathrm{ind}_B^{P(\alpha)}(K_\lambda)$ and use the notations from 4.2(4). The subspace $H_\alpha^0(\lambda)^- = \sum_{i=1}^n Kv_i$ (the negative part) of $H_\alpha^0(\lambda)$ is a B-submodule, and Kv_n is a B-submodule of $H_\alpha^0(\lambda)^-$. Let us denote the quotient by $H_\alpha^0(\lambda)^m = H_\alpha^0(\lambda)^-/Kv_n$ (the middle part). We get two exact sequences of B-modules

(1) $0 \to H_\alpha^0(\lambda)^- \to H_\alpha^0(\lambda) \to K_\lambda \to 0$

and

(2) $0 \to K_{s_\alpha\lambda} \to H_\alpha^0(\lambda)^- \to H_\alpha^0(\lambda)^m \to 0$.

For $n \geq 2$ we can also consider $H_\alpha^0(\lambda-\alpha)$. Let us denote the basis of $H_\alpha^0(\lambda-\alpha)$ as in 4.2(4) by $\tilde v_0, \tilde v_1, \ldots, \tilde v_{n-2}$. The map $v_i \mapsto (n-i)\tilde v_{i-1}$ is easily checked to induce a homomorphism $H_\alpha^0(\lambda)^m \to H_\alpha^0(\lambda-\alpha)$ of B-modules. Let us denote the kernel, image and cokernel of this homomorphism by $N_\alpha(\lambda)$, $I_\alpha(\lambda)$, and $Q_\alpha(\lambda)$. So we have exact sequences

(3) $0 \to N_\alpha(\lambda) \to H_\alpha^0(\lambda)^m \to I_\alpha(\lambda) \to 0$

and

(4) $0 \to I_\alpha(\lambda) \to H_\alpha^0(\lambda-\alpha) \to Q_\alpha(\lambda) \to 0$.

Obviously $\mathrm{ch}\, H_\alpha^0(\lambda)^m = \mathrm{ch}\, H_\alpha^0(\lambda-\alpha)$, hence $\mathrm{ch}\, N_\alpha(\lambda) = \mathrm{ch}\, Q_\alpha(\lambda)$. A simple consideration shows (remember $n = \langle\lambda,\alpha^\vee\rangle$)

(5) $\quad \operatorname{ch} N_\alpha(\lambda) = \operatorname{ch} Q_\alpha(\lambda) = \sum_{\substack{r\in \mathbb{N} \\ 0<rp<n}} e(s_\alpha\lambda+rp\alpha).$

Let us tensor the exact sequences (1)-(4) with $K_{-\rho}$, apply ind_B^G and look at the corresponding long exact sequences. As $H_\alpha^0(\lambda)$ and $H_\alpha^0(\lambda-\alpha)$ are $P(\alpha)$-modules, we get from 5.1(1) that

$$R^\cdot \operatorname{ind}_B^G(H_\alpha^0(\lambda)\otimes K_{-\rho}) = 0 = R^\cdot \operatorname{ind}_B^G(H_\alpha^0(\lambda-\alpha)\otimes K_{-\rho}).$$

So (1), (4) yield isomorphisms

$$H^i(\lambda-\rho) \simeq R^{i+1}\operatorname{ind}_B^G(H_\alpha^0(\lambda)^-\otimes K_{-\rho})$$

and

$$R^i\operatorname{ind}_B^G(Q_\alpha(\lambda)\otimes K_{-\rho}) \simeq R^{i+1}\operatorname{ind}_B^G(I_\alpha(\lambda)\otimes K_{-\rho}).$$

We insert this into the long exact sequences arising from (2) and (3) and then replace λ by $\lambda+\rho$. In this way we get long exact sequences for each $\lambda\in X(T)$ and $\alpha\in S$ with $\langle\lambda+\rho,\alpha^\vee\rangle \geq 0$:

(6) $\quad \ldots \to H^i(s_\alpha\cdot\lambda) \to H^{i-1}(\lambda) \to R^i\operatorname{ind}_B^G(H_\alpha^0(\lambda+\rho)^m\otimes K_{-\rho}) \to \ldots$

and (for $\langle\lambda+\rho,\alpha^\vee\rangle \geq 2$)

(7) $\quad \ldots \to R^i\operatorname{ind}_B^G(N_\alpha(\lambda+\rho)\otimes K_{-\rho}) \to R^i\operatorname{ind}_B^G(H_\alpha^0(\lambda+\rho)^m\otimes K_{-\rho})$
$\quad\quad \to R^{i-1}\operatorname{ind}_B^G(Q_\alpha(\lambda+\rho)\otimes K_{-\rho}) \to \ldots$

Observe: If $\langle\lambda+\rho,\alpha^\vee\rangle = 0$, then $H^\cdot(s_\alpha\cdot\lambda) = H^\cdot(\lambda) = 0$, if $\langle\lambda+\rho,\alpha^\vee\rangle = 1$, then $H_\alpha^0(\lambda+\rho)^m = 0$.

<u>5.3</u> For any $\beta\in R$ and $n\in\mathbb{Z}$ let $s_{\beta,n}$ be the affine reflection $s_{\beta,n}(\lambda) = \lambda-(\langle\lambda,\beta^\vee\rangle-n)\beta = s_\beta(\lambda)+n\beta$ on $X(T)$ or $X(T)\otimes\mathbb{Q}$. We shall extend the "dot" convention $(w\cdot\lambda = w(\lambda+\rho)-\rho)$ also to maps of the form $s_{\beta,n}$ or to products of such maps.

Define an order relation $\uparrow$ on $X(T)$ through:

(1) $\mu\uparrow\lambda \Longleftrightarrow \mu=\lambda$ <u>or there are</u> $r\in\mathbb{N}$ <u>and</u> $\mu=\mu_0,\mu_1,\ldots,\mu_r = \lambda\in X(T)$ <u>and</u> $\alpha_i\in R$, $m_i\in\mathbb{Z}$ <u>with</u> $\mu_{i-1} = s_{\alpha_i,m_ip}\cdot\mu_i$ <u>and</u> $\mu_{i-1}\leq\mu_i$ <u>for all</u> i $(1\leq i\leq r)$.

Let us now state the strong linkage principle:

(2) Let $\mu \in X(T)_+$ and $\lambda \in X(T)$ with $<\lambda+\rho,\alpha^\vee> \geq 0$ for all $\alpha \in S$. If $L(\mu)$ is a composition factor of some $H^i(w\cdot\lambda)$ with $i \in \mathbb{N}$ and $w \in W$, then $\mu \uparrow \lambda$.

This theorem can be proved using induction over λ. We may assume the result for all $\lambda' \in -\rho+X(T)_+$ with $\lambda' < \lambda$. Consider for all $w \in W$ and $\alpha \in S$ with $w^{-1}\alpha > 0$ the long exact sequences 5.2(6),(7) with $w\cdot\lambda$ instead of λ. For any weight μ_1 of some $N_\alpha(w\cdot\lambda+\rho) \otimes K_{-\rho}$ or $Q_\alpha(w\cdot\lambda+\rho) \otimes K_{-\rho}$ there is a unique $\lambda_1 \in W\cdot\mu_1 \cap (-\rho+X(T)_+)$. Now 5.2(5) and Satz 9 in [26] imply $\lambda_1 < \lambda$ and $\lambda_1 \uparrow \lambda$. By induction any composition factor of some $R^i\mathrm{ind}_B^G(N_\alpha(w\cdot\lambda+\rho) \otimes K_{-\rho})$ or $R^i\mathrm{ind}_B^G(Q_\alpha(w\cdot\lambda+\rho) \otimes K_{-\rho})$, hence by 5.2(7) also of each $R^i\mathrm{ind}_B^G(H_\alpha^O(w\cdot\lambda+\rho)^m \otimes K_{-\rho})$ has the form $L(\mu)$ with $\mu \uparrow \lambda$ and $\mu < \lambda$.

This shows: Any counter-example $L(\mu)$ to (2) or $L(\lambda)$ will never show up in the kernel or cokernel of one of the maps $H^{i+1}(s_\alpha w\cdot\lambda) \to H^i(w\cdot\lambda)$ arising from 5.2(6).

In order to go on with (2), look at first at all $H^i(w\cdot\lambda)$ with $i < \ell(w)$ and use induction on $\ell(w)$ for fixed $\ell(w)-i$. The case $\ell(w) = 0$ (i.e. $w = 1$) is trivial. For $w \neq 1$ there is $\alpha \in S$ with $w^{-1}\alpha < 0$. Then we get a map $H^i(w\cdot\lambda) \to H^{i-1}(s_\alpha w\cdot\lambda)$ from 5.2(6) which any counter-example and also any factor of type $L(\lambda)$ has to survive. As $\ell(s_\alpha w) = \ell(w)-1$ we can now use induction. For $i > \ell(w)$ a similar, descending induction works. This settles the case for $i \neq \ell(w)$ and also for $\lambda \notin X(T)_+$ as in that case $w\cdot\lambda = ws_\beta\cdot\lambda$ for a suitable β and $\ell(w) \neq \ell(ws_\beta)$. We get also

(3) If $\lambda \in X(T)_+$ and if $L(\mu)$ is a composition factor of $H^i(w\cdot\lambda)$ with $w \in W$ and $i \in \mathbb{N}$, $i \neq \ell(w)$, then $\mu \uparrow \lambda$ and $\mu < \lambda$.

We have still to look at all $H^{\ell(w)}(w\cdot\lambda)$ for $\lambda \in X(T)_+$. Set $n = |R^+| = \ell(w_0)$. We can form a sequence of homomorphisms starting at $H^n(w_0\cdot\lambda)$, passing through $H^{\ell(w)}(w\cdot\lambda)$ and ending in $H^O(\lambda)$. Each single homomorphism arises from 5.2(6). We know $L(\lambda) = \mathrm{soc}_G H^O(\lambda)$ and also $L(\lambda) = H^n(w_0\cdot\lambda)/\mathrm{rad}_G H^n(w_0\cdot\lambda)$ by Serre duality (3.4(1)). The character formula 3.6(3) implies that $L(\lambda)$ occurs with multiplicity one as a composition factor of $H^O(\lambda)$. Hence any non-zero homomorphism $H^n(w_0\cdot\lambda) \to H^O(\lambda)$ has kernel $\mathrm{rad}_G H^n(w_0\cdot\lambda)$ and image $L(\lambda)$. Any counter example to (2) would have to appear in this image by our previous arguments, so there is no counter-example. We get at the same time:

(4) <u>If</u> $\lambda \in X(T)_+$, <u>then</u> $L(\lambda)$ <u>occurs with multiplicity 1 in all</u> $H^{\ell(w)}(w\cdot\lambda)$ <u>with</u> $w \in W$.

<u>5.4</u> Consider $\lambda,\mu \in X(T)_+$ with $\mu\uparrow\lambda$. Let us call μ <u>close</u> to λ if and only if:

a) There is no $\alpha \in R^+$ with $\mu\uparrow\lambda-p\alpha$,

b) For all $\lambda' \in X(T)$ with $\mu\uparrow\lambda'\uparrow\lambda$ we have $\lambda' \in X(T)_+$.

More or less the same arguments as in 5.3 show

(1) <u>Let</u> $\lambda,\mu \in X(T)_+$ <u>with</u> $\mu\uparrow\lambda$ <u>and</u> μ <u>close to</u> λ. <u>Then</u> $L(\mu)$ <u>is no composition factor of any</u> $H^i(w\cdot\lambda)$ <u>with</u> $i \neq \ell(w)$. <u>It occurs in</u> $H^{\ell(w)}(w\cdot\lambda)$ <u>for any</u> $w \in W$ <u>with the same multiplicity as in</u> $H^0(\lambda)$. <u>This multiplicity is positive</u>.

Induction over $\lambda-\mu$ shows that $L(\mu)$ occurs always with the same multiplicity in the kernel as in the cokernel of the maps $H^i(s_\alpha w\cdot\lambda) \to H^{i-1}(w\cdot\lambda)$ arising from 5.2(6), that this multiplicity is 0 for $i \neq \ell(s_\alpha w)$ and at least once non-zero for some w. For a maximal μ one gets more precisely:

(2) <u>Let</u> $\lambda \in X(T)_+$. <u>Consider</u> $\mu \in X(T)$ <u>maximal for</u> $\mu\uparrow\lambda, \mu < \lambda$. <u>If</u> $\mu \in X(T)_+$ <u>and if</u> $\mu \neq \lambda-p\beta$ <u>for all</u> $\beta \in R^+$, <u>then</u> $L(\mu)$ <u>occurs with multiplicity 1 in</u> $H^0(\lambda)$.

<u>5.5</u> Denote by W_p the group generated by all $s_{\beta,mp}$ with $\beta \in R$ and $m \in \mathbb{Z}$. It is called the <u>affine Weyl group</u> corresponding to p and can also be described as the semi-direct product $W \ltimes \mathbb{Z}R$ where any $\nu \in \mathbb{Z}R$ operates as the translation by $p\nu$.

Let us call $\lambda,\mu \in X(T)$ <u>linked</u> if and only if there is some $w \in W_p$ with $\mu = w\cdot\lambda$. Obviously $\mu\uparrow\lambda$ implies that μ is linked to λ.

Take $\lambda,\mu \in X(T)_+$, consider the exact sequence

$$0 \to L(\lambda) \to H^0(\lambda) \to H^0(\lambda)/\mathrm{soc}_G H^0(\lambda) \to 0,$$

cf. 3.3(2), and apply $\mathrm{Hom}_G(L(\mu),?)$ to it. We have by 4.6(1) isomorphisms $\mathrm{Ext}^i_G(L(\mu),H^0(\lambda)) \simeq \mathrm{Ext}^i_B(L(\mu),K_\lambda)$. If $\lambda < \mu$, then any exact sequence $0 \to K_\lambda \to M \to L(\mu) \to 0$ of B-modules has to split: Choose $v \in M_\mu$ with a non-zero image in $L(\mu)$, then $Kv \oplus \bigoplus_{\nu<\mu} M_\nu$ is a B-stable complement to K_λ in M. So $\mathrm{Ext}^1_G(L(\mu),H^0(\lambda)) = 0$ for $\lambda < \mu$. Furthermore $\mathrm{Hom}_G(L(\mu),H^0(\lambda)) = 0$ for $\lambda \neq \mu$ whereas for $\lambda = \mu$

the map $\operatorname{Hom}_G(L(\mu),L(\lambda)) \to \operatorname{Hom}_G(L(\mu),H^0(\lambda))$ is bijective. So we get:

(1) <u>If</u> $\lambda \not< \mu$, <u>then</u> $\operatorname{Ext}^1_G(L(\mu),L(\lambda)) \simeq \operatorname{Hom}_G(L(\mu),H^0(\lambda)/\operatorname{soc}_G H^0(\lambda))$.

Using duality and an automorphism of G inducing $-w_0$ on T we get

(2) $\operatorname{Ext}^1_G(L(\mu),L(\lambda)) \simeq \operatorname{Ext}^1_G(L(\lambda),L(\mu))$

for all $\lambda,\mu \in X(T)$. So we get from (1) and 5.3(2):

(3) <u>Any</u> $\lambda,\mu \in X(T)^+$ <u>with</u> $\operatorname{Ext}^1_G(L(\lambda),L(\mu)) \neq 0$ <u>are linked</u>.

More generally, the <u>linkage principle</u> holds:

(4) <u>Let</u> M <u>be an indecomposable</u> <u>G-module and let</u> $L(\lambda),L(\mu)$ <u>be composition factors of</u> M. <u>Then</u> λ,μ <u>are linked</u>.

Consider the smallest equivalence relation on $X(T)_+$ such that λ,μ are linked if $\operatorname{Ext}^1_G(L(\lambda),L(\mu)) \neq 0$. The equivalence classes are called the blocks of G. The block of some $\mu \in X(T)_+$ is by (3) contained in $W_p \cdot \mu \cap X(T)_+$. More precisely, one has:

(5) <u>Let</u> $\mu \in X(T)_+$. <u>If there is</u> $\beta \in R$ <u>with</u> $p \nmid \langle \mu+\rho,\beta^\vee\rangle$, <u>then</u> $W_p \cdot \mu \cap X(T)_+$ <u>is the block of</u> μ.

Now we get all blocks via:

(6) <u>If</u> $\underline{B}$ <u>is a block as in</u> (5) <u>then all</u> $\{(p^r-1)\rho+p^r\lambda \mid \lambda \in \underline{B}\}$ <u>with</u> $r \in \mathbb{N}$ <u>are blocks</u>.

6. <u>The Translation Principle</u>

<u>References</u>: [25],[27],[28],[29],[6],[36]

<u>6.1</u> The group W_p acts on $X(T)\otimes\mathbb{Q}$ (or on $X(T)\otimes\mathbb{R}$) as a reflection group, hence defines a system of alcoves and facets. We shall always use the "dot" operation $w\cdot\lambda = w(\lambda+\rho)-\rho$ so that the reflection hyperplanes are the $\{\lambda \in X(T)\otimes\mathbb{Q} \mid \langle\lambda+\rho,\beta^\vee\rangle = mp\}$ for all $\beta \in R$ and $m \in \mathbb{Z}$. Two elements of $X(T)\otimes\mathbb{Q}$ belong to the same <u>facet</u>, if and only if they lie on the same side of each of these hyperplanes. Hence a facet is a non-empty set of the form

$$F = \{\lambda \in X(T)\otimes\mathbb{Q} \mid n_\alpha p = \langle\lambda+\rho,\alpha^\vee\rangle \quad \text{for all } \alpha \in R_0^+(F),$$
$$n_\alpha p < \langle\lambda+\rho,\alpha^\vee\rangle < (n_\alpha+1)p \quad \text{for all } \alpha \in R_1^+(F)\}$$

for suitable $n_\alpha \in \mathbb{Z}$ and a disjoint decomposition $R^+ = R_0^+(F) \cup R_1^+(F)$. In this situation we set

$$\bar{F} = \{\lambda \in X(T) \otimes \mathbb{Q} \mid n_\alpha p = \langle\lambda+\rho,\alpha^\vee\rangle \quad \text{for all } \alpha \in R_0^+(F),$$
$$n_\alpha p \le \langle\lambda+\rho,\alpha^\vee\rangle \le (n_\alpha+1)p \quad \text{for all } \alpha \in R_1^+(F)\}$$

and

$$\hat{F} = \{\lambda \in X(T) \otimes \mathbb{Q} \mid n_\alpha p = \langle\lambda+\rho,\alpha^\vee\rangle \quad \text{for all } \alpha \in R_0^+(F),$$
$$n_\alpha p < \langle\lambda+\rho,\alpha^\vee\rangle \le (n_\alpha+1)p \quad \text{for all } \alpha \in R_1^+(F)\}.$$

Obviously $\bar{F}$ is the <u>closure</u> of F. We call $\hat{F}$ the <u>upper closure</u> of F. Both $\bar{F}$ and $\hat{F}$ are unions of facets.

An <u>alcove</u> is a facet F where $R_0^+(F) = \emptyset$. The group W_p permutes the alcoves simply transitively. One special alcove is

$$C = \{\lambda \in X(T) \otimes \mathbb{Q} \mid 0 \le \langle\lambda+\rho,\beta^\vee\rangle < p \quad \text{for all } \alpha \in R^+\}.$$

(Then $\bar{C}_{\mathbb{Z}}$ from 4.4(1) is $\bar{C} \cap X(T)$.) The closure of any alcove is a fundamental domain for W_p. Hence $\bar{C}_{\mathbb{Z}}$ is a fundamental domain for W_p operating on $X(T)$.

A <u>wall</u> is a facet F with $|R_0^+(F)| = 1$. It is therefore contained in a unique reflection hyperplane. Let s_F be the corresponding reflection.

For any alcove C_0 the reflections s_F with F running through the walls with $F \subset \bar{C}_0$ generate W_p and $(W_p, \{s_F \mid F \subset \bar{C}_0\})$ is a Coxeter system. In the case $C_0 = C$ we denote this set of reflections by Σ. It consists of all s_α with $\alpha \in S$ and of those $s_{\beta,p}$ where β is the largest short root of some component of R.

For any $\lambda \in X(T) \otimes \mathbb{Q}$ its stabilizer $W_p^0(\lambda)$ in W_p is generated by all reflections $s \in W_p$ with $s\cdot\lambda = \lambda$. More precisely, if $\lambda \in \bar{C}$, then $(W_p^0(\lambda), \Sigma^0(\lambda))$ is a Coxeter system where $\Sigma^0(\lambda) = \{s \in \Sigma \mid s\cdot\lambda = \lambda\}$.

<u>6.2</u> For any $\lambda \in X(T)$ and any G-module M let $\mathrm{pr}_\lambda M$ be the largest G-submodule such that all its composition factors have the form $L(w\cdot\lambda)$ with $w \in W_p$. (We have obviously $\mathrm{pr}_{w\cdot\lambda} M = \mathrm{pr}_\lambda M$ for all $w \in W_p$.) Now 5.5(4) implies

(1) $M = \bigoplus_{\lambda \in \bar{C}_{\mathbb{Z}}} pr_\lambda M.$

This is immediate if $\dim M < \infty$ and follows in general from the local finiteness of M. We may of course replace $\bar{C}_{\mathbb{Z}}$ by any other system of representatives for $X(T)/W_p$. We have a canonical bijection for all G-modules M, M':

(2) $\mathrm{Hom}_G(M,M') \cong \prod_{\lambda \in \bar{C}_{\mathbb{Z}}} \mathrm{Hom}_G(pr_\lambda M, pr_\lambda M').$

This shows not only that each pr_λ is an exact functor, but also (together with (1)) that the category of all G-modules is the direct product of all $\underline{\underline{M}}_\lambda$ with $\lambda \in \bar{C}$. (We define $\underline{\underline{M}}_\lambda$ for each $\lambda \in X(T)$ as the category of all G-modules M with $pr_\lambda M = M$.)

The strong linkage principle implies obviously for all $\lambda, \mu \in X(T)$ and $i \in \mathbb{N}$

(3) $$pr_\lambda H^i(\mu) = \begin{cases} H^i(\mu) & \text{if } \mu \in W_p \cdot \lambda, \\ 0 & \text{otherwise.} \end{cases}$$

For any finite dimensional G-module V we have a unique expression

$$\mathrm{ch}\, V = \sum_{\mu \in X(T)_+} (V:\mu)\chi(\mu)$$

with $(V:\mu) \in \mathbb{Z}$. Then we get for all λ

$$\mathrm{ch}\, pr_\lambda V = \sum_{\mu \in W_p \cdot \lambda} (V:\mu)\chi(\mu).$$

<u>6.3</u> Define for any $\lambda, \mu \in \bar{C}_{\mathbb{Z}}$ the translation functor T_λ^μ from {G-modules} to itself (or from $\underline{\underline{M}}_\lambda$ to $\underline{\underline{M}}_\mu$) through

(1) $T_\lambda^\mu V = pr_\mu(pr_\lambda(V) \otimes L(\nu_1))$

for all G-modules V, where ν_1 is the only element in $W(\mu-\lambda) \cap X(T)_+$. As pr_λ and pr_μ are exact functors, so is T_λ^μ. Using $L(\nu_1)^* = L(-w_0\nu_1)$ and $\{-w_0\nu_1\} = W(\lambda-\mu) \cap X(T)_+$ we get easily

(2) <u>For all</u> $\lambda, \mu \in \bar{C}_{\mathbb{Z}}$ <u>the functors</u> T_λ^μ <u>and</u> T_μ^λ <u>are adjoint to each other</u>.

<u>6.4</u> Let us apply T_λ^μ at first to G-modules of the form $H^0(\lambda')$. Because of 6.2(3) we may restrict ourselves to the case $\lambda' = w \cdot \lambda$

for some $w \in W_p$. Take ν_1 as in 6.3(1). Then the definition and the tensor identity imply

$$T_\lambda^\mu H^0(w\cdot\lambda) = pr_\mu(L(\nu_1) \otimes ind_B^G(K_{w\cdot\lambda}))$$
$$= pr_\mu \circ ind_B^G(L(\nu_1) \otimes K_{w\cdot\lambda}).$$

The B-module $M = L(\nu_1) \otimes K_{w\cdot\lambda}$ has a composition series $M = M_0 \supset M_1 \supset M_2 \supset \ldots$ with each M_i/M_{i+1} of the form $K_{\nu+w\cdot\lambda}$ with ν running through the weights of $L(\nu_1)$. As $ch\ L(\nu_1)$ is W-invariant we can write these factors also as $K_{w\cdot(\nu(i)+\lambda)}$ where $\{\nu(i) | i \geq 0\}$ are the weights of $L(\nu_1)$ counted with their multiplicity. We get long exact sequences

$$\ldots \to pr_\mu R^j ind_B^G M_{i+1} \to pr_\mu R^j ind_B^G M_i \to pr_\mu H^j(w\cdot(\nu(i)+\lambda)) \to \ldots$$

Only the $\nu(i)$ with $\lambda+\nu(i) \in W_p\cdot\mu$ contributes something. They can be determined explicitly

(1) <u>If</u> $\lambda, \mu \in \bar{C}_{\mathbf{Z}}$ <u>and</u> $\nu_1 \in X(T)_+ \cap W(\mu-\lambda)$, <u>then</u> $W_p\cdot\mu \cap \{\lambda+\nu | L(\nu_1)_\nu \neq 0\} = \{w'\cdot\mu | w' \in W_p^0(\lambda)\}$. <u>For all</u> ν <u>occurring one has</u> $\dim L(\nu_1)_\nu = 1$.

So the relevant weights of $L(\nu_1) \otimes K_{w\cdot\lambda}$ are the $ww'\cdot\mu$ with $w' \in W_p^0(\lambda)$. We may assume $ww'\cdot\lambda = w\cdot\lambda \in X(T)_+$, hence $\langle ww'\cdot\lambda+\rho, \alpha^\vee\rangle \geq 0$ for all $\alpha \in S$ and $H^j(K_{ww'\cdot\mu}) = 0$ for all $i > 0$. This shows:

(2) <u>Let</u> $\lambda, \mu \in \bar{C}_{\mathbf{Z}}$ <u>and</u> $w \in W_{p_0}$ <u>with</u> $w\cdot\lambda \in X(T)_+$. <u>Then</u> $T_\lambda^\mu H^0(w\cdot\lambda)$ <u>has a filtration with factors</u> $H^0(ww'\cdot\mu)$ <u>with</u> w' <u>running through a system of representatives for</u> $W_p^0(\lambda)/(W_p^0(\lambda) \cap W_p^0(\mu))$.

If $W_p^0(\lambda) \subset W_p^0(\mu)$ then obviously $T_\lambda^\mu H^0(w\cdot\lambda) \cong H^0(w\cdot\mu)$. In this situation the discussion above gives even more information:

(3) <u>Let</u> $\lambda, \mu \in \bar{C}_{\mathbf{Z}}$, <u>let</u> F <u>be the facet with</u> $\lambda \in F$. <u>If</u> $\mu \in \bar{F}$, <u>then</u> $T_\lambda^\mu H^i(w\cdot\lambda) \cong H^i(w\cdot\mu)$ <u>for all</u> $w \in W_p$ <u>and</u> $i \in \mathbb{N}$.

The next simple case is the following

(4) <u>Let</u> $\lambda, \mu \in \bar{C}_{\mathbf{Z}}$ <u>and</u> $s \in \Sigma$. <u>Suppose</u> $\Sigma^0(\lambda) = \{s\}$ <u>and</u> $\Sigma^0(\mu) = \emptyset$. <u>Then we have for all</u> $w \in W_p$ <u>with</u> $ws\cdot\mu < w\cdot\mu$ <u>a long exact sequence of G-modules</u>

$$0 \to H^0(ws\cdot\mu) \to T_\lambda^\mu H^0(w\cdot\lambda) \to H^0(w\cdot\mu) \to H^1(ws\cdot\mu)$$
$$\ldots \to H^i(ws\cdot\mu) \to T_\lambda^\mu H^i(w\cdot\lambda) \to H^i(w\cdot\mu) \to \ldots$$

<u>If</u> $w\cdot\lambda \in X(T)_+$, <u>then we have an exact sequence</u>

$$0 \to H^0(ws\cdot\mu) \to T_\lambda^\mu H^0(w\cdot\lambda) \to H^0(w\cdot\mu) \to 0.$$

<u>6.5</u> We want to apply the T_λ^μ now to simple G-modules. Let us assume that we are in the situation of 6.4(3). If $w\cdot\lambda \in X(T)_+$ for some $w \in W_p$, then $L(w\cdot\lambda)$ is the image of some homomorphism $H^n(w_0w\cdot\lambda) \to H^0(w\cdot\lambda)$ where $n = |R^+|$. As T_λ^μ is exact, the module $T_\lambda^\mu L(w\cdot\lambda)$ has to be the image of some homomorphism $T_\lambda^\mu H^n(w_0w\cdot\lambda) \to T_\lambda^\mu H^0(w\cdot\lambda)$, i.e. $H^n(w_0w\cdot\mu) \to H^0(w\cdot\mu)$. If $w\cdot\mu \notin X(T)_+$, then $T_\lambda^\mu L(w\cdot\lambda) = 0$ as $H^0(w\cdot\mu) = 0$. If $w\cdot\mu \in X(T)_+$, then $T_\lambda^\mu L(w\cdot\lambda)$ is either 0 or $L(w\cdot\mu)$. As $L(w\cdot\mu)$ is a composition factor of $H^0(w\cdot\mu) \cong T_\lambda^\mu H^0(w\cdot\lambda)$, it has to occur in some $T_\lambda^\mu L(w'\cdot\lambda)$ for some composition factor $L(w'\cdot\lambda)$ of $H^0(w\cdot\lambda)$. The same argument as above implies $T_\lambda^\mu L(w'\cdot\lambda) \in \{0, L(w'\cdot\mu)\}$, hence $w'\cdot\lambda = w\cdot\mu$ and $w' \in wW_p^0(\mu)$. Therefore the following case is trivial:

(1) <u>If</u> $\lambda,\mu \in \bar{C}_{\mathbb{Z}}$ <u>belong to the same facet, then</u> $T_\lambda^\mu L(w\cdot\lambda) \cong L(w\cdot\mu)$ <u>for all</u> $w \in W_p$ <u>with</u> $w\cdot\lambda \in X(T)_+$.

For such λ,μ it is now easy to show that T_λ^μ is an equivalence of categories $\underline{\underline{M}}_\lambda \to \underline{\underline{M}}_\mu$. This is one version of the <u>translation principle</u>: "The category $\underline{\underline{M}}_\lambda$ depends only on the facet to which λ belongs."

In general one can show, using 5.4(1):

(2) <u>Let</u> $\lambda,\mu \in \bar{C}_{\mathbb{Z}}$ <u>and</u> $w \in W_p$ <u>with</u> $w\cdot\lambda \in X(T)_+$. <u>Let</u> F <u>be the facet with</u> $w\cdot\lambda \in F$ <u>and suppose</u> $w\cdot\mu \in \bar{F}$. <u>Then</u>

$$T_\lambda^\mu L(w\cdot\lambda) \cong \begin{cases} L(w\cdot\mu) & \underline{\text{if}}\ \ w\cdot\mu \in \hat{F}, \\ 0 & \underline{\text{otherwise}}. \end{cases}$$

<u>6.6</u> Let us denote by $[V:L]$ the multiplicity of any simple module L as a composition factor of a module V. Using the exactness of T_λ^μ we get from 6.5(2) and 6.4(3):

(1) <u>Let</u> $\lambda,\mu \in \bar{C}_{\mathbb{Z}}$ <u>and</u> $w,w' \in W_p$ <u>with</u> $w\cdot\lambda \in X(T)_+$. <u>Let</u> F <u>be the facet with</u> $w\cdot\lambda \in F$ <u>and suppose</u> $w\cdot\mu \in \bar{F}$. <u>Then we have for all</u> $i \in \mathbb{N}$

$$[H^i(w'\cdot\lambda):L(w\cdot\lambda)] = [H^i(w'\cdot\mu):L(w\cdot\mu)].$$

If $W_p^0(\lambda)$ is strictly contained in $W_p^0(\mu)$, then different w'

will lead in (1) to different $w'\cdot\lambda$, but the same $w'\cdot\mu$. This gives equalities for certain multiplicities of $L(w\cdot\lambda)$. The simplest case is the following:

(2) <u>Let</u> $\lambda \in \bar{C}_{\mathbb{Z}}$ <u>with</u> $\Sigma^0(\lambda) = \emptyset$. <u>Let</u> $w \in W_p$ <u>and</u> $s \in \Sigma$ <u>with</u> $w\cdot\lambda \in X(T)_+$ <u>and</u> $w\cdot\lambda < ws\cdot\lambda$. <u>Then we have for all</u> $w' \in W_p$ <u>and</u> $i \in \mathbb{N}$:

$$[H^i(w'\cdot\lambda):L(w\cdot\lambda)] = [H^i(w's\cdot\lambda):L(w\cdot\lambda)].$$

One simply has to show that there is some $\mu \in \bar{C}_{\mathbb{Z}}$ with $\Sigma^0(\mu) = \{s\}$. This is an elementary exercise. (The fact that there is some $\lambda \in \bar{C}_{\mathbb{Z}}$ with $\Sigma^0(\lambda) = \emptyset$ implies that p cannot be too small.)

<u>6.7</u> Let $\lambda \in \bar{C}_{\mathbb{Z}}$ with $\Sigma^0(\lambda) = \emptyset$. For any $w \in W_p$ with $w\cdot\lambda \in X(T)_+$ there are unique integers $a(w,w')$ for all $w' \in W_p$ with $w'\cdot\lambda \in X(T)_+$ such that

(1) $\operatorname{ch} L(w\cdot\lambda) = \sum_{w'} a(w,w')\chi(w'\cdot\lambda)$.

One simply has to invert the matrix $[H^0(w'\cdot\lambda):L(w\cdot\lambda)]$ which is unipotent lower triangular. It follows from 6.6(1) that the $a(w,w')$ are independent of λ and that for all $\mu \in \bar{C}$:

(2) $$\sum_{w'} a(w,w')\chi(w'\cdot\mu) = \begin{cases} \operatorname{ch} L(w\cdot\mu) & \underline{\text{if}}\ \ w\cdot\mu \in w\cdot C, \\ 0 & \underline{\text{otherwise}}. \end{cases}$$

As in 6.6(2) the second part yields relations between different $a(w,w')$.

In [38] there is a conjecture how to express the $a(w,w')$ in terms of Kazhdan-Lusztig polynomials for W_p as long as $\langle w(\lambda+\rho),\alpha^\vee\rangle < p(p-\langle\rho,\alpha^\vee\rangle+1)$ for all $\alpha \in R \cap X(T)_+$. In general one should probably replace $L(w\cdot\lambda)$ by some other modules as Kato's proof [32] of the compatibility of this conjecture with Steinberg's tensor product theorem indicates. There are some other conjectures which imply by [8] and [9] the conjecture above mentioned.

It is not difficult to show that $a(w,w') = \sum_i (-1)^i \dim \operatorname{Ext}^i_G(L(w\cdot\lambda), H^0(w'\cdot\lambda))$. One may ask how these dimensions are related to the Kazhdan-Lusztig polynomials. In some very special cases they can be identified with coefficients in these polynomials: combine [11] and [31].

References

[1] H.H. Andersen: Cohomology of line bundles on G/B, Ann. Scient. Éc. Norm. Sup. (4) 12 (1979), 85-100

[2] H.H. Andersen: The first cohomology group of a line bundle on G/B, Invent. math. 51 (1979), 287-296

[3] H.H. Andersen: The strong linkage principle, J. reine angew. Math. 315 (1980), 53-59

[4] H.H. Andersen: The Frobenius morphism on the cohomology of homogeneous vector bundles on G/B, Ann. of Math. 112 (1980), 113-121

[5] H.H. Andersen: On the structure of Weyl modules, Math. Z. 170 (1980), 1-14

[6] H.H. Andersen: On the structure of the cohomology of line bundles on G/B, J. Algebra 71 (1981), 245-258

[7] H.H. Andersen: Extensions of modules for algebraic groups, Amer. J. Math. 106 (1984), 489-504

[8] H.H. Andersen: An inversion formula for the Kazhdan-Lusztig polynomials for affine Weyl groups, Adv. in Math.

[9] H.H. Andersen: Filtrations of cohomology modules for Chevalley groups, Ann. Scient. Éc. Norm. Sup. (4) 16 (1983), 495-528

[10] H.H. Andersen: Schubert varieties and Demazure's character formula

[11] H.H. Andersen, J.C. Jantzen: Cohomology of induced representations for algebraic groups, Math. Ann. 269 (1984), 487-525

[12] E. Cline, B. Parshall, L. Scott: Induced modules and affine quotients, Math. Ann. 230 (1977), 1-14

[13] E. Cline, B. Parshall, L. Scott: On injective modules for infinitesimal algebraic groups I, Proc. London Math. Soc.

[14] E. Cline, B. Parshall, L. Scott, W. van der Kallen: Rational and generic cohomology, Invent. math. 39 (1977), 143-163

[15] S. Donkin: The blocks of a semi-simple algebraic group , J. Algebra 67 (1980), 36-53

[16] S. Donkin: Tensor products and filtrations for rational representations of algebraic groups

[17] W. Haboush: Homogeneous vector bundles and reductive subgroups of reductive algebraic groups, Amer. J. Math. 100 (1978), 1123-1137

[18] W. Haboush: Central differential operators on split semi-simple groups over fields of positive characteristic, pp. 35-85 in: Séminaire d'Algèbre P. Dubreil et M.-P. Malliavin (Proc. 1979), Lecture Notes in Mathematics 795, Berlin/Heidelberg/New York 1980 (Springer)

[19] W. Haboush: A short proof of the Kempf vanishing theorem, Invent. math. 56 (1980), 109-112

[20] R. Hartshorne: Algebraic Geometry, New York/Heidelberg/Berlin 1977 (Springer)

[21] J.E. Humphreys: Modular representations of classical Lie algebras and semisimple groups, J. Algebra 19 (1971), 51-79

[22] J.E. Humphreys: Linear Algebraic Groups, Berlin/Heidelberg/New York 1975 (Springer)

[23] J.E. Humphreys, J.C. Jantzen: Blocks and indecomposable modules for semisimple algebraic groups, J. Algebra 54 (1978), 494-503

[24] B. Iversen: The geometry of algebraic groups, Advances in Math. 20 (1976), 57-85

[25] J.C. Jantzen: Zur Charakterformel gewisser Darstellungen halbeinfacher Gruppen und Lie-Algebren, Math. Z. 140 (1974), 127-149

[26] J.C. Jantzen: Darstellungen halbeinfacher Gruppen und kontravariante Formen, J. reine angew. Math. 290 (1977), 157-199

[27] J.C. Jantzen: Über das Dekompositionsverhalten gewisser modularer Darstellungen halbeinfacher Gruppen und ihrer Lie-Algebren, J. Algebra 49 (1977), 441-469

[28] J.C. Jantzen: Weyl modules for groups of Lie type, pp. 291-300 in M. Collins (ed.), Finite Simple Groups II (Proc. Durham 1978), London/New York 1980 (Academic Press)

[29] J.C. Jantzen: Darstellungen halbeinfacher Gruppen und ihrer Frobenius-Kerne, J. reine angew. Math. 317 (1980), 157-199

[30] V. Kac, B. Weisfeiler: Coadjoint action of a semisimple algebraic group and the center of the enveloping algebra in characteristic p, Indag. math. 38 (1976), 136-151

[31] S. Kato: Spherical functions and a q-analogue of Kostant's weight multiplicity formula, Invent. math. 66 (1982), 461-468

[32] S. Kato: On the Kazhdan-Lusztig polynomials for affine Weyl groups, Adv. in Math. 55 (1985), 103-130

[33] G. Kempf: Linear systems on homogeneous spaces, Ann. of Math. 103 (1976), 557-591

[34] G. Kempf: The Grothendieck-Cousin complex of an induced representation, Adv. in Math. 29 (1978), 310-396

[35] M. Koppinen: On the composition factors of Weyl modules, Math. Scand. 51 (1982), 212-216

[36] M. Koppinen: On the translation functors for a semisimple algebraic group, Math. Scand. 51 (1982), 217-226

[37] S. Lang: Algebra (2nd ed.), Reading, Mass. 1984 (Addison-Wesley)

[38] G. Lusztig: Some problems in the representation theory of finite Chevalley groups, pp. 313-317 in: B. Cooperstein, G. Mason (eds.), The Santa Cruz Conference on Finite Groups (1979), Proc. Symp. Pure Math. 37, Providence, R.I. 1980 (Amer. Math. Soc.)

[39] V.B. Mehta, A. Ramanathan: Frobenius splitting and cohomology vanishing for Schubert varieties (to appear)

[40] J. O'Halloran: Cohomology of a Borel subgroup of a Chevelley group (to appear)

[41] S. Ramanan, A. Ramanathan: Projective normality of flag varieties and Schubert varieties, Invent. math. 79 (1985), 217-224

[42] C.S. Seshadri: Line bundles on Schubert varieties (to appear)

[43] T.A. Springer: Linear Algebraic Groups, Boston/Basel/Stuttgart 1981 (Birkhäuser)

DIFFERENTIAL ALGEBRAIC GROUPS

E.R. Kolchin
Columbia University

The lectures will attempt to describe the general theory of differential algebraic groups that has been developed in recent years in analogy with and as a generalization of the older theory of algebraic groups. Limitations of time make it necessary to omit a number of topics and to give broad descriptions instead of proofs.

I have given Professor Tuan at least one copy of each of several references, and I hope that these can be consulted by anyone interested in any of the details. The main reference is the manuscript of my forth-coming book [8]; two earlier works that may prove helpful on occasion are my 1973 book [6] and my paper [7]. Also included are reprints of five papers by P.J. Cassidy [1-5], one paper by J. Kovacic [9], and one paper by W.Y. Sit [10], all bearing on results in the theory that I shall not have time to describe.

These lectures are intended as a selective survey of the subject.

Algebraic groups

An algebraic group is a group, defined by a system of algebraic equations, such that the group law and group symmetry (inverse) are rational mappings. Thus, the additive group G_a, the multiplicative group G_m, the general linear group $GL(n)$, the special linear group $SL(n)$, and the upper triangular group $T(n)$ all are algebraic groups. So, too, is the elliptic curve $W(g_2, g_3)$ in the projective plane, given by the equation

$$y_0 y_2^2 - (y_1^3 - g_2 y_0^2 y_1) - g_3 y_0^3) = 0,$$

where g_2, g_3 are constants with $g_2^3 - 27g_3^2 \neq 0$. In general, when the coefficients in the algebraic equations and in the rational mappings are in a given field K, we say that the algebraic group is defined over K, or is a K-group.

Examples of differential algebraic groups.

Algebraic equations form a special case of algebraic differential equations. Therefore it is natural to try to generalize the idea of algebraic group to one of differential algebraic group. Of course, whatever defintion we use, every algebraic group should be a differential algebraic group. More generally, however, if G is one of the algebraic groups mentioned above, and Σ is a system of algebraic differential equations in the coordinates of the elements of G having the property

$$a,b \in G \text{ and } a,b \text{ are solutions of } \Sigma$$
$$\Longrightarrow ab,\ a^{-1} \text{ are solutions of } \Sigma,$$

then the set of elements of G that are solutions of Σ should be a differential algebraic group. Thus, for example, if Λ is a linear differential operator, the set L_Λ of solutions of the linear differential equation $\Lambda y = 0$ should be a differential algebraic subgroup of the additive group G_a.

The differential algebraic setting

Before we go further, the setting for our differential equations must be made clear. In what follows, all rings are commutative, with 1, and all fields are of characteristic 0.

Suppose given a ring R and a finite set Δ that operates on R as a commuting family of derivation operators:

$$\delta(a+b) = \delta a + \delta b,\ \delta(ab) = \delta a \cdot b + a\delta b,\ \delta\delta' a = \delta' \delta a$$
$$(a,b \in R,\quad \delta,\delta' \in \Delta).$$

Then we say that R is a differential ring (relative to Δ), or that R is a Δ-ring. When the ring is a field, we have a Δ-field. (In general, we use the prefix Δ- as a synonym for "differential" or "differentially".) We put $m = \operatorname{Card}\Delta$ and denote the elements of Δ by $\delta_1,\ldots,\delta_m$. The free commutative monoid Θ generated by Δ, the elements of which are the expressions $\delta_1^{e_1}\cdots\delta_m^{e_m}$ $(e_1 \in \mathbb{N},\ldots,e_m \in \mathbb{N})$, operates on any Δ-ring in an obvious way. The definitions of Δ-subring and homomorphism of Δ-rings are as expected.

If R is a Δ-subring of a Δ-ring R', and $s = (s_j)_{j \in J}$ is a family of elements of R', there is a unique smallest Δ-subring of R' containing the elements of R and every s_j; this is the Δ-<u>ring generated</u> by s over R, denoted by $R\{s\}$ (or $R\{s\}_\Delta$ if necessary). As a ring, $R\{s\}$ coincides with $R[(\theta s_j)_{\theta \in \Theta, j \in J}]$. If F is a Δ-subfield of a Δ-field F', and s is now a family of elements of F', the Δ-<u>field generated</u> by s over F is defined in a similar way; it is denoted by $F\langle s\rangle$ and is called, also, the <u>extension</u> of F generated by s. We say that s is a Δ-<u>algebraically dependent</u> over R if the family $\Theta s = (\theta s_j)_{\theta \in \Theta, j \in J}$ is algebraically dependent over R; in the contrary case, we say that s is Δ-<u>algebraically independent</u> over R or that s is a family of Δ-<u>interdeterminates</u> over R. For any set J, there exists a family $(y_j)_{j \in J}$ of Δ-indeterminates; the elements of $R\{(y_j)_{j \in J}\}$ are called Δ-<u>polynomials</u> in $(y_j)_{j \in J}$ over R.

Let G be an extension of the Δ-field F and $(y_1,\ldots,y_n)$ be a finite family of Δ-indeterminates. If $u = (u_1,\ldots,u_n)$ is a family of elements of G, there exists a unique homomorphism

$$F\{y_1,\ldots,y_n\} \longrightarrow G$$

over F with $y_j \rightarrow u_j\ (1 \leq j \leq n)$; for each Δ-polynomial $P \in F\{y_1,\ldots,y_n\}$, the image of P is denoted by $P(u)$. When $P(u) = 0$, we say that P <u>vanishes</u> at u; the kernel $\mathfrak{p}$ of the homomorphism, which is called the <u>defining</u> Δ-<u>ideal</u> of u over F, is a prime Δ-ideal of $F\{y_1,\ldots,y_n\}$. If Σ is a subset of $\mathfrak{p}$, we say that u is a solution of the system of Δ-equations

$$P = 0\ (P \in \Sigma)$$

or that u is a <u>zero</u> of Σ. If also $u' = (u'_1,\ldots,u'_n)$ is a family of elements of an extension of F, and if $\mathfrak{p}'$ is its defining Δ-ideal over F, the following two conditions are equivalent to each other:

1) $\mathfrak{p} \subset \mathfrak{p}'$;

2) there exists a surjective homomorphism $\mathcal{F}\{u\} \to \mathcal{F}\{u'\}$ over $\mathcal{F}$ with $u_j \mapsto u'_j$ $(1 \leq j \leq n)$.

When these conditions are satisfied we say that u' is a Δ-<u>specialization</u> of u over $\mathcal{F}$ and we write $u \to u'$ (or $u \xrightarrow[\mathcal{F}]{} u'$ or $u \xrightarrow{\Delta} u'$ or $u \xrightarrow[\mathcal{F}]{\Delta} u'$, when necessary). When $\mathfrak{p} = \mathfrak{p}'$, the above homomorphism is an isomorphism and hence extends to an isomorphism $\mathcal{F}\langle u\rangle \to \mathcal{F}\langle u'\rangle$, we then say that the Δ-specialization $u \to u'$ is <u>generic</u> and write $u \longleftrightarrow u'$.

It will be convenient, in discussing Δ-equations over $\mathcal{F}$, to have an extension $\mathcal{U}$ of $\mathcal{F}$ so large that the solutions can always be taken in $\mathcal{U}$. More precisely, we require that for any finitely generated extension $\mathcal{F}'$ of $\mathcal{F}$ in $\mathcal{U}$ and any finitely generated extensions $\mathcal{G}$ of $\mathcal{F}'$ whatsoever, there exist an embedding of $\mathcal{G}$ into $\mathcal{U}$ over $\mathcal{F}'$. Such an extension $\mathcal{U}$ of $\mathcal{F}$ is said to be <u>universal</u> (over $\mathcal{F}$). That $\mathcal{F}$ always has a universal extension is a grand exercise in Zorn's lemma [6].

In what follows, <u>we fix a universal extension</u> $\mathcal{U}$ <u>of</u> $\mathcal{F}$. <u>All Δ-fields that we discuss, except those for which the contrary is stated or is obvious, will be Δ-subfields of</u> $\mathcal{U}$ <u>over which</u> $\mathcal{U}$ <u>is universal</u>. The field of <u>constants</u> of $\mathcal{U}$ (the set of elements $u \in \mathcal{U}$ such that $\delta_i u = 0$ $(1 \leq i \leq m)$) will be denoted by $\mathcal{K}$. It is obvious that $\mathcal{U}$ is not universal over $\mathcal{K}$.

Δ- $\mathcal{F}$-<u>groups</u>

If G is any one of the Δ-algebraic groups described above, or is any other "concrete" Δ-algebraic group defined over the Δ-field $\mathcal{F}$, we have the following "Δ- $\mathcal{F}$-data" on G:

for each $x \in G$, the extension $\mathcal{F}\langle x\rangle$;

the binary relation $x \to x'$ on G of Δ-specialization over $\mathcal{F}$;

for each $(x,x') \in G^2$ with $x \leftrightarrow x'$, the isomorphism

$$s_{x',x}\colon \mathcal{F}\langle x\rangle \to \mathcal{F}\langle x'\rangle.$$

The group G, the extensions $\mathcal{F}\langle x\rangle$, the relation $x \rightarrow x'$, and the isomorphisms $S_{x',x}$ have certain formal properties. We use these formal properties as axioms for an abstract definition of Δ-$\mathcal{F}$-group.

This definition carries with it natural definitions of Δ-$\mathcal{F}$-subgroup and Δ-$\mathcal{F}$-homomorphism of Δ-$\mathcal{F}$-groups. The latter is a mapping $f: G \rightarrow G'$ between Δ-$\mathcal{F}$-groups that is a group homomorphism of their underlying groups and satisfies the conditions

$$\mathcal{F}\langle x\rangle \supset \mathcal{F}\langle f(x)\rangle,$$

$$x \rightarrow x' \implies f(x) \rightarrow f(x'),$$

$$x \leftrightarrow x' \implies S_{x',x} \text{ extends } S_{f(x'),f(x)}.$$

Homogeneous Δ-$\mathcal{F}$-spaces

For any group G, there is the notion of homogeneous space for G; this is a set M together with a transitive action of G on M, that is, a mapping $M \times G \rightarrow M$, for which we usually use the notation $(v,x) \mapsto vx$, that satisfies the identities

$$v(x_1x_2) = (vx_1)x_2, \quad v1 = v, \quad vG = M.$$

When G is a Δ-$\mathcal{F}$-group, we define, by axioms analogous to some of the axioms for Δ-$\mathcal{F}$-group, a companion notion of homogeneous Δ-$\mathcal{F}$-space for G and a corresponding notion of Δ-$\mathcal{F}$-homomorphism of homogeneous Δ-$\mathcal{F}$-spaces for G. When the action of G on the homogeneous Δ-$\mathcal{F}$-space M for G is simply transitive and some further axioms are satisfied, M is said to be a principal homogeneous Δ-$\mathcal{F}$-space for G. G itself is a principal homogeneous Δ-$\mathcal{F}$-space for G. An interesting problem, addressed later, is that of classifying, up to Δ-$\mathcal{F}$-isomorphism, the principal homogeneous Δ-$\mathcal{F}$-spaces for a given Δ-$\mathcal{F}$-group G.

$\mathcal{F}$-components

A subset V of a homogeneous Δ-$\mathcal{F}$-space M consisting of all the Δ-specialization over $\mathcal{F}$ of a fixed $v \in M$ is said to be $\mathcal{F}$-irreducible, and v is said to be an $\mathcal{F}$-generic element of V. The union of finitely many $\mathcal{F}$-irreducible subsets of M is said to be a Δ-$\mathcal{F}$-set; when none of these $\mathcal{F}$-irreducible sets is contained in another, they are unique, being the maximal $\mathcal{F}$-irreducible subsets of the Δ-$\mathcal{F}$-set (called its $\mathcal{F}$-components). One of the early results is that M itself is an F-set of which the F-components are pairwise disjoint. Since M can be G, it follows that G has a unique F-component, which we denote by $G°$, that contains the neutral element 1 of G. This $G°$ is a normal Δ-F-subgroup of G of finite index.

Varying $\mathcal{U}$ or F or Δ

In building the general theory there are a huge number of constructions to make and propositions to prove. Some of these are obvious for concrete Δ-F-groups, but require substantial effort in general.

For example, consider an extension $\mathcal{U}_0$ of F in $\mathcal{U}$, over which $\mathcal{U}$ need not be universal, such that $\mathcal{U}_0$ is universal over F. For any Δ-F-group G relative to $\mathcal{U}$, abstract or concrete, the set $G_{\mathcal{U}_0}$, consisting of the elements $x \in G$ with $F\langle x\rangle \subset \mathcal{U}_0$, has an easily described natural structure of Δ-F-group relative to $\mathcal{U}_0$. Conversely, if G_0 is a Δ-F-group relative to $\mathcal{U}_0$, we can associate to G_0, in a canonical way, a Δ-F-group G relative to $\mathcal{U}$ such that $G_{\mathcal{U}_0} = G_0$. In the concrete case this is evident, but in general, for axiomatically defined Δ-F-groups, this must be established.

Similarly, consider an extension $\mathcal{G}$ of F. A concrete Δ-F-group is obviously a Δ-$\mathcal{G}$-group (every algebraic Δ-equation that has coefficients in F has coefficients in $\mathcal{G}$). For an abstract Δ-F-group this must be proved.

Finally, consider the set $F\Delta$ consisting of the derivation operators $\Sigma a_i\delta_i$ with $a_1,\ldots,a_m \in F$; this has obvious structures of vector space over F and of Lie ring. Let Δ' be a linearly independent subset of $F\Delta$ the elements of which commute with each other. Every Δ-field may be regarded as a Δ'-field, and it is easy to see that, as Δ'-fields, $\mathcal{U}$ is universal over F. A concrete Δ'-F-group (relative to $\mathcal{U}$) is obviously a Δ-F-group. For abstract Δ'-F-groups the result is still true, but it is far from obvious. (In the extreme case in which Δ' is empty, the result states that every F-group is a Δ-F-group.)

The proof, for each of these results, consists in specifying the basic data (the extensions, the binary relation, the isomorphisms), verifying the axioms, and showing that the new structure bears the desired relation to the given structure. This can be a long sometimes boring process, but is always elementary. Of course, all these results extend to homogeneous Δ-$\mathcal{F}$-spaces.

Quotients and products

Some results that are difficult in the concrete case benefit from the axiomatic approach. For example, if H is a normal Δ-$\mathcal{F}$-subgroup of the Δ-$\mathcal{F}$-group G, how do we make G/H a Δ-$\mathcal{F}$-group? The method in broad terms is the same: specify the Δ-$\mathcal{F}$-data on the group G/H, verify the axioms, and show that the canonical group homomorphism $\pi: G \longrightarrow G/H$ is a Δ-$\mathcal{F}$-homomorphism with the correct universal mapping property. When H is not normal, the method exhibits G/H as a homogeneous Δ-$\mathcal{F}$-space for G. Similarly, if G_1 and G_2 are Δ-$\mathcal{F}$-groups, we can define on $G_1 \times G_2$ a structure of Δ-$\mathcal{F}$-group that deserves to be called the direct product of the Δ-$\mathcal{F}$-groups, and when M_i is a homogeneous Δ-$\mathcal{F}$-space for G_i $(i = 1,2)$, we can define on $M_1 \times M_2$ a structure of homogeneous Δ-$\mathcal{F}$-space for $G_1 \times G_2$.

The Δ-Zariski topologies

As in algebraic geometry, it is possible and useful to introduce into our subject the language of topology. Let M be a homogeneous Δ-$\mathcal{F}$-space for a Δ-$\mathcal{F}$-group. As stated above, for any extension $\mathcal{G}$ of $\mathcal{F}$, M is a homogeneous Δ-$\mathcal{G}$-space, too, and therefore it makes sense to speak of Δ-$\mathcal{G}$-subsets of M. If $\mathcal{H}$ is an extension of $\mathcal{F}$ in $\mathcal{U}$ over which $\mathcal{U}$ is not universal, we define a Δ-$\mathcal{H}$-subset of M to be a Δ-$\mathcal{G}$-subset of M for some extension $\mathcal{G}$ of $\mathcal{F}$ with $\mathcal{G} \subset \mathcal{H}$. It turns out that the Δ-$\mathcal{H}$-subsets of M are the closed sets of a topology on M; we call it the Δ-Zariski topology relative to $\mathcal{H}$ (or just the Δ-$\mathcal{H}$-topology) on M. The finest of these topologies is the Δ-$\mathcal{U}$-topology which we usually call the Δ-Zariski topology (or Δ-topology). As in algebraic geometry, the Δ-topology is Noetherian, that is, every strictly decreasing sequence of Δ-closed subsets of M is finite. This is, at bottom, a consequence of the Ritt-Raudenbush basis theorem, according to which the perfect Δ-ideals of a Δ-polynomial ring $\mathcal{H}\{y_1,\ldots,y_n\}$ satisfy the ascending chain condition.

Δ-rational mappings

Further development of the theory requires a good definition of Δ-$\mathcal{F}$-mapping of a Δ-$\mathcal{F}$-set A into a Δ-$\mathcal{F}$-set B. In the concrete case and with A $\mathcal{F}$-irreducible this means a mapping f of a nonempty Δ-$\mathcal{F}$-open subset of A into B such that, for any $\mathcal{F}$-generic element x of A, the coordinates of $f(x)$ are expressible rationally over $\mathcal{F}$ in terms of the coordinates of x and their derivatives of various orders, f being defined at an element x' of A if the rational expressions can be chosen with the denominators not vanishing at x'. In general, we use a fairly complicated definition in terms of the Δ-$\mathcal{F}_a$-data of the homogeneous Δ-$\mathcal{F}$-spaces containing A and B and of their respective Δ-$\mathcal{F}$-groups (where $\mathcal{F}_a$ denotes the algebraic closure of $\mathcal{F}$).

For any extension $\mathcal{G}$ of $\mathcal{F}$, it turns out that every Δ-$\mathcal{F}$-mapping of A into B is a Δ-$\mathcal{G}$-mapping. If $\mathcal{H}$ is an extension of $\mathcal{F}$ in $\mathcal{U}$ with $\mathcal{U}$ not universal over $\mathcal{H}$, we define a Δ-$\mathcal{H}$-mapping of A into B to be a Δ-$\mathcal{G}$-mapping for some extension $\mathcal{G}$ of $\mathcal{F}$ with $\mathcal{G} \subset \mathcal{H}$; such a $\mathcal{G}$ can always be taken finitely generated over $\mathcal{F}$. Every Δ-$\mathcal{F}$-homomorphism of Δ-$\mathcal{F}$-groups or of homogeneous Δ-$\mathcal{F}$-spaces is a Δ-$\mathcal{F}$-mapping, as are the group law $\mu_G\colon G \times G \to G$ of a Δ-$\mathcal{F}$-group G given by $(x,y) \mapsto xy$ and its group symmetry $\iota_G\colon G \to G$ given by $x \mapsto x^{-1}$; so are the action $\mu_M\colon M \times G \to M$ of a homogeneous Δ-$\mathcal{F}$-space M for G given by $(v,x) \mapsto vx$ and, when M is principal, the mapping $\psi_M\colon M \times M \to G$ given by $(v,w) \mapsto v^{-1}w$.

If f is a Δ-$\mathcal{F}$-mapping of A into B, then f is Δ-$\mathcal{F}$-continuous (that is, is continuous relative to the Δ-$\mathcal{F}$-topologies), and the domain of definition of f is a Δ-$\mathcal{F}$-open Δ-$\mathcal{F}$-dense subset of A. If also g is a Δ-$\mathcal{F}$-mapping of B into a Δ-$\mathcal{F}$-set C such that g is defined at $f(u)$ whenever u is an $\mathcal{F}$-generic element of A, then there exists a unique Δ-$\mathcal{F}$-mapping h of A into C such that $h(u) = g(f(u))$ for all such u. We call this h the <u>generic composite</u> of g and f, and denote it by $g \,\square\, f$. When there exists a Δ-$\mathcal{F}$-mapping f' of B into A such that $f' \square f$ exists and equals id_A and $f \,\square\, f'$ exists and equals id_B, we say that f is <u>generically invertible</u> and that f' is its generic inverse; when this is the case, and $u \in A$ has the property that f is defined at u and f' is defined at $f(u)$, we say that f is <u>bidefined</u> at u; the set of all such u is a Δ-$\mathcal{F}$-open Δ-$\mathcal{F}$-dense subset of A called the <u>domain of bidefinition of</u> f.

Δ-rational functions

For Δ-$\mathcal{H}$-mappings of the Δ-$\mathcal{H}$-set A into G_a we have special terminology and notation. Such a Δ-$\mathcal{H}$-mapping is called a Δ-$\mathcal{H}$-function on A; the set of all Δ-$\mathcal{H}$-functions on A is denoted by $\mathfrak{F}_{\mathcal{H}}(A)$, and for $\mathfrak{F}_{\mathcal{U}}(A)$ we generally write just $\mathfrak{F}(A)$. The set of elements of $\mathfrak{F}_{\mathcal{H}}(A)$ resp. $\mathfrak{F}(A)$ that are defined at a given element $u \in A$ is denoted by $\mathfrak{F}_{\mathcal{H},u}(A)$ resp. $\mathfrak{F}_u(A)$. $\mathfrak{F}(A)$ has a natural structure of Δ-ring (indeed, of Δ-algebra over $\mathcal{U}$), of which $\mathfrak{F}_{\mathcal{F}}(A)$ and $\mathfrak{F}_u(A)$ are Δ-subrings. When A is $\mathcal{F}$-irreducible, then $\mathfrak{F}_{\mathcal{F}}(A)$ is a Δ-field (an extension of $\mathcal{F}$, not in $\mathcal{U}$, of course). When A is irreducible (that is, is $\mathcal{G}$-irreducible for every extension $\mathcal{G}$ of $\mathcal{F}$ or, what is equivalent, is $\mathcal{F}_a$-irreducible), then $\mathfrak{F}(A)$ is an extension of $\mathcal{U}$; in this case, $\mathcal{U}$ and $\mathfrak{F}_{\mathcal{F}}(A)$ are linearly disjoint over $\mathcal{F}$ and $\mathcal{U}.\mathfrak{F}_{\mathcal{F}}(A) = \mathfrak{F}(A)$, and for any $u \in A$, $\mathfrak{F}_u(A)$ is a local Δ-ring with field of quotients $\mathfrak{F}(A)$.

Now consider a Δ-$\mathcal{F}$-mapping of A into a Δ-$\mathcal{F}$-set B. If f is generically surjective (that is, f maps the set of $\mathcal{F}$-generic elements of A onto the set of $\mathcal{F}$-generic elements of B) then $\psi \circ f$ exists for every $\psi \in \mathfrak{F}(B)$ and hence f induces an injective mapping $f^*: \mathfrak{F}(B) \to \mathfrak{F}(A)$. This f^* is a homomorphism of Δ-rings over $\mathcal{U}$. If f is not generically surjective, this is no longer the case, but when $u \in A$ and f is defined at u, then f induces a homorphism $f_u^*: \mathfrak{F}_{f(u)}(B) \to \mathfrak{F}_u(A)$.

Principal homogeneous Δ-$\mathcal{F}$-spaces and constrained cohomology

One of the interesting problems in the theory of Δ-$\mathcal{F}$-groups is that of classifying, up to Δ-$\mathcal{F}$-isomorphism, the principal homogeneous Δ-$\mathcal{F}$-spaces for a given Δ-$\mathcal{F}$-group G.

Recall that G itself can be considered a principal homogeneous Δ-$\mathcal{F}$-space for G. Since G has an element rational over $\mathcal{F}$, namely the element $1 \in G$, every principal homogeneous Δ-$\mathcal{F}$-space for G that is Δ-$\mathcal{F}$-isomorphic to G has an element rational over $\mathcal{F}$. Conversely, if a principal homogeneous Δ-$\mathcal{F}$-space M for G has an element u with $\mathcal{F}\langle u\rangle = \mathcal{F}$, then the formula $x \mapsto ux \, (x \in G)$ defines a Δ-$\mathcal{F}$-isomorphism of G onto M.

Again, consider the set L_Λ of solutions of the Δ-equation

$$\Lambda y = 0$$

corresponding to a nonzero linear Δ-operator $\Lambda = \Sigma a_\theta \theta$ (where the sum extends over a finite set of derivative operators $\theta \epsilon \Theta$ and $a_\theta \epsilon \mathcal{F}$ for each θ); L_Λ is a Δ-$\mathcal{F}$-subgroup of G_a. For any element $a \epsilon \mathcal{F}$, the set $L_{\Lambda,a}$ of solutions of the Δ-equation

$$\Lambda y = a$$

has a natural structure of principal homogeneous Δ-$\mathcal{F}$-space for L_Λ. We can ask two questions.

(1) If $a,b \epsilon \mathcal{F}$, when are $L_{\Lambda,a}$, $L_{\Lambda,b}$ Δ-$\mathcal{F}$-isomorphic?

(2) Does there exist a principal homogeneous Δ-$\mathcal{F}$-space for L_Λ that is not Δ-$\mathcal{F}$-isomorphic to any $L_{\Lambda,a}$?

The answer to question (1) is easy: $L_{\Lambda,a}$ is Δ-$\mathcal{F}$-isomorphic to $L_{\Lambda,b}$ if and only if the equation $\Lambda y = b - a$ has a solution in $\mathcal{F}$, that is, $a \equiv b (\text{mod } \Lambda\mathcal{F})$. In other words, the set of Δ-$\mathcal{F}$-isomorphism classes of the $L_{\Lambda,a}$ with $a \epsilon \mathcal{F}$ is in bijective correspondence with $\mathcal{F}/\Lambda\mathcal{F}$. Thus these Δ-$\mathcal{F}$-isomorphism classes have something to do with the possibility of solving certain Δ-equations <u>in</u> $\mathcal{F}$. The answer to question 2 is less immediate.

We shall ultimately see, as a result of a general classification theorem, that the answer is negative. The theorem will classify the set $P(\mathcal{F},G)$ of Δ-$\mathcal{F}$-isomorphism classes of principal homogeneous Δ-$\mathcal{F}$-spaces for a Δ-$\mathcal{F}$-group G in terms of a new kind of cohomology set $H^1_\Delta(\mathcal{F},G)$. This <u>constrained</u> cohomology set is somewhat analogous to the Galois cohomology set $H^1(K,G)$ of a field K in a K-group G (see e.g. [1]). The constrained cohomology is based on the notion of <u>constrained closure</u> of a Δ-field, a notion in differential algebra originated, curiously enough, by logicians, namely the late Abraham Robinson and his followers, Lenore Blum and Saharon Shelah (who used the term "differential closure"). A constrained closure of a Δ-field is somewhat analogous to an algebraic closure of a field, and has many of the same properties (see [7].) In particular, $\mathcal{F}$ has a constrained closure in $\mathcal{U}$, unique up to a Δ-isomorphism over $\mathcal{F}$; however,

$\mathcal{U}$ may contain more than one constrained closure of $\mathcal{F}$, indeed, one can very well contain another. Also, for any Δ-$\mathcal{F}$-set B, the set of elements of B that are rational over a given constrained closure of $\mathcal{F}$ is dense in B relative to the Δ-topology. In what follows, we fix a constrained closure $\mathcal{F}^+$ of $\mathcal{F}$ in $\mathcal{U}$ (over which $\mathcal{U}$ need not be universal), and put $A = \mathrm{Aut}(\mathcal{F}^+/\mathcal{F})$.

For any finitely generated extension $\mathcal{G}$ of $\mathcal{F}$ in $\mathcal{F}^+$, the set $\mathrm{Iso}(\mathcal{G}/\mathcal{F})$ of all isomorphisms of $\mathcal{G}$ over $\mathcal{F}$ onto an extension of $\mathcal{F}$ in $\mathcal{U}$ has a natural structure of Δ-$\mathcal{F}$-set. If $\mathcal{H}$ is another such extension, with $\mathcal{G} \subset \mathcal{H}$, we have the obvious mappings

$$r: \mathrm{Iso}(\mathcal{H}/\mathcal{F}) \twoheadrightarrow \mathrm{Iso}(\mathcal{G}/\mathcal{F}), \qquad i: \mathrm{Iso}(\mathcal{H}/\mathcal{G}) \longrightarrow \mathrm{Iso}(\mathcal{H}/\mathcal{F})$$

of restriction and inclusion. It turns out that r is a Δ-$\mathcal{F}$-mapping and i is a Δ-$\mathcal{G}$-mapping. Similarly, we have the restriction mapping

$$A \longrightarrow \mathrm{Iso}(\mathcal{G}/\mathcal{F}),$$

the image of which is the set $\mathrm{Iso}_{\mathcal{F}^+}(\mathcal{G}/\mathcal{F})$ consisting of the elements of $\mathrm{Iso}(\mathcal{G}/\mathcal{F})$ that are rational over $\mathcal{F}^+$, and the diagram

$$\begin{array}{ccc} & A & \\ \swarrow & & \searrow \\ \mathrm{Iso}(\mathcal{H}/\mathcal{F}) & \xrightarrow{\;r\;} & (\mathrm{Iso}(\mathcal{G}/\mathcal{F}) \end{array}$$

is commutative. It follows that A may be regarded as a projective limit

$$A = \varprojlim_{\mathcal{G}} \mathrm{Iso}_{\mathcal{F}^+}(\mathcal{G}/\mathcal{F})$$

A <u>constrained</u> 1-<u>cocycle</u> of $\mathcal{F}$ in the Δ-$\mathcal{F}$-group G is defined to be a mapping $\Phi: A \to G$ such that:

(a) There exist a finitely generated extension $\mathcal{G}$ of $\mathcal{F}$ and an everywhere defined Δ-$\mathcal{G}$-mapping $\Phi_{\mathcal{G}}: \mathrm{Iso}(\mathcal{G}/\mathcal{F}) \to G$ such that $\Phi(\sigma) = \Phi(\sigma|\mathcal{G})$ for every $\sigma \in A$.

(b) $\Phi(\sigma\tau) = \Phi(\sigma)\cdot\sigma\Phi(\tau)$ for all $\sigma, \tau \in A$.

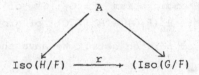

The subsequent definitions are as expected. Letting $Z^1_\Delta(F,G)$ denote the set of all constrained 1-cocycles of F in G, we say that two such cocycles Φ, Ψ are <u>cohomologous</u> if there exists an $x \in G_{F+}$ such that $\Psi(\sigma) = x^{-1}\Phi(\sigma)\sigma x \quad (\sigma \in A)$; this is an equivalence on $Z^1_\Delta(F,G)$. The <u>constrained cohomology set</u> of F in G, denoted by $H^1_\Delta(F,G)$, is the set of equivalence classes. It is a pointed set.

We can now state our classification theorem in the following form.

Theorem. We have a canonical bijection $P(F,G) \to H^1_\Delta(F,G)$.

To describe the bijection, consider any principal homogeneous Δ-F-space M for G. There exists an element $u \in M_{F+}$. The formula $\sigma \to u^{-1}\sigma u$ defines a mapping $\Phi_{M,u}: A \to G$. It is not very hard to show that when N is another principal homogeneous Δ-F-space for G and $v \in N_{F+}$, then N is Δ-F-isomorphic to M if and only if $\Phi_{N,v}$ is cohomologous to $\Phi_{M,u}$. Thus, we have an injection $P(F,G) \to H^1_\Delta(F,G)$. The final step is to prove that, for any $\Phi \in Z^1_\Delta(F,G)$, there exist M, u as above with $\Phi = \Phi_{M,v}$. This part of the proof is very long.

In order to compute $H^1_\Delta(F,G)$ in some cases, it is useful to put $H^0_\Delta(F,G) = G_F$. Every Δ-F-homomorphism $f: G \to G'$ of Δ-F-groups induces homomorphisms f^0: $H^0_\Delta(F,G) \to H^0_\Delta(F,G)$ of groups and $f^1: H^1_\Delta(F,G) \to H^1_\Delta(F,G')$ of pointed sets. We have the following results.

Theorem. Let

$$1 \longrightarrow N \xrightarrow{j} G \xrightarrow{\pi} G' \longrightarrow 1$$

be a short exact sequence of Δ-F-homomorphisms of Δ-F-groups. There exists a connecting homomorphism $\delta: H^0_\Delta(F,G') \to H^1_\Delta(F,G)$ *such that the sequence*

$$1 \longrightarrow H^0_\Delta(F,N) \xrightarrow{j^0} H^0_\Delta(F,G) \xrightarrow{\pi^0} H^0_\Delta(F,G') \xrightarrow{\delta} H^1_\Delta(F,N) \xrightarrow{j^1} H^1_\Delta(F,G) \xrightarrow{\pi^1} H^1_\Delta(F,G')$$

is exact

Theorem. When G is an F-group, then $H^1_\Delta(F,G)$ *is canonically isomorphic to the Galois cohomology set* $H^1_\Delta(F,G)$.

The latter theorem shows, for example, that every principal homogeneous Δ-F-space for $GL(n)$ is Δ-F-isomorphic to $GL(n)$, because it is known that $H^1(F, GL(n)) = 1$. The former theorem shows, when applied to the exact sequence

$$0 \longrightarrow L_\Lambda \longrightarrow G_a \xrightarrow{\Lambda} G_a \longrightarrow 0,$$

that $H^1(F, L_\Lambda) \approx F/\Lambda F$ and every principal homogeneous Δ-F-space for L_Λ is Δ-F-isomorphic to some $L_{\Lambda,a}$.

Another application of constrained cohomology is to the classification of Δ-F-vector spaces over the universal Δ-field $\mathcal{U}$. We are going to see that every such Δ-F-vecotor space is Δ-F-isomorphic to $\mathcal{U}^n$ for some n. More precisely, we have the following result.

<u>Theorem</u>. <u>Let</u> V <u>be a</u> Δ-F-<u>vector space over</u> $\mathcal{U}$. <u>Then the</u> Δ-<u>dimension of</u> V <u>is also its dimension as a vector space over</u> $\mathcal{U}$, <u>and</u> V <u>has a basis that is rational over</u> F.

Indeed, for any finite family $(u_1,\ldots,u_k)$ of elements of V_{F^+} that are linearly independent over $\mathcal{U}$, the formula $(\alpha_1,\ldots,\alpha_k) \mapsto \Sigma\, \alpha_i u_i$ defines an injective Δ-F^+-homomorphism $\mathcal{U}^k \to V$ of Δ-F^+-vector spaces over $\mathcal{U}$. The image is a Δ-F^+-vector subspace of V of Δ-dimension k, so that $k \leqslant \Delta\text{-dim}\, V$. If this image is not V, its complement in V, which is Δ-open, contains an element u_{k+1} of the Δ-dense set V_{F^+}. It follows that V has a basis $u = (u_1,\ldots,u_n)$ with $n = \Delta\text{-dim}\, V$ that is rational over F^+.

For any automorphism $\sigma \in A$, the family $\sigma u = (\sigma u_1, \ldots, \sigma u_n)$ also is a basis of V, rational over F^+, so that there exists a matrix $a(\sigma) \in GL_{F^+}(n)$ such that $\sigma u = u a(\sigma)$. It can be shown that the mapping $a\colon A \to GL(n)$ given by the formula $\sigma \mapsto a(\sigma)$ is an element of $Z^1_\Delta(F, GL(n))$. Since $H^1_\Delta(F, GL(n)) = 1$ (see above), it follows that there exists an element $b \in GL_{F^+}(n)$ such that $a(\sigma) = b^{-1}\sigma b$ $(\sigma \in A)$. Hence, for the basis $v = ub^{-1}$ of V,

$$\sigma v = \sigma u \sigma b^{-1} = u a(\sigma) \sigma b^{-1} = u b^{-1} = v \quad (\sigma \in A).$$

This implies that the basis v of V is rational over F, and completes the proof.

The computation of $H^1_\Delta(F, G)$ in other cases would permit other applications.

Lie Theory

We conclude with a brief indication of how the mechanism of Lie theory can be introduced into our subject.

(a) Other Δ-$\mathcal{F}$-structures.

The first step is to define Δ-$\mathcal{F}$-structures other than groups and their homogeneous spaces. This is fairly straight-forward. A Δ-$\mathcal{F}$-ring, for example, is a ring R, together with Δ-$\mathcal{F}$-group structure on the additive group R, such that the ring multiplication $R \times R \rightarrow R$ is a Δ-$\mathcal{F}$-mapping. Thus, the field K of constants of $\mathcal{U}$ is a Δ-$\mathcal{F}$-ring (indeed, is a Δ-$\mathbb{Q}$-field). For any Δ-$\mathcal{F}$-ring R, a Δ-$\mathcal{F}$-module over R is a module V over the ring R, together with a Δ-$\mathcal{F}$-group structure on the additive group V, such that the external law of composition $V \times R \rightarrow V$ of the module is a Δ-$\mathcal{F}$-mapping. We shall be interested in the case in which R is K, in which case V is called a Δ-$\mathcal{F}$-vector space over K. Finally, we shall need the notion of Δ-$\mathcal{F}$-Lie algebra over K, which is defined as expected.

(b) Tangent spaces.

Consider an irreducible Δ-$\mathcal{F}$-set V and an element $v \in V$. The definition of tangent space is analogous to the definition in algebraic geometry. The Δ-algebra $\mathfrak{F}_v(V)$ over $\mathcal{U}$, consisting of the Δ-rational functions on V that are defined at v, is now (because V is irreducible) a local Δ-ring. A tangent vector to V at v is a local Δ-derivation of $\mathfrak{F}_v(V)$ at v, that is, is a $\mathcal{U}$-linear mapping $T: \mathfrak{F}_v(V) \rightarrow \mathcal{U}$ such that

$$T(\varphi\psi) = T\varphi \cdot \psi(v) + \varphi(v)T\psi \quad (\varphi,\psi \in \mathfrak{F}_v(V)),$$

$$T(\delta\varphi) = \delta(T\varphi) \qquad (\varphi \in \mathfrak{F}_v(V),\ \delta \in \Delta).$$

The set $\mathfrak{T}_v(V)$ of tangent vectors to V at v has an obvious structure of vector space over K. Because it is usually infinite dimensional, we need more structure on it to keep it under control. It turns out that for a "large enough" extension $\mathcal{G}$ of $\mathcal{F}\langle v\rangle$, $\mathfrak{T}_v(V)$ has a natural structure of Δ-$\mathcal{G}$-vector space over K. "Large enough" means that there must exist a generically invertible Δ-$\mathcal{G}$-mapping of V into a Δ-$\mathcal{G}$-subset of some affine space $\mathcal{U}^n$ that is bidefined at v, or as we shall say, that v is Δ-$\mathcal{G}$-affine in V. ("Most" elements of V

are Δ-$\mathcal{F}$-affine in V; there exists a finitely generated extension $\mathcal{G}$ of $\mathcal{F}$ such that every element of V is Δ-$\mathcal{G}$-affine in V.) We call $\mathfrak{T}_v(V)$, with its structure of Δ-$\mathcal{G}$-vector space over $\mathcal{K}$, the <u>tangent space</u> to V at v.

As remarked before, any Δ-$\mathcal{F}$-mapping f of V into an irreducible Δ-$\mathcal{F}$-set V', such that f is defined at v, induces a homomorphism $f_v^*: \mathfrak{F}_{f(v)}(V') \longrightarrow \mathfrak{F}_v(V)$. For any $T \in \mathfrak{T}_v(V)$, the composite mapping $T \circ f_v^*$ is a tangent vector to V' at $f(v)$. Therefore the formula $T \mapsto T \circ f_v^*$ defines a mapping

$$f_v^{**} : \mathfrak{T}_v(V) \to \mathfrak{T}_{f(v)}(V').$$

When v and $f(v)$ are Δ-$\mathcal{G}$-affine in V and V', respectively, then f_v^{**} is a Δ-$\mathcal{G}$-homomorphism of Δ-$\mathcal{G}$-vector spaces over $\mathcal{K}$.

(c) The Lie algebra

The Lie algebra of a connected Δ-$\mathcal{F}$-group G, or more generally of a principal homogeneous Δ-$\mathcal{F}$-space V for G, is defined in terms of the invariant Δ-derivations on V. A Δ-<u>derivation</u> on V is a $\mathcal{U}$-linear mapping $D: \mathfrak{F}(V) \longrightarrow \mathfrak{F}(V)$ such that

$$D(\varphi\psi) = D\varphi\cdot\psi + \varphi D_\psi, \quad D(\delta\varphi) = \delta(D\varphi)$$

for all $\varphi, \psi \in \mathfrak{F}(V)$ and $\delta \in \Delta$. For any $x \in G$, the mapping $\rho_x: V \to V$, given by the formula $v \mapsto vx$, is a Δ-$\mathcal{F}\langle x\rangle$-mapping, and $\rho_{x^{-1}}$ is its inverse; hence ρ_x induces an automorphism $\rho_x{}^*$ of $\mathfrak{F}(V)$ over $\mathcal{U}$. Therefore the mapping $\rho_x{}^{**}(D) = \rho_{x^{-1}}{}^* \circ D \circ \rho_x{}^*$ is again a Δ-derivation on V. D is <u>invariant</u> when $\rho_x{}^{**}(D) = D (x \in G)$. The set $\mathfrak{L}_\Delta(V)$ of invariant Δ-derivations on V has an obvious structure of Lie algebra over $\mathcal{K}$. This tends to be infinite dimensional, but there is a natural structure on $\mathfrak{L}_\Delta(V)$ of Δ-$\mathcal{F}$-Lie algebra over $\mathcal{K}$. The difficulty here is not to specify the Δ-$\mathcal{F}$-data on $\mathfrak{L}_\Delta(V)$, but rather to verify all the appropriate axioms. We call $\mathfrak{L}_\Delta(V)$, with its structure of Δ-$\mathcal{F}$-Lie algebra over $\mathcal{K}$, the <u>Lie algebra</u> of V.

For any $D \in \mathfrak{L}_\Delta(V)$ and $v \in V$, it turns out out that $D\mathfrak{F}_v(V) \subset \mathfrak{F}_v(V)$. Therefore the formula $\varphi \mapsto (D\varphi)(v)$ defines a mapping

$\mathfrak{F}_V(V) \to U$ which is easily seen to be a tangent vector to V at v and which we denote by D_v. The formula $D \mapsto D_v$ thus gives a canonical mapping $\mathcal{L}_\Delta(V) \to \mathcal{T}_v(V)$ which turns out to be an isomorphism of vector spaces over K. It is this isomorphism that allows us to complete the proof that $\mathcal{L}_\Delta(V)$ is a Δ-F-vector space over K, and then to conclude, for any extension G of $F\langle v\rangle$ such that v is Δ-G-affine in V, that the mapping is a Δ-G-isomorphism of Δ-G-vector spaces over K.

Consider a Δ-F-mapping f of V into a principal homogeneous Δ-F-space V', with f defined at v. If G is now large enough so that v is Δ-G-affine in V and $f(v)$ is Δ-G-affine in V', then we see that f induces a Δ-G-homomorphism $f_v^{\#}: \mathcal{L}_\Delta(V) \to \mathcal{L}_\Delta(V')$ of Δ-G-vector spaces over such that the diagram

$$\begin{array}{ccc} \mathcal{L}_\Delta(V) & \xrightarrow{\;f_v^{\#}\;} & \mathcal{L}_\Delta(V') \\ \downarrow & & \downarrow \\ \mathcal{T}_v(V) & \xrightarrow[\;f_v^{**}\;]{} & \mathcal{T}_{f(v)}(V') \end{array}$$

is commutative (the vertical arrows being the canonical ones). The functoriality of the Lie algebra $\mathcal{L}_\Delta(V)$ comes from the following corollary of a result about "crossed" Δ-F-homomorphisms:

<u>If</u> $f: V \to V'$ <u>is a "relative"</u> Δ-F-<u>homomorphism of principal homogeneous</u> Δ-F-<u>spaces, then</u> $f_v^{\#}$ <u>is independent of</u> v (and hence may be denoted by $f^{\#}$) <u>and is a</u> Δ-F-<u>homomorphism of</u> Δ-F-<u>Lie algebras over</u> K.

(d) Logarithmic derivations

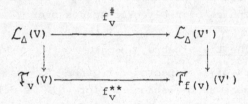

Let V be a principal homogeneous Δ-F-space for a connected Δ-F-group G. Let H be an extension of F with U not necessarily universal over H, and let ε be a Δ-derivation of H into U over F. Put $H_\varepsilon = \operatorname{Ker}\varepsilon$. For any $v \in V_H$, the formula $\varphi \mapsto \varepsilon(\varphi(v))$ defines a mapping $\mathfrak{F}_{H_\varepsilon}(V) \to U$; when v is Δ-H_ε-affine in V, this can be extended to a unique tangent vector to V at v which, by the above, is D_v for a unique $D \in \mathcal{L}_\Delta(V)$. We denote this D by $\ell\varepsilon_V(v)$ or by $\ell\varepsilon(v)$ and call it the <u>logarithmic derivative of</u> v <u>on</u> V <u>relative to</u> ε. The invariant derivation $\ell\varepsilon(v)$ on V is characterized by the condition

$$\ell\varepsilon(v)_v\varphi = \varepsilon(\varphi(v)) \quad (\varphi \in \mathfrak{F}_{H_\varepsilon, V}(V)).$$

Let V_ε denote the set of elements of V_H that are Δ-H_ε-affine in V. The formula $v \to \ell\varepsilon(v)$ defines a mapping $\ell\varepsilon = \ell\varepsilon_V\colon V_\varepsilon \to \mathcal{L}_\Delta(V)$ that we call the <u>logarithmic derivation on</u> V <u>relative to</u> ε. It is easy to see that $\ell\varepsilon(v) = 0$ if and only if $v \in V_{H_\varepsilon}$. The key property of logarithmic derivations is given by the following result.

<u>Theorem</u>. <u>Let the notation be as above, let</u> $v \in V_H$, $x \in G_H$ <u>and suppose that</u> v, x, vx <u>are</u> Δ-H_ε<u>-affine in</u> V, G, V, <u>respectively. Let</u> $\lambda_v\colon G \to V$ <u>denote the</u> Δ-$F\langle v\rangle$<u>-mapping defined by the formula</u> $y \mapsto vy \; (y \in G)$. <u>Then</u>

$$\ell\varepsilon_V(vx) = \ell\varepsilon_V(v) + \lambda_v^{\#}(\ell\varepsilon_G(x))$$

When $V = G$ and G is commutative, this equation reduces to

$$\ell\varepsilon_G(yx) = \ell\varepsilon_G(y) + \ell\varepsilon_G(x).$$

Logarithmic derivations relative to ε are most useful when $H = U$ and U is universal over F as a differential field relative to the set $\Delta(\varepsilon) = \Delta \cup \{\varepsilon\}$ of derivation operators. When that is the case, and ι_G denotes the symmetry mapping $x \mapsto x^{-1}$ of G, then $\ell\varepsilon_G \circ \iota_G$ is a surjective everywhere defined crossed $\Delta(\varepsilon)$-F-homomorphism of G into $\mathcal{L}_\Delta(G)$, and for every Δ-F-subgroup H of G, $\ell\varepsilon_G(H) = \mathrm{in}_{G,H}{}^{\#}(\mathcal{L}_\Delta(H^\circ))$, where $\mathrm{in}_{G,H}$ denotes the inclusion mapping $H \to G$ (which is a Δ-F-homomorphism, of course). Also then ε induces a derivation $\varepsilon^{\#}$ of the Lie algebra $\mathcal{L}_\Delta(G)$ for which

$$\lambda_x^{\#}(\varepsilon^{\#}\lambda_{x^{-1}}{}^{\#}(D)) = \varepsilon^{\#}D + [\ell\varepsilon(x), D] \quad (x \in G,\ D \in \mathcal{L}_\Delta(G)),$$

and when W is a Δ-closed vector subspace of the Δ-F-vector space $\mathcal{L}_\Delta(G)$ over K, a necessary and sufficient condition that W be Δ-U_ε-closed in $\mathcal{L}_\Delta(G)$ is that $\varepsilon^{\#}W \subset W$.

(e) The Lie-Cassidy-Kovacic method.

For the most effective application of logarithmic derivations, we need a Δ-derivation ε of $\mathcal{U}$ over $\mathcal{F}$ such that $\mathcal{U}$ is universal as a $\Delta(\varepsilon)$-field extension of $\mathcal{F}$. The Lie-Cassidy-Kovacic method satisfies this need and makes it possible to use logarithmic derivations to the same end that the logarithmic and exponential maps are used in classical Lie theory.

Fix an element ε of an overset of Δ with $\varepsilon \notin \Delta$ and put $\Delta(\varepsilon) = \Delta \cup \{\varepsilon\}$. The operation of Δ on $\mathcal{U}$ can be extended to one of $\Delta(\varepsilon)$ on $\mathcal{U}$ by putting $\varepsilon u = 0$ $(u \in \mathcal{U})$. This makes $\mathcal{U}$(as well as every Δ-subfield of $\mathcal{U}$) a $\Delta(\varepsilon)$-field. Now fix a universal extension $\mathfrak{U}$ of the $\Delta(\varepsilon)$-field $\mathcal{U}$. Of course, $\mathfrak{U}$ is a universal extension of the Δ-field $\mathcal{U}$, too. Let $\mathfrak{K}$ denote the field of constants of $\mathfrak{U}$ as a Δ-field.

Now, to every Δ-$\mathcal{F}$-group G relative to the universal Δ-field $\mathcal{U}$ there is a canonically associated Δ-$\mathcal{F}$-group $\mathfrak{G}$ relative to the universal Δ-field $\mathfrak{U}$ such that $G = \mathfrak{G}_{\mathcal{U}}$. To a Δ-$\mathcal{F}$-homomorphism $f: G \to H$ of Δ-$\mathcal{F}$-groups there exists a unique Δ-$\mathcal{F}$-homomorphism $\mathfrak{f}: \mathfrak{G} \longrightarrow \mathfrak{H}$ of their canonically associated Δ-$\mathcal{F}$-groups such that $f = \mathfrak{f}|G$. When G and H are connected, then so too are $\mathfrak{G}$ and $\mathfrak{H}$; also $\mathcal{L}_\Delta(G)$ and $\mathcal{L}_\Delta(H)$ can be canonically identified with $\mathcal{L}_{\Delta,\mathcal{U}}(\mathfrak{G})$ and $\mathcal{L}_{\Delta,\mathcal{U}}(\mathfrak{H})$, and then $f^\# = \mathfrak{f}^\#|\mathcal{L}_\Delta(G)$.

This permits us to prove theorems about G and $\mathcal{L}_\Delta(G)$ by proving them instead for $\mathfrak{G}$ and $\mathcal{L}_\Delta(\mathfrak{G})$.

(f) Completing the theory.

There remains the task of forging the connection between properties of a connected Δ-$\mathcal{F}$-group and those of its Lie algebra. Using logarithmic derivations and the Lie-Cassidy-Kovacic method, we can obtain the following result.

Theorem. *Let* $H \xrightarrow{f} G \xrightarrow{g} G'$ *be a sequence of* Δ-$\mathcal{F}$-*homomorphisms of connected* Δ-$\mathcal{F}$-*groups. A necessary and sufficient condition that* $\mathrm{Im}(f) = \mathrm{Ker}(g)^\circ$ *is that the induced sequence*

$$\mathcal{L}_\Delta(H) \xrightarrow{f^\#} \mathcal{L}_\Delta(G) \xrightarrow{g^\#} \mathcal{L}_\Delta(G')$$

be exact.

The Lie algebras of the connected Δ-algebraic subgroups of $\mathcal{F}$ do not immediately reflect the inclusion relations among these subgroups; indeed, the way we have defined them, these Lie algebras are in general not even subsets of $\mathcal{L}_\Delta(G)$. However, for any connected say Δ-G-subgroup H of G, the inclusion mapping $in_{G,H}: H \to G$, which is a Δ-G-homomorphism, induces an injective Δ-G-homomorphism $in_{G,H}^{\#} : \mathcal{L}_\Delta(H) \longrightarrow \mathcal{L}_\Delta(G)$. Its image is a Δ-G-Lie subalgebra of $\mathcal{L}_\Delta(G)$ that we denote by $\mathcal{L}_\Delta^G(H)$ and call the Lie algebra of H in G. Using logarithmic derivations and the Lie-Cassidy-Kovacic method, we can prove the following result.

Let G be a connected Δ-F-group.

(a) The formula $H \to \mathcal{L}_\Delta^G(H)$ defines an injective mapping of the set of connected Δ-closed subgroups of G into the set of Δ-closed Lie subalgebras of $\mathcal{L}_\Delta(G)$.

(b) For any two connected Δ-closed subgroups H_1, H_2 of G,

$$H_1 \subset H_2 \iff \mathcal{L}_\Delta^G(H_1) \subset \mathcal{L}_\Delta^G(H_2).$$

(c) For any family $(H_i)_{i \in I}$ of connected Δ-closed subgroups of G,

$$\mathcal{L}_\Delta^G((\bigcap H_i)^\circ) = \bigcap \mathcal{L}_\Delta^G(H_i).$$

Other links between Δ-F-groups and their Lie algebras require the following partial analog of a result of Chevalley about linear algebraic groups ("Theorie des Groupes de Lie", vol. II, "Groups Algébriques". Hermann, Paris, 1951; see p. 172).

Theorem. Let G be a connected Δ-F-group, and V and V' be Δ-F-vector subspaces of $\mathcal{L}_\Delta(G)$ such that $V' \subset V$. Put

$$H = \{x \in G \mid \tau_x^{\#}(D) - D \in V' \ (D \in V)\},$$

$$L = \{X \in \mathcal{L}_\Delta(G) \mid [X,D] \in V' \ (D \in V)\}.$$

(a) H is a Δ-F-subgroup of G, L is a Δ-F-Lie subalgebra of $\mathcal{L}_\Delta(G)$, and $\mathcal{L}_\Delta^G(H^\circ) \subset L$.

(b) If V' is 0 or V, then $\mathcal{L}_\Delta^G(H^\circ) = L$.

The proof of this theorem uses logarithmic derivations and the Lie-Kovacic-Cassidy method. In the light of Chevalley's result, it is natural to ask whether the extra hypothesis in part (b) is needed. I have not been able to settle this question.

Using this theorem, we can prove a whole sequence of results of which we mention just three (in the statement of which G is a connected Δ-F-group and N is a connected Δ-F-subgroup of G.)

1. N is normal in G if and only if $\mathcal{L}_\Delta^G(N)$ is an ideal of $\mathcal{L}_\Delta(G)$.

2. If N is normal in G and $C_G(N)$ denotes the commutant of N in G, then $\mathcal{L}_\Delta^G(C_G(N)^\circ)$ is the commutant of $\mathcal{L}_\Delta^G(N)$ in $\mathcal{L}_\Delta(G)$. Hence, in particular, if $C(G)$ denotes the center of G, then $\mathcal{L}_\Delta^G(C(G)^\circ)$ is the center of $\mathcal{L}_\Delta(G)$.

3. G is solvable (resp. nilpotent) if and only if $\mathcal{L}_\Delta(G)$ is solvable (resp. nilpotent).

References

1. P.J. Cassidy. Differential algebraic groups, Amer. J. Math. 94 (1972), 891-954.

2. P.J. Cassidy. The differential rational representation algebra on a linear differential algebraic group, J. Algebra 37(1975), 222-238.

3. P.J. Cassidy. Unipotent differential algebraic groups, in "Contributions to Algebra" (H. Bass, P.J. Cassidy, and J. Kovacic, eds.), Academic Press, New York, 1977, pp. 83-115.

4. P.J. Cassidy. Differential algebraic Lie algebras, Trans. Amer. Math. Soc. 247(1979), 247-273.

5. P.J. Cassidy. Differential algebraic group structures on the plane, Proc. Amer. Math. Soc. 80 (1980), 210-214.

6. E.R. Kolchin. "Differential Algebra and Algebraic Groups." Academic Press, New York, 1973.

7. E.R. Kolchin. Constrained extensions of differential fields, Adv. in Math. 12(1974), 141-170.

8. E.R. Kolchin. "Differential Algebraic Groups." Academic Press, Orlando, 1985.

9. J. Kovacic. Constrained cohomology, in "Contributions to Algebra" (H. Bass, P.J. Cassidy, and J. Kovacic, eds.), Academic Press, New York, 1977, pp. 251-266.

10. W.Y. Sit. Differential algebraic subgroups of SL(2) and strong normality in simple extensions, Amer. J. Math. 97 (1975), 627-695.

CONJUGACY CLASSES IN ALGEBRAIC GROUPS

T.A. Springer
Mathematisch Instituut
Budapestlaan 6
Utrecht

1. INTRODUCTION, GENERALITIES

1.1. These lectures contain a review of results about conjugacy classes in linear algebraic groups. In the more recent developments there has been some emphasis on geometric aspects. I shall try to bring out these aspects.
General references for the subject-matter of these lectures are [5], [24], [31].
We assume the reader to be familiar with the theory of linear algebraic groups (exposed in [4], [15], [27]). We shall recall results from that theory as we go along.
We begin by fixing some notation, to be used throughout.

1.2. We denote by k an algebraically closed field, of arbitrary characteristic p, and by G a linear algebraic group over k, with identity component G°. If $x \in G$ we denote by x_s and x_u the semi-simple and unipotent part in the Jordan decomposition $x = x_s x_u$.
B and T will denote a Borel subgroup of G and a maximal torus contained in it. N denoting the normalizer of T in G, the Weyl group of (G,T) is W = N/T, it is a finite group operating on T. If G is reductive the root system $\Phi = \Phi(G,T)$ is a finite subset of $X = \mathrm{Hom}(T, \mathbb{G}_m)$, which is a root system in the sense of [9 ,Ch.VI], in the subspace of the vector space $X \otimes_{\mathbb{Z}} \mathbb{Q}$ spanned by Φ.
The Borel group B defines an order on Φ, we denote by Φ^+ the corresponding set of positive roots and by Δ the basis of Φ defined by Φ^+.
The Lie algebra of G will be denoted by the corresponding lower case gothic letter $\mathfrak{g}$.

If ℓ is a subfield of k and G is defined over ℓ we denote by $G(\ell)$ its group of ℓ-rational points. An important case is that of a finite field ℓ (in which case we usually take k to be an algebraic closure of ℓ). Then $G(\ell)$ is a <u>finite group of Lie type</u>. Another special case is $k = \mathbb{C}$, $\ell = \mathbb{R}$. Assuming G to be connected and reductive, $G(\mathbb{R})$ is an example of a "real reductive Lie group".

1.3. The topic of these lectures is the study of conjugacy classes in a connected reductive linear algebraic group G. Ideally, the purpose of this study should be (a) to parametrize, in some way, the conjugacy classes in G, (b) to describe the structure of the centralizers of elements in G. The same questions can be studied in groups $G(\ell)$.
An example is the case $G = GL_n$, where the theory of Jordan normal forms supplies a parametrization. Also, the answer to (b) is known in that case (the details can be found in [6 ,p.250-251]. See also no. 5.
Nowadays, for any type of simple algebraic group G over an algebraically closed field, a parametrization of conjugacy classes is available, and also a description of centralizers. So, one could say that (a) and (b) have been achieved. However, the answers are partly in the form of tables, and one would really like to have an a priori understanding of the matter, of a less "experimental" kind. I want to concentrate on this sort of understanding.

1.4. Semi-simple and unipotent elements
We first recall a general reduction based on the Jordan decomposition in G. We denote by $C(x) = C_G(x) = \{yxy^{-1} | y \in G\}$ the conjugacy class of x in G and by $Z(x) = Z_G(x)$ the centralizer of x in G.
The following result is an easy consequence of the properties of the Jordan decomposition in G.

Lemma. *Let* $x,y \in G$.
(i) *If* x *is conjugate to* y *then* x_s *is conjugate to* y_s;
(ii) *Let* $x_s = y_s$. *Then* x *is conjugate to* y *if and only if* x_u *is conjugate to* y_u *in* $Z_G(x_s)$;
(iii) $Z_G(x) = Z_{Z(x_s)}(x_u)$.
The lemma shows that the study of conjugacy classes and centralizers splits up into the study of:
conjugacy classes of semi-simple elements,
centralizers of semi-simple elements,
unipotent elements and their centralizers.

1.5. In the study of conjugacy classes in groups of rational points $G(\ell)$ (in the situation of 1.2) the following questions arise:
(a) if C is a conjugacy class in G, give conditions for the intersection $C \cap G(\ell)$ to be non-empty, (b) if $C \cap G(\ell) \neq \emptyset$ it is a union of conjugacy classes in $G(\ell)$, describe these. In other words, describe the "fusion" in G of conjugacy classes in $G(\ell)$.
If ℓ is finite (and k is an algebraic closure $\overline{\ell}$), these questions can

be answered satisfactorily, and we shall describe the answers briefly. Let F be the Frobenius automorphism of $k = \overline{\ell}$ over ℓ (so $Fa = a^{|\ell|}$). Then F operates on G, since G is assumed to be defined over ℓ. The basic result about algebraic groups over finite fields is <u>Lang's theorem</u>, which states that if G is <u>connected</u>, any element of G is of the form $x(Fx)^{-1}$, for some $x \in G$.
Another formulation is as follows. Let $H^1(\ell,G)$ be the set of classes in G for the equivalence relation: $x \sim y$ if $x = zy(Fz)^{-1}$ for some $z \in G$. Then $H^1(\ell,G)$ has one element.
Using Lang's theorem it is not hard to give answers to (a) and (b). They are contained in the following result.

<u>Proposition</u>. (i) *If* G *is connected then* $C \cap G(\ell) \neq \emptyset$ *if and only if* $FC = C$ *i.e. is and only if* C *is defined over* ℓ;
(ii) *Let* $x \in C \cap G(\ell)$ *and put* $Z = Z_G(x)$. *The number of conjugacy classes of* $G(\ell)$ *contained in* $C \cap G(\ell)$ *equals the number of elements of* $H^1(\ell, Z/Z^\circ)$.
For a discussion of these matters we refer to [5 , part E].

For arbitrary ℓ, the situation is more tricky and there are very few results of a general nature.
Part (ii) of the proposition shows that it is desirable to have informations on the groups Z/Z°. In particular, one would like to have criteria for Z to be connected.

2. SEMI-SIMPLE ELEMENTS

2.1. We shall first discuss conjugacy classes of semi-simple elements, and their centralizers. We assume that G is connected and reductive. We need some more notation. If $\alpha \in \Phi$ there exist an isomorphism x_α of the additive group $\mathbb{G}_a \simeq k$ onto a closed subgroup U_α of G such that

$$t x_\alpha(\xi) t^{-1} = x_\alpha(\alpha(t)\xi) \quad (t \in T,\ \xi \in k).$$

The U_α with $\alpha \in \Phi^+$ generate the unipotent part U of B.

Denote by Q the subgroup of the character group X generated by Φ. Then G is semi-simple if and only if Q has finite index in X. This index $i(G)$ is then bounded by a constant (depending only on Φ). G is <u>simply connected</u> if $(X:Q)$ is maximal and <u>adjoint</u> if $X = Q$.

2.2. We first recall some results about semi-simple elements of G.

2.2.1. Lemma. (i) *Any semi-simple element is conjugate to an element in* T;
(ii) *If* $x,y \in T$ *are conjugate in* G *there is* $w \in W$ *with* $y = w.x$.
The proof of (ii) uses the Bruhat decomposition in G. (If $y = gxg^{-1}$, write $g = unu'$ with $u,u' \in U$, $n \in N$, and use that in such a decomposition n is unique.)

So, if C is a semi-simple conjugacy class, the intersection $C \cap T$ is a W-orbit in T. We now define a map $\chi: G \to T/W$ (the set of W-orbits in T) by $\chi(x) = C(x_s) \cap T$.
This is, in fact, a morphism of algebraic varieties. G is an affine algebraic variety by definition, let $k[G]$ be its algebra of regular functions. Also, the orbit space T/W can be viewed as an affine algebraic variety whose algebra $k[T/W]$ is the algebra $k[T]^W$ of W-invariants in the algebra $k[T]$ of our torus T. To prove our claim that χ is a morphism we have to show that there is an algebra homomorphism $\chi^*: k[T]^W \to k[G]$ such that

(1) $(\chi^* f)(x) = f(\chi(x))$ $(f \in k[T]^W, x \in G)$.

To define χ^* we introduce the subalgebra R of $k[G]$ of class functions $f \in k[G]$ (such that $f(xyx^{-1}) = f(y)$ for all $x,y \in G$).

2.2.2. Lemma. *The restriction map* $R \to k[T]$ *defines an isomorphism* $R \simeq k[T]^W$.
The most serious part is the proof that $k[T]^W$ is in the image of the restriction map. For this one uses representation theory (see [31, p. 87]).

Taking now χ^* to be the composite of the restriction map $R \to k[G]$ and the inverse of the isomorphiam of 2.2.2, the property (1) follows by observing that $f(x) = f(x_s)$, for all $f \in R$[loc.cit.,p.89]. The main results about the structure of the algebra $k[T]^W$ is the following one [loc.cit.,p.87].

2.2.3. Proposition. *If* G *is semi-simple and simply connected then* $k[T]^W$ *is isomorphic to a polynomial algebra in* r = rank G *indeterminates*.
So T^W is isomorphic to affine r-space $\mathbb{A}^r$, where r = rank G = dim T.

Example. If $G = SL_n$ (which is semi-simple and simply connected, of rank n-1) identify $\mathbb{A}^{n-1}$ with the space of monic polynomials in one indeterminate t of degree n and constant term 1. Then $\chi: G \to \mathbb{A}^{n-1}$ can be given by $\chi(x) = \det(t+x)$.

We call $\chi: G \to T/W$ the <u>adjoint quotient map</u> or morphism.

<u>Remark</u>. The map $x \mapsto x_s$ is in general not a morphism of G into itself (as one sees already in SL_2).

2.3. <u>Centralizers of semi-simple elements</u>

We have the following information about centralizers of semi-simple elements.

<u>Theorem</u>. *Let* $t \in T$.

(i) *The centralizer* $Z_G(t)$ *is reductive. Its identity component* $Z_G(T)^o$ *is generated by* T *and the* U_α *with* $\alpha(t) = 1$;

(ii) *If* G *is semi-simple and simply connected then* $Z_G(t)$ *is connected.*

One ingredient of the proof is Bruhat's lemma. The proof of (ii) is reduced via Bruhat's lemma to a statement about the action of W on T (see [5 ,p.201-203]).

(ii) is a connectedness result as mentioned at the end of 1.5. It is a useful one for the theory of the finite groups of Lie type $G(\ell)$ of Lie type of 1.5. In fact, if G is as in (ii), defined over the finite field of characteristic p, then (ii) implies that the number of p-regular classes in $G(\ell)$ equals $|\ell|^{\text{rank } G}$ (see [29]).

2.4. <u>Study of the adjoint quotient map</u>

The geometric study of the adjoint quotient map χ can be viewed as the more refined theory of the Jordan decomposition in reductive groups. We discuss here a number of results, following [22]. Notations being as in 2.2, we denote by $\bar{t}$ the image in T/W of $t \in T$. Also, we denote by $V = V_G$ the set of unipotent elements in G. We shall call V_G the <u>unipotent variety</u> of G. The name is justified by the following lemma.

2.4.1. <u>Lemma</u>. $V = \chi^{-1}(\bar{e})$. *Hence* V *is a closed subset of* G.

We have already observed that $f(x) = f(x_s)$ for all $x \in G$, $f \in R$, whence $f(x) = f(e)$ is unipotent. Conversely, if $f(x) = f(e)$ for all $f \in R$ then x is unipotent. (This is obvious if one views G as a closed subgroup of some GL_n.)

If $t \in T$ we put $V(t) = V_{Z(t)}$. It is known that $V_{Z(t)^o} = V_{Z(t)}$ (i.e. any unipotent element commuting with t lies in the identity component of $Z(t)$).

2.4.2. Lemma. *There is a G-equivariant isomorphism* $\alpha_t: G\times^{Z(t)}V(t) \tilde{\to} \chi^{-1}(\bar{t})$. Here the fibre product $G\times^{Z(t)}V(t)$ is the quotient of $G\times V(t)$ by the $Z(t)$-action $z.(g,v) = (gz^{-1},zv)$. G acts on the left-hand side via translations in the first factor and by inner automorphisms on the right-hand side. The quotient exists by general theorems (see [4 , §6]). It is a fibre space over $G/Z(t)$, in the sense of algebraic geometry. Let $g*v$ be the image of (g,v) under the quotient map. Then $\alpha_t(g*v) = gtvg^{-1}$ is as required [22,p.30].

The preceding lemma shows that $\chi^{-1}(\bar{t})$ is "built up" from $G/Z(t)$ and $V(t)$. One extreme case is that t is regular, in which case $Z(t)^{\circ} = T$ and $V(t) = e$. The other extreme is $t = e$, $V(t) = V$. We shall say that $x \in G$ is regular if $\dim Z(x) = r$ (= dim T).

The main results about the adjoint quotient map are given in the following theorem (see [22,p.31], where G is assumed to be semi-simple, however). We put $G_t = \chi^{-1}(\bar{t})$ $(t \in T)$. We denote by G' the derived group of G which is connected, semi-simple.

2.4.3. Theorem. (i) G_t *is irreducible, of dimension* dim G-r;
(ii) G_t *is a union of finitely many conjugacy classes;*
(iii) G_t *contains exactly one class of semi-simple elements, viz.* C(t). *It is the only closed class in* G_t, *and it is contained in the closure of any other class in* G_t;
(iv) G_t *contains exactly one class of regular elements. It is open and dense in* G_t *and its completement has codimension* ≥ 2;
(v) *If* G' *is simply connected and* char k *does not divide* i(G') *then* χ *is a flat morphism and the fibers* G_t *are normal varieties;*
(vi) *If* char k *does not divide* i(G') *then the smooth points of* G_t *are precisely the regular elements.*
By the previous lemma, (i)-(iv) have only to be proved for $V = G_e$. Then (ii) is the result that there are only finitely many unipotent conjugacy classes. This is fairly easy if one imposes some mild conditions on p = char k (which gives, in particular, the case p = 0). In all generality this finiteness result is quite difficult. We shall come back to these matters in 3.1.
The proof of the statements (i) and (iii) (for V) is rather easy. (iv) comes from a detailed study of regular unipotent elements, made by Steinberg [30]. The statements (v) and (vi) are also proved via the results

of that study. They are discussed in [22,p.32] in the case that G is semi-simple, i.e. $G = G'$. In the general case we have, S denoting the identity component of the center of G, a homomorphism of algebraic groups $G' \times S \to G$, whose kernel is a finite group contained in the center, of order prime to p. (v) and (vi) for G are then easily deduced from the same statements for G'.

As a consequence of part (iii) of the theorem we have the following geometric description of semi-simple elements.

2.4.4. Corollary. $x \in G$ *is semi-simple if and only if its conjugacy class* $C(x)$ *is closed.*

(vi) implies the following characterization of the smooth fibers G_t.

2.4.5. Corollary. G_t *is smooth if and only if* t *is regular.*

The set of $\bar{t} \in T/W$ such that G_t is non-smooth is the discriminant of χ, which is a hypersurface in T/W.
In the example $G = SL_n$ (see 2.2) the discriminant is indeed given by the vanishing of a discriminant polynomial, as is readily seen.

2.5. The Lie algebra case
There are similar results for the Lie algebra $\mathfrak{g}$ of G. Here one has an additive decomposition $x = x_s + x_n$. Again, the orbit of x_s (for the adjoint action of G) intersects the Lie algebra $\mathfrak{t}$ of T in a W-orbit, whence an adjoint quotient map $\chi: \mathfrak{g} \to \mathfrak{t}/W$. The preceding results, in particular those of theorem 2.4.3, carry over, possibly under some stronger restrictions on p. These matters are discussed in [22,p.36-42], to which we refer for further details. As a complement, it should be added that the analogue of the finiteness statement (ii) is always true. This follows from the result, proved more recently (see [14]), that the number of nilpotent orbits in $\mathfrak{g}$ is finite. Over $\mathbb{C}$ the results like 2.4.3 in the Lie algebra case were proved already by Kostant (see [17]).

We have to mention a result which connects unipotent elements in G with nilpotent elements in $\mathfrak{g}$. We denote by V the unipotent variety of G, and by N the nilpotent variety of .
If p is a prime number we say that p is good for the root system Φ if the following holds: (a) Φ is irreducible. Then p is arbitrary in type

A_ℓ, $p \neq 2$ in types B_ℓ, C_ℓ, D_ℓ; $p \neq 2,3$ in types G_2, F_4, E_6, E_7, $p \neq 2,3,5$ in type E_8. (b) Φ is arbitrary. Then p is required to be good for all irreducible components of Φ.
We say that the characteristic p of k is good for G, if either $p = 0$ or p is good for Φ. We say that p is <u>very good</u> for G, if it is good and if p does not divide the integers n+1 such that the root system Φ has an irreducible component of type A_n.

2.5.1. <u>Theorem</u>. *Let* G *be semi-simple and simply connected. Assume* p *to be good for* G. *There is a* G*-equivariant isomorphism* $\varphi: V \to N$.
See [25] (a better proof is given in [3]).

2.6. <u>Adjoint quotient map for parabolic subgroups</u>

Let G,T,B be as before and let P be a parabolic subgroup of G containing B. We denote by U_P the unipotent radical of P and by L the Levi subgroup of P containing T. Then P is the semi-direct product of L and U_P (in the sense of algebraic groups) and L is connected, reductive. We denote by W_L the Weyl group of L relative to T. So we have an adjoint quotient map

$$\chi_L : L \to T/W_L .$$

We denote by χ_P the composite map $P \to P/U_P \tilde{\to} L \overset{\chi_L}{\to} T/W_L$
Define $W_P = N_P(T)/T$. It is easily seen that $W_P \simeq W_L$. Thus we have defined a morphism $\chi_P: P \to T/W_P$. Since W_P is a subgroup of W, we have a canonical projection $\psi_P: T/W_P \to T/W$.

2.6.1. <u>Lemma</u>. *The following diagram commutes.*

$$\begin{array}{ccc} P & \hookrightarrow & G \\ {\scriptstyle \chi_P}\downarrow & & \downarrow{\scriptstyle \chi} \\ T/W_P & \xrightarrow{\psi_P} & T/W \end{array}$$

This is tantamount to the following statement: if $x \in L$, $y \in U_P$ then $(xy)_s$ is conjugate in G to x_s. Now $(xy)_s$ is conjugate in P to an element $t \in T$ and by projection onto L one sees that t is conjugate in L to x_s, whence the statement.

Next consider the fibre product $G \times^P P$, the quotient of $G \times P$ by the P-action $x(g,y) = (gx^{-1}, xyx^{-1})$. We denote by $g * y$ the image in $G \times^P P$ of (g,y). Observing that χ_P is constant on conjugacy classes of P, we can define a morphism $\theta_P: G \times^P P \to T/W_P$ by $\theta_P(g * y) = \chi_P(y)$.

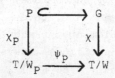

Also, define $\phi_P\colon G \times^P P \to G$ by $\phi_P(g * y) = gyg^{-1}$. The previous lemma now implies that we have a commutative diagram

$$\begin{array}{ccc} G \times^P P & \xrightarrow{\ \phi_P\ } & G \\ \theta_P \downarrow & & \downarrow \chi \\ T/W_P & \xrightarrow{\ \psi_P\ } & T/W \end{array}$$

Notice that ϕ_P is G-equivariant, G operating by left translations in $G \times^P P$ and by conjugation in G.

2.6.2. Induction of conjugacy classes

We keep the same notations. Let C be a conjugacy class in L. Then $\theta_P(G * CU_P)$ is one point in $T/W_P = T/W_L$, namely $\chi_L(C)$. It follows that $\phi_P(G * CU_P)$ is contained in a fiber G_t of χ. Since $\phi_P(G * CU_P)$ is G-invariant and irreducible, it follows from 2.4.3(ii) that there is a unique conjugacy class D of G which intersects $\phi_P(G \times CU_P)$ densely. We write $D = \mathrm{Ind}_{L,P}^G C$ and we call D the class in G obtained by induction from C. This induction procedure was introduced by Lusztig and Spaltenstein for the case of unipotent classes [19]. We shall come back later to that case in 4.5.

We now discuss some properties of the varieties and morphism which we just introduced. Let $\mathcal{P} = G/P$, this is a projective variety on which G operates by left translations. Let $\alpha\colon G \times \mathcal{P} \to \mathcal{P} \times \mathcal{P}$ be the morphism with $\alpha(x,\xi) = (x.\xi,\xi)$. We denote by $\Delta_{\mathcal{P}}$ the diagonal in $\mathcal{P} \times \mathcal{P}$ and by $pr_G, pr_{\mathcal{P}}$ the projection maps of $G \times \mathcal{P}$.

2.6.3. Lemma. (i) *There is an isomorphism* $\beta\colon G \times^P P \tilde{\to} \alpha^{-1}(\Delta_{\mathcal{P}})$ *such that* $pr_G \circ \beta = \phi_P$;
(ii) *The morphism* $pr_{\mathcal{P}} \circ \beta$ *makes* $G *^P P$ *into a locally trivial fiber space over* $\mathcal{P}$, *whose fibers are isomorphic to* P.

Define β by $\beta(g * x) = (gxg^{-1}, gP)$. We then have the properties of (i), as is easily checked. (ii) now follows by using that G is a locally trivial fiber space over $\mathcal{P}$, which is a consequence of a version of Bruhat's lemma.

2.6.4. Proposition. (i) ϕ_P *is a proper morphism;*
(ii) *For* $x \in G$, $\phi_P^{-1}x$ *is isomorphic to the fixed point variety* $\mathcal{P}_x$ *of* x *in* $\mathcal{P}$;
(iii) *If the derived group* L_1 *of* L *is simply connected and if* char k *does not divide* $i(L_1)$ *then* θ_P *is a flat morphism.*

Since P is a projective variety, pr_G is proper. This implies (i), via 2.6.3 (i). Similarly, we have (ii). The last point follows from 2.6.3 (ii) and 2.4.3 (v).

It is known that L_1 is simply connected if G_1 is so, see [5, p. 207].

2.6.5. Lemma. *If P,Q are parabolic subgroups containing B with $Q \subset P$ there is a commutative diagram of morphisms*

$$\begin{array}{ccc} G \times^Q Q & \xrightarrow{\phi_{P,Q}} & G \times^P P \\ \theta_R \downarrow & & \downarrow \theta_P \\ T/W_Q & \xrightarrow{\psi_{P,Q}} & T/W_P, \end{array}$$

where $\phi_{P,Q}$ and $\psi_{P,Q}$ are the obvious maps.

2.6.6. The simultaneous resolution of χ.

Of particular interest is the case P = B. We then write $\mathcal{B} = G/B$, $\tilde{G} = G \times^B B$, and we drop the indices B.

So there is a commutative diagram

$$\begin{array}{ccc} \tilde{G} & \xrightarrow{\varphi} & G \\ \theta \downarrow & & \downarrow \chi \\ T & \xrightarrow{\psi} & T/W . \end{array}$$

Now (iii) implies that θ is flat (in fact, $L_1 = \{e\}$ in this case). θ is even smooth: it follows by using 2.6.3 (ii) that the fibers $\theta^{-1}(t)$ are smooth and θ, being flat, is a smooth morphism. If we work over $\mathbb{C}$ our varieties have structures of complex varieties, and than θ is a locally trivial C^∞-fiber bundle.

The last diagram is Grothendieck's simultaneous resolution of the morphism χ. We shall come back to it later (4.1.2), when we discuss the resolution of the unipotent variety. Its study is of great interest for the finer aspects of the theory of conjugacy classes. For future use we notice the following. The notations are obvious.

2.6.7. Lemma. *Let $x = x_s x_u \in G$ and put $Z = Z(x_s)^\circ$. Then $\varphi_G^{-1}(x) \simeq \simeq G \times^Z \varphi_Z^{-1}(x_u)$.*

There is an analogous construction in the case of the Lie algebra $\mathfrak{g}$, leading to a diagram

$$\begin{array}{ccc} \tilde{\mathfrak{g}} & \xrightarrow{\varphi} & \mathfrak{g} \\ \theta\downarrow & & \downarrow\chi \\ \mathfrak{t} & \xrightarrow{\psi} & \mathfrak{t}/W \end{array}$$

with similar properties. We refer to [22, p. 58] (where the case that G is semi-simple is discussed).

2.6.8. An example

As an example of these simultaneous resolutions we discuss the Lie algebra situation, for $G = SL_2$ and char $k \neq 2$.
We take T,B to be the subgroup of diagonal resp. upper triangular matrices. We identify $\mathcal{B} = G/B$ with the projective line $\mathbb{P}^1$, the G-action being given by $\begin{pmatrix} x & y \\ z & t \end{pmatrix}.\xi = \frac{x\xi+y}{z\xi+t}$. Both $\mathfrak{t}$ and $\mathfrak{t}/W$ can be identified with k, such that $\psi a = a^2$. The Lie algebra $\mathfrak{g}$ consists of the 2×2-matrices $\begin{pmatrix} a & b \\ c & -a \end{pmatrix}$, and $\chi\begin{pmatrix} a & b \\ c & -a \end{pmatrix} = a^2 + bc$.
Now $\tilde{\mathfrak{g}}$ can be viewed as the subvariety of $\mathfrak{g} \times \mathbb{P}^1$ consisting of the $\left(\begin{pmatrix} a & b \\ c & -a \end{pmatrix},\xi\right)$ with $c\xi^2 - 2a\xi - b = 0$, and then

$$\varphi\left(\begin{pmatrix} a & b \\ c & -a \end{pmatrix},\xi\right) = \begin{pmatrix} a & b \\ c & -a \end{pmatrix}, \theta\left(\begin{pmatrix} a & b \\ c & -a \end{pmatrix}\right) = a^2 + bc.$$

3. UNIPOTENT ELEMENTS

We keep the notations of no. 2. We shall assume, moreover, that G is semi-simple. We shall discuss here a number of results on the classification of unipotent elements.

3.1. As before, let $V = V_G$ be the unipotent variety. We first collect a number of properties which are already contained in no. 2.

Theorem. (i) V *is an irreducible affine algebraic variety, of dimension* dim G - rk G;
(ii) *If* G *is simply connected and* char k *does not divide* i(G) *then* V *is normal;*
(iii) *The number of* G*-orbits on* V *(i.e. the number of unipotent conjugacy classes of* G) *is finite.*
These results are contained in 2.4.3 and constitute the core of that result.
The finiteness statement (iii) can be proved rather easily if char k is good for G (see [5 , p. 106]) The general case is more difficult,

and the proof involves either l-adic cohomology or the quite laborious work of an explicit classification (see [24, I,4] for further references).

3.2. Classification

3.2.1. In a classification of unipotent elements one would like to have a parametrization of the unipotent conjugacy classes. There are some finer aspects about which a classification should also provide information:
(a) The structure of the centralizer $Z_G(x)$ of a unipotent element x. Of interest are also the finite groups $A(x) = Z_G(x)/Z_G(x)^o$ (as is shown already by prop. 1.5).
(b) There is a natural order on the set of unipotent conjugacy classes, or on the set of unipotent elements: $x \leq y$ if and only if $C(x) \subset \overline{C(y)}$.
At present, there is available for each type of simple group $(A_n, \ldots, G_2)$ an explicit classification of the unipotent conjugacy classes. This is discussed in [24, I.2], to which we refer to further references. (In [loc.cit.] also the case of a non-connected group G is discussed.)

Example (see also no. 5). Let $G = PGL_n$. The unipotent conjugacy classes are parametrized by partitions $\lambda = (\lambda_1, \lambda_2, \ldots)$ of n, as follows from the theory of Jordan normal forms. The structure of centralizers can be described (see [5, part E) in particular $A(x) = \{e\}$ for all unipotent x.
Let $x \in G$ be unipotent and let $\lambda(x) = (\lambda_1, \lambda_2, \ldots)$ be the corresponding partition. So $\lambda_1 \geq \lambda_2 \geq \ldots$ and $\lambda_1 + \lambda_2 + \lambda \ldots \doteq n$. We order the partitions by the "natural" ordering: if $\lambda = (\lambda_1, \lambda_2, \ldots)$, $\mu = (\mu_1, \mu_2, \ldots)$ then $\lambda \leq \mu$ if $\lambda_1 + \lambda_2 + \ldots + \lambda_i \leq \mu_1 + \mu_2 + \ldots + \mu_i$ for all $i \geq 1$.
Then the order of unipotent elements is described by: $x \leq y$ if and only if $\lambda(x) \leq \lambda(y)$, see e.g. [13].

In the case of the other classical groups (types B_n, C_n, D_n) one has results of a similar combinatorial flavour. In the exceptional types the classification results rely on tables.
There are several principles whoch one can use in the description of a classification. We shall briefly review them.

3.2.2. Classification principles
We mention three such principles.
(1) Here one uses Th. 2.5.1, which shows that in good characteristics

the classification of unipotent elements in G is equivalent to the classification of nilpotent elements in the Lie algebra.
Let $x \in \mathfrak{g}$ be nilpotent and $\neq 0$. Under more severe restrictions on char k the theorem of Jacobson-Morozov applies, which states that there exist $y,h \in \mathfrak{g}$ with $[x,y] = h$, $[h,x] = 2x$, $[h,y] = -2y$. (In other words, x lies in a subalgebra of $\mathfrak{g}$ isomorphic to $\mathfrak{sl}_2$.) We may assume that $h \in \mathfrak{t}$. Now $\mathfrak{t}$ is the dual of the vector space $X \otimes_{\mathbb{Z}} k$ (notations as in 1.1), so a root $\alpha \in \Phi$ defines a linear function $\tilde{\alpha} = \alpha \otimes 1$ on $\mathfrak{t}$. One shows that (when x has been replaced by a suitable conjugate) one has $\tilde{\alpha}(h) = 0,1,2$ for all $\alpha \in \Delta$. Now associate to the conjugacy class of x a <u>weighted Dynkin diagram</u>, obtained by putting the integer $\tilde{\alpha}(h)$ in the node of the Dynkin diagram defined by the simple root α. One shows that these weighted Dynkin diagrams parametrize the nilpotent conjugacy classes.
For the simple classical types the possible weighted Dynkin diagrams are given in [5 , Part E, Ch. IV]. For the exceptional types they are listed in Dynkin's paper [11], where this classification principle was first introduced in the case $k = \mathbb{C}$.
The drawback of a classification along these lines is that it is restricted to the case that char k is so large that the theorem of Jacobson-Morozov holds.
(2) A parabolic subgroup $P = L.U_P$ is called <u>distinguished</u> if $\dim U_P/U_P' = \dim L'$, the prime denoting commutator subgroups. Consider the set $\mathcal{D}$ of pairs (M,Q) where M is a Levi subgorups of a parabolic subgroup of G, and Q a distinguishes parabolic subgroup of M. G operates on $\mathcal{D}$. If $(M,Q) \in \mathcal{D}$ there is a unique unipotent class C in G such that $C \cap U_Q$ is dense in U_Q. This defines a map of the set of G-orbits of $\mathcal{D}$ to the set of unipotent conjugacy classes of G. This map was introduced by Bala and Carter and they showed that, under some restrictions on char k, it is a bijection [2]. Subsequently it has been shown that this is true as soon as char k is good [21].
(3) The following procedure for obtaining a classification is quite general. A unipotent element x in a connected reductive group G is called <u>semi-regular</u> if all semi-simple elements of its centralizer $Z_G(x)$ are central in G.
Consider the set S of pairs (H,x) consisting of a connected reductive subgroup of our semi-simple group G, of the same rank as G, and a semi-regular unipotent element of H. It is obvious that any unipotent element x occurs in such a pair (but perhaps for several non-conjugate groups H). So there is a map of the set of G-orbits in S onto the set of unipotent conjugacy classes in G.

This leads to a description of unipotent classes. Here one has to know (G being given): (a) the possible H, (b) the semi-regular classes of any semi-simple subgroup. For $k = \mathbb{C}$ these data are already contained in [11].
Now (a) is a matter about root systems and (b) is tractable because the number of semi-regular classes is quite small.
For an example how this classification can be used we refer to the discussion of the exceptional types in [24, Ch. I, no. 2]. There one also finds a description of the semi-regular classes in exceptional groups, and in [loc.cit. Ch. IV] one finds explicit data on the ordering of conjugacy classes, using the procedure described here.

As a consequence of the classification results one obtains a description of the possible finite groups A(x).

3.2.3. Proposition. *Assume that* G *is simple and that* char k *is good. The non-trivial groups which occur are isomorphic to one of the following:* $(\mathbb{Z}/2\mathbb{Z})^a$, S_3, S_4, S_5.
This is established in [1] (over $\mathbb{C}$).

4. A RESOLUTION OF THE UNIPOTENT VARIETY

4.1. We return to the situation of 2.6.6. We put $\tilde{V} = \theta^{-1}e$. It is clear from the definitions that $\tilde{V} \simeq G \times^B U$ and that we have a morphism $\varphi: \tilde{V} \to \overline{V}$, which is G-equivariant (for the obvious G-actions).

4.1.1. Theorem. *If* char k *does not divide* i(G) *then* φ *is a resolution of singularities of* V.
This means that $\tilde{V}$ is smooth, φ is proper and that, V_s denoting the set of smooth (= nonsingular) points of V, the morphism φ induces an isomorphism $\varphi^{-1}V_s \simeq V_s$. The first two points appeared already in 2.6.6. The proof of the last point requires results about regular unipotent elements (as is to be expected from 2.4.3 (iv)). For a proof of the theorem we refer to [32].

4.1.2. Corollary. *Assume that* G *simply connected and that* char k *does not divide* i(G). *Then the diagram of* 2.6.6 *is a simultaneous resolution of the morphism* χ.
This means: φ is proper, θ is smooth, ψ is finite and surjective and for all $t \in T$ the morphism induced by $\varphi: \theta^{-1}(t) \to \chi^{-1}(\overline{t})$ is a

resolution of singularities of $\chi^{-1}(\bar{t})$.
For a proof of the corollary see [22, 4.4].

There is a similar situation for the nilpotent variety N, isomorphic to V_G in good characteristics (2.5.1). Putting $\tilde{N} = G \times^B \mathfrak{b}$ we have a morphism $\rho: \tilde{N} \to N$.

4.1.3. Proposition. *If* char k *is very good then* ρ *is a resolution of singularities of* N.
See [loc.cit., 4.7].

Example. Let $G = SL_2$ and char $k \neq 2$, as in 2.6.8. We then have $N = \left\{\begin{pmatrix} a & b \\ c & -a\end{pmatrix} \middle| a^2+bc = 0\right\}$ and $\tilde{N} = \left\{\left(\begin{pmatrix} a & b \\ c & -a\end{pmatrix}, \xi\right) \in N \times \mathbb{P}^1 \mid c\xi^2-2a\xi-b = 0,\right.$ $\left. c-2a\xi^{-1}-b\xi^{-2} = 0\right\}$, and $\varphi\left(\begin{pmatrix} a & b \\ c & -a\end{pmatrix}, \xi\right) = \begin{pmatrix} a & b \\ c & -a\end{pmatrix}$.

4.2. The resolution of N as a moment map
There is a nice geometric interpretation of the morphism ρ (see [7, §2]).

We first recall the definition of a moment map. Let X be an algebraic variety and G an algebraic group acting on X. We denote by T*X the cotangent bundle of X, whose points are pairs (x,ξ) where $x \in X$ and ξ lies in the cotangent space $(T_xX)^*$ of X at x (the dual of the tangent space T_xX at x). There is a morphism $\mu: T^*X \to \mathfrak{g}^*$ of algebraic varieties, the moment map associated to X and G. It is defined as follows. If $x \in X$, let $\phi_x: g \mapsto g.x$ be the orbit map $G \to X$ and $d\phi_x: \mathfrak{g} \to T_xX$ its differential at e. Then, $\langle\ ,\ \rangle$ denoting the canonical pairing between $\mathfrak{g}$ and its dual $\mathfrak{g}^*$, for $\eta \in \mathfrak{g}$, $\xi \in \mathfrak{g}^*$, $x \in X$,

$$\langle\eta,\mu(x,\xi)\rangle = \langle d\phi_x(\eta),\xi\rangle.$$

Example. If X = G and G acts by inner automorphisms then $T^*X \simeq G \times \mathfrak{g}^*$ and the moment map can be viewed as the map $(x,\xi) \mapsto (Ad^*(x)-1)\xi$ (Ad^* denoting the contragradient of the adjoint representation).

4.2.1. Proposition. *Let* G *be connected semi-simple and assume the Killing form on* $\mathfrak{g}$ *to be non-degenerate. There is an isomorphism* $\alpha: \tilde{N} \xrightarrow{\sim} T^*\mathcal{B}$, *such that there is a commutative diagram*

$$\begin{array}{ccc} \tilde{N} & \xrightarrow{\varphi} & N \hookrightarrow \mathfrak{g} \\ \alpha \wr\downarrow & & \swarrow \kappa \\ T^*\mathcal{B} & \xrightarrow{\mu} & \mathfrak{g}^* \end{array}$$

where κ is the isomorphism defined by the Killing form.
We have (cf. 2.6.3) that $\tilde{N}$ is isomorphic to $\{(x,gB) \in G \times \mathcal{B} \mid Ad(g)^{-1}x \in \mathfrak{u}\}$. On the other hand, the tangent space to $\mathcal{B}$ at gB is isomorphic to $\mathfrak{b}/Ad(g)\mathfrak{b}$, so the cotangent space is isomorphic to the subspace $(Ad(g)\mathfrak{b})^{\perp}$ of $\mathfrak{g}^*$ orthogonal to $Ad(g)\mathfrak{b}$. Via the Killing form this can be identified with $Ad(g)\mathfrak{u}$. The existence of α follows readily from these observations, and the proposition follows, from the definitions of φ and μ.

4.3. <u>The fibers of</u> φ
From now on we only assume G to be reductive. If $x \in V$ the fiber $\varphi^{-1}x$ is isomorphic to the fixed point set $\mathcal{B}_x$ of x in $\mathcal{B}$ (see 2.6.4 (ii)). The projective varieties $\mathcal{B}_x$ are of considerable interest. We shall discuss some of their elementary properties. First we introduce some notations and preliminaries. G operates on $\mathcal{B} \times \mathcal{B}$ and one version of Bruhat's lemma says that the G-orbits are parametrized by the elements of the Weyl group W. More precisely, putting

$$O(w) = \{(gB,g'B) \in \mathcal{B} \times \mathcal{B} \mid g^{-1}g' \in BwB\},$$

we have a decomposition into G-orbits

$$\mathcal{B} \times \mathcal{B} = \coprod_{w \in W} O(w)$$

(in particular, O(w) is the diagonal). All O(w) are locally closed.
We say that $\xi, \eta \in \mathcal{B}$ are <u>in position</u> w if $(\xi,\eta) \in O(w)$.
As before, let Δ be the basis of the root system Φ defined by B. For $\alpha \in \Delta$ define P_α to be the subgroup of G generated by B and $U_{-\alpha}$ (see 2.1). This is a parabolic subgroup containing B such that P_α/B is a projective line. Any set gP_α/B in $\mathcal{B}$ is called a line of type α in $\mathcal{B}$, or a line of type s if $s = s_\alpha$, the reflection in W defined by α. We then write $P_s = P_\alpha$.
Put $S = \{s_\alpha \mid \alpha \in \Delta\}$. This is a set of generators for W. We denote by ℓ the length function on W defined by S. The Bruhat order on W is denoted by $<$.
Let $\varphi_{G\times G} : \widetilde{V \times V} \to V \times V$ be the resolution of the unipotent variety

of $G \times G$. We put $Y = \varphi_{G\times G}^{-1}(\Delta_N)$ (where Δ_N is the diagonal of $N \times N$). Then $\varphi_{G\times G}$ defines a proper surjective morphism $\Phi\colon Y \to V$. On the other hand, the projection $\widetilde{V \times V} \to \mathcal{B} \times \mathcal{B}$ (see 2.6.3 (ii)) induces a surjective morphism $\Pi\colon Y \to \mathcal{B} \times \mathcal{B}$. We shall identify Y with the variety of all $(x, gB, g'B) \in V \times \mathcal{B} \times \mathcal{B}$ such that $x \in gBg^{-1} \cap (g')B(g')^{-1}$. Then Φ and Π correspond to the projection on the first factor (resp. the product of the last two factors). Also, Φ and Π are G-equivariant morphisms.

4.3.1. <u>Lemma</u>. (i) $Y_w = \Pi^{-1}(O(w))$ *is an irreducible variety of dimension* $2\dim \mathcal{B}$ *and* Y *is the disjoint union* $Y = \coprod_{w\in W} Y_w$;

(ii) *The closures* $\overline{Y}_w$ *are the irreducible components of* Y.

The lemma follows readily from the last description of Y.

<u>Remark</u>. There is a similar variety for N. In that case $\overline{Y}_w$ is nothing but the conormal variety of $\mathcal{B} \times \mathcal{B}$ along $O(w)$.

4.3.2. <u>Lemma</u>. *Let* $(x, \xi, \eta) \in \Pi^{-1}(O(w))$. *Assume* $s \in S$ *to be such that* $ws < w$. *Then,* L *denoting the line of type* s *through* η, *we have* $(x, \xi, L) \subset Y$.

To prove this we may assume $\xi = B$, $\eta = nB$ were $n \in N$ represents w. Let $s = s_\alpha$ $(\alpha \in \Delta)$. We have to show that for all $g \in P_\alpha$ we have $x \in B \cap ngBg^{-1}n^{-1}$. This follows from $B \cap nBn^{-1} \subset B \cap ngBg^{-1}n^{-1}$, which is proved by an easy computation involving the one parameter groups x_β $(\beta \in \Phi)$ introduced in 2.1.

Now let $x \in V$. Denote by $C(x)$ its conjugacy class and by $Z(x)$ its centralizer. We put $e(x) = \frac{1}{2}(\dim Z(x) - \mathrm{rk} G)$.

4.3.3. <u>Proposition</u>. (i) $\varphi^{-1}x$ *is connected;*

(ii) $\dim \varphi^{-1}x \leq e(x)$.

Let $\xi, \eta \in \varphi^{-1}x$ and assume $(x, \xi, \eta) \in \Pi^{-1}(O(w))$. If $w = e$ we have $\xi = \eta$. If $w \neq e$, let $s \in S$ be such that $ws < w$. Then, L being as in 4.3.2, there is a unique $\zeta \in L$ such that ξ, ζ are in position ws. We thus have connected η and ζ by a projective line lying in $\mathcal{B}_x$. Now (i) follows by an induction on $\ell(w)$.

Let Γ and Γ' be two components of $\varphi^{-1}x$. Then $\{x\} \times \Gamma \times \Gamma' \subset Y$ so $Y_{x,\Gamma,\Gamma'} = \bigcup_{g\in G} (\{x\}\times\Gamma\times\Gamma')$ is also contained in Y and is an irreducible contructible subset. Now $Z(x)$ stabilizes $\varphi^{-1}(x)$, and its identity component fixes all components. So $A(x) = Z(x)/Z(x)^\circ$ operates on the set of components. (It is easy to see that $Y_{x,\Gamma,\Gamma'} = Y_{x,\Gamma_1,\Gamma_1'}$ if and only there is $a \in A(x)$ such that $\Gamma_1 = a.\Gamma, \Gamma_1' = a.\Gamma'$.)

Now $\dim Y_{x,\Gamma,\Gamma'} = \dim C(x) + \dim \Gamma + \dim \Gamma'$. Since $Y_{x,\Gamma,\Gamma'}$ must be contained in an irreducible component of Y, we conclude from 4.3.1 that $\dim Y_{x,\Gamma,\Gamma'} \leq 2 \dim \mathcal{B}$. Taking $\Gamma = \Gamma'$ we get

$$2 \dim \Gamma \leq 2 \dim \mathcal{B} - \dim C(x),$$

which is equivalent to the inequality of (ii).

4.3.4. There is another way to formulate this inequality. A surjective morphism $f: X \to Y$ of irreducible algebraic varieties is said to be <u>small</u> (resp. <u>semi-small</u>) if

$$\operatorname{codim}\{y \in Y \mid \dim f^{-1}y \geq i\} > 2i$$

for all $i > 0$ (resp. $\geq 2i$ for all $i \geq 0$). Such morphisms have good homological properties (see [12, p. 120-121] and [8 , §1]). Then 4.3.3 (ii) is tantamount to saying that φ is a semi-small morphism.

We now consider the extended morphism $\varphi: \tilde{G} \to G$ of 2.6.6.

4.3.5. <u>Corollary</u>. (i) *For all* $x \in G$ *we have* $\dim \varphi^{-1}x \leq \frac{1}{2}(\dim Z(x) - \operatorname{rk} G)$; (ii) $\varphi: \tilde{G} \to G$ *is a small morphism.*
The first point follows from 2.6.7 and 4.3.3 (ii). To prove the second one one can proceed as follows. If $x \in G$ then $Z_G(x_s)^\circ$ is a connected reductive subgroup of G whose connected center contains x_s and which contains x_u, There are only finitely many conjugacy classes of such subgroups. Fix a connected reductive subgroup H containing T. One proves by applying 4.3.3 (ii) to the derived group of H that the set of $x \in G$ such that $Z_G(x_s)^\circ$ is conjugate to H and that $\dim Z_G(x) - \operatorname{rk} G \geq 2i$ has dimension $\leq \dim G - \operatorname{rk} H' - 2i$, and one sees that the codimension of this set is $\geq \operatorname{rk} H' + 2i > 2i$, if $\operatorname{rk} H' \neq 0$. Since $\operatorname{rk} H' = 0$ implies that $x = x_s$ is regular semi-simple, the assertion (ii) follows (a similar procedure gives a partition of G into finitely many locally closed smooth G-invariant pieces, see [18, p. 216].)

Let again $x \in V$ be a unipotent element, with centralizer $Z(x)$ and conjugacy class $C(x)$. As before, U denotes the unipotent part of B. There is a connection between the components of $\varphi^{-1}x$ and those of $U \cap C(x)$. To establish it, consider the closed subset of G

$$S(x) = \{y \in G \mid y^{-1}xy \in B\}.$$

Clearly, $S(x)$ is a union of double cosets $Z(x)gB$. Now $\beta y = y * (y^{-1}xy)$

defines a surjective morphism $S(x) \to \varphi^{-1}x$ whose fibers are isomorphic to B and $\gamma y = y^{-1}xy$ defines a surjective morphism $S(x) \to U \cap C(x)$ whose fibers are isomorphic to $Z(x)$. These remarks and 4.3.3 (ii) imply the following.

4.3.6. Lemma. (i) *If Γ is a component of $\mathcal{B}_x$ then $\gamma(\beta^{-1}\Gamma)$ is one of* $U \cap C(x)$. *Each component of* $U \cap C(x)$ *is of this form and we have* $\gamma(\beta^{-1}\Gamma) = \gamma(\beta^{-1}\Gamma')$ *if and only if there is* $a \in A(x)$ *with* $\Gamma' = a.\Gamma$;
(ii) *All components of* $U \cap C(x)$ *have dimension* $\leq \frac{1}{2} \dim C(x)$.

Now let $s \in S$ and denote by P_s the corresponding parabolic subgroup containing B. We denote its unipotent radical by U_s. We identify $\varphi^{-1}x$ with the fixed point set $\mathcal{B}_x$ of x in $\mathcal{B}$ (2.6.4 (ii)).

4.3.7. Lemma. (i) *Let* $y \in S(x)$ *and let Γ be a component of $\varphi^{-1}x$ passing through βy. Then the line of type s through βy lies on Γ if and only if* $\gamma y \in C(x) \cap U_s$;
(ii) *Let Γ be a component of $\varphi^{-1}x \simeq \mathcal{B}_x$. The set of points $\xi \in \mathcal{B}_x$ such that the line of type s passing through ξ is contained in Γ is either all of Γ or a closed subset whose components have codimension 1. The first case prevails if and only if $\gamma(\beta^{-1}\Gamma) \subset C(x) \cap U_s$.*
Let $s = s_\alpha$, $\alpha \in \Delta$. The first point follows from the following result, which is readily established using standard formulas (see [27, Ch. 11]): if $y \in U$ and if $x_{-\alpha}(t)yx_{-\alpha}(t)^{-1} \in U$ for all $t \in k$ then $y \in U_s$. Then (ii) follows readily.

4.3.8. Lemma. *All components of $\varphi^{-1}x$ have the same dimension.*
This is proved as follows (after Spaltenstein, see [24, p. 47]). From the proof of 4.3.3 (i) we see that given $\xi, \eta \in \varphi^{-1}x \simeq \mathcal{B}_x$, there exists $s_i \in S$ $(0 \leq i \leq h)$ and lines L_i of type s_i, such that $\xi \in L_0$, $\eta \in L_h$, $L_i \cap L_{i+1} \neq \emptyset$ $(0 \leq i \leq h-1)$. It follows that it suffices to prove the following: if Γ is a component of $\varphi^{-1}x$ of maximal dimension and L a line of type s (for some $s \in S$) which intersects Γ, then L lies in a component of $\varphi^{-1}x$ of maximal dimension. This is clear if we have the first case of 4.3.7 (ii). In the second case it follows from 4.3.7 (ii) that there is an irreducible subvariety Γ_1 of Γ of codimension 1 such that all lines of type s through points of Γ_1 lie on $\varphi^{-1}x$ and that L is such a line. But then the set of those lines is contained in a component of $\varphi^{-1}x$ of maximal dimension and we have proved the desired assertion.

In the next lemma $\dot{w}$ denotes a representative of $w \in W$ in N.

4.3.9. Lemma. *The following assertions are equivalent:*
(a) *All components of* $\varphi^{-1}x$ *have dimension* $e(x)$,
(b) *There exists* $w \in W$ *such that* $C(x)$ *intersects* $U \cap \dot{w}U\dot{w}^{-1}$ *in a dense subset.*

It follows from 4.3.3 (ii) and 4.3.8 that (a) holds as soon as there is one component Γ of $\varphi^{-1}x$ of dimension equal to $e(x)$. In that case it follows from the proof of 4.3.3 (ii) that there is $w \in W$ such that, with the notation of 4.3.1 and 4.3.3, the set $Y_{x,\Gamma,\Gamma}$ intersects a component Y_w densely. It is a straightforward verification that this is equivalent to (b). The argument may be reversed, proving the equivalence of (a) and (b).

4.3.10. Theorem. *For any* $x \in V$ *we have that all components of* $\varphi^{-1}x$ *have dimension* $e(x)$.

This is proved by Spaltenstein in [24, p. 163]. The proof involves the classification. We shall say below in 4.5 a bit more about some ingredients of the proof.

4.3.11. Corollary. *All components of* $C(x) \cap U$ *have dimension* $\frac{1}{2}\dim C(x)$.

This follows by 4.3.6.

For $k = \mathbb{C}$ this corollary can be interpreted in a different way. We then have a similar statement for nilpotent elements $X \in \mathfrak{g}$, as follows by using 2.5.1. On the conjugacy class $C(X)$ of $X \in \mathfrak{g}$ there exists a $\mathfrak{g}$ G-equivariant symplectic structure with the following property. Let F be the Killing form on $\mathfrak{g}$ (G is assumed to be semi-simple). The tangent space T_X to $C(X)$ at X is isomorphic to $\mathfrak{g}/\mathfrak{z}$, where $\mathfrak{z}$ is the centralizer if X in $\mathfrak{g}$. Now the non-degenerate alternating form on T_X defined by the symplectic structure is induced by the (degenerate) alternating form $(Y,Z) \mapsto F(Y,[X,Z])$ on $\mathfrak{g}$.

Then the analogue of 4.3.11 can be reformulated as follows: all components of $C(X) \cap U$ are Lagrangian submanifolds of $C(X)$, for the symplectic structure.

4.4. Components and the Weyl group.

If $x \in V$ let $\Sigma(x)$ be the set of components of $\varphi^{-1}x$. Then, as in the proof of 4.3.3 (ii), the group $A(x)$ operates on $\Sigma(x)$ and (diagonally) on $\Sigma(x) \times \Sigma(x)$. We denote by $A(x)\backslash\Sigma(x) \times \Sigma(x)$ the set of orbits, and by $\sigma(x)$ the permutation of this set induced by permutation of the two factors. The symbol $\coprod'$ below denotes disjoint union, taken over a set of representatives of the set $G\backslash V$ of unipotent conjugacy classes.

4.4.1. Proposition. *There is a bijection* $\rho: W \xrightarrow{\sim} \coprod' A(x)\backslash\Sigma(x) \times \Sigma(x)$. *We have* $\rho(w^{-1}) = \sigma(\rho(w))$, *where* s *is defined by the* $\sigma(x)$.
By 4.3.10 it follows from the proof of 4.3.3 (ii) that the components Y_w of Y coincide with the varieties $\overline{Y}_{x,\Gamma,\Gamma'}$ (x running through a set of representatives). The bijection ρ then comes from the identification of the Y_w with the $\overline{Y}_{x,\Gamma,\Gamma'}$. The second point readily follows.

The projections on first factors define a surjective map $\rho_1: W \to \coprod' A(x)\backslash\Sigma(x)$. There is a similar map defined by projections on second factors, which by 4.4.1 is $w \mapsto \rho_1(w^{-1})$. Now by 4.3.6 (i) we can identify $A(x)\backslash\Sigma(x)$ with the set of components $\tilde{\Sigma}(x)$ of $U \cap C(x)$, and we have a surjective map $\tilde{\rho}: W \to \coprod' \tilde{\Sigma}(x)$. The next result gives a geometric description of ρ.

4.4.2. Corollary. $\overline{\tilde{\rho}(w)} = \overline{\bigcup_{b \in B} b(U \cap \dot{w}U\dot{w}^{-1})b^{-1}}$
We have $Y_w = \overline{Y}_{x,\Gamma,\Gamma'}$, as above. This is a closed subset of $V \times \mathcal{B} \times \mathcal{B}$. The assertion follows by intersecting it with $V \times \{B\} \times \mathcal{B}$, and projecting on the first factor, using the descriptions of Y_w and $Y_{x,\Gamma,\Gamma'}$.

The composite of $\tilde{\rho}$ and the obvious map $\tilde{\Sigma}(x) \to G\backslash V$ defines a surjective map $\mu: W \to G\backslash V$. It follows from 4.4.2. that μ can be described as follows.

4.4.3. Lemma. $\mu(w)$ *is the conjugacy class* C *which intersects* $U \cap \dot{w}U\dot{w}^{-1}$ *densely*.

We define an equivalence relation on W by $w \sim w'$ if $\tilde{\rho}(w) = \tilde{\rho}(w')$. This relation has the property of the next proposition. We shall see below (5.4) how this property leads to a combinatorial description of the relation in the case $G = GL_n$.

4.4.4. Proposition. *Let* $s,t \in S$ *and* $w \in W$ *be such that* $sts = tst$ *and* $tsw < sw < w < tw$. *Then* $sw \sim w$.
Here $<$ denotes again the Bruhat order on W.
Let $\alpha,\beta \in \Delta$ be such that $s = s_\alpha$, $t = s_\beta$. Then $sts = tst$ is equivalent to $s_\alpha\beta = s_\beta\alpha = \alpha+\beta$.
As before, we denote for $\gamma \in \Phi$ by U_γ the one parameter subgroup defined by γ (2.1). If $x \in W$ we put $U_x = U \cap \dot{x}U\dot{x}^{-1}$. One knows that U_x is generated by the U_γ with $\gamma > 0$ (i.e. $\gamma \in \Phi^+$) and $x^{-1}\gamma > 0$.
One concludes from this that U_{sw} is generated by U_w and U_α and that U_{tsw} is generated by U_w, U_α and U_β. Also, $U_{\alpha+\beta} \not\subset U_w$, as

$w^{-1}(\alpha+\beta) = w^{-1}s_\alpha\beta < 0$, for $U_\beta \not\subset U_{s_\alpha w}$

We now claim that if $U_\gamma \subset U_w$ we have $U_{ts\gamma} \subset U_{sw}$. First note that $s\gamma > 0$, $ts\gamma > 0$. Otherwise we had either $\gamma = \alpha$ or $\gamma = \alpha+\beta$, which is impossible since $U_\alpha \not\subset U_w$, $U_{\alpha+\beta} \not\subset U_w$. (We used here that if $\gamma > 0$, $\alpha \in \Delta$ and $s_\alpha\gamma < 0$ we have $\gamma = \alpha$, see e.g. [27, p. 216].)

Let $W_{\alpha,\beta}$ be the subgroup of W generated by $s=s_\alpha$ and $t=s_\beta$. One knows that we can write uniquely $w = xy$ with $x \in W_{\alpha,\beta}$ and $y^{-1}\alpha > 0$, $y^{-1}\beta > 0$ (see [9, p. 37, ex. 3]) and one concludes from our assumptions that $x = st$. So we have $\gamma > 0$, $w^{-1}\gamma = y^{-1}ts\gamma > 0$ and we have to show that $w^{-1}sts\gamma = y^{-1}s\gamma > 0$. Let $y^{-1} = s_1 \ldots s_1$ with $s_i \in S$ be a shortest expression, then $y^{-1}s = s_1 \ldots s_1 s$ is one of $y^{-1}s$, since $y^{-1}\alpha > 0$. If $y^{-1}s\gamma < 0$ we had $\gamma = ss_1 \ldots s_i s_{i-1}$ for some $i \geqslant 2$ [27, p.216], in which case $y^{-1}ts\gamma = \gamma^{-1}ts_1 \ldots s_i s_{i-1} < 0$, by the same result, applied to the shortest expression $s_1 \ldots s_1 t$ of $y^{-1}t$. This contradiction implies our claim. We deduce from it that $\dot{t}\dot{s}U_w\dot{s}^{-1}\dot{t}^{-1} \subset U_{sw}$. If C is the unipotent conjugacy class intersecting U_w densely, we conclude that $C \cap U_{sw}$ has codimension $\leqslant 1$ in U_{sw}. Now $U_{\alpha+\beta} \subset U_{sw}$ (since $w^{-1}s(\alpha+\beta) = w^{-1}\beta > 0$) and $U_{\alpha+\beta} \not\subset \dot{t}\dot{s}U_w\dot{s}^{-1}\dot{t}^{-1}$ (since $s_\beta s_\alpha(\alpha+\beta) = \beta < 0$), moreover $U_{\alpha+\beta} = \dot{s}\,U_\beta\dot{s}^{-1}$. We conclude from these facts that C must also intersect U_{sw} densely. But then 4.4.2 implies that $\overline{\tilde{\rho}(w)} \subset \overline{\tilde{\rho}(sw)}$. Since both sets are irreducible of the same dimension the proposition follows.

4.4.5. The previous proposition is a special case of a general result due to Spaltenstein [24, p. 139-140]. We have singled out the special case because we shall use only the property of 4.4.4 in the discussion of the case of the symmetric groups in no. 5.

We now describe the more general result alluded to above. Let $S_1 \in S$ be a subset and let $W_1 \subset W$ be the subgroup generated by S_1. To S_1 there corresponds a unique parabolic subgroup $P = LU_P \supset B$, where L is the Levi subgroup containing T, such that W_1 is the Weyl group of (L,T) and that S_1 is defined by the Borel group $B \cap L$ of L.

Let $\sim_1$ be the equivalence relation on W_1 defined above.

Let W' be the set of $w' \in W$ such that $l(w's) > l(w')$ for all $s \in S_1$. One knows [9, p. 37, ex. 3] that any $w \in W$ can be written uniquely as $w = w_1w'$, with $w_1 \in W_1$, $w' \in W'$. Then:

4.4.6. <u>Proposition</u>. *Let* $x,y \in W_1$, $w' \in W'$. *If* $x \sim_1 y$ *then* $xw' \sim yw'$. For the proof see [24, p. 139-140].

We finally mention another property of the equivalence relation on W. If $w \in W$ let $R(w) = \{s \in S \mid l(ws) < l(w)\}$.

4.4.7. Proposition. *If* $w \sim x$ *then* $R(w) = R(x)$.
This follows from 4.3.2 and 4.3.7 (ii) (see also [loc.cit., p. 73]).

4.5. Dimension of components and parabolic subgroups
Fix a parabolic subgroup $P \supset B$. We use the notations of 4.4.5. Also, $\tilde{w}$ resp. $\tilde{w}_1$ denote the longest elements of W, resp. W_1.

4.5.1. Proposition. *Let* $x \in V_L$. *Assume that* $w \in W_1$ *is such that* $C_L(x)$ *intersects* $U \cap \dot{w}U\dot{w}^{-1} \cap L$ *densely. Then, putting* $y = w\tilde{w}_1\tilde{w}$, *we have that* $C_G(x)$ *intersects* $U \cap \dot{y}U\dot{y}^{-1}$ *densely.*
By an easy reasoning involving the one parameter groups U_α one shows that $U \cap \dot{y}U\dot{y}^{-1} = U \cap \dot{w}U\dot{w}^{-1} \cap L$, whence the result [24, p. 68].

We next recall the definition of induction of conjugacy classes, in the unipotent case (see 2.6.2). If C is a unipotent class in L there is a unique conjugacy class D in G such that D intersects CU_P densely. Then $D = \mathrm{Ind}_{L,P}^G C$, as one readily sees (notation of 2.6.2). Let $x \in C$, $y \in D$. If $C = \{e\}$ then D is the class intersection U_P densely. Such a unipotent conjugacy class is called a Richardson class (defined by P), and its elements are Richardson elements.
One can show that $\mathrm{Ind}_{L,P}^G C$ depends only on L and C, and not on the choice of the parabolic subgroup P with Levi group L (see [19]).

Now let $x \in C$, $y \in D$. The next result is a special case of [24, p. 60, prop. 3.2].

4.5.2. Proposition. *Assume that* $\dim \mathcal{B}_x^L = \frac{1}{2}(\dim Z_L(x) - \mathrm{rk} L)$. *Then* $\dim \mathcal{B}_y^G = \frac{1}{2}(\dim Z_G(y) - \mathrm{rk}\, G)$. *Moreover* $\dim \mathcal{B}_x^L = \dim \mathcal{B}_y^G$.
Viewing (as we may) $\mathcal{B}^G$ resp. $\mathcal{B}^P = P/B$ as the set of Borel subgroups of G resp. P, it is clear that $\mathcal{B}_x^L \simeq \mathcal{B}_x^P \subset \mathcal{B}_y^G$, whence $\dim Z_L(x) \leq \dim Z_G(y)$, by 4.3.5 (since rk G = rk L).
On the other hand, we have by 4.3.6 (ii)

$$\tfrac{1}{2} \dim C_G(y) \geq \dim D \cap U \geq \dim CU_P \cap U = \dim C \cap U \cap L + \dim U_P =$$
$$= \tfrac{1}{2} \dim C_L(x) + \dim U_P = \tfrac{1}{2} \dim G - \tfrac{1}{2} \dim Z_L(x).$$

From this one deduces that $\dim Z_L(x) \geq \dim Z_G(y)$, and the proposition follows.

4.5.3. Corollary. *If* D *is the Richardson class defined by* P *then* $\dim \mathcal{B}_y^G = \frac{1}{2}(\dim Z_G(x) - \mathrm{rk}\, G) = \dim \mathcal{B}^L$. *Moreover*, $\mathcal{B}^P$ *is a component of* $\mathcal{B}_y^G$.

We can now comment on the proof of 4.3.10. By induction, using 4.5.1 and 4.5.2, we are reduced to proving 4.3.10 in the case that x is a semi-regular unipotent element (defined in 3.2.2) in a simple group G whose conjugacy class is not induced from a proper parabolic subgroup. The number of possible conjugacy classes of such elements is very small (in fact there are none at all if char $k \neq 2,3$), and they can be dealt with. See [24, p. 163] for the details. The proof given in [loc.cit.] is quite complicated, as it involves the classification of unipotent elements. An a priori proof would be desirable.

If char k is good we have the classification of Bala-Carter (see 3.2.2) and in that case 4.5.1 implies 4.3.9 (b), whence 4.3.10. So in this case the proof is rather easy.

5. GL_n

5.1. As an illustration of the results of the preceding section we discuss in more detail the case $G = GL_n$. We now take B to be the subgroup of upper triangular matrices and for T the subgroup of diagonal matrices. Then N is the group of $n \times n$-matrices which have in each row and each column exactly one non-zero entry. The Weyl group $W = N/T$ is isomorphic to the symmetric group S_n. The set of generators S of W correspond to the permutations s_i with $s_i(j) = j$ for $j \neq i, i+1$ and $s_i(i) = i+1$ $(1 \leq i \leq n-1)$.

Moreover, we have $l(ws_i) > l(w)$ if and only if $w(i) < w(i+1)$.

The group GL_n operates in $V = k^n$. A complete flag in V is a set of distinct subspaces $(V_i)_{1 \leq i \leq n}$, with $V_i \subset V_{i+1}$ $(1 \leq i \leq n-1)$. Clearly, $\dim V_i = i$. These flags can be viewed as the points of the quotient variety $\mathcal{B} = G/B$, whcih we now also call the flag variety. If $x \in G$ then the variety $\mathcal{B}_x$ can be viewed as the set of flags fixed by x.

5.2. Partitions and unipotent elements

5.2.1. We shall have to use some combinatorial notions. We recall that a partition $\lambda = (\lambda_1, \lambda_2, \ldots)$ of n is a finite sequence of non-negative integers with $\lambda_1 \geq \lambda_2 \geq \ldots$ and $\Sigma\lambda_i = n$. We put $n = |\lambda|$. The λ_i are the parts of λ. We do not distinguish between partitions which differ only in the number of zeros.

We associate to the partition λ a diagram. This can be defined to be the set of points $(-i,j)$ in the plane, with i,j integers > 0 and $1 \leq j \leq \lambda_i$. We prefer to view the diagram as formed by rows of squares of lengths $\lambda_1, \lambda_2, \ldots$ starting at the top. We identify λ with its diagram. The dual λ' of λ is obtained by interchanging rows and columns of λ. We put $n(\lambda) = \sum_{i \geq 1} (i-1)\lambda_i$. Then, $\lambda' = (\lambda_1', \lambda_2', \ldots)$ denoting the dual, we have $n(\lambda) = \frac{1}{2} \sum_{i \geq 1} \lambda_i'(\lambda_i'-1)$.
For all these facts see [20, Ch. I].

5.2.2. Let $x \in G$ be unipotent. By the theory of Jordan normal forms, we can associate to x a partition $\lambda = (\lambda_1, \lambda_2, \ldots)$, the λ_i being the sizes of the Jordan blocks of x. Then the partitions λ of n parametrize the unipotent conjugacy classes.
We denote by $x_\lambda \in G$ the direct sum of consecutive unipotent Jordan blocks of sizes $\lambda_1, \lambda_2, \ldots$ So the x_λ represent the unipotent classes. We denote by Z_λ and C_λ the centralizer and conjugacy class of x_λ.

5.2.3. Lemma. (i) *If* $x \in G$ *then* $Z_G(x)$ *is connected;*
(ii) $\dim Z_\lambda = n + 2n(\lambda)$;
(iii) *If* $\lambda \neq (n)$ *then* Z_λ *contains a non-central torus.*
These facts are contained in the results of [5, part E].

5.2.4. Proposition. *If* $x \in G$ *is unipotent then* $\dim \mathcal{B}_x = \frac{1}{2}(\dim Z_G(x) - n)$.
This follows from 4.3.9, 4.5.1 and 5.2.3 (iii). (So the proof of 4.3.10 is rather easy in the case of GL_n.)
Notice that $\dim \mathcal{B}_{x_\lambda} = n(\lambda)$.

Given a unipotent element $x \in G$ the dual of its partition is described as follows, as one sees by using a Jordan basis of V relative to x.

5.2.5. Lemma. $\dim \operatorname{Ker}(x-1)^i = \lambda_1' + \lambda_2' + \ldots + \lambda_i'$.

5.3. Standard tableaux and components

5.3.1. A tableau τ is a diagram together with a labelling of its squares by distinct strictly positive integers. The set of integers involved in τ is the content of τ. The underlying partition (or diagram) λ is the shape of τ.
A tableaux is standard if its labels increase in the rows and columns of the diagram. We denote by T_n the set of standard tableaux of content

$\{1,2,\ldots,n\}$ and by $T_\lambda \subset T_n$ the subset of standard tableaux of shape λ (so $|\lambda| = n$). Notice that if $\tau \in T_n$, we obtain a standard tableau $\bar{\tau} \in T_{n-1}$ by deleting the square with label n. This square of the shape λ of τ is a corner of λ, i.e. a square such that deletion gives again a diagram.
If $\tau \in T_\lambda$ and $|\lambda| = n$, let τ_i be the column of λ which contains the label i of τ. Then τ is completely determined by the sequence $(\tau_1,\ldots,\tau_n)$.

5.3.2. Now let $x \in G$ be unipotent and put $X = B_x$. We denote by λ the associated partition of n. We define now a surjective map $t: X \to T_\lambda$, after Spaltenstein [23].
First a lemma, whose easy proof is omitted.

5.3.3. Lemma. *Let* $W_i = \mathrm{Ker}(x-1) \cap \mathrm{Im}(x-1)^{i-1}$.
(i) $\dim W_i = \lambda'_i$;
(ii) $Z_G(x)$ *operates transitively on* $W_i - W_{i+1}$.
(As before, λ' is the dual of λ).

If $F = (V_i)_{1\leq i\leq n}$ is a flag in X then $V_1 \subset W_1$. So we have a morphism π of X to the projective space $\mathbb{P}(W_1)$, which sends F to V_1. Then x defines a unipotent automorphism $\bar{x}$ of V/V_1, whose diagram $\bar{\lambda}$ is λ with a corner deleted, and $\bar{F} = (V_i/V_1)_{2\leq i\leq n}$ defines a flag fixed by $\bar{x}$. By induction we may assume that we have already associated to $\bar{F}$ a standard tableau $t(\bar{F}) \in T_{\bar{\lambda}}$. We define $t(F) \in T_\lambda$ to be the standard tableau τ obtained from $t(\bar{F})$ giving the only square which is in λ but not in $\bar{\lambda}$ the label n. Then, with the notation of 5.3.1 we have $\tau_n = i$ if and only if $V_1 \subset W_i - W_{i+1}$.

5.3.4. Proposition. (i) $X = \coprod_{\sigma\in T_\lambda} t^{-1}(\tau)$;

(ii) *If* $\tau \in T_\lambda$ *then* $t^{-1}(\tau)$ *is an irreducible locally closed subset of* X, *which is smooth. Its dimension is* $n(\lambda)$;
(iii) *The closures* $\overline{t^{-1}(\tau)}$ $(\tau \in T_\lambda)$ *are the irreducible components of* X.
The first point is clear. The second point follows by deducing from 5.3.3 (ii) that $\pi^{-1}(\mathbb{P}(W_i)-\mathbb{P}(_{i+1}))$ is a locally trivial fibre space over $\mathbb{P}(W_i)-\mathbb{P}(W_{i+1})$ for the Zariski topology, whose fibers are isomorphic to the variety of flags in a space V/V_1 fixed by a unipotent element $\bar{x}$ as above. Then (iii) is a direct consequence of (i) and (ii). (In [loc.cit.] one finds somewhat more precise results.)

From the fact that the components of $\mathcal{B}_x = X$ are described by standard tableaux, we now infer, using 4.4.1 and 5.2.3 (i) the following combinatorial result.

5.3.5. Corollary. *There is a bijection* $\rho: S_n \xrightarrow{\sim} \coprod_{|\lambda|=n} T_\lambda \times T_\lambda$. *If* $\rho(w) = (\tau_1,\tau_2)$, *where* $\tau_1,\tau_2 \in T_\lambda$ *then* $\rho(w^{-1}) = (\tau_2,\tau_1)$.

Moreover, defiming the equivalence relation $\sim$ on S_n as in 4.4, we see:

5.3.6. Corollary. *The number of equivalence classes equals the number of elements of* T_n.

5.3.7. Corollary. *If* $x,w \in S_n$ *and* $x \sim w$, $x^{-1} \sim w^{-1}$ *then* $x = w$.
This follows from 5.3.5.

5.4. Equivalence relations on S_n.

5.4.1. The equivalence relation $\sim$ just recalled has the property of 4.4.4. This can be described differently in the case of S_n. To do this we identify $w \in S_n$ with the ordered set of integers $(a_1,\ldots,a_n)$, where $a_i = w^{-1}(i)$ $(1 \leq i \leq n)$. Let $s = s_i \in S$. As recalled in 5.1, we have $sw < w$ if and only if $a_{i+1} < a_i$.
If s and t are as in 4.4.4 we have either $s = s_i$ or $t = s_{i+1}$ or $s = s_{i+1}$, $t = s_i$ for some $i = 1,\ldots,n-1$. The property of 4.4.4 is now translated the following one of ordered sets of integers.

5.4.2. Property. *Let* a,b,c *be integers with* $a < b < c$. *Then the following operations preserve equivalence classes: replacing three consecutive integers* cab *(resp.* bac*) by* acb *(resp.* bca*).*

Now let $\tau \in T_n$ be a standard tableau. We associate to it the permutation $w(\tau)$ obtained by reading the rows of τ consecutively, starting at the bottom. So, if $\tau = \begin{smallmatrix}1&2&3\\4\end{smallmatrix}$ then $w(\tau) = (4123)$. We denote by $\sim_1$ the minimal equivalence relation on S_n (viewed as a subset of $S_n \times S_n$) with the property of 4.4.4 (or of 5.4.2).

5.4.3. Lemma. *Let* $a_1,\ldots,a_s,b$ *be integers with* $a_1 < a_2 < \ldots < a_s$, $b < a_s$. *Let* $i \geq 1$ *be the first index such that* $b < a_i$. *The following operation preserves equivalence classes for* $\sim_1$: *replacing a set of consecutive integers* $a_1a_2\ldots a_sb$ *by* $a_ia_1\ldots a_{i-1}ba_{i+1}\ldots a_s$.
This follows by a repeated application of the properties of 5.4.2.

5.4.4. Proposition. *For each* $w \in S_n$ *there is* $\tau \in T_n$ *such that* $w \sim_1 w(\tau)$.
It is convenient to consider here arbitrary ordered sets $(a_1,\dots,a_n)$ of integers > 0, which are not necessarily permutations of $(1,\dots,n)$. Also, we can associate to a standard tableau τ with arbitrary content an ordered set of integers $w(\tau)$, by the same rule as before.
Now let $w = (a_1,\dots,a_n) \in S_n$. By induction on n we may assume that there is a standard tableau τ_1 with content $\{1,\dots,n\} - \{a_n\}$ such that $w(\tau) = (a_1,\dots,a_{n-1})$. If $a_{n-1} < a_n$ the proposition is proved. If $a_n < a_{n-1}$ let λ_1 be the length of the first row of τ_1. If $i \geq 0$ is the first integer such that $a < a_{n-\lambda_1+i}$ then, using the previous lemma, we see that $w \sim (a_1,\dots,a_{n-\lambda_1-1}, a_{n-\lambda_1+i}, a_{n-\lambda_1},\dots,a_{n-\lambda_1+i-1}, a_n, a_{n-\lambda_1+i+1},\dots,a_{n-1})$. Using induction, we see that $(a_1,\dots,a_{n-\lambda_1-1}, a_{n-\lambda_1+i})$ is equivalent for $\sim_1$ with a $w(\tau_2) = (b_1,\dots,b_{n-\lambda_1})$, say. One then checks that $(b_1,\dots,b_{n-\lambda_1}, a_{n-\lambda_1},\dots,a_{n-\lambda_1+i-1}, a_n, a_{n-\lambda_1+i+1},\dots,a_{n-1})$ is of the form $w(\tau)$, which proves the proposition.

The proposition is due to Knuth [16]. The inductive proof given here can be viewed as giving the algorithm of [loc.cit.] to associate with any set of distinct integers >0 a standard tableau.

5.4.5. Corollary. *The equivalence relations* $\sim$ *and* $\sim_1$ *on* S_n *coincide.*
By 5.4.4 the number of equivalence classes for $\sim_1$ is at most the number of elements of T_n. Now the minimality of $\sim_1$ and 5.3.6 imply the corollary.

5.4.6. Corollary. *Any equivalence class for* $\sim$ *contains a unique element of the form* $w(\tau)$.
If $w \in S_n$ let $\tau(w) \in T_n$ be such that $w \sim w(\tau(w))$.

5.4.7. Theorem. *The map* $w \to (\tau(w), \tau(w^{-1}))$ *defines a bijection* $S_n \tilde{\to} \coprod_{|\lambda|=n} T_\lambda \times T_\lambda$
By 5.3.7 and 5.4.6 the map is injective. Since, by 5.3.5 the sets S_n and $\coprod_{|\lambda|=n} T_\lambda \times T_\lambda$ have the same number of elements, the theorem follows.
The map in question defines the Robinson-Schensted correspondence. We have established here this correspondence using the geometric results of no. 4. For more details about it see for example [10, part 1].

6. REPRESENTATIONS OF THE WEYL GROUP

6.1. In no. 4 we have discussed some elementary geometric results about the simultaneous resolution $\varphi: \tilde{G} \to G$ and its Lie algebra analogue $\tilde{\mathfrak{g}} \to \mathfrak{g}$ (see 2.6.6). Using the tools of algebraic topology one can make a more refined analysis of these resolutions. A thorough discussion of this would be beyond the scope of these lectures. We shall only discuss some aspects, which lead to a better a priori insight into the classification of unipotent conjugacy classes.
The notations are as in the preceding sections. For simplicity, we assume that G is connected, semi-simple adjoint and that the base field is the field of complex numbers $\mathbb{C}$.

6.2. Now G and $\tilde{G}$ are complex algebraic varieties, and hence have a structure of complex manifold. The morphism $\varphi: \tilde{G} \to G$ of 2.6.6 is then a proper morphism of such manifolds. If $x \in G$, the fiber $\varphi^{-1}x$ is a complex projective variety, isomorphic to the fixed point set $\mathcal{B}_x$ of x in $\mathcal{B}$ (2.6.4 (ii)). We shall have to consider the cohomology groups with rational coefficients $H^i(\mathcal{B}_x,\mathbb{Q})$. These are finite dimensional vector spaces over $\mathbb{Q}$, which are zero if $i < 0$ or $i > 2e(x)$, where $e(x) =$ $= \dim \mathcal{B}_x = \frac{1}{2}(\dim Z(x) - \mathrm{rk}\, G)$ (see 4.3 and 2.6.7). One knows that each component C of $\mathcal{B}_x$ (which has dimension e(x) by 4.3.8, 4.3.10 and 2.6.7) defines an element [C] of $H^{2e(x)}(\mathcal{B}_x,\mathbb{Q})$ and that these elements form a basis of $H^{2e(x)}(\mathcal{B}_x,\mathbb{Q})$.
Let T,B and W be as before. The crucial point is now that one can define, for each $x \in G$, a representation of the Weyl group W in the cohomology spaces $H^i(\mathcal{B}_x,\mathbb{Q})$. The most interesting case is when x is unipotent. Such representations were first introduced in [26], in the Lie algebra situation, over the algebraic closure of a finite field (working in l-adic cohomology). Subsequently, various other constructions of these representations were given. We shall now comment a bit on an elegant construction in the context of intersection cohomology, which is due to G. Lusztig and has been exploited by Borho and MacPherson [6]. We refer to [28] for further references about these matters.

6.3. First notice that the groups $H^i(\mathcal{B}_x,\mathbb{Q})$, for varying $x \in G$, can be assembled in a sheaf $R^i\varphi_*\mathbb{Q}$, the ith higher direct image of the constant sheaf $\mathbb{Q}$ on $\tilde{G}$. These sheaves are the cohomology sheaves of a complex of sheaves of finite dimensional vector spaces denoted by $\mathbb{R}\varphi\mathbb{Q}$. The construction mentioned above now leads to a W-action on $\mathbb{R}\varphi\mathbb{Q}$. (A correct treatment of these matters requires the introduction of the appropriate

derived category.)
It is easy to deal with the Weyl group actions if one restricts to the set G_{reg} of regular semi-simple elements of G (which is a dense Zariski-open subset of G). Let $T_{reg} = T \cap G_{reg}$. Define a morphism $\alpha: G/T \times T_{reg} \to \varphi^{-1}G_{reg}$ by $\alpha(gT,t) = g * t$ (notation of 2.6).

6.4. Lemma. (i) α *is an isomorphism;*
(ii) $\varphi: \varphi^{-1}G_{reg} \to G_{reg}$ *is a Galois covering with group* W.
(i) follows from the observation that an element in $T_{reg}.U$ is conjugate in B to a unique element in T_{reg}. As a consequence of (i), the morphism of (ii) can be viewed as the morphism $G/T \times T_{reg} \to G_{reg}$ sending (gT,t) to gtg^{-1}. That this is a Galois covering is readily seen. The covering group W operates by $w.(gT,t) = (g\dot{w}^{-1}T, \dot{w}t\dot{w}^{-1})$ ($\dot{w}$ denoting a representive of $w \in W$).

6.5. By the lemma, we have that if $x \in G_{reg}$ the inverse image $\varphi^{-1}x$ is a finite set of points, so $H^i(\mathcal{B}_x,\mathbb{Q}) = 0$ if $i \neq 0$. Moreover, the Galois action of W then induces a representation of W on $H^0(\mathcal{B}_x,\mathbb{Q})$, which is isomorphic to the regular representation over $\mathbb{Q}$. More precisely, there is a W-action on the restriction of $R^0\varphi_*\mathbb{Q}$ to G_{reg} inducting these W-actions in the stalks $H^0(\mathcal{B}_x,\mathbb{Q})$. It amounts to the same to say that there is a W-action on the restriction of $\mathbb{R}\varphi\mathbb{Q}$ to G_{reg}. We denote by L the restriction of $R^0\varphi_*\mathbb{Q}$ to G_{reg}, it is a "local system" on G_{reg}.
It is here that the machinery of intersection cohomology comes in. This gives an "extension" of the local sytem L on G_{reg} to a complex $\mathbb{I}C^{\cdot}(G,L)$ on all of G, characterized by certain axioms (see [12, p. 120]). The W-action on L is inherited by $\mathbb{I}C^{\cdot}(G,L)$. Now the fact that φ is a small morphism (see 4.3.5) allows one to prove that $\mathbb{R}\varphi\mathbb{Q}$ satisfies these axioms [loc.cit., p. 121], whence $\mathbb{I}C^{\cdot}(G,L) = \mathbb{R}\varphi\mathbb{Q}$. This implies that there is a W-action on $\mathbb{R}\varphi\mathbb{Q}$.

6.6. We assume that $x \in G$ is unipotent. The centralizer $Z(x)$ operates on $\mathcal{B}_x$, and hence on $H^i(\mathcal{B}_x,\mathbb{Q})$, and the connected centralizer $Z(x)^0$ acts trivially. Hence the finite group $A(x) = Z(x)/Z(x)^0$ operates linearly in $H^i(\mathcal{B}_x,\mathbb{Q})$ and it turns out that the actions of $A(x)$ and W in $H^i(\mathcal{B}_x,\mathbb{Q})$ commute. So we have a representation of the direct product $A(x) \times W$ in these spaces.
Now put $V_x = H^{2e(x)}(\mathcal{B}_x,\mathbb{Q})$ (the top cohomology group) and for any character λ of an absolutely irreducible representation of $A(x)$ over $\mathbb{Q}$, let $V_{x,\lambda}$ be the λ-isotopic subspace of V_x. Then $A(x) \times W$ acts linearly in $V_{x,\lambda}$, let $\lambda \times \chi_{x,\lambda}$ be the character of the corresponding

representation of $A(x) \times W$. It should be pointed out that the fact that in the next theorem one can work over $\mathbb{Q}$ uses the classification result 3.2.3.
The following theorem gives a sort of parametrization of the irreducible characters of W.

6.7. Theorem. *If $V_{x,\lambda} \neq 0$ then $\chi_{x,\lambda}$ is an absolutely irreducible character of* W. *Any irreducible character of* W *is of this form, the pair (x,λ) being unique up to conjugacy.*
A proof in the context sketched in 6.5 is given in [6].

Example. $G = PSL_n(C)$. Using 5.3.2(i) one sees that only $\lambda = 1$ occurs here. So the V_x are irreducible W-modules, and the representations of the symmetric group are parametrized by conjugacy classes of unipotent elements in $SL_n(\mathbb{C})$.

We notice the following consequence of the theorem.

6.8. Corollary. *All irreducible representations of a Weyl group are defined over* $\mathbb{Q}$.

6.9. If Γ is a finite group we denote by $\Gamma^\vee$ the set of its irreducible characters. We then see from 6.7 that certain conjugacy classes of pairs (x,λ) of the theorem are parametrized by $W^\vee$. In fact, not all $\lambda \in A(x)^\vee$ need occur (i.e. it may happen that $V_{x,\lambda} = 0$).
This remark shows that it may be reasonable to look at conjugacy classes of pairs (x,λ), instead of conjugacy classes of unipotent elements. It turns out that, in fact, there is a good a priori description of conjugacy classes of pairs. Thus the goal of an a priori parametrization of unipotent classes has been achieved, to some extent. The relevant results are due to G. Lusztig [18] and we shall briefly describe them. The proofs of [loc.cit.] rely heavily on the use of the recently developed new tools in algebraic topology (intersection homology and perverse sheaves). It seems that, at the moment, these tools are indispensable for the finer theory of conjugacy classes. (The same is true for the theory of characters of finite groups of Lie type.)

6.10. We now only assume G to be connected, reductive. Let $P = LU_P$ be a parabolic subgroup of G (notations as in 2.6). Denote by V_G and V_L the unipotent varieties of G resp. L. For $x \in V_G$, $y \in V_L$ put

$$Y_{x,y} = \{gZ_L(y)^\circ U_P \mid g \in G, g^{-1}xg \in U_P\}.$$

Then $\dim Y_{x,y} \leq \frac{1}{2}(\dim Z_G(x) - \dim Z_G(y)) = d$, say. This is a generalization of 4.3.3 (ii) (in which equality need not always hold). More general results of this kind are established in [loc.cit., §1].

The component group $A_G(x)$ operates on the finite set $\Sigma_{x,y}$ of components of $Y_{x,y}$ of dimension d. (If P = B and y = 1 then $\Sigma_{x,y}$ is the set of components of $\mathcal{B}_x$.)

An irreducible representation (over $\mathbb{C}$) of $A_G(x)$ is called cuspidal if it does not occur in any permutation representation on $\Sigma_{x,y}$, for P,y as before, with $P \neq G$.

6.11. Theorem. *Fix a character* χ *of the group of components of the centre of* G. *There is at most one* G*-conjugacy class of pairs* (x,λ), *with* $x \in V_G$ *and* λ *a complex irreducible character of* $A_G(x)$, *such that* λ *is cuspidal.*

This is a consequence of the main result of [loc.cit.]. The result proved there is more general, in that the ground field is only assumed to have good characteristic. In [loc.cit.] the classification is given of the possible cases with G quasi-simple.

Moreover, Lusztig shows that to any pair (x,λ) one can attach a triple (L,y,λ'), unique up to G-conjugacy, consisting of a Levi subgroup L of some parabolic subgroup, an element $y \in V_L$ and a cuspidal character λ' of $A_L(y)$. The conjugacy classes of pairs (x,λ) giving rise to a fixed triple (L,y,λ') can be put in a bijective correspondence with the irreducible characters of the group N_GL/L (which happens to be a Coxeter group).

These results give an a priori description of the conjugacy classes of pairs (x,λ). As already pointed out, the proofs rely very much on the use of algebraic topology.

REFERENCES

[1] A.V. Alexeevsky, Component groups of centralizers for unipotent elements in semisimple algebraic groups (in Russian), Trudy Tbilissk. Math. Inst. 62 (1979), 5-28.

[2] P. Bala, R.W. Carter, Classes of unipotent elements in simple algebraic groups, Math. Proc. Camb. Phil. Soc. 79 (1976), 401-425 and 80 (1976), 1-18.

[3] P. Beardsly, R.W. Richardson, Etale slices for algebraic transformation groups in characteristic p, to appear.

[4] A. Borel, Linear algebraic groups, New York, Benjamin, 1969.

[5] A. Borel et al. Seminar on algebraic groups and related finite groups, Lect. Notes in Math. no. 131, Springer-Verlag, 1970.

[6] W. Borho, R. MacPherson, Représentations de groupes de Weyl et homologie d'intersection pour les variétés nilpotentes, C.R. Acad. Sc. Paris 292 (1981).

[7] W. Borho, J.-L. Brylinski, Differential operators on homogeneous spaces I, Inv. Math. 69 (1982), 437-476.

[8] W. Borho, R. MacPherson, Partial resolutions of nilpotent varieties, in: Analyse et topologie sur les espaces singuliers Astérisque vol.101-102, p. 23-74, Soc. Math. Fr. 1983.

[9] N. Bourbaki, Groupes et algèbres de Lie, chap. IV,V,VI, Paris, Hermann, 1968.

[10] Combinatoire et représentation du groupe symétrique (Strasbourg 1976), Lect. Notes in Math. no. 579, Springer-Verlag, 1977.

[11] E.B. Dynkin, Semisimple subalgebras of semisimple Lie algebras, Amer. Math. Soc. Transl. Ser. 2, 6 (1957), 111-245 (= Math. Sbornik N.S. 30 (1952), 349-462).

[12] M. Goresky, R. MacPherson, Intersection homology II, Inv. Math. 71 (1983), 77-129.

[13] W. Hesselink, Singularities in the nilpotent scheme of a classical group, Trans. Amer. Math. Soc. 222 (1976), 1-32.

[14] D.F. Holt, N. Spaltenstein, Nilpotent orbits of exceptional Lie algebras over algebraically closed fields of bad characteristic, to appear.

[15] J.E. Humphreys, Linear algebraic groups, Springer-Verlag, 1975.

[16] D.E. Knuth, Permutation matrices and generalised Young tableaux, Pac.J. Math. 34 (1970), 709-727.

[17] B. Kostant, Lie group representations in polynomial rings, Amer. J. Math. 85 (1963), 327-404.

[18] G. Lusztig, Intersection cohomology complexes on a reductive group, Inv. Math. 75 (1984), 205-273.

[19] G. Lusztig, N. Spaltenstein, Induced unipotent classes, J. London Math. Soc. 19 (1979), 41-52.

[20] I.G. Macdonald, Symmetric functions and Hall polynomials, Oxford Univ. Press, 1979.

[21] K. Pommerening, Uber die unipotenten Klassen reduktiver Gruppen II, J. Alg. 65 (1980), 373-398.

[22] P. Slodowy, Simple singularities and simple algebraic groups, Lect. Notes in Math. no. 815, Springer-Verlag, 1980.

[23] N. Spaltenstein, The fixed point set of a unipotent transformation on the flag manifold, Proc. Kon. Ak. v. Wet. Amsterdam, Ser. A, 79 (1976), 452-456.

[24] N. Spaltenstein, Classes unipotentes et sous-groupes de Borel, Lect. Notes in Math. no. 946, Springer-Verlag, 1982.

[25] T.A. Springer, The unipotent variety of a semisimple group, in: Algebraic geometry (Bombay Colloquium 1968), p. 373-391, Oxford Univ. Press, 1969.

[26] T.A. Springer, Trigonometric sums, Green functions of finite groups and representations of Weyl groups, Inv. Math. 36 (1976), 173-207.

[27] T.A. Springer, Linear algebraic groups, (2nd ed.) Birkhäuser, 1981.

[28] T.A. Springer, Quelques applications de la cohomologie d'intersection, (sém. Bourbaki no. 582), Astérisque nr. 92-93, p. 249-273, Soc. Math. Fr., 1982.

[29] R. Steinberg, Representations of algebraic groups, Nagoya Math. J. 22 (1963), 33-56.

[30] R. Steinberg, Regular elements of semisimple algebraic groups, Publ. Math. I.H.E.S. no. 25 (1965), 49-80.

[31] R. Steinberg, Conjugacy classes in algebraic groups, Lect. Notes in Math. no. 366, Springer-Verlag, 1974.

[32] R. Steinberg, On the desingularization of the unipotent variety, Inv. Math. 36 (1976), 209-224.

SOME FINITE GROUPS WHICH APPEAR AS Gal L/K, WHERE $K \subseteq Q(\mu_n)$

John G. Thompson

The series of six lectures given by the author at the meeting dealt largely with the author's paper with the above title which appeared in Journal of Algebra 89, 437-499 (1984). The Contents of that paper are listed below.
The remainder of this contribution deals with related topics, and includes remarks designed to facilitate the reading of the J.Algebra paper.

Table of Contents

REGULAR GALOIS EXTENSIONS OF $Q(x)$

Let $\overline{Q}$ be a fixed algebraic closure of Q. It is a straightforward exercise to check that if x is an indeterminate, then every subfield of $\overline{Q}(x)$ which contains $Q(x)$ is of the form $E(x)$ for some subfield E of $\overline{Q}$. This is a handy fact in dealing with extensions of $Q(x)$.

An extension field R of $Q(x)$ is said to be <u>regular</u> if and only if $R \cap \overline{Q} = Q$, that is, R is regular precisely when every algebraic number in R is rational.

The object of this note is to prove the following result.

Theorem. <u>Suppose</u> $1 \to A \to B \to C \to 1$ <u>is an exact sequence of finite groups such that</u> A <u>is Abelian, and</u> B <u>splits over</u> A. <u>If</u> F <u>is a regular Galois extension of</u> $Q(x)$ <u>and</u> gal $F/Q(x) = C$, <u>then there is a regular Galois extension</u> E <u>of</u> $Q(x)$ <u>such that</u> $E \supset F$, gal $E/Q(x) \cong B$, <u>and such that in this isomorphism, the fixed field of</u> A <u>is</u> F.

To prove the theorem, several preliminaries are in order. We give ourselves a Galois extension K of Q(x) such that [K:Q(x)] is finite. Let

$$G = \text{gal } K/Q(x).$$

If G_o is a subgroup of G, let K_{G_o} denote the fixed subfield of G_o, and let $\mathcal{N}_{G_o}$ be the usual norm map:

$$\mathcal{N}_{G_o} : K \to K_{G_o}$$

$$\kappa \longmapsto \prod_{g \in G_o} \kappa \circ g = \mathcal{N}_{G_o}(\kappa) .$$

Similarly, when $G_o \triangleleft G$, let

$$\mathcal{N}_{G/G_o} : K_{G_o} \to Q(x),$$

$$\kappa \longmapsto \prod_{g \in G/G_o} \kappa \circ g = \mathcal{N}_{G/G_o}(\kappa).$$

This, too, is the usual norm map since

$$G/G_o = \text{gal } K_{G_o}/Q(x).$$

If F is a subfield of K which contains Q(x), let G_F be the subgroup of G whose fixed field is F.

Let L be the algebraic closure of Q in K, let $\mathcal{O}$ be the ring of integers of L, and let

$$H = \text{gal } L/Q \cong \text{gal } L(x)/Q(x).$$

Then, setting $N = G_{L(x)}$, there is an exact sequence

$$1 \to N \to G \xrightarrow{r} H \to 1,$$

where r is the restriction map from K to L(x). Set $\mathcal{N}_H = \mathcal{N}_{G/N}$.

Let y be a primitive element for K over L(x), and let

$p(Y) \in L(x)[Y]$ be the monic irreducible polynomial which satisfies $p(y) = 0$. Then $\{y \circ \eta | \eta \in N\}$ is the set of roots of $p(Y)$ in K, and

(1) $$p(Y) = \prod_{\eta \in N} (Y - y \circ \eta).$$

I argue that $p(Y)$ is irreducible in $\overline{Q}(x)[Y]$. If false, then there is a non empty proper proper subset N_o of N such that the elementary symmetric functions of $\{y \circ \eta | \eta \in N_o\}$ are in E(x) for some subfield E of $\overline{Q}$. By the remark in the first paragraph, this forces $K \cap \overline{Q} \supset L$, against the definition of L. So $p(Y)$ is absolutely irreducible.

Theorem 1. *Suppose* m *is a natural number and* $\zeta_m \in K$, *where* ζ_m *is a primitive* m^{th} *root of unity. Let*

$$M_m = K^{\times}/K^{\times m}, \quad A_m = Z/mZ.$$

View M_m *as a* $A_m G$*-module, via*

$$(a,\xi) \longmapsto \prod_{g \in G} (a \circ g)^{n(g)} = a^{\xi},$$

$$a \in M_m, \quad \xi = \sum_{g \in G} n(g)g \in A_m G.$$

Then for each natural number ℓ, M_m *has a submodule* $\overline{U} = U_{\ell}/K^{\times m}$ *which is free of rank* ℓ, *and such that*

$$U_{\ell} \cap L^{\times} = L^{\times m}.$$

Proof. Let

$$V = \{z \in K^{\times} | z^m \in L(x)\}.$$

Then V is a multiplicative group containing $L(x)^{\times}$, and $V/L(x)^{\times}$ is of exponent dividing m. Since $\zeta_m \in L$, $L(x)(V)$ is a Kummer extension of $L(x)$, and

$$G_{L(x)(V)} = [N,N].N^m .$$

Moreover, $N/[N,N].N^m$ and $V/L(X)^\times$ are isomorphic finite Abelian groups.

Choose elements $v_1,\dots,v_s \in V$ such that $\{v_1L(x)^\times,\dots,v_sL(x)^\times\}$ is a basis for $V/L(x)^\times$. If the order of $v_iL(x)^\times$ in $V/L(x)^\times$ is m_i, then

$$v_i^{m_i} = \lambda_i \prod_j q_{ij}^{rij} ,$$

where $\lambda_i \in L^\times$, the q_{ij} are irreducible polynomials in $\mathcal{O}[x]$, and each r_{ij} is a non zero integer. Let

$$R = \prod_{i,j} \mathcal{N}_H(q_{ij}) ,$$

Then $R \in Z[x]$.

Let I be the integral closure of Q[x] in K. By Hilbert's Irreducibility Theorem, there is an integer q such that

(i) $x - q$ does not divide R.

(ii) $I/(x-q)I = k$ is a Galois extension of Q with $[k:Q] = [K:Q(x)]$.

By a fundamental property of number fields, there are ℓ distinct primes $p_1,\dots,p_\ell$, each of which is unramified in k and each of which splits completely in k. Let $P = p_1p_2\cdots p_\ell$, and let $\mathfrak{p}_i$ be a prime ideal in the ring of integers of k which divides p_i. By the Approximation Theorem, there are algebraic integers ξ_i in k such that

(i) $\xi_i \in \mathfrak{p}_i$, $\xi_i \notin \mathfrak{p}_i^2$.

(ii) If $\mathfrak{q}$ is any prime ideal distinct $\mathfrak{p}_i$ which divides P, then $\xi_i \notin \mathfrak{q}$.

Next, I argue that there is f_i in I such that

(i) $\xi_i = f_i + (x-q)I$.

(ii) $\mathcal{N}_N(f_i)$ is relatively prime to R in L[x].

To see this, I use the fact that $x - q$ and R are relatively prime.

There are a, b $\in L[x]$ such that

$$a(x-q) + bR = 1 ,$$

and so $I = (x-q)I + RI$. Thus, RI maps surjectively onto k. Pick $f_i' \in RI$ with $\xi_i = f_i' + (x-q)I$. Hence, as $R \in Q[x]$, we have $f_i' \circ g \in RI$ for all $g \in G$, so in particular, if $|N| = n$, then

$$\prod_{g\in N} (Y + f_i' \circ g) = Y^n + c_{i,1} Y^{n-1} +\ldots+ c_{i,n} \in L[x,Y] ,$$

where $c_{i,1},\ldots,c_{i,n} \in RL[x]$. Set $f_i = x-q + f_i'$. Then

$$\mathcal{N}_N(f_i) = (x-q)^n + c_{i,1}(x-q)^{n-1} +\ldots+ c_{i,n}$$

is relatively prime to R , and $\xi_i = f_i + (x-q)I$, as desired.

Set

$$\eta_i = f_i K^{\times m} .$$

Suppose $\alpha_1,\ldots,\alpha_\ell \in A_m G$, and

(*)
$$\prod_{i=1}^{\ell} \eta_i^{\alpha_i} = 1 \text{ in } M_m .$$

Let $\alpha_i = \sum_{g\in G} n_i(g)g$, $n_i(g) \in A_m$. Choose natural numbers $\tilde{n}_i(g) \in n_i(g)$ for all i,g , and set

$$\tilde{\alpha}_i = \sum_{g\in G} \tilde{n}_i(g).g .$$

From (*), we get

(**)
$$\prod_{i=1}^{\ell} f_i^{\tilde{\alpha}_i} = f^m$$

for some $f \in K^\times$. The left hand side is in I, which is integrally closed in K , and so $f \in I$. Take the image of (**) in k, obtaining

$$\prod_{i=1}^{\ell} \xi_i^{\tilde{\alpha}_i} = \xi^m$$

in k . The stabilizer in G of each γ_i is 1 , and so

$$\nu_{\gamma_i} \circ g\left(\prod_{i=1}^{\ell} \xi_i^{\tilde{\alpha}_i}\right) = \tilde{n}_i(g) = m\nu_{\gamma_i} \circ g(\xi) .$$

Here ν_γ is the usual valuation. So $\tilde{n}_i(g) \equiv 0 \pmod m$ for all i,g , and each α_i is 0 in A_mG . Let $\tilde{U}_\ell$ be the ZG-submodule of $K^\times$ generated by the f_i and $K^{\times m}$. We conclude that $\overline{U}_\ell = U_\ell/K^{\times m}$ is a free A_mG-module of rank ℓ . Moreover, if U^*_ℓ is the ZG-module generated by the f_i alone, then $\overline{U}_\ell \cong U^*_\ell/U^*_\ell \cap K^{\times m}$ as ZG-modules, and in addition the image of $U^*_\ell/U^*_\ell \cap K^{\times m}$ in $k^\times/k^{\times m}$ is isomorphic to $\overline{U}_\ell$. This is helpful in handling the remaining difficulty in the proof of Theorem 1.

Let D be a non zero rational integer such that

$$S = D \prod_i \mathcal{N}_G(f_i) \in Z[x] .$$

We must examine the polynomials p(Y) mentioned in (1). We say that the element Y of K is <u>good</u> if and only if y is a primitive element for K over Q(x) (not merely over L(x)) , and in addition, y is in the integral closure of Z[x] in K . Good elements exist, since if y is any primitive element for K over Q(x) , then cy is good for suitable $c \in Z[x]$. Choose a good y_o and set

$$y = PRSy_o .$$

Then p(Y), the polynomial in (1), is of the form

$$p(Y) = Y^n + c_1Y^{n-1} + \ldots + c_n ,$$

where $c_1,\ldots,c_n$ lie in PRS [x]. By the Irreducibility Theorem, there are integers $q_1,\ldots,q_\ell$ such that

(i) $q_i \equiv 1 \pmod P$, $1 \le i \le \ell$.

(ii) $\{p(q_i)\circ \eta \mid 1 \le i \le \ell , \eta \in H\}$ is a set of $\ell|H|$ polynomials in L[x] which are irreducible and pairwise relatively prime.

For each i, set $g_i = (q_i - y).f_i$. Note that $\mathcal{N}_N(q_i - y) = p(q_i)$. In particular, each $p(q_i)$ is relatively prime to RS in $L[x]$. Moreover, the image of $q_i - y$ in k is an integer of k which is $\equiv 1 \pmod{P}$. Let U_ℓ be the ZG-module generated by the g_i and $K^{\times m}$. If $\mathfrak{P}$ is any prime ideal of the ring of integers of k which divides P, then the image of f_i and the image of g_i in k have the same valuation at $\mathfrak{P}$, and this guarantees that $U_\ell/K^{\times m} = \bar{U}$ is free of rank ℓ.

Now suppose $\lambda \in L^\times$ and $\lambda K^{\times m} \in \bar{U}$. Then in particular, $\lambda K^{\times m}$ is a fixed point of N on $\bar{U}$. As $\bar{U}$ is a free $A_m N$- module, every fixed point is of the form $\mathcal{N}_N(u)K^{\times m}$ for some $u \in U_\ell$. Thus, there are elements $\beta_1, \ldots, \beta_\ell \in ZH$ such that

$$(+) \qquad \prod_{i=1}^{\ell} \mathcal{N}_N(g_i)^{\beta_i} = \lambda f^m$$

for some $f \in K^\times$. The left hand side is in $L(x)^\times$, and so is λ, so $f^m \in L(x)$, whence $f \in V$. Thus

$$f = vh, \quad v \in \langle v_1, \ldots, v_s \rangle, \quad h \in L(x)^\times .$$

By construction of the g_i, the only term in the left hand side of (+) which is not relatively prime to $p(q_i) \circ \eta$ $(\eta \in H)$ is $\mathcal{N}_N(g_i)^{\beta i}$, and

$$\nu_{p(q_i) \circ \eta}(\mathcal{N}_N(g_i)^{\beta_i}) = \text{coefficient of } \eta \text{ in } \beta_i .$$

Now $\nu_{p(q_i) \circ \eta}(\lambda v^m) = 0$, since $p(q_i)$ is relatively prime to R in $L[x]$, so we get that each β_i is in mZH. Now (+) implies that $\lambda \in K^{\times m}$, and as L is algebraically closed in K, $\lambda \in L^{\times m}$, as required. The proof of Theorem 1 is complete.

It is now easy to complete the proof of the Theorem. Suppose C and F are as indicated. Let m be the exponent of A. Let $K = F(\zeta_m)$, $G = \text{gal } K/Q(x)$. Then $G = C \times H$, where $H = G_F = \text{gal } K/F$. View A as a G-module by letting H act trivially.

Let ℓ be the number of generators of A as G-module. We may assume that $A \neq 1$, otherwise there is nothing to prove. So $\ell > 0$.

Choose U_ℓ as in Theorem 1. Note that since F is regular, the algebraic closure of Q in K is $\mathbb{Q}(\zeta_m)$, as $Q(\zeta_m,x)$ is a subfield K_o of K which is maximal subject to $K_o \cap F = Q(x)$. Let $E_o = K(\sqrt[m]{U_\ell})$. Observe that by the condition $U_\ell \cap L = L^{\times m}$ in Theorem 1, $Q(\zeta_m)$ is the set of all algebraic numbers in $\sqrt[m]{U_\ell}$. Since E_o is a Kummer extension of K, $Q(\zeta_m)$ is the algebraic closure of Q in E_o. Let $A_o = \text{Gal } E_o/K$. Since $\bar{U}$ is a free $A_m G$-module of rank ℓ, so is A_o. (See Remark 1 which follows the proof of the Corollary.) Moreover, since $H^2(G,A_o) = 0$, there is a subgroup $G_o = G$ which is a complement to A_o in gal $E_o/Q(x)$. Using this isomorphism, we abuse notation slightly by taking $G = G_o$. More precisely, each element of G has a unique extension to an automorphism of E_o which lies in G_o. With this convention, gal $E_o/Q(x) = A_o.G$, $A_o \cap G = 1$. Let

$$\pi: A_o \to A$$

be a G-homomorphism of A_o onto A, and let $B_o = \ker \pi$. Then H acts trivially on A_o/B_o, so B_oH/B_o is a direct factor of A_oG/B_o. Let E be the fixed field of B_oH, so that $A_oG/B_oH = \text{gal } E/Q(x)$, and the fixed field of A_oH/B_oH is F. Moreover, $A_oG/B_oH/A_oH/B_oH = A_oG/A_oH = G/H = C$, and as ZC-module, $A_oH/B_oH = A_o/B_o = A$. Visibly, B_oG/B_oH is a complement to A_oH/B_oH in A_oG/B_oH, so $A_oG/B_oH = B$ via an isomorphism which carries A_oH/B_oH to A. The proof of the Theorem is complete.

There is a corollary which is worth recording.

Corollary. <u>Suppose G is a finite group and one of the following holds</u>:

(i) G <u>is solvable and every Sylow subgroup is Abelian</u>.

(ii) G <u>is nilpotent of class</u> 2.

<u>Then there is a regular Galois extension</u> E <u>of</u> Q(x) <u>such that</u>

$$\text{gal } E/Q(x) = G .$$

Proof. (i) If G is Abelian, we take $A = B = G$, $C = 1$, and apply the Theorem. We may assume that G is non Abelian. Define $G_1 = G$, $G_{n+1} = [G_n, G_n]$, $n = 1,2,\ldots,$ and let n be the smallest integer such that G_{n+1} is Abelian. Let C_n be a Carter subgroup of G_n. As all Sylow subgroups of G are Abelian, while C_n is nilpotent, it follows that C_n is Abelian. Since all Carter subgroups of C_n are self normalizing in G_n, and as any two Carter subgroups are conjugate in the group they generate, it follows that $G_n = C_n G_{n+1}$. Since $G_{n+1} \cap Z(G_n) = 1$, it follows that C_n is complement to G_{n+1} in G. So we are done by induction on the derived length of G.

(ii) We may assume that G is a p-group. Let m, n be natural numbers, set $q = p^m$, and define by generators and relations the group $G(n,q)$, as follows:

(a) if $p \neq 2$, then

$$G(n,q) = \langle x_1,\ldots,x_n \mid x_i^q = [x_i, x_j]^q = 1,$$

$$[x_i, x_j, x_k] = 1,\ 1 \le i,j,k \le n\rangle.$$

(b) if $p = 2$, then

$$G(n,q) = \langle x_1,\ldots,x_n \mid x_i^{2q} = [x_i, x_j]^q = 1,$$

$$[x_i, x_j, x_k] = 1,\ 1 \le i,j,k \le n\rangle .$$

We check that $|G(n,q)| = q^{n+\binom{n}{2}} d_n$, where $d_n = 1$ if $p \neq 2$ and $d_n = 2^n$ if $p = 2$. There is a visible homomorphism of $G(n+1,q)$ onto $G(n,q)$ obtained by collapsing x_{n+1} to 1. The kernel of this homomorphism is

$K_{n+1} = \langle x_{n+1}, [x_{n+1},x_1], [x_{n+1},x_2],\ldots,[x_{n+1},x_n]\rangle$, which is Abelian. The corresponding extension obviously splits, as $\langle x_1,\ldots,x_n\rangle$ is a complement in $G(n+1,q)$ to K_{n+1}. As $G(1,q)$ is Abelian, we see that for each n,m, there is a regular Galois extension of $Q(x)$ whose Galois group is $G(n,q)$. As every p-group of class 2 is a homomorphic image of some $G(n,q)$, (ii) follows.

Remark 1. The fact that A_o is isomorphic to $\overline{U}$ is well known. Pick $\alpha \in A_o$, $u \in U_\ell$. Choose v in E_o with $v^m = u$. Then $v\circ\alpha = v.\zeta_m^a$ for some $a \in A_m$. This produces for us an element $\hat{\alpha}$ of $\mathrm{Hom}(\overline{U},A_m)$, namely, the one which sends $uK^{\times m}$ to a, for one checks that this is well defined, and that $\overline{u}_1\overline{u}_2.\hat{\alpha} = \overline{u}_1\hat{\alpha} + \overline{u}_2\hat{\alpha}$. On the other hand, $\mathrm{Hom}(\overline{U},A_m)$ is a left ZG-module via $\overline{u}(g\circ f) = (\overline{u}\circ g)f(\overline{u} \in \overline{U}, g \in G, f \in \mathrm{Hom}(\overline{U},A_m))$. Of course, A_o is a left ZG-module via

$$G \times A_o \to A_o, \ (g,\alpha) \mapsto g\alpha g^{-1} .$$

Finally, A_m itself is a ZG-module via

$$g \circ 1 = r \text{ if } \zeta_m\circ g^{-1} = \zeta_m^r .$$

The pieces fit together by the map

$$\phi: A_o \to A_m \otimes_Z \mathrm{Hom}(\overline{U},A_m) ,$$

$$\alpha \mapsto 1 \otimes \hat{\alpha}.$$

With these definitions, one checks directly that φ is a ZG-isomorphism, and in the case at hand,

$$A_m \otimes \mathrm{Hom}(\overline{U},A_m) \cong \mathrm{Hom}(\overline{U},A_m) \cong \overline{U} ,$$

all these modules being free $A_m G$-modules of rank ℓ. The Tate twist of tensoring with Z/mZ is part of the machinery of Kummer extensions.

Remark 2. Let $\mathcal{R}$ be the smallest family of isomorphism classes of finite groups which satisfies the following conditions:

(i) $1 \in \mathcal{R}$.

(ii) If $G \in \mathcal{R}$ and $H \triangleleft G$, then $G/H \in \mathcal{R}$.

(iii) If $1 \to A \to B \to C \to 1$ is an exact sequence for which A is Abelian and is complemented in B, and $C \in \mathcal{R}$, then $B \in \mathcal{R}$.

Since the family $\mathcal{S}$ of all finite soluble groups satisfies these conditions, it follows that $\mathcal{R} \subseteq \mathcal{S}$. It is not difficult to check that $SL(2,3) \notin \mathcal{R}$. In any case, if $G \in \mathcal{R}$, then G is the Galois group of a regular Galois extension of $Q(x)$.

The results in this paper have also been obtained by David Saltman.

January 1984

Remarks (on Lectures 1-5)

Soon after completing my paper on Galois groups, I learned from J-P. Serre and other algebraic geometers that the function theoretic portion of my paper could be greatly shortened by appealing to Riemann's Existence Theorem. However, I believe my approach is still of considerable worth, for several reasons. It does not appeal to Riemann's theorem, and indeed can be derived from work of Legendre and Puiseux, and therefore can be read by relatively inexperienced mathematicians. Also, the very concreteness of my approach gives an explicit quality to the constructions.

I decided to give 5 lectures on Galois groups rather than 4 (as originally planned), and concentrated in the fifth lecture on the rather diverse simple groups which have been shown in recent months to be Galois groups over Q.

BILINEAR FORMS IN CHARACTERISTIC p AND THE FROBENIUS-SCHUR INDICATOR

(Lecture 6)

1. Introduction

If G is a finite group and M is an irreducible CG-module, then there is a nonsingular G-invariant form on M which is one of the following:

(+) symmetric.

(-) skew symmetric.

(O) Hermitian.

Moreover, if χ is the character afforded by M, and

$$\varepsilon(\chi) = \frac{1}{|G|} \sum_{g \in G} \chi(g^2)$$

is the Frobenius-Schur indicator of χ, the three cases correspond to the values of 1, - 1, 0 of $\varepsilon(\chi)$; $\varepsilon(\chi) = 1$ if and only if χ is the character afforded by a RG-module, and $\varepsilon(\chi) = 0$ if and only if $\chi(g) \notin R$ for some g in G. This result [1] has been of use in determining the character tables of various groups. Conversely, if we know the character table of G, and if we also know the squaring map, then, given M, we can decide which of the three possibilities occurs. More precisely, deciding which of (+), (-), (O) holds is reduced to a calculation based on the character table and the squaring map.

The object of this note is to show that an analogous result holds for irreducible kG-modules, where k is a splitting field for G of characteristic $\neq 0,2$. The precise situation in characteristic 2 is elusive and seems to require considerations which go beyond the arguments used here.

In order to state the relevant result, which is given in section 4 some notation and hypotheses are needed, as well as a preliminary result which is based on an idea of Witt.

(1.1) K is an algebraic number field which is a splitting field for G; $\mathfrak{p}$ is a prime ideal of the ring of the integers of K which contains the rational prime p, and $p \neq 2$;

$\mathcal{O}$ is the completion of the ring of $\mathfrak{p}$-integers of K;

$\bar{K}$ is the field of fractions of $\mathcal{O}$;

π is a generator for the maximal ideal of $\mathcal{O}$.

$k = \mathcal{O}/\pi\mathcal{O}$ is a splitting field for G.

If M is a finitely generated kG-module, ϕ_M is the Brauer character associated to M via $\mathfrak{p}$. If χ is an ordinary irreducible character of G, and ϕ is the Brauer character of an irreducible kG-module, then $d(\chi,\phi)$ is the corresponding decomposition number.

2. The Witt kernel of a EG-module.

In this section, E denotes an arbitrary field and M is a finitely generated EG-module. We assume also that M is equipped with a non singular G-invaraint bilinear form $\langle\ ,\ \rangle$, with values in E, which is either symmetric or skew symmetric.

Let

$$\mathcal{M} = \{M_o \mid M_o \text{ is a EG-submodule of M and } \langle M_o , M_o\rangle = 0\}.$$

Obviously, $\{0\} \in \mathcal{M}$, and $\mathcal{M}$ is partially ordered by inclusion. If $M_o \in \mathcal{M}$, then $M_o^{\perp}/M_o$ inherits a non singular form given by

$$(m_o + M_o , m'_o + M_o)_{M_o} = \langle m_o , m'_o\rangle \quad (m_o, m_o \in M_o^{\perp}) .$$

If M_1 is a maximal element of $\mathcal{M}$, we observe that

(2.1) $M_1^{\perp}/M_1$ is a completely reducible EG-module, and the restriction of $(\ ,\ )_{M_1}$ to any EG-submodule of $M_1^{\perp}/M_1$ is non singular.

Definition. The Witt kernel of M is $M_1^{\perp}/M_1$, where M_1 is a maximal element of $\mathcal{M}$.

A priori, this definition presents us with competing candidates for the Witt kernel. However, we have

Lemma 2.1. If M_1, M_2 are maximal elements of $\mathcal{M}$, then there is a EG-isomorphism

$$\phi: M_1^{\perp}/M_1 \to M_2^{\perp}/M_2 .$$

such that

$$(m_1 , m_1')_{M_1} = (\phi(m_1), \phi(m_1'))_{M_2} , \quad m_1, m_1' \in M_1^{\perp}/M_1 .$$

Proof. Since M_1, M_2 are in $\mathcal{M}$, we have

(2.2) $\quad M_1 \subseteq M_1^{\perp}, \; M_2 \subseteq M_2^{\perp}$.

Set

(2.3) $\quad N = M_1 \cap M_2$.

By (2.2) we have

(2.4) $\quad M_1^{\perp} \subseteq N^{\perp}, \; M_2^{\perp} \subset N^{\perp}$.

Set $\overline{M}_i = M_i/N$, $i = 1, 2$, and define

(2.5) $\quad \psi: \overline{M}_1 \times \overline{M}_2 \to E,$

$\quad \psi(\overline{m}_1, \overline{m}_2) = (\overline{m}_1, \overline{m}_2)_N.$

Suppose $\overline{m}_1 \in \overline{M}_1$ and $(\overline{m}_1, \overline{M}_2) = 0$. The set of elements $\overline{m}_1$ of $\overline{M}_1$ which satisfy this condition is a EG-submodule $\tilde{M}_1/N$ of $\overline{M}_1$.

Set $M_2^* = \tilde{M}_1 + M_2$. By definition of ψ, together with (2.2), we have $M_2^* \in \mathcal{M}$. By maximality of M_2, this forces $M_2 = \tilde{M}_2$, that is, $\tilde{M}_1 \subseteq M_2$, so by (2.3), $\tilde{M}_1 \subseteq N$, whence $\tilde{M}_1/N = \{0\}$. By symmetry, we conclude that ψ is an exact pairing, and so, as EG-modules,

(2.6) $\quad \overline{M}_1 \cong \mathrm{Hom}_E(\overline{M}_2, E), \; \overline{M}_2 \cong \mathrm{Hom}_E(\overline{M}_1, E)$.

Moreover, since ψ is an exact pairing and $M_i \in \mathcal{M}$, we have

(2.7) $\quad M_1^{\perp} \cap M_2 = N , \; M_1 \cap M_2^{\perp} = N$.

Next, we argue that

(2.8) $N^{\perp} = M_1^{\perp} + M_2 = M_1 + M_2^{\perp}$.

By symmetry, together with (2.2) and (2.4), it is enough to show that $N^{\perp} \subseteq M_1^{\perp} + M_2$. Pick $x \in N^{\perp}$. This gives us a homomorphism $< , x >$ of M_1 into E whose kernel contains N, so by (2.6) there is y in M_2 such that

$$<m_1 , x> = <m_1 , y> , m_1 \in M_1 .$$

Hence $x = x-y+y \in M_1^{\perp} + M_2$, and (2.8) follows.

If $m_1 , m_1' \in M_1^{\perp} / M_1$, say $m_1 = x + M_1$, $m_1' = x' + M_1$, then by (2.8) we have

$$x = u_1 + v, x' = u_1 + v' , u_1, u_1' \in M_1 , v, v' \in M_2^{\perp} .$$

Since $x, x' \in M_1^{\perp}$, we have

$$<x , u_1'> = <u_1 , x'> = 0.$$

Since $M_1 \in \mathcal{M}$, we also have $<u_1 , u_1'> = 0$.

Hence, $<x , x'> = <x, u_1'> + <x, v'> = <x, v'> =$

$<u_1 + v , v'> = <u_1 , v'> + <v, v'>$. Moreover,

$$v' = x' - u_1 , \text{ so } <u_1 , v'> = <u_1 , x'> - <u_1 , u_1'> = 0 - 0,$$

and so

(2.9) $\langle x , x' \rangle = \langle v , v' \rangle$.

Define

$$\phi: M_1^{\perp} / M_1 \rightarrow M_2^{\perp} / M_2$$

$$\phi(m_1) = v + M_2 , m_1 = x + M_1 , x = u_1 + v .$$

This is well defined, since

$$x = u_1 + v = \tilde{u}_1 + \tilde{v} , u_1 , \tilde{u}_1 \in M_1 , \tilde{v} , v \in M_2^{\perp} ,$$

implies that $v - \tilde{v} = \tilde{u}_1 - u_1 \in M_2^{\perp} \cap M_1 = N$, by (2.7) , so that $v - \tilde{v} \in M_2$, by (2.3). Visibly, ϕ is a EG-homomorphism. If $v \in M_2$, then $v = x - u_1 \in M_1^{\perp} \cap M_2 = N \subseteq M_1$, so $x \in u_1 + M_1 = M_1$ whence ϕ is an injection. By (2.8) , ϕ is also surjective, so is an isomorphism. Since $(m_1, m'_1)_{M_1} = \langle x, x' \rangle$, and $(\phi(m_1), \phi(m'_1))_{M_2} = \langle v, v' \rangle$, (2.9) completes the proof.

3. G-lattices.

In this section we use the notation and the hypotheses introduced in (1.1), and suppose that V is an irreducible $\bar{K}G$-module whose character χ is real-valued. Since K is a splitting field, so is $\bar{K}$, and there is a non singular G-invariant bilinear form $< , >$ on V, with values in $\bar{K}$, which is symmetric or skew-symmetric. By a G-lattice in V we mean a finitely generated G-submodule L of V such that $\bar{K}L = V$. Let $\mathcal{L} = \mathcal{L}_V$ be the family of G-sublattices. If $L \in \mathcal{L}$, then L^* denotes the dual lattice defined by

$$L^* = \{\ell \in V \mid < L, \ell > \subseteq \mathcal{O}\}.$$

Let $\mathcal{L}_i$ denote the set of G-lattices L with $<L,L> \subseteq \mathcal{O}$. If L is any element of $\mathcal{L}$, there is an integer n such that $\pi^n L \in \mathcal{L}_i$. Obviously, $\mathcal{L}_i$ is partially ordered by inclusion and if $L_1, L_2 \in \mathcal{L}_i$ with $L_1 \subseteq L_2$, then $L_2 \subseteq L_1^*$. Since L_1^*/L_1 is finite, every element of $\mathcal{L}_i$ is contained in a maximal element of $\mathcal{L}_i$. In the following discussion, L denotes a fixed maximal element of $\mathcal{L}_i$.

Lemma 3.1.

(i) $\pi L^* \subseteq L$.

(ii) Let $M = L^*/L$. There is a non singular G-invariant form $<,>_M$ on M, with values in k, defined as follows: if $m_1, m_2 \in M$, $m_i = x_i + L$ then $<m_1, m_2>_M$ = image in k of $\pi <x_1, x_2>$.

Proof. Let h be the smallest integer ≥ 0 such that $\pi^h L^* \subseteq L$. If $h \leq 1$, then (i) holds. Suppose $h \geq 2$.

Let $L_1 = L + \pi^{h-1}L^*$. Then $L_1 \in \mathcal{L}$. Moreover, if $u_1, u_2 \in L_1$, say

$u_1 = \ell_1 + \pi^{h-1}\ell_1^*$, $u_2 = \ell_2 + \pi^{h-1}\ell_2^*$, $\ell_1, \ell_2 \in L$, $\ell_1^*, \ell_2^* \in L^*$, then

$$<u_1, u_2> = <\ell_1, \ell_2> + \pi^{h-1} (<\ell_1, \ell_2^*> + <\ell_1^*, \ell_2>) + \pi^{h-2} <\pi^h \ell_1^*, \ell_2^*> \in \mathcal{O},$$

by definition of L^* and of h. Thus, $L_1 \in \mathcal{L}_i$. Since $L \subseteq L_1$, this violates the maximality of L. So (i) holds.

If $\ell_1^*, \ell_2^* \in L^*$, then $\pi \ell_1^* \in L$, so $< \pi\ell_1^*, \ell_2^* > \in \mathcal{O}$. Since $<L, L^*>$ and $<L^*, L>$ are contained in $\mathcal{O}$, and since π is a generator for the maximal ideal of $\mathcal{O}$, it follows that $< , >_M$ is well defined. To see that this form is non singular, suppose $\ell^* \in L^*$ and $<\ell^*, L^*> \subseteq \mathcal{O}$. Then $\ell^* \in L^{**} = L$, so $\ell^* + L = 0$ in M. This is (ii).

Lemma. 3.2. $\{\ell \in L \mid <\ell, L> \subseteq \pi\mathcal{O}\} = \pi L^*$.

Proof. By Lemma 3.1(i), $\pi L^* \subseteq L$. By definition of L^*, $\pi L^* \subseteq \{\ell \in L \mid <\ell, L> \subset \pi\mathcal{O}$. Thus it suffices to show that if $\ell \in L$ and $<\ell, L> \subseteq \pi\mathcal{O}$, then $\ell \in \pi L^*$. This clear, since $<\frac{1}{\pi}\ell, L> \subseteq \mathcal{O}$, so that by definition of L^*, we have $\frac{1}{\pi}\ell \in L^*$.

Now we consider $L^*/\pi L^* = A$, and the kG-submodule $B = L/\pi L^*$ of A. We have an exact sequence of kG-modules,

(3.1) $0 \to B \to A \to C \to 0,$

where $C = L^*/L$. Furthermore on C, Lemma 3.1 gives us a non singular form, and Lemma 3.2 gives us a non singular form on B. These two forms are of the same type, that is, both are symmetric or both are skew symmetric, and the type is given by $\varepsilon(\chi)$.

Lemma 3.3 _Suppose_ P _is an irreducible_ kG-_module and_

(i) _the Brauer character_ ϕ _of_ P _is real-valued_.

(ii) $d(\chi, \phi)$ _is odd_.

Let B_1, C_1 _be the Witt kernels of_ B, C, _respectively_. _Then the multiplicity of_ P _in_ $B_1 \oplus C_1$ _is odd_.

Proof. By hypothesis, the multiplicity of P in $L^* / \pi L^*$ is odd. Since ϕ is real-valued, P and $Hom_k(P, k)$ are isomorphic kG-modules, and if B_o is a kG-submodule of B which is maximal subject to $\langle B_o, B_o \rangle_B = 0$, then the multiplicity of P in B_o equals the multiplicity of P in $B/B_o^{\perp}$, so the parity of the multiplicity of P in B is the parity of the multiplicity of P in B_1. The same argument applies to C, and the lemma follows.

4. Self-dual irreducible kG-modules.

Let M be an irreducible self-dual kG-module.

Let $\rho: G \to GL(M)$ be a matrix representation of G with respect to some k-basis of M. Then there is a non singular matrix X such that

$$(4.1) \qquad {}^t\rho(g)^{-1} = X^{-1}\rho(g)X, \quad g \in G,$$

where t is the transpose map. Applying the transpose inverse map to (4.1) gives us

$$(4.2) \quad \rho(g) = {}^tX \, {}^t\rho(g)^{-1} \, . \, {}^tX^{-1},$$

and if we use (4.1) in (4.2) we get that $X . {}^tX^{-1}$ centralizes $\rho(G)$. Since k is a splitting field, we have $X . {}^tX^{-1} = cI$ for some c in k^X, that is, $X = c\,{}^tX$.

Applying the transpose map to this equation gives us $c^2 = 1$, so $c = 1$ or -1. Set $\varepsilon(M) = c$. Note that by (4.1) we have (4.3) $X = \rho(g) . X . {}^t\rho(g)$.

If $c = 1$, (4.3) produces for us a non singular symmetric G-invariant bilinear form on M, and if $c = 1$, we get a skew symmetric form.

We can now state the main result of this note.

Theorem. _If ϕ is the Brauer character of the self-dual irreducible kG-module M, there is an irreducible K-character χ of G such that_

(i) χ _is real valued._

(ii) $d(\chi, \phi)$ _is odd._

<u>Moreover, if</u> χ <u>is any irreducible</u> K-<u>character which satisfies</u>

(i) <u>and</u> (ii), <u>then</u> $\varepsilon(\chi) = \varepsilon(M)$.

Proof. It suffices to show that χ exists, the equality $\varepsilon(\chi) = \varepsilon(M)$ being a consequence of the construction in section 3.

Let B be the block containing ϕ, let $\{\chi_1,\ldots,\chi_m\}$ be all the irreducible K-characters in B, and let $\{\phi_1,\ldots,\phi_n\}$ be all the irreducible Brauer characters in B. Let

$$d_{ij} = d(\chi_i,\phi_j),\ 1 \le i \le m, \le 1 \le j \le n,$$

$$c_{rs} = \sum_{i=1}^{m} d_{ir}\, d_{is}\ ,$$

$$C = (c_{rs})\ ,\ 1 \le r,\ s \le n.$$

Then det C is a power of p, and in particular, is odd.

Choose notation so that $n = 2n_1 + n_2$, where $\{\phi_1,\phi_2\ ,\ \{\phi_3,\phi_4\},\ldots,\{\phi_{2n_1 . 1},\phi_{2n_1}\}$ are pairs of complex conjugates, and $\phi_{2n_1+1},\ldots,\phi_n$ are real valued. By hypothesis, $n_2 \ne 0$. Write C in block form

$$C = \begin{pmatrix} C_0 & C_2 \\ {}^tC_2 & C_1 \end{pmatrix},$$

where C_0 is $2n_1 \times 2n_1$ and C_1 is $n_2 \times n_2$.

Lemma 4.1. det C_1 is <u>odd</u>.

Proof. For i = 1, 2,...,m, let P_i be the projective indecomposable kG-module whose socle has Brauer character ϕ_i, and let Φ_i be the Brauer character of P_i.
Then

$$c_{ij} = (\Phi_i,\Phi_j)\ .$$

Let σ be the permutation matrix (and also the permutation) which interchanges $2i - 1$ and $2i$ for $i = 1,2,\ldots,n_1$, and fixes $2n_1 + 1,\ldots,n$. S ince

$$\overline{\Phi}_i = \Phi_{i+1}, \quad i = 1,3,\ldots,2n_1 - 1,$$

$$\overline{\Phi}_i = \Phi_i, \quad i = 2n_1 + 1,\ldots,n,$$

we get

$$\sigma^{-1} C \sigma = C .$$

Let $\tilde{S}_n$ be the set of all permutations of $\{1,2,\ldots,n\}$ which do not leave invariant $\{2n_1 + 1,\ldots,n\}$. Then

$$\det C = \det C_o \cdot \det C_1 + \sum_{\tau \in \tilde{S}_n} sg(\tau) c_{1\tau(1)} c_{2\tau(2)} \cdots c_{n\tau(n)} .$$

Since $\tilde{S}_n$ is the complement in S_n of the centralizer of σ, it follows that $\sigma^{-1} \tilde{S}_n \sigma = \tilde{S}_n$, and σ has no fixed points on $\tilde{S}_n$. Since $\sigma^{-1} C \sigma = C$, we have

$$c_{ij} = c_{\sigma(i)\,\sigma(j)} .$$

Pick $\tau \varepsilon \tilde{S}_n$, and set $\tau' = \sigma\tau\sigma$. Then $\tau' \neq \tau$. Also,

$$\prod_{i-1}^{n} c_{i\tau(i)} = \prod_{i=1}^{n} c_{\sigma(i)\,\sigma\tau(i)} = \prod_{i=1}^{n} c_{\sigma(i)\,\tau'\sigma(i)} =$$

$$\prod_{i=1}^{n} c_{i\tau'(i)}$$

and so $\det C \equiv \det C_o \cdot \det C_1 \pmod 2$. The lemma follows.

Lemma 4.2 <u>For each</u> $j = 2n_1 + 1,\ldots,n$, <u>there is</u> $i \;\varepsilon \{1,2,\ldots,m\}$ <u>such that</u> d_{ij} <u>is odd and such that</u> χ_i <u>is real-valued</u>.

Proof. Let $m = 2m_1 + m_2$, where notation is chosen so that $\{\chi_1, \chi_2\},\ldots,\{\chi_{2m_1-1}, \chi_{2m_1}\}$ are pairs of complex conjugates, while $\chi_{2m_1+1},\ldots,\chi_m$ are real valued.

Suppose $d_{ij} \equiv 0 \pmod 2$, for all $i = 2m_1 + 1,\ldots, m$.
Then for each $k \in \{1,2\ldots,n\}$, we have

$$c_{jk} = \sum_{i=1}^{m} d_{ij}\, d_{ik} \equiv \sum_{i=1}^{2m_1} d_{ij}\, d_{ik} \pmod 2.$$

On the other hand, $\overline{\Phi}_j = \Phi_j$, and if $k \in \{2n_1 + 1,\ldots, n\}$, then $\overline{\phi}_k = \phi_k$, and so

$$d_{ij} = d_{i+1j}, \quad d_{ik} = d_{i+1k}, \quad i = 1,3,\ldots,2m_1-1,$$

whence $c_{jk} \equiv 0 \pmod 2$, $k \in \{2n_1 + 1,\ldots,n\}$. This means that some row of C_1 consists of even integers. This violates Lemma 4.1. This contradiction completes the proof of Lemma 4.2 and the theorem is proved.

The result in this paper has also been obtained by Wolfgang Willems.

[1] G. Frobenius and I. Schur, Über die reellen Darstellungen der endlichen Gruppen, collected works of Frobenius vol. 3, p. 355.

May 1984

<u>Remarks</u> (on Lecture 6)

In recent years, a great deal of work by a small number of people in Cambridge (Conway, Norton, Parker, Thackray,...) has been devoted to the determination of the irreducible representations in characteristic p of various simple groups. The techniques are largely ad hoc and have relied heavily on computing machines. The theorem proved in my lecture was arrived at empirically by Richard Parker, and it can be said that the theorem stems from the computer, although my proof makes no allusion to or use of the computer.

Department of Pure Mathematics and Mathematical Statistics
University of Cambridge, Cambridge CB2 1SB, England, U.K.

HOMOMORPHISMS FROM LINEAR GROUPS OVER DIVISION RINGS TO ALGEBRAIC GROUPS

Yu Chen
Institute of Systems Science, Academia Sinica
Beijing 100080, China

The problem of homomorphisms of classical groups is one of the central problems in the theory of classical groups. Since O. Schreier and B.L.van der Waerden's paper [13] appeared in 1928, the problem of isomorphisms of classical groups have been solved to a great extent. It was not until 1970's that people began to consider some general problems the so-called "abstract" homomorphisms of algebraic groups, A.Borel & J.Tits [3] determined first the abstract homomorphisms between simple isotropic algebraic groups. Then B.Weisfeiler gave a series of further results. The aim of present paper is to determine the homomorphisms from linear groups over division rings to algebraic groups, which is one of the open problems proposed by B.Weisfeiler in the Conference on "Abstract homomorphisms of algebraic groups" [10] held in Pennsylvania State University, USA in 1979. In next paper, the homomorphisms from unitary groups over division rings to algebraic groups will be studied.

The present paper is submitted in partial fulfillment of the requirements for the Doctor Degree in the Institute of Systems Science, Academia Sinica under the guidance of Professor Zhe-xian Wan, to whom the author likes to express his sincere thanks. He also wishes to express his gratitude to Professor Boris Weisfeiler for his valuable suggestions and kind guidance which will remain in his memory forever.

1. Preliminaries

1.1 Let K be an algebraically closed field and k be a subfield of K. The algebraic varieties in this paper are affine varieties over K. Suppose k_s is the separable algebraic closure of k and $k_s \subseteq K$, denote by $Gal(k_s/k)$ the Galois group of k_s over k. Let X be an affine variety defined over k, i.e. a k-variety, denote the set of k-rational points in X by X(k). If G is an algebraic group, $R_u(G)$ stands for its unipotent radical. Moreover, if G is a reductive k-group splitting over k and P is a k-parabolic subgroup of G, then $R_u(P)$ splits over k [4, §4.6]. Suppose H is a subset of G, denote the normalizer, centralizer and closure of H in G by $N_G(H)$, $C_G(H)$ and $\overline{H}$ respectively. Denote by [G,G] the commutator subgroup of G.

1.2. Let G be a reductive k-group splitting over k, then it contains a maximal k-torus T splitting over k [3,§8.2]. Suppose B is a Borel subgroup of G, which contains T and splits over k. If S is the set of involutions of $N_G(T)/T$, which is finite in number and generates $N_G(T)/T$, the root system, positive root system and simple root system relative to Tits system $(G,B,N_G(T),S)$ are denoted by Φ(or $\Phi(T)$), Φ^+ and Δ respectively. Let $U_a (a \in \Phi)$ be the root subgroup relative to the root a, then $\{T, U_a (a \in \Phi)\}$ is a "donnée de déploiment" [4.§2.8]. If G is a semi-simple algebraic group, $G=G_1 \cdot\ldots\cdot G_s$ is an almost direct product of some simple algebraic groups (which are not necessarily simple as abstract groups), then $\Phi = \bigcup_{i=1}^{s} \Phi_i$, where Φ_i is the root system of G_i, relative to its maximal torus $T \cap G_i$. Moreover, $\Phi_i \perp \Phi_j$, $\forall i \neq j$ and $\Phi^+ = \bigcup_{i=1}^{s} \Phi_i^+$, where $\Phi_i^+ = \Phi^+ \cap \Phi_i$. $\Delta = \bigcup_{i=1}^{s} \Delta_i$, where $\Delta_i = \Phi_i \cap \Delta$ is the simple root system of G_i.

1.3. Lemma. Suppose G,G' are simple algebraic groups splitting over k and G' is of adjoint type. $\{T, U_a\}$ $(a \in \Phi)$ and $\{T', U_{a'}\}$ $(a' \in \Phi')$ are k-splitting données of G and G' repsectively. Δ and Δ' are simple root systems of Φ and Φ' respectively. If $\sigma: \Delta \to \Delta'$ is an isomorphism of simple root systems, then there is a k-isogeny $\pi: G \to G'$ such that $\pi(U_a) = U_{\sigma(a)}$, $\forall a \in \Delta$. Moreover, π is unique up to k-isomorphisms.
Proof. Cf. [4.§2.13] and [16 §2.6.1]. ∎

1.4. Let D be a division ring, denote the group consisting of non-zero elements of D by D*. Recall that
$SL_n(D) = \{X \in Mat_{n\times n}(D) \mid \det X = 1 \in D^*/[D^*,D^*]\}$ [8] and $PSL_n(D) = SL_n(D)/C(SL_n(D))$ where $n \geq 2$. If $(a_{ij}) \in SL_n(D)$, $a_{ij} \in D$, $1 \leq i,j \leq n$, denote by $[a_{ij}]$ the image of (a_{ij}) under canonical homomorphism $SL_n(D) \longrightarrow PSL_n(D)$.
Let G be a simple algebraic group defined over k, $\alpha: SL_n(D) \longrightarrow G(k)$ be a homorphism with Zariski dense image. We give the following remarks first.
(i) If $|D| < \infty$, $\alpha(SL_n(D))$ is finite, it follows that dim G=0. Moreover, G={1} since G is connected. In this case, α is trival. Thus in following we may always suppose $|D| = \infty$.
(ii) In case D is commutative, α has been determined in [3]. Henceforth we only consider the case that D is not commutative.
(iii) If M is a subset of $SL_n(D)$, denote the closure of $\alpha(M)$ by $\bar{\alpha}(M)$. We have $[\alpha(SL_n(D), \alpha(SL_n(D))] = \alpha(SL_n(D))$ since $[SL_n(D), SL_n(D)] = SL_n(D)$ [8], hence

$$[G,G]=[\bar\alpha(SL_n(D)),\ \bar\alpha(SL_n(D))]=\bar\alpha(SL_n(D))=G.$$

Inparticular G is not solvable.

1.5. Lemma. If F' is a maximal subfield of D and $|D|=\infty$, then $|F'|=\infty$.
Proof. F' is an extension field of G. If $|F|$ is infinite, so is $|F'|$. If $|F|<\infty$, we have $\dim_F D=\infty$ since $|D|=\infty$, hence $|F':F|=\infty$ [11.VIII Th 8,1.7] which means $|F'|=\infty$. ∎

2. Some Properties of $SL_n(D)$ and Algebraic Groups

2.1. We consider first some useful "Bruhat" decompositions of $SL_n(D)$. Suppose $1\leq i\leq n-1$. Let

$$Z_i=\left\{\begin{pmatrix}A & 0\\ 0 & B\end{pmatrix}\in SL_n(D)\;\middle|\; A\in GL_{n-i}(D),\ B\in GL_i(D)\right\}$$

$$U_i=\left\{\begin{pmatrix}I_{n-i} & B\\ 0 & I_i\end{pmatrix}\in SL_n(D)\;\middle|\; B\in Mat_{(n-i)\times i}(D)\right\}$$

$$U_i^-=\left\{\begin{pmatrix}I_{n-i} & 0\\ B & I_i\end{pmatrix}\in SL_n(D)\;\middle|\; B\in Mat_{i\times(n-i)}(D)\right\}$$

$$Z_0=\{\operatorname{diag}(b_1,\dots,b_n)\in SL_n(D)\mid b_j\in D^*,\ j=1,\dots,n\}$$

$$U_0=\left\{\begin{pmatrix}1 & b_{12} & \cdots & b_{1n}\\ 0 & 1 & \ddots & \vdots\\ \vdots & \vdots & \ddots & b_{n-1,n}\\ 0 & 0 & \cdots & 1\end{pmatrix}\in SL_n(D)\;\middle|\;\forall b_{ij}\in (D),\ 1\leq i<j\leq n\right\}$$

$$U_0^-=\left\{\begin{pmatrix}1 & 0 & \cdots & 0\\ b_{21} & 1 & \ddots & \vdots\\ \vdots & \ddots & \ddots & 0\\ b_{n1} & \cdots & b_{n,n-1} & 1\end{pmatrix}\in SL_n(D)\;\middle|\;\forall b_{ij}\in D,\ 1\leq j<i\leq n\right\}$$

<u>Lemma</u>. For any i, such that $0\leq i\leq n-1$, we have

$$SL_n(D)=\bigcup_{j=1}^{m} w_jU_i^-Z_iU_i,$$

where w_j's are permutation matrices and $0<m<\infty$.

Proof. Suppose $1\leq i\leq n-1$, $\begin{pmatrix}A_1 & A_2\\ A_3 & A_4\end{pmatrix}\in SL_n(D)$, where $A_1\in Mat_{(n-i)\times(n-i)}$, $A_2\in Mat_{(n-i)\times i}$, $A_3\in Mat_{i\times(n-i)}$ and $A_4\in Mat_{i\times i}$. If $A_1\in GL_{n-i}(D)$, then

$$\begin{pmatrix}A_1 & A_2\\ A_3 & A_4\end{pmatrix}=\begin{pmatrix}I_{n-i} & 0\\ A_3A_1^{-1} & I_i\end{pmatrix}\begin{pmatrix}A_1 & 0\\ 0 & A_4-A_3A_2\end{pmatrix}\begin{pmatrix}I_{n-i} & A_1^{-1}A_2\\ 0 & I_i\end{pmatrix}\in U_i^-Z_iU_i.$$

If $A_1\notin GL_{n-i}(D)$, there is a permutation matrix $w\in SL_n(D)$, such that $w\begin{pmatrix}A_1\\ A_3\end{pmatrix}=\begin{pmatrix}B_1\\ B_3\end{pmatrix}$ where $B_1\in GL_{n-i}(D)$. Hence we have the decompositions in lemma in case $1\leq i\leq n-1$. Now, suppose $i=0$ and $n>2$, we prove the lemma by induction on n. Suppose $\begin{pmatrix} a & B \\ C & Q \end{pmatrix} \in SL_n(D)$, where $C=\begin{pmatrix} c_1 \\ \vdots \\ c_{n-1} \end{pmatrix} \in Mat_{(n-1)\times 1}(D)$, $B=(b_1 \cdots b_{n-1}) \in Mat_{1\times(n-1)}(D)$, $a \in D$. Denote by E_{ij} the matrix with 1 in the (i,j) position, 0 in other positions. If $a\neq 0$, then

$$\begin{pmatrix} a & B \\ C & Q \end{pmatrix} = \prod_{i=1}^{n-1} (I_n + c_i a^{-1} E_{i+1}) \begin{pmatrix} a & 0 \\ 0 & Q_1 \end{pmatrix} \prod_{j=1}^{n-1} (I_n + a^{-1} b_j E_{1,j+1})$$

where $Q_1 = Q - Ca^{-1}B \in GL_{n-1}(D)$. Hence there exist $W_1 \in SL_{n-1}(D)$ and $W_2 = \mathrm{diag}\ (1,\ldots,d) \in GL_{n-1}(D)$ such that $Q_1 = W_1 W_2$. By inductive assumption, $W_1 \in w' U_0'^{-} Z_0' U_0'$, where $w' \in SL_{n-1}(D)$ is a permutation matrix, Z_0' is the subgroup of $SL_{n-1}(D)$ consisting of diagonal matrices, $U_0'(U_0)$ is the subgroups of $SL_{n-1}(D)$ consisting of inferior (upper) triangular matrices with diagonal elements 1. It follows that

$$\begin{pmatrix} a & B \\ C & Q \end{pmatrix} \in U_1^- \begin{pmatrix} 1 & 0 \\ 0 & w' \end{pmatrix} U_0^- Z_0 U_1 \begin{pmatrix} 1 & 0 \\ 0 & w_2 \end{pmatrix} U_0 = \begin{pmatrix} 1 & 0 \\ 0 & w' \end{pmatrix} U_0^- Z_0 U_0.$$

If $a=0$, then $C\neq 0$, there is a permutation matrix $w \in SL_n(D)$, such that $\begin{pmatrix} a & B \\ C & Q \end{pmatrix} = w \begin{pmatrix} a_1 & b_1 \\ c_1 & Q_1 \end{pmatrix}$ where $a_1 \neq 0$. We also have the decompositions required. Finiteness of m comes from that of permutation matrices of $SL_n(D)$. ∎

__2.2. Lemma__. Suppose H is an algebraic group, I an index set, $\{H_i\}_{i\in I}$ a family of abstract unipotent subgroups of H and $[H_i, H_j] = \{1\}$, $\forall i,j \in I$. Then the closed subgroup $\overline{\langle H_i \mid \forall i \in I \rangle}$ generated by all H_i, $i \in I$, is unipotent.

Proof. Take H to be a closed subgroup of $GL_n(K)$. Let T_n be the subgroup of $GL_n(K)$ consisting of upper triangular matrices and U_n be the subgroup of T_n, consisting of matrices which have diagonal elements 1. Since $\langle H_i \mid \forall i \in I \rangle$ is a commutative subgroup, there is $g \in GL_n(K)$ such that $g \langle H_i \mid \forall i \in I \rangle g^{-1} \subsetneq T_n$. Inparticular, $gH_i g^{-1} \subsetneq U_n$, $\forall i \in I$ since H_i is unipotent, so that $g \langle H_i \mid \forall i \in I \rangle g^{-1} \subsetneq U_n$, which means $\langle H_i \mid \forall i \in I \rangle$ is unipotent, hence, so is its closure $\overline{\langle H_i \mid \forall i \in I \rangle}$. ∎

__2.3. Lemma__. Suppose H is an algebraic group and $H_1,\cdots,H_n$ are abstract subgroups of H. If $H_j \subseteq N_H(H_i)$, $\forall 1\leq i<j\leq n$, then

$$\overline{H_1 \cdot H_2 \cdots H_n} = \overline{H}_1 \cdot \overline{H}_2 \cdots \overline{H}_n.$$

Inparticular, if H_i is closed for every i, $1\leq i\leq n$, then $H_1 \cdots H_n$ is also closed.

Proof. We have $H_1 \cdots H_n \subsetneq \overline{H}_1 \cdots \overline{H}_n \subsetneq \overline{H_1 \cdots H_n}$, it's enough to show that

$\bar{H}_1 \cdots \bar{H}_n$ is closed. Suppose $\bar{H}_i = \bigcup_{\ell_i=1} h_{i\ell_i} \bar{H}_i^{\circ}$, where $h_{i\ell_i} \in \bar{H}_i$, $i=1,\ldots,n$ then

$$\bar{H}_1 \cdots \bar{H}_n = \bigcup_{\ell_1=1}^{m_1} \cdots \bigcup_{\ell_n=1}^{m_n} h_{1\ell_1} \bar{H}_1^{\circ} \cdots h_{n\ell_n} \bar{H}_n^{\circ} = \bigcup_{\ell_1=1}^{m_1} \cdots \bigcup_{\ell_n=1}^{m_n} h_{1\ell_1} \cdots$$

$$h_{n\ell_n} \bar{H}_1^{\circ} \cdots \bar{H}_n^{\circ}.$$

Since $\bar{H}_1^{\circ} \cdots \bar{H}_n^{\circ}$ is closed, so is $\bar{H}_1 \cdots \bar{H}_n$. ▮

2.4. Lemma. Suppose H is an algebraic group. H_1, H_2 are abstract unipotent subgroups of H and $H_2 \subseteq N_H(H_1)$. then $\bar{H}_1 \bar{H}_2$ is a closed unipotent subgroup of H.

Proof. Take H to be a closed subgroup of $GL(V)$, where V is a vector space over K of dimension n. We first show that there exists a vector which is fixed under the action of $H_1 H_2$. Let V' be the subspace of V consisting of vectors which are fixed under the action of H_1. Since H_1 is unipotent, $\dim V' \geqslant 1$. Suppose $v \in V'$, $h_1 \in H_1$, $h_2 \in H_2$, then

$$h_1 h_2 v = h_2 h_2^{-1} h_1 h_2 v = h_2 v,$$

therefore $h_2 v \in V'$, hence $H_2 V' = V'$. Since H_2 is unipotent, there is a vector $v_1 \in V'$ which is fixed by H_2. It is obvious that v_1 is fixed by $H_1 H_2$. Now $H_1 H_2$ acts on $V/{kv_1}$. Since $H_1 H_2$ is unipotent in $GL(V/{kv_1})$, we can find $\bar{v}_2 \in V/{kv_1}$ which is fixed by $H_1 H_2$, where $v_2 \in V$. Following the above process, we get a basis $\{v_1, \ldots, v_n\}$ of V and a composition series of subspaces of V, which is stable under action of $H_1 H_2$

$$\{0\} = V_0 \subset V_1 \cdots \subset V_n = V$$

where $V_i = \bigoplus_{i=1}^{i} kv_1$, $i=1,\ldots,n$, such that $H_1 H_2 v_i \subset v_i + V_{i-1}$. In other words, $H_1 H_2$ is conjugate to a subgroup of U_n which consists of upper triangular matrices with diagonal elements 1. Hence $\bar{H}_1 \bar{H}_2$ is unipotent.

Corollary. Suppose H is an algebraic group, $H_1, \ldots, H_n$ are unipotent abstract subgroups of H and $H_j \subseteq N_H(H_i)$, $\forall 1 \leqslant i < j \leqslant n$, then $\bar{H}_1 \cdots \bar{H}_n$ is a closed unipotent subgroup of H. ▮

2.5. Lemma. Suppose H is an algebraic group, H_1 and H_2 are connected subgroups of H, then $\dim H_1 + \dim H_2 - \dim \overline{H_1 H_2} > 0$ provided $\dim(H_1 \cap H_2) > 0$.

Proof. Consider the variety $H_1 \times H_2$ which is irreducible. Define a map $\mu: H_1 \times H_2 \longrightarrow H$ by $\mu(h_1, h_2) = h_1 h_2$, where $h_i \in H_i$, $i=1,2$. μ is a morphism of varieties. $\overline{\mu(H_1 \times H_2)} = \overline{H_1 H_2}$ is also an irreducible affine variety. It follows from [12.I.§8] that there is an open subset U of $H_1 H_2$ such that for any $x \in U \cap H_1 H_2$ we have $\dim \mu^{-1}(x) = \dim(H_1 \times H_2) - \dim \overline{H_1 H_2}$. Suppose

$x=x_1x_2 \in U \cap H_1H_2$, where $x_i \in H_i$, $i=1,2$. Define maps

$$\lambda_{x_1}: \quad H_1 \longrightarrow H_1$$
$$h_1 \longmapsto x_1h_1, \quad \forall h_1 \in H_1$$

$$\rho_{x_2}: \quad H_2 \longrightarrow H_2$$
$$h_2 \longmapsto h_2^{-1}x_2, \quad \forall h_2 \in H_2$$

Both λ_{x_1} and ρ_{x_2} are isomorphisms of varieties, so is $\lambda_{x_1}\times\rho_{x_2}: H_1\times H_2 \longrightarrow H_1 \times H_2$. Consequently, we have

$$\dim(\lambda_{x_1}\times\rho_{x_2})^{-1}\mu^{-1}(x_1x_2)=\dim\mu^{-1}(x_1x_2)=\dim H_1+\dim H_2-\dim\overline{H_1H_2}.$$

Let $N=\{(h,h) \in H_1 \times H_2 \mid \forall h \in (H_1 \cap H_2)\}$. It is obvious that both N and $(H_1 \cap H_2)\times(H_1 \cap H_2)$ are closed subvarieties of $H_1 \times H_2$. Moreover, $\dim N=\dim H_1 \cap H_2$. Since $\dim H_1 \cap H_2 > 0$ and $N \subseteq (\lambda_{x_1}\times\rho_{x_2})^{-1}\mu^{-1}(x_1x_2)$, it follows that

$$\dim H_1+\dim H_2-\dim\overline{H_1H_2} \geqslant \dim N > 0. \quad \blacksquare$$

<u>2.6. Lemma</u>. Suppose H is an algebraic group, $\alpha: SL_n(D) \longrightarrow H$ is a homomorphism, $n \geqslant 2$, then $\bar{\alpha}(SL_n(D))$ is connected.
Proof. Suppose $H=\bar{\alpha}(SL_n(D))$, we show that $H=H^o$. Consider composition of homomorphisms

$$SL_n(D) \xrightarrow{\ \alpha\ } H \xrightarrow{\ \pi\ } H/H^o.$$

where π is the canonical homomorphism. $\ker \pi\alpha = \alpha^{-1}(H^o)$, so that $\alpha^{-1}(H^o)$ is either contained in $C(SL_n(D))$ or equal to $SL_n(D)$. The former case is impossible, for otherwise, $\pi\alpha$ would induce an injection

$SL_n(D)/\alpha^{-1}(H^o) \longrightarrow H/H^o$, which implies that $|SL_n(D)/\alpha^{-1}(H^o)| \leqslant |H/H^o| < \infty$

but, on the other hand, $|D| = \infty$ means $|SL_n(D)/\alpha^{-1}(H^o)| \geqslant |SL_n(D)/C(SL_n)| = \infty$. This is a contradiction. Thus $H=H^o$. $\blacksquare$

<u>2.7. Lemma</u>. Let H be an algebraic group and $\alpha: SL_n(D) \longrightarrow H$ be a nontrivial homomorphism, then $\bar{\alpha}(U_i)$ and $\bar{\alpha}(U_i)$, $i=0,\ldots,n-1$, are connected unipotent unipotent subgroups of H.
Proof. For $d \in D^*$, denote the maximal subfield of D containing d by F_d. If $1 \leqslant i,j \leqslant n$ and $i \neq j$, let

$Z_{ij,d}:=\{\mathrm{diag}(b_1,\ldots,b_n)\in SL_n(D)\mid b_\ell=1,\ \ell\neq i,j,\ 1\leq \ell\leq n,\ b_i,b_j\in F_d\}$

$U_{ij}:=\{I_n+bF_{ij}\mid \forall\, b\in D\}$; $U_{ij,d}:=U_{ij}\cap SL_n(F_d)$.

It is obvious that $Z_{ij,d}\subseteq N_{SL_n(F_d)}(U_{ij,d})$. Denote $W_{ij,d}=Z_{ij,d}\ltimes U_{ij,d}$. We have

$$[W_{ij,d},\ W_{ij,d}]\subseteq U_{ij,d}. \tag{2.7.1}$$

We show that $\bar{\alpha}(U_i)$ is connected and unipotent in four steps: (i) $|Z_{ij,d}\cap\alpha^{-1}(\bar{\alpha}(W_{ij,d})^{o})|$ is infinite. (ii) $\bar{\alpha}(U_{ij,d})$ is a connected unipotent subgroup of H for $1\leq i<j\leq n$. (iii) $\bar{\alpha}(U_{ij})$ is a connected unipotent subgroup of H for $1\leq i<j\leq n$. (iv) $\bar{\alpha}(U\ )$ is a connected unipotent subgroup of H for $0<r<n-1$. The proof of $\bar{\alpha}(U_i)$ being connected and unipotent are similar.

(i) $\bar{\alpha}(W_{ij,d})^{o}\cap\alpha(W_{ij,d})$ is a normal subgroup of $\alpha(W_{ij,d})$, therefore $\alpha^{-1}(\bar{\alpha}(W_{ij})^{o}\cap\alpha(W_{ij,d}))$ is a normal subgroup of $W_{ij,d}\ker\alpha$. If $|Z_{ij}\cap\alpha^{-1}(\bar{\alpha}(W_{ij,d})^{o})|$ is finite, it follows from $|F|=\infty$ and $|Z_{ij,d}|=\infty$ that

$$\left|\, Z_{ij,d}\big/ Z_{ij,d}\cap\alpha^{-1}(\bar{\alpha}(W_{ij,d})^{o})\,\right|=\infty \tag{2.7.2}$$

from which a contradiction will be deduced. In fact, since

$$Z_{ij,d}\cap\alpha^{-1}(\bar{\alpha}(W_{ij,d})^{o})=Z_{ij,d}\cap\alpha^{-1}(\bar{\alpha}(W_{ij,d})^{o}\cap\alpha(W_{ij,d}))$$

the canonical homomorphism

$$W_{ij,d}\ker\alpha\longrightarrow W_{ij,d}\ker\alpha\big/\alpha^{-1}(\bar{\alpha}(W_{ij,d})^{o}\cap\alpha(W_{ij,d}))$$

restricted to $Z_{ij,d}$ induces an injection $Z_{ij,d}\big/Z_{ij,d}\cap\alpha^{-1}(\bar{\alpha}(W_{ij,d})^{o})\longrightarrow$ $W_{ij,d}\ker\alpha\big/\alpha^{-1}(\bar{\alpha}(W_{ij,d})^{o}\cap\alpha(W_{ij,d}))$ therefore, by (2.7.2),

$$W_{ij,d}\ker\alpha\big/\alpha^{-1}(\bar{\alpha}(W_{ij,d})^{o}\cap\alpha(W_{ij,d}))$$

is infinite. But, on the other hand, consider the composition of homomorphisms

$$W_{ij,d}\ker\alpha\xrightarrow{\ \alpha\ }\bar{\alpha}(W_{ij,d})\xrightarrow{\ \pi\ }\bar{\alpha}(W_{ij,d})\big/\bar{\alpha}(W_{ij,d})^{o}$$

where π is the canonical homomorphism. Evidently, the kernel is

$\alpha^{-1}(\bar{\alpha}(W_{ij,d})^{\circ} \cap \alpha(W_{ij,d}))$. It follows that

$$\left| W_{ij,d}^{\ker\alpha} \Big/ \alpha^{-1}(\bar{\alpha}(W_{ij,d})^{\circ} \cap \alpha(W_{ij,d})) \right| \leqslant \left| \bar{\alpha}(W_{ij,d}) \Big/ \bar{\alpha}(W_{ij,d})^{\circ} \right| < \infty$$

(i) is proved.

(ii) It follows from (i) there is $g = \begin{pmatrix} 1 & & & & \\ & \ddots & & & \\ & & b & & \\ & & \ddots & b^{-1} & \\ & & & & \ddots & \\ & & & & & 1 \end{pmatrix} \in$

$i \quad j$

$Z_{ij,d} \cap \alpha^{-1}(\bar{\alpha}(W_{ij,d})^{\circ})$ where $b^2 \neq 1$. For any $c \in F_d$, we have (2.7.3)

$$\begin{pmatrix} 1 & & \\ & \ddots & 1 \cdots c \\ & & 1 \\ & & \ddots \\ & & & 1 \end{pmatrix} = \begin{pmatrix} 1 & \\ & \ddots b \\ & b^{-1} \\ & \ddots \\ & & 1 \end{pmatrix} \begin{pmatrix} 1 & \\ & \ddots 1 \cdots (1-b^2)^{-1}c \\ & 1 \\ & \ddots \\ & & 1 \end{pmatrix} \begin{pmatrix} 1 & \\ & \ddots b \\ & b^{-1} \\ & \ddots \\ & & 1 \end{pmatrix}^{-1} \begin{pmatrix} 1 & \\ & \ddots 1 \cdots (1-b^2)^{-1}c \\ & 1 \\ & \ddots \\ & & 1 \end{pmatrix}^{-1}$$

hence $U_{ij,d} = [g, U_{ij,d}]$. Moreover

$$\bar{\alpha}(U_{ij,d}) = [\alpha(g), \bar{\alpha}(U_{ij,d})] = [\bar{\alpha}(W_{ij,d})^{\circ}, \bar{\alpha}(W_{ij,d})] \subseteq \bar{\alpha}(W_{ij,d})^{\circ}.$$

Further more,

$$\bar{\alpha}(U_{ij,d}) \subseteq \overline{[\bar{\alpha}(W_{ij,d})^{\circ}, \bar{\alpha}(W_{ij,d})^{\circ}]} = [\bar{\alpha}(W_{ij,d})^{\circ}, \bar{\alpha}(W_{ij,d})^{\circ}].$$

On the other hand, $[W_{ij,d} W_{ij,d}] = U_{ij,d}$ by (2.7.1) & (2.7.3) so that

$$[\bar{\alpha}(W_{ij,d})^{\circ}, \bar{\alpha}(W_{ij,d})^{\circ}] \subseteq \bar{\alpha}[W_{ij,d}, W_{ij,d}] = \bar{\alpha}(U_{ij,d})$$

hence

$$\bar{\alpha}(U_{ij,d}) = [\bar{\alpha}(W_{ij,d})^{\circ}, \bar{\alpha}(W_{ij,d})^{\circ}] \qquad (2.7.4)$$

which means that $\bar{\alpha}(U_{ij,d})$ is connected. In particular, since $W_{ij,d}$ is solvable, so is $\bar{\alpha}(W_{ij,d})^{\circ}$, it follows that $\bar{\alpha}(U_{ij,d})$ is unipotent by (2.7.4)

(iii) For any $d \in D^*$, there is a connected unipotent subgroup $\bar{\alpha}(U_{ij,d})$ by (ii). Consider the connected closed subgroup $<\bar{\alpha}(U_{ij,d}) | \forall d \in D^*>$ of H [1.§2.1]. For $b, d \in D^*$, $[U_{ij,d}, U_{ij,d}] = \{I_n\}$, hence $[\bar{\alpha}(U_{ij,b}), \bar{\alpha}(U_{ij,d})] = \{1\}$. It follows from the corollary (2.4) that $<\bar{\alpha}(U_{ij,d}) | \forall d \in D^*>$ is unipotent. It is obvious that $<\bar{\alpha}(U_{ij,d}) | \forall d \in D^*> \subseteq \bar{\alpha}(U_{ij})$. On the other hand,

$$\alpha(U_{ij}) \subseteq <\alpha(U_{ij,d}) | \forall d \in D^*> \subseteq <\bar{\alpha}(U_{ij,d}) | \forall d \in D^*>.$$

Hence $\overline{\alpha}(U_{ij})=\langle\overline{\alpha}(U_{ij,d})\mid \forall d\in D^*\rangle$. (iii) is proved.

(iv) Since $U_\gamma=\langle U_{ij}\mid U_{ij}\subseteq U_\gamma;\ 1\leq i<j\leq n\rangle$, $\gamma=0,1,\dots,n-1$, similar to the proof of (iii) we have $\overline{\alpha}(U_\gamma)=\langle\overline{\alpha}(U_{ij})\mid U_{ij}\subseteq U_\gamma,\ 1\leq i<j\leq n\rangle$. $\gamma=0,1,\dots,n-1$. hence $\overline{\alpha}(U_\gamma)$ is connected. To prove $\overline{\alpha}(U_\gamma)$, $0\leq r\leq n-1$, is unipotent, it is enough to show that $\overline{\alpha}(U_0)$ is unipotent since $U_\gamma\subseteq U_0$. Let $U_{(i)}=\prod_{j=i+1}^{n} U_{ij}$, $i=1,\dots,n-1$. It follows from (ii), (2.3) & (2.4) that

$$\overline{\alpha}(U_{(i)})=\overline{\alpha}(U_{i,i+1})\,\overline{\alpha}(U_{i,i+2})\cdots\overline{\alpha}(U_{in}).$$

therefore $\overline{\alpha}(U_{(i)})$ is connected and unipotent. Furthermore, $\overline{\alpha}(U_{(i)})\subseteq N_H(\overline{\alpha}(U_{(i)})$, $\forall\ 1\leq j<i\leq n-1$, since $U_0=U_{(1)}U_{(2)}\cdots U_{(n-1)}$ and $U_{(i)}\subseteq N_{SL_n(D)}(U_j)$, $\forall\ 1\leq j<i\leq n-1$. Consequently, by (2.3) & corollary (2.4), we have

$$\overline{\alpha}(U_0)=\overline{\alpha}(U_{(1)})\,\overline{\alpha}(U_{(2)})\cdots\overline{\alpha}(U_{(n-1)}) \tag{2.7.5}$$

and $\overline{\alpha}(U_0)$ is unipotent. ∎

<u>2.8. Lemma</u>. Let H be an algebraic group, $\alpha:SL_n(D)\longrightarrow H$ a homomorphism with Zariski dense image, then $\overline{\alpha}(U_i)\overline{\alpha}(Z_i)^o\overline{\alpha}(U_i^-)$ is dense in H, where $0\leq i\leq n-1$.

Proof. By (2.1), we have

$$H=\overline{\alpha}\Big(\bigcup_{j=1}^{m} w_jU_iZ_iU_i\Big)=\bigcup_{j=1}^{m}\alpha(w_j)\overline{\alpha(U_i)\alpha(Z_i)\alpha(U_i)}$$

since H is connected (2.6), it follows that

$$H=\overline{\alpha(U_i)\alpha(Z_i)\alpha(U_i^-)}=\overline{\overline{\alpha}(U_i)\overline{\alpha}(Z_i)^o\overline{\alpha}(U_i^-)}.\ ∎$$

<u>2.9. Lemma</u>. Let $P_i=Z_iU_i$, $P_i^-=U_i^-Z_i$, $i=0,1,\dots,n-1$. Suppose H is a reductive algebraic group, $\alpha:SL_n(D)\longrightarrow H$ is a homomorphism with Zariski dense image, then

(i). $\overline{\alpha}(Z_i)$, $\overline{\alpha}(P_i)$ and $\overline{\alpha}(P_i^-)$ are connected subgroups of H.

(ii). $\overline{\alpha}(P_i)$ and $\overline{\alpha}(P_i^-)$ are opposite parabolic subgroups of H with common Levi-component $\overline{\alpha}(Z_i)$.

(iii). $R_u(\overline{\alpha}(P_i))=\overline{\alpha}(U_i)$, $R_u(\overline{\alpha}(P_i^-))=\overline{\alpha}(U_i^-)$.

Proof. We divide the proof into three parts; (1) $\overline{\alpha}(Z_i)^o$ is reductive, (2) $\overline{\alpha}(Z_i)^o\overline{\alpha}(U_i)$ and $\overline{\alpha}(U_i^-)\overline{\alpha}(Z_i)^o$ are opposite parabolic subgroups of

H with common Levi-component $\bar{\alpha}(Z_i)^o$, their unipotent radicals are $\bar{\alpha}(U_i)$ and $\bar{\alpha}(U_i^-)$ respectively. (3) $\bar{\alpha}(Z_i)=\bar{\alpha}(Z_i)^o$, $\bar{\alpha}(P_i^-)=\bar{\alpha}(U_i^-)\bar{\alpha}(Z_i^-)$, $\bar{\alpha}(P_i)=\bar{\alpha}(Z_i)\alpha(U_i)$.

(1) Let $\pi:\bar{\alpha}(Z_i)^o \longrightarrow \bar{\alpha}(Z_i)^o/R_u(\bar{\alpha}(Z_i)^o$ be the canonical homomorphism, $\tilde{B}_i$ and $\tilde{B}_i^-$ be opposite Borel subgroups of $\bar{\alpha}(Z_i)^o/R_u(\bar{\alpha}(Z_i)^o)$.

Let $\tilde{V}_i$ and $\tilde{V}_i^-$ be maximal unipotent subgroups of $\tilde{B}_i$ and $\tilde{B}_i^-$ respectively and $\tilde{T}_i=\tilde{B}_i\cap \tilde{B}_i^-$, then $\tilde{B}_i=\tilde{V}_i\rtimes\tilde{T}_i$, $\tilde{B}_i^-=\tilde{V}_i^-\rtimes\tilde{T}_i$, $\tilde{V}_i\cap\tilde{V}_i=\{1\}$. Denote $V_i=\pi^{-1}(\tilde{V}_i)$ and $V_i^-=\pi^{-1}(\tilde{V}_i^-)$, then V_i and V_i^- are maximal unipotent closed subgroups of $\bar{\alpha}(Z_i)^o$. Moreover, both V_i and V_i^- are connected since π is an open homomorphism. Obviously, $V_i\cap V_i^-=R_u(\bar{\alpha}(Z_i)^o)$. We are going to show that it is trivial. At first we have to prove that $\bar{\alpha}(U_i)V_iT_iV_i\bar{\alpha}(U_i)$ is dense in H, where T_i is a maximal torus of $\bar{\alpha}(Z_i)^o$ such that $\pi(T_i)=\tilde{T}_i$. Note that $\bar{\alpha}(U_i^-)V_iT_i$ and $V_i\cdot\bar{\alpha}(U_i)$ are connected closed subgroups of H since $\bar{\alpha}(Z_i)\subseteq N_H(\bar{\alpha}(U_i^-))\cap N_H(\bar{\alpha}(U_i))$. Let $M=\bar{\alpha}(U_i^-)V_iT_i\times V_i\bar{\alpha}(U_i)$ be a direct product of algebraic groups, define an action of M on H by

$$M\times H \longrightarrow H$$
$$((x,y),h)\longmapsto xhy^{-1}$$

The orbits are locally closed in H. In particular, $\bar{\alpha}(U_i^-)V_i^-T_iV_i\bar{\alpha}(U_i)$ is open in its closure. Since $V_i^-T_iV_i$ is an open subset of $\bar{\alpha}(Z_i)^o/R_u(\bar{\alpha}(Z_i)^o)$, $V_i^-T_iV_i=\pi^{-1}(\tilde{V}_i^-\tilde{T}_i\tilde{V}_i)$ is open in $\bar{\alpha}(Z_i)^o$, hence by (2.8) we have

$$\overline{\bar{\alpha}(U_i^-)V_i^-T_iV_i\bar{\alpha}(U_i)}=\overline{\bar{\alpha}(U_i^-)\bar{\alpha}(Z_i)^o\bar{\alpha}(U_i)} = H$$

so that

$$\dim H = \dim\bar{\alpha}(U_i^-)V_i^-T_iV_i\bar{\alpha}(U_i).$$

Furthermore since H is reductive, it is easy to deduce from the above that $\bar{\alpha}(U_i^-)V_i^-T_i$ and $T_iV_i\bar{\alpha}(U_i)$ are Borel subgroups of H with maximal torus T_i, maximal unipotent subgroup $\bar{\alpha}(U_i^-)V_i$ and $V_i\bar{\alpha}(U_i)$ respectively. Therefore

$$\dim H=\dim\bar{\alpha}(U_i^-)V_i+\dim T_i+\dim V_i\bar{\alpha}(U_i) \qquad (2.9.2)$$

Now if $V_i\cap V_i^-$ is not trivial, then

$$\dim(\bar{\alpha}(U_i^-)V_i^-T_i\cap V_i\bar{\alpha}(U_i))\geqslant \dim V_i^-\cap V_i > 0$$

consequently $\dim H > \dim \bar\alpha(U_i^-)V_i^-T_iV_i\bar\alpha(U_i)$ by (2.5), which contradicts to (2.9.1). Hence $R_u(\bar\alpha(Z_i)^o)=\{1\}$, i.e. $\bar\alpha(Z_i)^o$ is reductive.

(2) $\bar\alpha(Z_i)^o\bar\alpha(U_i^-)$ and $\bar\alpha(Z_i)^o\bar\alpha(U_i)$ contain Borel subgroups $T_iV_i\bar\alpha(U_i)$ and $\bar\alpha(U_i^-)V_i^-T_i$ respectively, they are parabolic subgroup of H. Obviously, $R_u(\bar\alpha(Z_i)^o\bar\alpha(U_i))\supseteq\bar\alpha(U_i)$ and $R_u(\bar\alpha(Z_i)^o\bar\alpha(U_i^-))\supseteq\bar\alpha(U_i^-)$, so that to show $\bar\alpha(Z_i)^o\bar\alpha(U_i)$ and $\bar\alpha(Z_i)^o\bar\alpha(U_i^-)$ are opposite it is enough to show that the Borel subgroups $T_iV_i\bar\alpha(U_i)$ and $\bar\alpha(U_i^-)V_i^-T_i$ are opposite, or equivialently,

$$\bar\alpha(U_i^-)V_i^- \cap V_i\bar\alpha(U_i)=\{1\} \tag{2.9.3}$$

Since $\bar\alpha(U_i^-)V_i^-\cap V_i\bar\alpha(U_i)$ is an unipotent subgroup of the reductive group H and T_i-stable, it is connected. It follows from (2.9.2) & (2.5) that $\dim(\bar\alpha(U_i^-)V_i^-\cap V_i\bar\alpha(U_i))=0$, from which comes (2.9.3). Since $R_u(\bar\alpha(Z_i)^o\bar\alpha(U_i))\cap\bar\alpha(Z_i)^o$ is unipotent and T_i-stable, it is connected and contained in $R_u(\bar\alpha(Z_i)^o)$ which is trivial as we have seen. Therefore $R_u(\bar\alpha(Z_i)^o\bar\alpha(U_i))=\bar\alpha(U_i)$. Similarly, $R_u(\bar\alpha(Z_i)^o\bar\alpha(U_i^-))= \alpha(U_i^-)$. Now $\bar\alpha(U_i)\cap\bar\alpha(Z_i)^o\bar\alpha(U_i^-)$ is connected since it is unipotent and T_i-stable. By (2.9.2) and (2.5), $\dim(\bar\alpha(U_i)\cap\bar\alpha(Z_i)^o\bar\alpha(U_i^-))=0$, hence $\bar\alpha(U_i)\cap\bar\alpha(Z_i)\bar\alpha(U_i^-)$ is trivial. It follows that $\bar\alpha(Z_i)^o\bar\alpha(U_i)\cap\bar\alpha(Z_i)^o\bar\alpha(U_i^-)= \bar\alpha(Z_i)^o$. Thus (2) is proved.

(3) Since $P_i=Z_iU_i$ and $P_i^-=U_i^-Z_i$, we have $\bar\alpha(P_i)=\bar\alpha(Z_i)\bar\alpha(U_i)$ and $\bar\alpha(P_i^-)= \bar\alpha(Z_i)\bar\alpha(U_i^-)$ by (2.3), Moreover

$$\bar\alpha(Z_i)\subseteq N_H(\bar\alpha(Z_i)^o)\cap N_H(\bar\alpha(U_i))\subseteq N_H(\bar\alpha(Z_i)^o\bar\alpha(U_i))=\bar\alpha(Z_i)^o\bar\alpha(U_i)$$

Similarly, $\bar\alpha(Z_i)\cap\bar\alpha(Z_i)^o\bar\alpha(U_i^-)$ Hence

$$\bar\alpha(Z_i)\subseteq\bar\alpha(Z_i)^o\ \bar\alpha(U_i)\cap\bar\alpha(Z_i)^o\bar\alpha(U_i^-)=\bar\alpha(Z_i)^o$$

Thus (3) is proved. ▮

2.10. Lemma. Let H be a reductive algebraic group defined over k, $\alpha:SL_n(D)\longrightarrow H(k)$ a homomorphism with Zariski dense image. Then there is a finite separable extension field k' of k such that

(i) $\bar\alpha(Z_i)$, $\bar\alpha(U_i)$ and $\bar\alpha(U_i)$ split over k', $0\leq i\leq n-1$.

(ii) if B_i and B_i^- are opposite Borel subgroups of $\bar\alpha(Z_i)$ and split over k', then $B=\bar\alpha(U_i)B_i$ and $B^-=\bar\alpha(U_i^-)B_i^-$ are opposite Borel subgroups of H and split over k'.

(iii) if H splits over k, one can take k'=k.

Proof. (i) There is a finite extension field k' of k such that H splits over k' [2§8.3], hence it is enough to show that if H splits over k then so do $\bar{\alpha}(Z_i)$, $\bar{\alpha}(Z_i)$ and $\bar{\alpha}(U_i^-)$. Now $\alpha(SL_n(D)) \subsetneq G(k)$, so that

$$\alpha(P_i) \subsetneq \bar{\alpha}(P_i) \cap H(k) = \bar{\alpha}(P_i)(k) \subsetneq \bar{\alpha}(P_i)$$

which means that $\bar{\alpha}(P_i)(k)$ is dense in $\bar{\alpha}(P_i)$. Obviously, $\bar{\alpha}(P_i)(k)$ is stable under the action of $Gal(^k s/_k)$, therefore $\bar{\alpha}(P_i)$ is a k-parabolic subgroup of H [1. AG.§14.4]. Similar $\bar{\alpha}(P_i^-)$ is also a k-parabolic subgroup of H. Since H splits over k, the Levi-component $\bar{\alpha}(Z_i)$, and the unipotent radical $\bar{\alpha}(U_i)$ ($\bar{\alpha}(U_i^-)$) of $\bar{\alpha}(P_i)$ ($\bar{\alpha}(P_i^-)$) split over k [4.§4.6].

(ii) Suppose H splits over k, as above. $T_i = B_i \cap B_i^-$ is a maximal torus of H and splits over k. Let V_i and V_i^- be maximal unipotent subgroups of B_i and B_i^- respectively, then V_i and V_i^- also split over k. It follows from the Proof (2) of lemma (2.9) that $B = T_i V_i \bar{\alpha}(U_i)$ and $B^- = \bar{\alpha}(U_i^-) V_i T_i$ are opposite Borel subgroups of H. Since $\bar{\alpha}(U_i)$ is normalized by B_i and splits over k, $B = B_i \bar{\alpha}(U_i)$ also splits over k. Similarly $B^- = B_i^- \bar{\alpha}(U_i^-)$ splits over k.

(iii) Obvious. ∎

<u>2.11. Lemma</u>. Let H be a simple algebraic group, $\alpha : SL_n(D) \longrightarrow H$ a homomorphism with Zariski dense image. Then $\bar{\alpha}(P_i)$, $1 \leq i \leq n-1$, is a maximal parabolic subgroup of H.

Proof. $R_u(\bar{\alpha}(P_i)) = \bar{\alpha}(U_i) \subsetneq C_H(\bar{\alpha}(U_i))$ since U_i is an abelian group. Use [3 §4.8] to deduce that $\bar{\alpha}(P_i)$ is maximal. ∎

<u>2.12. Lemma</u>. Let H be a reductive algebraic group and T be a maximal torus of H. Let $\Phi(H,T)$ be the root system of H relative to T. Suppose $V_1, \ldots, V_n$ are unipotent closed T-stable (i.e. normalized by T) subgroups of H such that $V_i \subsetneq \bigcap_{j=1}^{i-1} N_H(V_j)$, $1 < i \leq n$, and $V_i \cap V_{i+1} \cdots V_n = \{1\}$, $1 \leq i < n$. If a root subgroup U_γ ($\gamma \in \Phi(H,T)$) $\subset V_1 \cdots V_n$, then U_γ is contained in exactly one of V_i's.

Proof. V_i is connected since it is unipotent and T-stable. Hence $V_1 \cdots V_n$ is also an unipotent connected subgroup of H. Let $\Psi = \{\beta \in \Phi(H,T) \mid U_\beta \subset V_1 \cdots V_n\}$, then the Lie algebra $L(V_1 \cdots V_n) = \bigoplus_{\beta \in \Psi} \mathfrak{g}_\beta$ where $\mathfrak{g}_\beta = L(U_\beta)$. Furthermore, let $\Psi_i = \{\beta \in \Phi(H,T) \mid U_\beta \subset V_i\}$, then

$$L(V_1 \cdots V_n) = \bigoplus_{i=1}^{n} L(V_i) = \bigoplus_{i=1}^{n} \Big(\bigoplus_{\beta \in \Psi_i} \mathfrak{g}_\beta \Big)$$

hence $\Psi=\bigcup_{i=1}^{n}\Psi_i$ and $\Psi_i\cap\Psi_j=\Psi$, $\forall\, i\neq j$. If $U_\gamma < V_1\ldots V_n$, then $L(U_\gamma)=\mathfrak{g}_\gamma\subset\bigoplus_{\beta\in\Psi}\mathfrak{g}_\beta$ so that γ belongs to exactly one of Ψ_i's, from which, the lemma follows. ∎

Note. The Proofs of (2.7) & (2.9) in some way follow the idea of Borel & Tits [3].

3. Homomorphisms From $SL_2(D)$ and $PSL_2(D)$ to Algebraic Groups

In this section, let G be a simple algebraic group splitting over k and $\alpha: SL_2(D)\longrightarrow G(k)$ be a homomorphism of algebraic group with Zariski dense image.

3.1. For convenience we simplify some notations used in §2. Denote $U=U_0=U_1$, $U^-=U_0^-=U_1^-$, $Z=Z_0=Z_1$, $P=P_0=P_1$. Let T be a maximal torus of $\bar{\alpha}(Z)$ and split over k, V and V^- be maximal connected unipotent subgroups of $\bar{\alpha}(Z)$ such that $B=TV\bar{\alpha}(U)$ and $B^-=TV^-\bar{\alpha}(U^-)$ are opposite Borel subgroup of G and split over k (2.10(ii)). Denote the root system, positive root system and simple root system of G relative to B and T by Φ, Φ^+ and Δ respectively. Suppose $a_1,a_2\in\Phi$, there is a unique positive integer r such that $a_2-ra_1\in\Phi$, $a_2-(r+1)a_1\notin\Phi$. Denote $N_{a_1,a_2}=\pm(r+1)$

3.2. Lemma. Let

$$Z'=\left\{\begin{pmatrix}u & 0\\ 0 & 1\end{pmatrix}\in SL_2(D)\ \middle|\ \forall\, u\in [D^*,D^*]\right\}$$

$$Z''=\left\{\begin{pmatrix}1 & 0\\ 0 & v\end{pmatrix}\in SL_2(D)\ \middle|\ \forall\, v\in [D^*,D^*]\right\}$$

then

(i) $\bar{\alpha}(Z')^0\subsetneq C_G(\bar{\alpha}(Z'')^0)$

(ii) $[\bar{\alpha}(Z),\bar{\alpha}(Z)]=\bar{\alpha}(Z')^0\bar{\alpha}(Z'')^0$

(iii) $\bar{\alpha}(Z')^0$ and $\bar{\alpha}(Z'')^0$ are normal subgroups of $[\bar{\alpha}(Z),\bar{\alpha}(Z)]$. In particular $\bar{\alpha}(Z')^0\cap\bar{\alpha}(Z'')^0$ is contained in the centre of $[\bar{\alpha}(Z),\bar{\alpha}(Z)]$.

Proof. $Z'\subsetneq C_{SL_2(D)}(Z'')$, so that $\bar{\alpha}(Z')\subsetneq C_G(\bar{\alpha}(Z''))$, which means (i). Since $[Z,Z]=Z'Z''$ and $\bar{\alpha}(Z)$ is connected, $[\bar{\alpha}(Z),\bar{\alpha}(Z)]=\bar{\alpha}(Z')\bar{\alpha}(Z'')=\bar{\alpha}(Z')^0\bar{\alpha}(Z'')^0$. (iii) comes from (i) & (ii) directly. ∎

3.3. Remark. We proved in (2.9) that $\bar{\alpha}(Z)$ is reductive. If $\bar{\alpha}(Z)$ is a torus, then $\bar{\alpha}(Z')$ is trivial. Since α is non-trivial, $Z'\subsetneq\ker\alpha\subsetneq C(SL_2(D))$, which means that D is a field. Therefore in the following we consider only the case that $\bar{\alpha}(Z)$ is not a torus (1.4.(ii)). In this case,

$[\bar{\alpha}(Z), \bar{\alpha}(Z)]=G_1\cdots G_r$ is an almost direct product of simple algebraic groups. By (3.2), both $\bar{\alpha}(Z')^o$ and $\bar{\alpha}(Z")^o$ are almost direct products of some G_i's, $1\leq i\leq r$. There is no G_i contained in $\bar{\alpha}(Z')^o \cap \bar{\alpha}(Z")^o$ because $|\bar{\alpha}(Z')^o \cap \bar{\alpha}(Z")^o| \leq |\{\text{Centre of } [\bar{\alpha}(Z),\bar{\alpha}(Z)]\}|<\infty$. Suppose $\bar{\alpha}(Z')^o=G_1\ldots G_q$, $\bar{\alpha}(Z")^o=G_{q+1}\ldots G_r$. Let $w=\begin{pmatrix} 0 & 1\\ -1 & 0\end{pmatrix} \in SL_2(D)$, then $wZ'w^{-1}=Z"$, and consequently, $\alpha(w)\bar{\alpha}(Z')^o\alpha(w)^{-1}=\bar{\alpha}(Z")^o$ hence $r=2q$.

3.4. Lemma. $\bar{\alpha}(Z')^o$ and $\bar{\alpha}(Z")^o$ are simple algebraic groups.

Proof. Let $T,\Phi\ \Delta$ be as in (3.1). Denote the root system of $[\bar{\alpha}(Z),\bar{\alpha}(Z)]$ relative to its maximal torus $T\cap[\bar{\alpha}(Z),\bar{\alpha}(Z)]$ by Φ'. $\Phi'\subset\Phi$. Let Φ_i be the root system of G_i (3.3) relative to its maximal torus $T\cap G_i$, $1\leq i\leq r$, then $\Phi'=\bigcup_{i=1}^{r}\Phi_i$ and $\Phi_i\perp\Phi_j$, $\forall i\neq j$, $1\leq i,j\leq r$. In particular, $\bigcup_{i=1}^{q}\Phi_i$ ($\bigcup_{i=q+1}^{r}\Phi_i$) is the root system of $\bar{\alpha}(Z')^o$ ($\bar{\alpha}(Z")^o$) relative to its maximal torus $T\cap\bar{\alpha}(Z')^o$ ($T\cap\bar{\alpha}(Z")^o$). $\Delta\cap\Phi'$ is a simple root system of Φ'. $\bar{\alpha}(P)$ is a maximal standard parabolic subgroup relative the Δ since $\bar{\alpha}(P)\supset B$, so that $R_u(\bar{\alpha}(P))$ contain exactly one simple root subgroup, say $U_a (a\in\Delta)$. Then $\Delta\cap\Phi'=\Delta\setminus\{a\}$. Moreover, $\Delta_i=\Delta\setminus\{a\}\cap\Phi$ is a simple root system of Φ_i, hence irreducible. We have

$$\Delta=\{a\}\cup(\bigcup_{i=1}^{r}\Delta_i);\ \Delta_i\perp\Delta_j,\ \forall i\neq j,\ 1\leq i,j\leq r \tag{3.4.1}$$

consider the Dynkin diagram of Δ. There are at most three irreducible components connected with root a since Δ is irreducible, hence $r\leq 3$. In fact $r=2$, $q=1$ by (3.3), which means $\bar{\alpha}(Z')^o$ and $\bar{\alpha}(Z")^o$ are simple algebraic groups. ∎

3.5. Theorem. Suppose G is a simple algebraic group and $\alpha:SL_2(D)\longrightarrow G$ is a homomorphism with Zariski dense image, then G is of type A_{2n-1}, where $n\in\mathbb{Z}^+$, $n>1$.

Proof. According to the Proof (3.4), $\Delta=\Delta_1\cup\Delta_2\cup\{a\}$, where Δ_1 and Δ_2 are simple root systems of $\bar{\alpha}(Z')^o$ and $\bar{\alpha}(Z")^o$ respectively. Therefore, the Dynkin diagram of Δ must be as follows

$$\Delta_1 \text{—— —— ——}\underset{a}{\circ}\text{—— —— ——}\Delta_2$$

Now, since there is an isomorphism of algebraic between $\bar{\alpha}(Z')^o$ and $\bar{\alpha}(Z")^o$ (3.3), $\Delta_1=\Delta_2$. Inparticular $|\Delta_1|=|\Delta_2|$ and $|\Delta|=2|\Delta_1|+1>3$ (3.3), Consequently, Δ is not of type A_1,A_2,B_2,C_2,G_2 and F_4. Furthermore considering the Dynkin diagram of Δ, one can find out that the following cases are impossible for Δ:

B_ℓ & C_ℓ ($\ell \geq 3$) 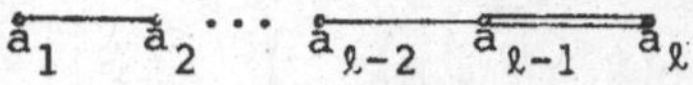

$a \neq a_2$ if $\ell = 3$

D_ℓ ($\ell > 3$) 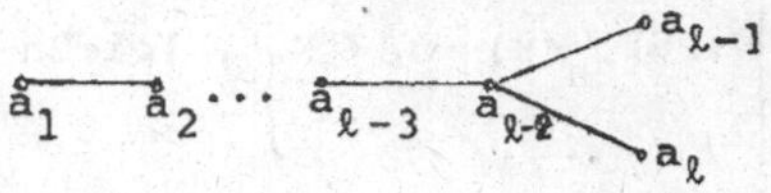

$a \neq a_4$ if $\ell = 7$ (3.5.1)

as well as E_ℓ ($\ell = 6,7,8$) & F_4. If G is of type B_3 with Dynkin diagram

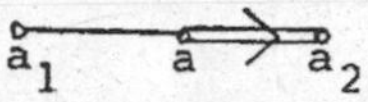

then

$$R_u(\bar{\alpha}(P)) = U_a U_{a+a_1} U_{a+a_2} U_{a+a_1+a_2} U_{a+a_1+2a_2} U_{a+2a_2} U_{a_1+2a+2a_2} \quad [6]$$

which is not commutative since $[U_a, U_{a+a_1+2a_2}] = U_{a_1+2a+2a_2}$. This contradicts to $R_u(\bar{\alpha}(P)) = \bar{\alpha}(U)$ being abelian. If G is of type C_3 with Dynkin diagram

Δ: 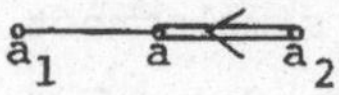

then

$$R_u(\bar{\alpha}(P)) = U_a U_{a+a_1} U_{a+a_2} U_{a+a_1+a_2} U_{2a+a_2} U_{a_1+a_2+2a} U_{2a_1+2a+a_2}. \quad [6]$$

Since $N_{a,a+a_1+a_2} = \pm 1$, $[U_a, U_{a+a_1+a_2}] \neq \{1\}$ [15. §3], which gives a contradiction. If G is of type D_7 with $a = a_4$ in Dynkin diagram (3.5.1), then

$R_u(\bar{\alpha}(P)) = \prod_{b \in \Phi^+ \setminus \phi} U_b$ (direct product of varieties), where $\phi = \{ \sum_{\substack{i=1 \\ i \neq 4}} \mathbb{Z} a_i \cap \Phi \}$

Now, $a+a_3, a+2a_5+a_6+a_7, 2a+a_3+2a_5+a_6+a_7 \in \Phi^+ \setminus \phi$, so that $[U_{a+a_3}, U_{a+2a_5+a_6+a_7}] \neq \{1\}$, which contradicts to $R_u(\bar{\alpha}(P)) = \bar{\alpha}(U)$ being abelian. Hence G is of type A_{2n-1}, $n = |\Delta_1| + 1 > 1$. ∎

<u>3.6. Corollary</u>. The simple root system Δ of G has Dynkin diagram

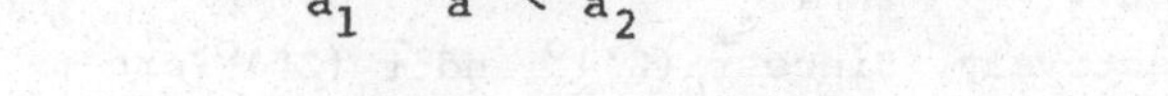

(3.6.1)

where $a = a_n$ is the only simple root such that the root subgroup U_a is contained in $R_u(\bar{\alpha}(P))$. Moreover, $\Delta_1 = \{a_1, \ldots, a_{n-1}\}$ is a simple root system of $\bar{\alpha}(Z')^\circ$ and $\Delta_2 = \{a_{n+1}, \ldots, a_{2n-1}\}$ is a simple root system of $\bar{\alpha}(Z'')^\circ$. ∎

<u>3.7. Remark</u>. Denote by B' and T' the subgroups of $PSL_{2n}(K)$ consisting of upper triangular matrices and diagonal matrices respectively. Let

$\Phi'(\Delta')$ be root system (simple root system) of $PSL_{2n}(k)$ relative to T' (and B'), suppose $\Delta'=\{a_1',\dots,a_{2n-1}'\}$, then the relative root subgroups are

$$U_{a_i'} = \left\{ \begin{bmatrix} 1 & & & \\ & \ddots & & \\ & & 1 & d \\ & & & 1 \\ & & & & \ddots \\ & & & & & 1 \end{bmatrix}^{i} \in PSL_{2n}(K) \;\middle|\; \forall d \in K \right\}, \quad 1 \leq i \leq 2n-1$$

$$U_{-a_i'} = \left\{ \begin{bmatrix} 1 & & & \\ & \ddots & & \\ & & 1 & \\ & & d & 1 \\ & & & & \ddots \\ & & & & & 1 \end{bmatrix} \in PSL_{2n}(K) \;\middle|\; \forall d \in K \right\}, \quad 1 \leq i \leq 2n-1$$
$$i$$

Since G splits over k, there is a k-isogeny $\pi: G \longrightarrow PSL_{2n}(k)$, such that $\pi(T)=T'$ and $\pi(U_{a_i})=U_{a_i'}$, $1\leq i\leq 2n-1$. Now because α has Zariski dense image, so does $\pi\alpha: SL_2(D) \longrightarrow PSL_{2n}(k)$. Hence the results for α we obtained before are also valid for $\pi\alpha$. Denote $\tau_\sigma = \pi\alpha$. Without loss of generality, denote $\Delta'(a_i', 1\leq i\leq 2n-1)$ by Δ $(a_i, 1\leq i\leq 2n-1)$. $\Delta_1 = \{a_1,\dots,a_{n-1}\}$ and $\Delta_2 = \{a_{n+1},\dots,a_{2n-1}\}$ are simple root systems of $\bar{\tau}_\alpha(Z')^o$ and $\bar{\tau}_\alpha(Z'')^o$ respectively. Since $\bar{\tau}_\alpha(Z')^o$ and $\bar{\tau}_\alpha(Z'')^o$ are simple algebraic groups, we have

$$\bar{\tau}_\alpha(Z')^o = \langle U_{\pm a_i} \mid a_i \in \Delta_1\rangle, \quad \bar{\tau}_\alpha(Z'')^o = \langle U_{\pm a_j} \mid a_j \in \Delta_2\rangle$$

hence

$$\bar{\tau}_\alpha(Z')^o = \left\{ \begin{bmatrix} A & 0 \\ 0 & I_n \end{bmatrix} \in PSL_{2n}(K) \;\middle|\; \forall A \in SL_n(K) \right\} \tag{3.7.1}$$

$$\bar{\tau}_\alpha(Z'')^o = \left\{ \begin{bmatrix} I_n & 0 \\ 0 & B \end{bmatrix} \in PSL_{2n}(K) \;\middle|\; \forall A \in SL_n(K) \right\} \tag{3.7.2}$$

<u>3.8. Lemma</u>. $\bar{\tau}_\alpha(Z')$ and $\bar{\tau}_\alpha(Z'')$ are connected subgroups of $PSL_{2n}(k)$.

Proof. $\bar{\tau}_\alpha(Z') \subsetneq \bar{\tau}_\alpha(Z')^o \bar{\tau}_\alpha(Z'')^o$ by (3.2), so that $\bar{\tau}_\alpha(Z') = (\bar{\tau}_\alpha(Z') \cap \bar{\tau}_\alpha(Z'')^o) \cdot \bar{\tau}_\alpha(Z')^o$. However, $\bar{\tau}_\alpha(Z') \cap \bar{\tau}_\alpha(Z'')^o \subsetneq C(\bar{\tau}_\sigma(Z'')^o)$ since $\bar{\tau}_\alpha(Z') \subsetneq C_G(\bar{\tau}_\alpha(Z'')^o)$. It follows from (3.7) that

$$C(\bar{\tau}_\alpha(Z'')^o) = \left\{ \begin{bmatrix} I_n & 0 \\ 0 & \varepsilon I_n \end{bmatrix} \in PSL_{2n}(K) \;\middle|\; \varepsilon \in K,\ \varepsilon^n = 1 \right\}$$

and

$$\begin{bmatrix} I_n & 0 \\ 0 & \varepsilon I_n \end{bmatrix} \begin{bmatrix} A & 0 \\ 0 & I_n \end{bmatrix} = \begin{bmatrix} \varepsilon^{n-1} A & 0 \\ 0 & I_n \end{bmatrix} \in \bar{\tau}_\alpha(Z')^o, \ \forall A \in SL_n(K).$$

which means that $(\bar{\tau}_\alpha(Z') \cap \bar{\tau}_\alpha(Z'')^o) \subsetneq \bar{\tau}_\alpha(Z')^o$. Consequently, $\bar{\tau}_\alpha(Z') = \bar{\tau}_\alpha(Z')^o$. Similarly $\bar{\tau}_\alpha(Z'') = \bar{\tau}_\alpha(Z'')^o$. ∎

3.9. Since $\bar{\tau}_\alpha(P)$ is a standard parabolic subgroup relative to Δ and its levi-component $\tau_\alpha(Z)$ contains root subgroups U_{a_i}, $i=1,\dots,n-1, n+1,\dots,2n-1$, we have

$$\bar{\tau}_\alpha(P) = \left\{ \begin{bmatrix} A & B \\ 0 & C \end{bmatrix} \in PSL_{2n}(K) \middle| \; A,C \in GL_n(K), \forall B \in Mat_{n\times n}(K) \right\}$$

$$\bar{\tau}_\alpha(P^-) = \left\{ \begin{bmatrix} A & 0 \\ B & C \end{bmatrix} \in SL_{2n}(K) \middle| \; A,C \in GL_n(K), \forall B \in Mat_{n\times n}(K) \right\}$$

In particular,

$$\bar{\tau}_\alpha(U) = R_u(\bar{\tau}_\alpha(P)) = \left\{ \begin{bmatrix} I_n & B \\ 0 & I_n \end{bmatrix} \in PSL_{2n}(K) \middle| \; \forall B \in Mat_{n\times n}(K) \right\}$$

$$\bar{\tau}_\alpha(U^-) = R_u(\bar{\tau}_\alpha(P^-)) = \left\{ \begin{bmatrix} I_n & 0 \\ C & I_n \end{bmatrix} \in PSL_{2n}(K) \middle| \; \forall C \in Mat_{n\times n}(K) \right\}$$

Let $b \in D^*$, suppose $\tau_\alpha\begin{pmatrix} 1 & b \\ 0 & 1 \end{pmatrix} = \begin{bmatrix} I_n & B \\ 0 & I_n \end{bmatrix}$, $\tau_\alpha\begin{pmatrix} 1 & 0 \\ -b^{-1} & 0 \end{pmatrix} = \begin{bmatrix} I_n & 0 \\ B_1 & I_n \end{bmatrix}$,

where $B, B_1 \in Mat_{n\times n}(k)$. From the identity

$$\begin{pmatrix} 0 & b \\ -b^{-1} & 0 \end{pmatrix} = \begin{pmatrix} 1 & b \\ 0 & 1 \end{pmatrix} \begin{pmatrix} 1 & 0 \\ -b^{-1} & 1 \end{pmatrix} \begin{pmatrix} 1 & b \\ 0 & 1 \end{pmatrix} \tag{3.9.1}$$

we have

$$\tau_\alpha\begin{pmatrix} 0 & b \\ -b^{-1} & 0 \end{pmatrix} = \begin{bmatrix} I_n & B \\ 0 & I_n \end{bmatrix} \begin{bmatrix} I_n & 0 \\ B_1 & I_n \end{bmatrix} \begin{bmatrix} I_n & B \\ 0 & I_n \end{bmatrix} = \begin{bmatrix} I_n + BB_1 & 2B+BB_1B \\ B_1 & I_n + B_1B \end{bmatrix} \tag{3.9.2}$$

On the other hand, $\tau_\alpha\begin{pmatrix} 0 & b \\ -b^{-1} & 0 \end{pmatrix} \bar{\tau}_\alpha(Z')^\circ \tau_\alpha\begin{pmatrix} 0 & b \\ -b^{-1} & 0 \end{pmatrix}^{-1} = \bar{\tau}_\alpha(Z'')^\circ$, so that $\tau_\alpha\begin{pmatrix} 0 & b \\ -b^{-1} & 0 \end{pmatrix}$ must be of the form $\begin{bmatrix} 0 & A \\ C & 0 \end{bmatrix}$, $A, C \in GL_n(k)$,

consequently, $B = -B_1^{-1}$. Furthermore, suppose $\tau_\alpha\begin{pmatrix} 1 & 1 \\ 0 & 1 \end{pmatrix} = \begin{bmatrix} I_n & Q \\ 0 & I_n \end{bmatrix}$ and

$\tau_\alpha\begin{pmatrix} 1 & 0 \\ -1 & 1 \end{pmatrix} = \begin{bmatrix} I_n & 0 \\ -Q^{-1} & I_n \end{bmatrix}$, then $Q \in GL_n(k)$ and $\tau_\alpha\begin{pmatrix} 0 & 1 \\ -1 & 0 \end{pmatrix} = \begin{bmatrix} 0 & Q \\ -Q^{-1} & 0 \end{bmatrix}$, so that

$$\beta : PSL_{2n}(K) \longrightarrow PSL_{2n}(K)$$

$$\begin{bmatrix} A & B \\ C & D \end{bmatrix} \longmapsto \begin{bmatrix} I_n & 0 \\ 0 & Q \end{bmatrix} \begin{bmatrix} A & B \\ C & D \end{bmatrix} \begin{bmatrix} I_n & 0 \\ 0 & Q \end{bmatrix}^{-1}$$

β is a k-isomorphism. Denote $\beta\tau_\alpha$ by σ_α, all the above results which are valid for τ_α is also valid for σ_α. In particular, $\sigma_\alpha\begin{pmatrix} 1 & 1 \\ 0 & 1 \end{pmatrix} =$

$\begin{bmatrix} I_n & I_n \\ 0 & I_n \end{bmatrix}$ $\sigma_\alpha\begin{pmatrix} 1 & 0 \\ -1 & 1 \end{pmatrix} = \begin{bmatrix} I_n & 0 \\ -I_n & I_n \end{bmatrix}, \sigma_\alpha\begin{pmatrix} 0 & 1 \\ -1 & 0 \end{pmatrix} = \begin{bmatrix} 0 & I_n \\ -I_n & 0 \end{bmatrix}$ Restricting σ_α to U and U^-, we have

$$\sigma_\alpha : U \longrightarrow \sigma_\alpha(U)$$
$$\begin{pmatrix} 1 & b \\ 0 & 1 \end{pmatrix} \longmapsto \begin{bmatrix} I_n & B \\ 0 & I_n \end{bmatrix}, \quad B \in \mathrm{Mat}_{n\times n}(k)$$

and

$$\sigma_\alpha : U^- \longrightarrow \sigma_\alpha(U^-)$$
$$\begin{pmatrix} 1 & 0 \\ b & 1 \end{pmatrix} \longmapsto \begin{bmatrix} I_n & 0 \\ B_1 & I_n \end{bmatrix}, \quad B_1 \in \mathrm{Mat}_{n\times n}(k)$$

which induce two homomorphisms of abelian groups,

$$\phi : \quad D \longrightarrow \mathrm{Mat}_{n\times n}(k)$$
$$b \longmapsto B$$

and

$$\phi : \quad D \longrightarrow \mathrm{Mat}_{n\times n}(k)$$
$$b \longmapsto B_1$$

<u>3.10. Lemma</u> $\phi = \phi'$. Moreover, $\phi(b^{-1}) = \phi(b)^{-1}$, $\forall b \in D^*$

Proof. Applying σ_α to both sides of the identity

$$\begin{pmatrix} 1 & 0 \\ b & 1 \end{pmatrix} = \begin{pmatrix} 0 & 1 \\ -1 & 0 \end{pmatrix}\begin{pmatrix} 1 & -b \\ 0 & 1 \end{pmatrix}\begin{pmatrix} 0 & -1 \\ 1 & 0 \end{pmatrix}$$

we obtain

$$\begin{bmatrix} I_n & 0 \\ B_1 & I_n \end{bmatrix} = \begin{bmatrix} 0 & I_n \\ -I_n & 0 \end{bmatrix}\begin{bmatrix} I_n & -B \\ 0 & I_n \end{bmatrix}\begin{bmatrix} 0 & -I_n \\ I_n & 0 \end{bmatrix} \quad \begin{bmatrix} I_n & 0 \\ B & I_n \end{bmatrix}$$

which means $B_1 = B$. hence $\phi = \phi'$. The second assertion of the lemma comes from (3.9.1) & (3.9.2). ∎

<u>3.11. Lemma</u>. $\phi(ab) = \phi(a)\phi(b)$, $\forall a,b \in [D^*,D^*]$.

Proof. By (3.9.1) & (3.9.2), $\sigma_\alpha\begin{pmatrix} 0 & b \\ -b^{-1} & 0 \end{pmatrix} = \begin{bmatrix} 0 & \phi(b) \\ -\phi(b)^{-1} & 0 \end{bmatrix}$,

so that

$$\sigma_\alpha\begin{pmatrix} b & 0 \\ 0 & b^{-1} \end{pmatrix} = \sigma_\alpha\begin{pmatrix} 0 & -1 \\ 1 & 0 \end{pmatrix}\sigma_\alpha\begin{pmatrix} 0 & b^{-1} \\ -b & 0 \end{pmatrix} = \begin{bmatrix} \phi(b) & 0 \\ 0 & \phi(b)^{-1} \end{bmatrix}, \forall b \in D^* \tag{3.11.1}$$

we prove first that $\sigma_\alpha\begin{pmatrix} b & 0 \\ 0 & 1 \end{pmatrix} = \begin{bmatrix} \phi(b) & 0 \\ 0 & I_n \end{bmatrix}$, $\forall b \in [D^*,D^*]$.

Suppose $\sigma_\alpha\begin{pmatrix} b & 0 \\ 0 & 1 \end{pmatrix} = \begin{bmatrix} B & 0 \\ 0 & I_n \end{bmatrix}$, $\sigma_\alpha\begin{pmatrix} 1 & 0 \\ 0 & b^{-1} \end{pmatrix} = \begin{bmatrix} I_n & 0 \\ 0 & B_1 \end{bmatrix}$ (cf. (3.7.1)

& (3.7.2)), then $\begin{bmatrix}\phi(b) & 0\\ 0 & \phi(b)^{-1}\end{bmatrix} = \begin{bmatrix}B & \\ & B_1\end{bmatrix}$ by (3.11.1),

hence $\begin{bmatrix}\phi(b)B^{-1} & 0\\ 0 & \phi(b)^{-1}B_1^{-1}\end{bmatrix} = [I_{2n}]$, which means that $B=\phi(b)$ and

$B_1=\phi(b)^{-I}$, where $\varepsilon \in k$ and $\varepsilon^{2n}=1$. Now,

$$\begin{pmatrix}b & 0\\ 0 & 1\end{pmatrix}=\begin{pmatrix}1 & -b\\ 0 & 1\end{pmatrix}\begin{pmatrix}1 & 0\\ b^{-1} & 0\end{pmatrix}\begin{pmatrix}b & 0\\ 0 & 1\end{pmatrix}\begin{pmatrix}1 & -1\\ 0 & 1\end{pmatrix}\begin{pmatrix}0 & 1\\ -1 & 0\end{pmatrix}, \forall b\in D^*$$

Applying σ_α to both sides of the above identity, we have

$$\begin{bmatrix}\phi(b) & 0\\ 0 & I_n\end{bmatrix}=\begin{bmatrix}I_n & -\phi(b)\\ 0 & I_n\end{bmatrix}\begin{bmatrix}I_n & 0\\ \phi(b)^{-1} & I_n\end{bmatrix}\begin{bmatrix}\phi(b) & 0\\ 0 & I_n\end{bmatrix}\begin{bmatrix}I_n & -I_n\\ 0 & I_n\end{bmatrix}\begin{bmatrix}0 & I_n\\ -I_n & 0\end{bmatrix}$$

$$=\begin{bmatrix}\phi(b) & 0\\ (\varepsilon-1)I_n & \varepsilon I_n\end{bmatrix}, \forall b\in [D^*,D^*].$$

It follows that

$$[I_{2n}]=\begin{bmatrix}\phi(b) & 0\\ 0 & I_n\end{bmatrix}^{-1}\begin{bmatrix}\phi(b) & 0\\ (\varepsilon-1)I_n & \varepsilon I\end{bmatrix}=\begin{bmatrix}\varepsilon^{-1}I_n & 0\\ (\varepsilon-1)I_n & I_n\end{bmatrix}, \forall b\in [D^*,D^*]$$

consequently, $\varepsilon=1$, $B=\phi(b)$ and $B_1=\phi(b)^{-1}$. Now, suppose $a,b\in[D^*,D^*]$,

since $\begin{pmatrix}ab & 0\\ 0 & 1\end{pmatrix}=\begin{pmatrix}a & 0\\ 0 & 1\end{pmatrix}\begin{pmatrix}b & 0\\ 0 & 1\end{pmatrix}$, we have

$$\begin{bmatrix}\phi(ab) & 0\\ 0 & I_n\end{bmatrix}=\begin{bmatrix}\phi(a) & 0\\ 0 & I_n\end{bmatrix}\begin{bmatrix}\phi(b) & 0\\ 0 & I_n\end{bmatrix}=\begin{bmatrix}\phi(a)\phi(b) & 0\\ 0 & I_n\end{bmatrix}$$

from which it follows that $\phi(ab)=\phi(a)\phi(b)$. |

<u>3.12. Lemma</u>. Suppose D is a non-commutative division ring, then

$$D=\left\{\sum_i \left(\sum_j \varepsilon_{ij}b_{ij}\right)^{-1} \,\middle|\, \varepsilon_{ij}=\pm 1,\ b_{ij}\in [D^*,D^*]\right\}$$

Proof. Let D_1 be the division ring generated by set $S=\{\sum_i(\sum_j\varepsilon_{ij}b_{ij})^{-1}|$ $\varepsilon_{ij}=\pm1$, $b_{ij}\in[D^*,D^*]\}$, then $D_1\subseteq D$. On the other hand, since D is generated by elements of $[D^*,D^*]$ [9.I.§11. Th3] which is contained in S, we have $D\subseteq D_1$, hence $D_1=D$. Now, we show that $S=D_1$. Obviously, S is closed under the addition of D. If $a_i,b_j\in[D^*,D^*]$, $\varepsilon_i,\varepsilon_j=\pm1$, then

$$\left(\sum_i\varepsilon_i a_i\right)^{-1}\left(\sum_j\varepsilon_j b_j\right)^{-1}=\left(\sum_i\sum_j\varepsilon_i\varepsilon_j b_j a_i\right)^{-1}\in S. \tag{3.12.1}$$

It follows that S is closed under the multiplication of D. Consider the inverses of the non-zero elements of S. Let $b=\sum_{i=j}^{n}(\sum_j\varepsilon_{ij}b_{ij})^{-1}$

$\in S$ and $b\neq 0$, we assert that $b=(\sum_\ell \varepsilon\varepsilon_\ell c_\ell)^{-1}(\sum_s \varepsilon_s d_s)$ for some $c_\ell, d_s \in [D^*,D^*]$, ε_ℓ, $\varepsilon_s=\pm 1$ hence

$$b^{-1}=(\sum_s \varepsilon_s d_s)^{-1}(\sum_\ell \varepsilon_\ell c_\ell)=\sum_\ell(\sum_s \varepsilon_\ell\varepsilon_s c_\ell^{-1}d_s)^{-1}\in S.$$

Clearly, the assertion holds for n=1. In case n=2, $b=(\sum_i \varepsilon_i b_i)^{-1}+(\sum_j \varepsilon_j b_j)^{-1}$, $a_i, b_j \in [D^*,D^*]$, $\varepsilon_i, \varepsilon_j=\pm 1$. Let $d=(\sum_i \varepsilon_i a_i)(\sum_j \varepsilon_j b_j)^{-1}(\sum_i \varepsilon_i a_i)^{-1}(\sum_j \varepsilon_j b_j)\in [D^*,D^*]$, then

$$(\sum_i \varepsilon_i a_i)(\sum_j \varepsilon_j b_j)^{-1}=d(\sum_j \varepsilon_j b_j)^{-1}(\sum_i \varepsilon_i a_i)=(\sum_j \varepsilon_j b_j')^{-1}(\sum_i \varepsilon_i a_i)$$

where $b_j'=b_j d^{-1}\in [D^*,D^*]$, therefore

$$(\sum_i \varepsilon_i a_i)^{-1}+(\sum_j \varepsilon_j b_j)^{-1}=(\sum_i \varepsilon_i a_i)^{-1}(1+(\sum_i \varepsilon_i a_i)(\sum \varepsilon_j b_j)^{-1})$$

$$=(\sum_i \varepsilon_i a_i)^{-1}(1+(\sum_j \varepsilon_j b_i)^{-1}(\sum_i \varepsilon_i a_i))$$

$$=(\sum_i \varepsilon_i a_i)^{-1}(\sum_j \varepsilon_j b_j')^{-1}(\sum_j \varepsilon_j b_j'+\sum_j \varepsilon_i a_i)$$

$$=(\sum_{i,j} \varepsilon_i \varepsilon_j b_j a_i)^{-1}(\sum \varepsilon_j b_j+\sum_i \varepsilon_i a_i). \qquad (3.12.2)$$

Suppose n >2 and the assertion holds for n-1, then

$$\sum_{i=1}^{n}(\sum_j \varepsilon_{ij} b_{ij})^{-1}=\sum_{i=1}^{n-1}(\sum_j \varepsilon_{ij} b_{ij})^{-1}+(\sum_j \varepsilon_{nj} b_{nj})^{-1}$$

$$=(\sum_\ell \varepsilon_\ell c_\ell)^{-1}(\sum_s \varepsilon s d_s)+(\sum_j \varepsilon_{nj} b_{nj})^{-1}, c_\ell, d_\ell \in [D^*,D^*], \varepsilon_\ell, \varepsilon_s=+1$$

$$=((\sum_\ell \varepsilon_\ell c_\ell)^{-1}+(\sum_j \varepsilon_{nj} b_{nj})^{-1}(\sum_s \varepsilon s d_s)^{-1})(\sum_s \varepsilon_s d_s)$$

$$=(\sum_r \varepsilon_r c_r')^{-1}(\sum_t \varepsilon_t d_t')(\sum_s \varepsilon_s d_s), \text{ by } (3.9.2)$$

$$=(\sum_r \varepsilon_r c_r')^{-1}(\sum_{s,t} \varepsilon_{st} d_s d_t').$$

Thus the assertion holds also for n. ∎

<u>3.13. Lemma</u>. The homomorphism of abelian groups $\phi: D \longrightarrow \mathrm{Mat}_{n\times n}(k)$, which is defined in (3.9), is a homomorphism of rings.

Proof. It is enough to show that $\phi(ab)=\phi(a)\phi(b)$, $\forall\, a,b\in D^*$. By (3.12) we can suppose $a=\sum_i(\sum_j \varepsilon_{ij} a_{ij})^{-1}$ and $b=\sum_\ell(\sum_s \varepsilon_{\ell s} b_{\ell s})^{-1}$, where $\varepsilon_{ij}, \varepsilon_{\ell s}=\pm 1$, $a_{ij}, b_{\ell s}\in [D^*,D^*]$.

$$(\sum_i(\sum_j \varepsilon_{ij} a_{ij})^{-1})(\sum_\ell(\sum_s \varepsilon_{\ell s} b_{\ell s})^{-1})=\sum_{i,\ell}(\sum_j \varepsilon_{ij} a_{ij})^{-1}(\sum_s \varepsilon_{\ell s} b_{\ell s})^{-1}$$

$$= \sum_{i,\ell} \left(\sum_{j,s} \varepsilon_{ij}\varepsilon_{\ell s} b_{\ell s} a_{ij} \right)^{-1}.$$

By (3.10) and (3.11) we have

$$\phi(ab) = \sum_{i,\ell} \left(\sum_{j,s} \varepsilon_{ij}\varepsilon_{\ell s}\phi(b_{\ell s})\phi(a_{ij}) \right)^{-1}$$

$$= \phi\left(\sum_i \left(\sum_j \varepsilon_{ij} a_{ij}\right)^{-1}\right)\phi\left(\sum_\ell \left(\sum_s \varepsilon_{\ell s} b_{\ell s}\right)^{-1}\right)$$

$$= \phi(a)\phi(b). \quad \blacksquare$$

3.14. Lemma. Let F be the centre of D, then $\phi: D \longrightarrow Mat_{n\times n}(k)$ induces an injection of fields $\phi^o: F \longrightarrow k$ such that $\phi(ta) = \phi^o(t).\phi(a)$, $\forall t \in F$, $\forall a \in D$.

Proof: Suppose $t \in F$, then $\begin{pmatrix} t & 0 \\ 0 & t^{-1} \end{pmatrix} \in C_{SL_2(D)}(Z')$, hence $\sigma_\alpha\begin{pmatrix} t & 0 \\ 0 & t^{-1} \end{pmatrix} =$

$\begin{bmatrix} \phi(t) & 0 \\ 0 & \phi(t)^{-1} \end{bmatrix} \in C_{PSL_{2n}(k)}(\overline{\sigma_\alpha}(Z')^o)$ (3.11.1) which means that

$$\begin{bmatrix} \phi(t) & 0 \\ 0 & \phi(t)^{-1} \end{bmatrix}\begin{bmatrix} A & 0 \\ 0 & I_n \end{bmatrix} = \begin{bmatrix} A & 0 \\ 0 & I_n \end{bmatrix}\begin{bmatrix} \phi(t) & 0 \\ 0 & \phi(t)^{-1} \end{bmatrix}, \quad \forall A \in SL_n(K).$$

i.e.

$$\begin{pmatrix} \phi(t) & 0 \\ 0 & \phi(t)^{-1} \end{pmatrix}\begin{pmatrix} A & 0 \\ 0 & I_n \end{pmatrix} = \begin{pmatrix} A & 0 \\ 0 & I_n \end{pmatrix}\begin{pmatrix} \phi(t) & 0 \\ 0 & \phi(t)^{-1} \end{pmatrix} \cdot \varepsilon I_{2n}, \quad \varepsilon \in K,$$

where $\varepsilon^{2n} = 1$. Therefore $\varepsilon\phi(t)^{-1} = \phi(t)^{-1}$, $\varepsilon = 1$. It follows that $\phi(t)A = A\phi(t)$, $\forall A \in SL_n(k)$, so that $\phi(t)$ is a scalar matrix, say, $\phi(t) = t'I_n$, where $t' \in k$. Since α is non-trivial, so is ϕ, and t' is uniquely determined by t. It is clear that the map $\phi^o: F \longrightarrow k$ defined by $\phi^o(t) = t'$ is injection of fields and $\phi(ta) = \phi^o(t)\phi(a)$, $\forall t \in F$, $a \in D$. $\blacksquare$

3.15. Theorem. Suppose G is a simple algebraic group defined over k, $\alpha: SL_2(D) \longrightarrow G(k)$ is a homomorphism with Zariski dense image, F is the centre of D. Then there is a finite separable extension field k' of k such that

(i) α induces an injection $\phi^o: F \longrightarrow k'$, inparticular, char F = char k.

(ii) Take F as a subfield of k', then α induces a homomorphism of F-algebras $\phi: D \longrightarrow Mat_{n\times n}(k')$, where $n = \operatorname{rank}\bar{\alpha}(Z')^o + 1$.

(iii)* D is a finite dimensional division algebra over F.

(iv) Define a map

$$\sigma_\alpha: SL_2(D) \longrightarrow PSL_{2n}(k')$$

*This statement is conjectured by B. Weisfeiler.

$$\begin{pmatrix} a & b \\ c & d \end{pmatrix} \longrightarrow \begin{bmatrix} \phi(a) & \phi(b) \\ \phi(c) & \phi(d) \end{bmatrix}, \qquad a,b,c,d \in D,$$

then σ_α is a homomorphism and, up to a k-isomorphism, there exists exactly one k'-isogeny $\pi: G \longrightarrow PSL_n(K)$ such that $\pi\alpha = \sigma_\alpha$.

(v) If G splits over k, one can take k'=k.

Proof. By [2.§8.3], there is a finite separable extension field k' of k such that G splits over k'. Therefore, it is enough to show that if G splits over k then (i) & (ii) holds, which, however, come from above lemmas. In particular, (iii) comes from (ii) by Kaplansky Theorem [11. Th.8.2.5] . Finally, since $SL_2(D) = \langle \begin{pmatrix} 1 & a \\ 0 & 1 \end{pmatrix}, \begin{pmatrix} 1 & 0 \\ b & 1 \end{pmatrix} \mid \forall a,b \in D \rangle$, σ_α defined in (iv) is just what we met in (3.9). |

3.16. Theorem. Suppose G is a reductive algebraic group. $\alpha: SL_2(D) \longrightarrow G$ is a homomorphism with Zariski dense image, then G is semisimple. Moreover, $G = G_1 \dots G_r$ is an almost direct product of simple algebraic groups of type A_{2n_i-1}, $i=1,\dots,r$, $n_i \in \mathbb{Z}^+$, $n_i > 1$.

Proof. Since $[SL_2(D), SL_2(D)] = SL_2(D)$, $[G,G] = G$, thus G is semisimple. If $G = G_1 \dots G_r$, where G_i, $i=1,\dots,r$, are simple algebraic groups, then there exists isogeny $\pi: G \to G' = G_1' \dots G_r'$, where G_i', $i=1,\dots,r$, are adjoint simple algebraic groups and G' is direct product of G_i's. Consider the composition of homomorphism

$$SL_2(D) \xrightarrow{\alpha} G \xrightarrow{\pi} G' \xrightarrow{P_i} G_i', \quad 1 \leq i \leq r.$$

where P_i is the projection of G' onto G_i'. Obviously, $P_i\pi\alpha$ is also a homomorphism with Zariski dense image, so that G_i', as well as G_i, is of type A_{2n_i-1} by (3.5). |

3.17. Remark. We may take $PSL_2(D)$ in stead of $SL_2(D)$ in theorems (3.5) (3.15) and (3.16), and the proofs are similar.

4. Homomorphisms From $SL_n(D)$ and $PSL_n(D)$ to Algebraic Groups

Let G be a simple algebraic group defined over k, $\alpha: SL_n(D) \longrightarrow G(k)$ a homomorphism with Zariski dense image, $n \geq 3$. In this section we use notations defined in §1 & §2.

4.1. Consider the following subgroups of $SL_n(D)$

$$Z_{11} = \left\{ \begin{pmatrix} A & \\ & 1 \end{pmatrix} \in SL_n(D) \;\middle|\; \forall A \in SL_{n-1}(D) \right\}$$

$$Z_{21}=\left\{\begin{pmatrix} B & \\ & I_2\end{pmatrix}\in SL_n(D)\,\middle|\,\forall B\in SL_{n-2}(D)\right\}$$

$$Z_{12}=\left\{\begin{pmatrix} I_{n-1} & \\ & d\end{pmatrix}\in SL_n(D)\,\middle|\,\forall d\in [D^*,D^*]\right\}$$

$$Z_{22}=\left\{\begin{pmatrix} I_{n-2} & \\ & C\end{pmatrix}\in SL_n(D)\,\middle|\,\forall C\in SL_2(D)\right\}$$

It is obvious that $Z_{11}\cong SL_{n-1}(D)$, $Z_{21}\cong SL_{n-2}(D)$, $Z_{22}\cong SL_2(D)$. $[Z_i,Z_i]=Z_{i1}\times Z_{12}$, $i=1,2$(cf.(2.11)). Furthermore, $\bar\alpha(Z_{11})=\bar\alpha(Z_{22})$ are connected by (2.6).

Lemma. (i) $\bar\alpha(Z_{i1})^o$ and $\bar\alpha(Z_{i2})^o$ are normal subgroups of $\bar\alpha(Z_i)$, $i=1,2$.
(ii) $\bar\alpha(Z_{i1})^o\subseteq C_G(\bar\alpha(Z_{i2})^o)$. $i=1,2$.
(iii) $[\bar\alpha(Z_i),\bar\alpha(Z_i)]=\bar\alpha(Z_{i1})^o\bar\alpha(Z_{i2})^o$, $i=1,2$. Moreover, $|\bar\alpha(Z_{i1})^o\cap\bar\alpha(Z_{12})^o|<\infty$.

Proof. Similar to (3.2). |

4.2. Lemma. $\bar\alpha(Z_{22})$ is a simple algebraic group.
Proof. Let $\Delta^{(2)}$ be a simple root system of G such that $\bar\alpha(P_2)$ is a standard parabolic subgroup relative to $\Delta^{(2)}$. Let θ be a simple root system of the Levi-component $\bar\alpha(Z_2)$ of $\bar\alpha(P_2)$ (2.9). $\Delta^{(2)}=\theta\cup\{a\}$, where a is the only simple root such that the root subgroup $U_a\subseteq R_u(\bar\alpha(P_2))$. Since $\bar\alpha(Z_2)$ is reductive, $[\bar\alpha(Z_2),\bar\alpha(Z_2)]=G_1\ldots G_r$ is an almost direct product of simple algebraic groups. It follows from (3.1) that $\bar\alpha(Z_{22})=G_{i_1}\ldots G_{i_s}$, $1\le i_1,\ldots,i_s\le r$, and $\bar\alpha(Z_{21})^o=G_{j_1}\ldots G_{j_t}$, $\{j_1,\ldots,j_t\}=\{1,\ldots,r\}\setminus\{i_1,\ldots,i_s\}$. Furthermore, taking simple root system Δ_i of G_i for each $1\le i\le r$ such that $\theta=\bigcup_{i=1}^{r}\Delta_i$, we obtain $r\le 3$ by a proof similar to (2.4). Since $n\ge 3$, $|Z_{21}|=\infty$ $\dim\bar\alpha(Z_{21})^o\ge 1$, so that $1\le s<r$. If $s=2$, then $t=1$ and Δ_{i_1}, Δ_{i_2} are simple root systems of type A by (3.16). Moreover $\Delta_{i_1}\ge 3$, $\Delta_{i_2}\ge 3$, $\Delta_{j_1}\ge 1$. Now, $\Delta^{(2)}=\Delta_{i_1}\cup\Delta_{i_2}\cup\{a\}$ and $\Delta_{i_1}\perp\Delta_{i_2}\perp\Delta_{j_1}$, it follows that the Dynkin diagram of $\Delta^{(2)}$ must be of the form

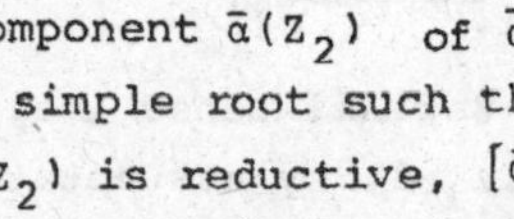

which is impossible. Hence s=1, $\bar\alpha(Z_{22})$ is a simple algebraic group. |

4.3. Lemma. Let $M_i=\{\mathrm{diag}\,(1,\ldots,b,\ldots,1)\in SL_n(D)\mid\forall b\in[D^*,D^*]\}$, $1\le i\le n$, then

(i) $\tilde{\alpha}(M_i)^o$ is a simple algebraic group of type A_m, $m>1$, $1\leq i\leq n$.

(ii) rank $\bar{\alpha}(M_i)^o$=rank$\bar{\alpha}(M_j)^o$, $1\leq i,j\leq n$.

Proof. $M_i \cong Z_{12}$ under an inner automorphism of $SL_n(D)$. Consequently, $\alpha(M_i)^o \cong \alpha(Z_{12})^o$, hence $\bar{\alpha}(M_i)^o$ is a simple algebraic group of type A_m by (4.2) & (3.4). (ii) is clear. |

4.4. Theorem. Let G be a simple algebraic group, $\alpha: SL_n(D) \longrightarrow G$ a homomorphism with Zariski dense image, then G is of type A_ℓ, $\ell>3$.

Proof. Apply induction on n. In case n=2, it is theorem (3.5). Suppose $n>2$. Let $\Delta^{(1)}$ be a simple root system of G such that $\bar{\alpha}(P_i)$ is a standard parabolic subgroup relative to $\Delta^{(1)}$ and θ_1 be a simple root system of Levi-component $\bar{\alpha}(Z_i)$ of $\bar{\alpha}(P_1)$. By (4.1), $[\bar{\alpha}(Z_1),\bar{\alpha}(Z_1)] = \bar{\alpha}(Z_{11})\bar{\alpha}(Z_{12})^o$ and $\bar{\alpha}(Z_{11})$ is a normal subgroup of $\bar{\alpha}(Z_1)$, $\bar{\alpha}(Z_1)$, as well as semisimple. Suppose $\bar{\alpha}(Z_{11})=G_1\ldots G_s$, where G_i, $1\leq i\leq s$, is a simple algebraic group. Let G'_i, $1\leq i\leq s$, is the adjoint simple algebraic group which is of same type as G_i. There is an isogeny $\pi: G \longrightarrow G'$. Consider the composition of homomorphisms

$$Z_{11} \xrightarrow{\alpha} G_1\ldots G_s \xrightarrow{\pi} \prod_{i=1}^{s} G'_i \xrightarrow{P_i} G'_i .$$

Since $P_i\pi\alpha$ is a homomorphism with Zariski dense image and $Z_{11}\cong SL_{n-1}(D)$, G'_i, as well as G_i, is of type A_{ℓ_i}, $\ell_i\geq 3$, by assumptation. similar to proof of (4.2) we have s=1 and $\bar{\alpha}(Z_{11})$ is a simple algebraic group of type A_{ℓ_1}. Lemma (4.1) enable us to take simple root systems $\Delta_1^{(1)}$ and $\Delta_2^{(2)}$ of $\tilde{\alpha}(Z_{11})$ and $\bar{\alpha}(Z_{12})^o$ respectively such that $\Delta^{(1)} = \Delta_1^{(1)} \cup \Delta_2^{(2)} \cup \{a_1\}$, where $\Delta_1^{(1)} \cup \Delta_2^{(1)} = \theta$ and a_1 is the only simple root such that the root subgroup $U_{a_1} \subset R_u(\bar{\alpha}(P_1))$. It follows from (4.3) and $\Delta_1^{(1)} \perp \Delta_2^{(1)}$ that the Dynkin diagram of $\Delta^{(1)}$ must be one of following cases:

A_ℓ $(\ell\geq 5)$

B_ℓ $(\ell\geq 5)$

C_ℓ $(\ell\geq 5)$

D_ℓ $(\ell\geq 5)$

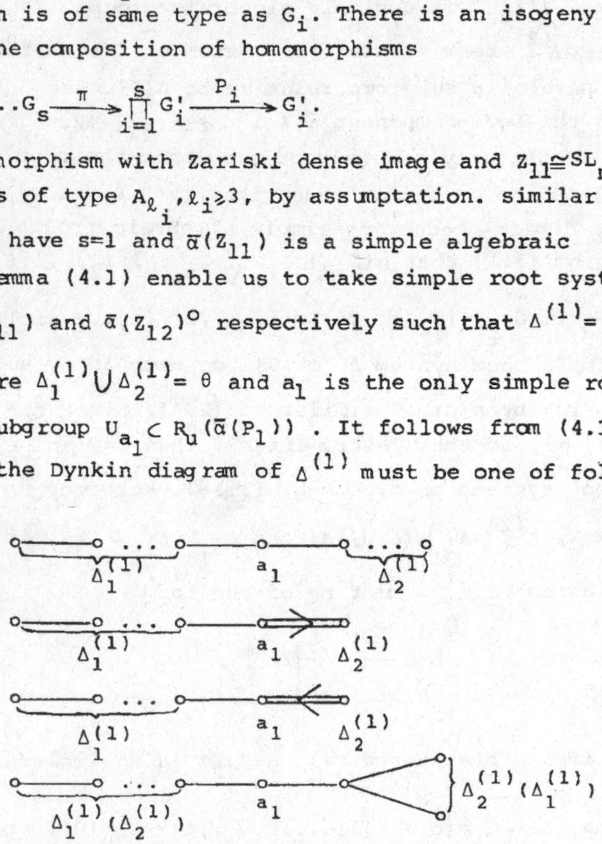

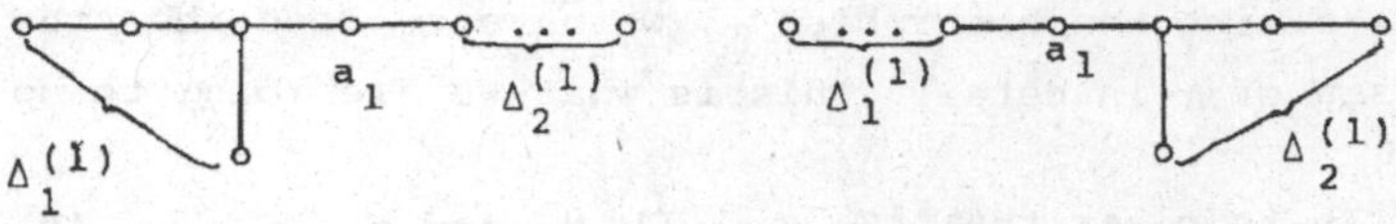

However, the unipotent radical $R_u(P_{\theta_1})$ of the standard parabolic subgroup P_{ℓ_1} is not commutative in cases B_ℓ, C_ℓ, D_ℓ and E_ℓ. Since $\bar{\alpha}(U_1)=R_u(\bar{\alpha}(P_1))$ is abelian, $\Delta^{(1)}$ must be of type A_ℓ. |

4.5. Theorem. Let G be a simple algebraic group, α: $SL_n(D)\longrightarrow G$ a homomorphism with Zariski dense image, the

$$\text{rank } G=n(\text{rank}\bar{\alpha}(M_i)^o+1)-1,\ 1\leqslant i\leqslant n.$$

Proof. Apply induction on n. If n=2, rank G=2m-1 by (3.5), where m= $\text{rank}\bar{\alpha}(M_i)^o+1$, i=1,2. Suppose n>2, let ℓ=rank G and m=$\text{rank}\bar{\alpha}(M_i)^o+1$= $\text{rank}\bar{\alpha}(Z_{12})^o+1$. By the proof of (4.4) the Dynkin diagram of $\Delta^{(1)}$ is

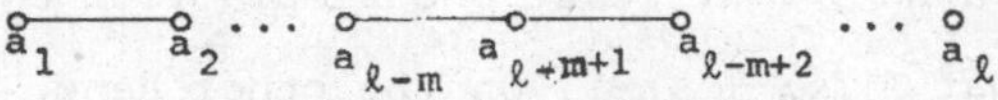

where $\{a_1,\dots,a_{\ell-m}\}$ is a simple root system of $\bar{\alpha}(Z_{11})$, $\{a_{\ell-m+2},\dots,a_\ell\}$ is a simple root system of $\bar{\alpha}(Z_{12})^o$ and $a_{\ell-m+1}$ is the only simple root such that the root subgroup $U_{a_{\ell-m+1}}\subset R_u(\bar{\alpha}(P_1))$. Since $Z_{11}\cong SL_{n-1}(D)$ and $\bar{\alpha}(Z_{11})$ is a simple algebraic group (cf. the proof of (4.4)), it follows from the assumption that

$$\ell-m=\text{rank}\,\bar{\alpha}(Z_{11})=(n-1)m-1$$

hence rank G= ℓ=nm-1. |

4.6. Remark. Suppose G splits over k, we have proved that $\bar{\alpha}(Z_o)$ and $\bar{\alpha}(U_o)$ split over k (2.10). Let T be a maximal torus of $\bar{\alpha}(Z_o)$ and split over k. Hence, T is also a maximal torus of G. If V is a maximal connected unipotent subgroup of $\bar{\alpha}(Z_o)$, then $B=TV\bar{\alpha}(U_o)$ is a k-split Borel subgroup of G(cf.(2.10)) Let S be the set of involutions of $N_G(T)/_T$, which generate $N_G(T)/_T$, then S is finte. The root system, positive root system and simple root system relative to the Tits system $(G,B,N_G(T),S)$ are denoted by Φ, Φ^+ and Δ respectively. Clearly, $\{T,U_a\}$ $(a\in\Phi)$ is a" donnée de déploiment". Let Φ_o, Φ_o^+ and Δ_o be the root system, positive root system and simple root system of $\bar{\alpha}(Z_o)$ relative to T and TV respectively, then $\Phi_o\subset\Phi$, $\Phi_o^+=\Phi^+\cap\Phi_o$, $\Delta_o=\Delta\cap\Phi_o$.

To determine the homorphism α, we have to describe the Dynkin diagram of Δ in detail. This is what we are going to do in (4.7)-(4.11).

<u>4.7</u>. It is clear that $[Z_o, Z_o] = \prod_{i=1}^{n} M_i$ and $M_i \subseteq C_{SL_n(D)}(M_j)$, $\forall i \neq j$, $1 \leq i, j \leq n$. We have

$$[\bar{\alpha}(Z_o), \bar{\alpha}(Z_o)] = \bar{\alpha}(M_1) \ldots \bar{\alpha}(M_n) = \bar{\alpha}(M_1)^o \ldots \bar{\alpha}(M_n)^o \qquad (4.7.1)$$

by connectivity of $\bar{\alpha}(Z_o)$, and $\bar{\alpha}(M_i)^o \subset C_G(\bar{\alpha}(M_j)^o)$, $\forall i \neq j$. Hence $\bar{\alpha}(M_i)^o \cap \bar{\alpha}(M_j)^o \subsetneq C(\bar{\alpha}(M_i)^o)$, which means $|\bar{\alpha}(M_i)^o \cap \bar{\alpha}(M_j)^o| < \infty$, $\forall i \neq j$ since $\bar{\alpha}(M_i)$ is a simple algebraic group. Moreover, $\bar{\alpha}(M_i)^o$ is a normal subgroup of $[\bar{\alpha}(Z_o), \bar{\alpha}(Z_o)]$, which enable us to choose simple root system Δ_i of $\bar{\alpha}(M_i)^o$ for every i, $1 \leq i \leq n$, such that $\bigcup_{i=1}^{n} \Delta_i = \Delta_o$. Obviously, $\Delta_i \perp \Delta_j$, $\forall i \neq j$, and Δ_i is irreducible and of type A_m for every i, $1 \leq i \leq n$. Now, consider the Dynkin diagram of Δ. For any pair (i,j) where $i \neq j$, there is, between Δ_i and Δ_j, at least one simple root contained in $\Delta \setminus \Delta_o$. It follows that $|\Delta \setminus \Delta_o| \geq n-1$. On the other hand, by (4.5),

$$|\Delta \setminus \Delta_o| = n(\text{rank}\bar{\alpha}(M_i)^o + 1) - 1 - n(\text{rank}\bar{\alpha}(M_i)^o) = n-1. \qquad (4.7.2)$$

Consequently, we have following

Lemma. Let $\{a_1, \ldots, a_{n-1}\} = \Delta \setminus \Delta_o$. There is a permutation $\sigma \in S_n$ such that the Dynkin diagram of Δ is as follows

$$(4.7.3)$$

<u>4.8</u>. Remark. Let $U_{[j]} = \prod_{i=1}^{j-1} U_{ij}$, $j=2, \ldots, n$. Since $U_o = U_{[n]} U_{[n-2]} \ldots U_{[2]}$, similar to (2.5), we have

$$\bar{\alpha}(U_o) = \bar{\alpha}(U_{[n]}) \bar{\alpha}(U_{[n-1]}) \ldots \bar{\alpha}(U_{[2]})$$

and

$$\bar{\alpha}(U_{[j]}) = \bar{\alpha}(U_{1j}) \bar{\alpha}(U_{2j}) \ldots \bar{\alpha}(U_{j-1,j}), \quad j=2, \ldots, n.$$

By (2.5), $U_{(i)} \subset U_{n-i}$ and $U_{(i+1)} \ldots U_{(n-1)} \subset Z_{n-i}$, hence $\bar{\alpha}(U_{(i)}) \subseteq \bar{\alpha}(U_{n-i})$; $\bar{\alpha}(U_{(i+1)}) \ldots \bar{\alpha}(U_{(n-1)}) \subseteq \bar{\alpha}(Z_{n-i})$. $\bar{\alpha}(U_{n-i})$ and $\bar{\alpha}(Z_{n-i})$ are unipotent radical and Levi-component of parabolic subgroup $\bar{\alpha}(P_{n-i})$ respectively, so that

$$\bar{\alpha}(U_{(i)}) \cap \bar{\alpha}(U_{(i+1)}) \ldots \bar{\alpha}(U_{(n-1)}) \subsetneq \bar{\alpha}(U_{n-i}) \cap \bar{\alpha}(Z_{n-i}) = \{1\}.$$

Inparticular, for $1 \leq i < j-1$ we have

$$\bar{\alpha}(U_{ij})\cap\bar{\alpha}(U_{i+1,j})\dots\bar{\alpha}(U_{j-1,j})\subseteq\bar{\alpha}(U_{(i)})\cap\bar{\alpha}(U_{(i+1)})\dots\bar{\alpha}(U_{(j-1)})=\{1\}$$

where $j=2,\dots,n$. Similar, for $j>2$.

$$\bar{\alpha}(U_{ij})\cap\bar{\alpha}(U_{i,j-1})\dots\bar{\alpha}(U_{i,2})\subset\bar{\alpha}(U_{[j]})\cap\bar{\alpha}(U_{[j-1]})\dots\bar{\alpha}(U_{[2]})=\{1\}$$

where $i=1,\dots,j-1$. It follows that

$$\bar{\alpha}(U_{[j]})=\prod_{i=1}^{j-1}\bar{\alpha}(U_{ij});\ \bar{\alpha}(U_{(i)})=\prod_{j=i+1}^{n}\bar{\alpha}(U_{ij}). \qquad (4.8.1)$$

Let T be the maximal torus given in (4.6), then $T\subset N_G(\bar{\alpha}U_{ij}))$ since $Z_o\subset N_{SL_n(D)}(U_{ij})$. In other words, $\bar{\alpha}(U_{ij})$ is a T-stable unipotent subgroup, so are $\bar{\alpha}(U_{(i)})$ and $\bar{\alpha}(U_{[j]})$. Moreover, if a root subgroup $U_\gamma\subseteq\bar{\alpha}(U_{(i)})$(or $\bar{\alpha}(U_{[j]})$), then it is contained exactly in one of $\bar{\alpha}(U_{ij})$'s, $j=i+1,\dots,n$ ($i=1,\dots,j-1$), by (2.12).

<u>4.9. Lemma</u>. Let Δ be the simple root system given in (4.6), then $\bar{\alpha}(U_{(i)})(\bar{\alpha}(U_{[j]}))$ contains exactly one root subgroup $U_{\beta_i}, \beta_i\in\Delta$ ($U_{\beta_j}, \beta_j\in\Delta$) for every i, $1\leqslant i\leqslant n-1$ (j, $2\leqslant j\leqslant n$).

Proof. we prove first that $\bar{\alpha}(U_{(i)})$ contains at most one simple root subgroup for any $1\leqslant i\leqslant n-1$. Otherwise, suppose $\gamma,\delta\in\Delta$, $\gamma\neq\delta$, such that both root subgroups U_γ and U_δ are contained in $\bar{\alpha}(U_i)$ for some $1<i\leqslant n-1$. Since $\bar{\alpha}(U_{(i)})\subset\bar{\alpha}(U_o)=R_u(\bar{\alpha}(P_o))$, $\gamma,\delta\in\Delta\setminus\Delta_o$. By (4.8) & (2.12), $U_\gamma\subset\bar{\alpha}(U_{ij})$ $i+1\leqslant j\leqslant n$, and $U_\delta\subset\bar{\alpha}(U_{i\ell})$, $i+1<\ell<n$. If $j=\ell$, then for any $t\in\{1,\dots,n\}\setminus\{i,j\}$ we have $\bar{\alpha}(M_t)^o\subset C_G(\bar{\alpha}(U_{ij}))\subset C_G(U_\gamma)\cap C_G(U_\delta)$ since $M_t\subset C_{SL_n(D)}(U_{ij})$, so that $\{\gamma,\delta\}\perp\Delta_t$, $\forall t\neq i,j$. However, this is impossible since, by Dynkin diagram (4.7.3), there are at most $n-3$ Δ_t's orthogonal to $\{\gamma,\delta\}$ If $j\neq\ell$, then $\{\gamma\}\perp\Delta_t$, $\forall t\neq i,j$, and $\{\delta\}\perp\Delta_s$ $\forall s\neq i,\ell$. In other words, in the Dynkin diagram (4.7.3), γ is just the root between Δ_i and Δ_j, δ is just the root between Δ_i and Λ_ℓ So the Dynkin diagram of Δ is

$$\circ\ \dots\ \Delta_j \text{---} \underset{\gamma}{\circ} \text{---} \Delta_i \text{---} \underset{\delta}{\circ} \text{---} \Delta_\ell \dots\ \circ$$

Suppose $\Delta_i=\{\beta_1,\dots,\beta_m\}$, then the root subgroup $U_{\beta_1+\dots+\beta_m}\subset\bar{\alpha}(M_i)^o$. Since $[M_i,U_{ij}]\subseteq U_{ij}$, $[\bar{\alpha}(M_i)^o, \bar{\alpha}(U_{ij})]\subset\bar{\alpha}(U_{ij})$, inparticular

$[U_{\beta_1+\dots\beta_m},U_\gamma]=U_{\alpha+\beta_1\dots+\beta_m}\subset\bar{\alpha}(U_{ij})$, hence

$$[\bar{\alpha}(U_{ij}),\ \bar{\alpha}(U_{i\ell})]\supseteq[U_{\alpha+\beta_1+\dots+\beta_m}, U_\delta]=U_{\gamma+\beta_1+\dots+\beta_m+\delta}.$$

This is a contradiction since $[U_{ij},U_{i\ell}]=\{I_n\}$ and $[\bar{\alpha}(U_{ij}),\bar{\alpha}(U_{i\ell})]=\{1\}$. Consequently, there is at most one simple root subgroup contained in $\bar{\alpha}(U_{(i)})$. If there is not any simple root subgroup contained in $\bar{\alpha}(U_{(i)})$

for some i, then there world be at least one $\bar{\alpha}(U_{(j)})$, $j \neq i$, which contains more than one simple root subgroups by (2.5.5) & (4.8). This contradicts to our above conclusion. Therefore $\bar{\alpha}(U_{(i)})$, $1 \leq i \leq n$ contains one and only one simple root subgroup. For $\bar{\alpha}(U_{[j]})$, $2 \leq j \leq n$, the proof is similar. |

4.10. Corollary. All the $n-1$ simple root subgroups of $\bar{\alpha}(U_o)$ (3.7.2) are contained in $\bigcup_{i=1}^{n-1} \bar{\alpha}(U_{i,i+1})$. More precisely, every $\bar{\alpha}(U_{i,i+1})$ ($i=1,\ldots,n-1$) contains a simple root subgroup.

Proof. We have, by (4.8.1) & (2.5.5).

$$\alpha(U_o) = \prod_{j_2=2}^{n} \bar{\alpha}(U_{1\,j_1}) \cdot \prod_{j_2=2}^{n} \bar{\alpha}(U_{2\,j_2}) \cdots \prod_{j_{n-2}=n-1}^{n} \bar{\alpha}(U_{n-2,j_{n-2}}) \bar{\alpha}(U_{n-1,n})$$

It follows from (2.12) & (4.7.2) that the $n-1$ simple root subgroups of $\bar{\alpha}(U_o)$ are contained in $\bigcup_{i,j} \bar{\alpha}(U_{ij})$, $1 \leq i < j \leq n$. Moreover, for every i, $1 \leq i \leq n$, there is exactly one $\bar{\alpha}(U_{ij})$ ($2 \leq j \leq n$) which contains a simple root subgroup and for every j, $2 \leq j \leq n$, there is exactly one $\bar{\alpha}(U_{ij})$ ($1 \leq i \leq n$) which contains a simple root subgroup by (4.9) & (2.12), which means that if both $\bar{\alpha}(U_{i_1 j_1})$ and $\bar{\alpha}(U_{i_2 j_2})$ contain simple root subgroups, then $i_1 \neq i_2$ & $j_1 \neq j_2$. Hence $\bar{\alpha}(U_{ij})$ contains a simple root subgroup only if $j = i+1$. |

4.11. Lemma. Let $a_i \in \Delta \setminus \Delta_o$ be the simple root such that the root subgroup $U_{a_i} \subset \bar{\alpha}(U_{i,i+1})$, $i=1,\ldots,n-1$, then the permutation σ in (3.7) is trivial, i.e. the Dynkin diagram of Δ is

where Δ_i is a simple root system of $\bar{\alpha}(M_i)^o$, $1 \leq i \leq n$.

Proof. It is clear $M_1 \subset \bigcap_{i=2}^{n} C_{SL_n(D)}(U_{(i)})$, so that

$\bar{\alpha}(M_1)^o \subset \bigcap_{i=2}^{n} C_G(\bar{\alpha}(U_{(i)}))$. $\Delta_1 \perp \{a_2,\ldots,a_{n-1}\}$ by (3.7), hence the Dynkin diagram of Δ is of form

$$\Delta_1 \overline{} \underset{a_1}{\circ} \overline{} \left\{ \bigcup_{i=2}^{n} \Delta_i \right\} \cup \{a_2,\ldots,a_{n-1}\}$$

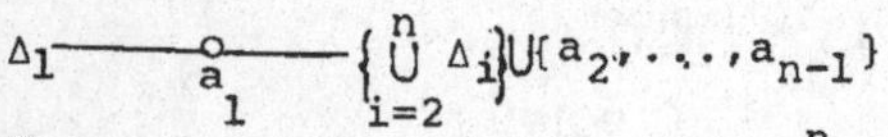

Since $U_{12} \subset \bigcap_{i=3}^{n} C_{SL_n(D)}(M_i)$, $U_{a_1} \subset \bar{\alpha}(U_{12}) \subset \bigcap_{i=3}^{n} C_G(\bar{\alpha}(M_i)^o)$, hence $\{a_1\} \perp \Delta_i$, $\forall 2 < i \leq n$. It follows that, except Δ_1, only Δ_2 connects with the root a_1, i.e. the Dynkin diagram of Δ is of form

$$\Delta_1 \underset{a_1}{\rule{1cm}{0.4pt}\!o\!\rule{1cm}{0.4pt}} \Delta_2 \rule{1cm}{0.4pt} \{\underset{i=3}{\overset{n}{\cup}} \Delta_i\} \cup \{a_2,\ldots,a_{n-1}\}.$$

Therefore, the lemma holds for n=3. For n>3, the Proof is similar. ∎

4.12. Lemma. Suppose G splits over k, then there is a k-isogeny $\pi:G \longrightarrow PSL_{nm}(k)$, where $m=\operatorname{rank}\bar{\alpha}(M_i)^O+1$, such that

$$\overline{\pi\alpha}(U_1) = \left\{ \begin{bmatrix} I_{(n-1)m} & B \\ 0 & I_m \end{bmatrix} \in PSL_{nm}(K) \,\middle|\, \forall\, B \in \operatorname{Mat}_{(n-1)m \times m}(K) \right\} \qquad (4.12.1)$$

$$\overline{\pi\alpha}(U_1^-) = \left\{ \begin{bmatrix} I_{(n-1)m} & B \\ C & I_n \end{bmatrix} \in PSL_{nm}(K) \,\middle|\, \forall\, C \in \operatorname{Mat}_{m \times (n-1)m}(K) \right\} \qquad (4.12.2)$$

$$\overline{\pi\alpha}(M_i)^O = \left\{ \begin{bmatrix} I_m & & & \\ & \ddots & & \\ & & A & \\ & & & \ddots & \\ & & & & I_m \end{bmatrix} \in PSL_{nm}(K) \,\middle|\, \forall\, A \in SL_m(K) \right\} \qquad (4.12.3)$$

$$\underset{\text{ith block}}{\underbrace{}}$$

where i=1,...,n.

Proof. Let T' and B' be subgroups of $PSL_{nm}(K)$ consisting of diagonal matrices and upper triangular matrices respectively. Let ϕ' be the root system relative to T', Denote the simple root system of ϕ' relative to B' by $\Delta'=\{a_1',\ldots,a_{nm-1}'\}$, where

$$U_{a_i'} = \left\{ \begin{bmatrix} 1 & & & & \\ & \ddots & & & \\ & & 1 & b & \\ & & & 1 & \\ & & & & \ddots \\ & & & & & 1 \end{bmatrix} \in PSL_{nm}(K) \,\middle|\, \forall\, b \in K,\ i=1,\ldots,nm-1. \right\}$$

It follows from (1.3), (4.6) & (4.11) that there is a k-isogeny $\pi:G \longrightarrow PSL_{nm}(K)$, which induces an isomorphism of simple root systems $\pi^{*-1}:\Delta' \longrightarrow \Delta$ such that

$$\pi^{*-1}(a_j') \begin{cases} \in \Delta_i, & \text{if } (i-1)m<j<im,\ i=1,\ldots,n \\ =a_i & \text{if } j=im,\ i=1,\ldots,n-1. \end{cases}$$

Since $\bar{\alpha}(P_o)$ is a standard parabolic subgroup relative to Δ and $\bar{\alpha}(P_1)\supset\bar{\alpha}(P_o)$, $\bar{\alpha}(P_1)$ is also a standard parablic subgroup relative to Δ. $R_u(\bar{\alpha}(P_1))=\bar{\alpha}(U_1)$ contains an unique simple root subgroup $U_{a_{n-1}}$ (3.10). Accordingly, $\overline{\pi\alpha}(P_1)$ is a standard parabolic subgroup relative to Δ' and $R_u(\overline{\pi\alpha}(P_1))$ contains an unique simple root subgroup $U_{a_{(n-1)m}'}$. Recall that π maps closed subgroups onto closed subgroups [7.§18] so that

$$\pi\bar{\alpha}(P_1)=\overline{\pi\alpha}(P_1)=\left\{ \begin{bmatrix} A & C \\ 0 & B \end{bmatrix} \in PSL_{nm}(K) \,\middle|\, A \in GL_{(n-1)m}(K),\ B \in GL_m(K), \right.$$

$$C \in \mathrm{Mat}_{(n-1)m\times m}(K) \Big\}$$

Now, (4.12.1) comes from $\overline{\pi\alpha}(U_1)=R_u(\overline{\pi\alpha}(P_1))$. Similar we have (4.12.2). Now, $\pi\bar{\alpha}(M_i)^o=\langle U_{\pm a_j'} \mid \forall (i-1)m<j<m\rangle$ since $\bar{\alpha}(M_i)^o=\langle U_{\pm a_j} \mid \forall a_j \in \Delta_i\rangle$, $1\leq i\leq n$, so that $\pi\bar{\alpha}(M_i)^o$ is connected and contained in $\overline{\pi\alpha}(M_i)^o$. On the other hand, since $|\bar{\alpha}(M_i)/\bar{\alpha}(M_i)^o|<\infty$, we have $|\overline{\pi\alpha}(M_i)/\pi\bar{\alpha}(M_i)^o|<\infty$, and hence $\overline{\pi\alpha}(M_i)^o \supseteq \pi\bar{\alpha}(M_i)$. Therefore $\overline{\pi\alpha}(M_i)^o=\overline{\pi\alpha}(M_i)^o$. Finally, (4.12.3) comes from the fact that $U_{\pm a_j'}$, $(i-1)m<j<im$, generate $\overline{\pi\alpha}(M_i)^o$. ∎

<u>4.13. Lemma</u>. $\overline{\pi\alpha}(M_i)$, $1\leq i\leq n$ is connected.

Proof. It is enough to show that $\overline{\pi\alpha}(M_n)=\overline{\pi\alpha}(Z_{12})$ is connected. Since $\pi\alpha$ is a homomorphism from $SL_n(D)$ to $PSL_{nm}(k)$ with Zariski dense image the preceding lemmas which hold for α are also hold for $\pi\alpha$. Inparticular, $\overline{\pi\alpha}(Z_1)$ is a Levi-component of $\overline{\pi\alpha}(P_1)$, so that

$$[\overline{\pi\alpha}(Z_1), \overline{\pi\alpha}(Z_1)]=\left\{\begin{bmatrix}A & 0\\ 0 & B\end{bmatrix} \in PSL_{nm}(K) \,\middle|\, \forall A \in SL_{(n-1)m}(K),\ B\in SL_m(K)\right\}.$$

It follows from (4.11) & (4.1) that $\{a_1,\dots,a_{n-2}\}\cup\left\{\bigcup_{i=1}^{n-1}\Delta_i\right\}$ is a simple root system of $\bar{\alpha}(Z_{11})$. Since $\bar{\alpha}(Z_{11})$ is a simple agebraic group, $\bar{\alpha}(Z_{11})=\langle U_{\pm a} \mid \forall a \in \{a_1,\dots,a_{n-2}\}\cup\left\{\bigcup_{i=1}^{n-1}\Delta_i\right\}\rangle$. Furthermore, by the proof of (4.12), we have

$$\overline{\pi\alpha}(Z_{11})=\langle U_{\pm a_j'} \mid \forall 1\leq j<(n-1)m\rangle=\left\{\begin{pmatrix}A & 0\\ 0 & I_m\end{pmatrix}\in PSL_{nm}(K) \,\middle|\, A\in SL_{(n-1)m}(K)\right\}$$

hence

$$C(\overline{\pi\alpha}(Z_{11}))=\left\{\begin{bmatrix}\varepsilon I_{(n-1)m} & 0\\ 0 & I_m\end{bmatrix}\in PSL_{nm}(K) \,\middle|\, \varepsilon^{(n-1)m}=1\right\}$$

Now, $[\overline{\pi\alpha}(Z_1),\overline{\pi\alpha}(Z_1)]=\overline{\pi\alpha}(Z_{11})\overline{\pi\alpha}(Z_{12})^o$ by (4.1), so that $\overline{\pi\alpha}(Z_{12})=(\overline{\pi\alpha}(Z_{12})\cap\overline{\pi\alpha}(Z_{11}))\overline{\pi\alpha}(Z_{12})^o$. Moreover, since $Z_{12}\subseteq C_{SL_n(D)}(Z_{11})$, $\overline{\pi\alpha}(Z_{12})\cap\pi\alpha(Z_{11})^o\subseteq C(\overline{\pi\alpha}(Z_{11}))\subset\overline{\pi\alpha}(Z_{12})^o$ by (4.12.3). Therefore $\overline{\pi\alpha}(Z_{12})=\overline{\pi\alpha}(Z_{12})^o$. ∎

<u>4.14. Remark</u>. From now on to (4.16), let G split over k. Consider the restriction of $\pi\alpha$ to the subgroup U_1

$$\pi\alpha|_{U_1}\quad U_1 \longrightarrow \pi\alpha(U_1)(k)$$

$$\begin{pmatrix}1 & & & b_1\\ & \ddots & & \vdots\\ & & 1 & b_{n-1}\\ & & & 1\end{pmatrix}\longmapsto\begin{bmatrix}I_m & & & B_1\\ & \ddots & & \vdots\\ & & I_m & B_{n-1}\\ & & & I_m\end{bmatrix},\ \forall b_i\in D,\ B_i\in \mathrm{Mat}_{m\times m}(k),\ 1\leq i<n.$$

It induces a homomorphism of abelian groups ϕ: $Mat_{(n-1)\times 1}(D) \longrightarrow Mat_{(n-1)m\times m}(k)$ such that $\phi\begin{pmatrix} b_1 \\ \vdots \\ b_{n-1}\end{pmatrix} = \begin{pmatrix} B_1 \\ \vdots \\ B_{n-1}\end{pmatrix}$. we claim that $\phi\begin{pmatrix} 0 \\ \vdots \\ b_i \\ \vdots \\ 0\end{pmatrix}$ is of form $\begin{pmatrix} 0 \\ \vdots \\ B_i \\ \vdots \\ 0\end{pmatrix}$, $1\leq i\leq n-1$. In fact, let $u_i = I_n + b_i E_{in}$, obviously, $u_i \in C_{SL_n(D)}(\prod_{j\neq i,n} M_j)\cap U_1$, so that $\pi\alpha(u_i)\in \overline{\pi\alpha}(M_1)\ldots\widehat{\overline{\pi\alpha}(M_i)}\ldots\overline{\pi\alpha}(M_{n-1})\cap \overline{\pi\alpha}(U_1)(k)$. By (4.12.3) & (4.12.4).

$$\overline{\pi\alpha}(M_1)\ldots\widehat{\overline{\pi\alpha}(M_i)}\ldots\overline{\pi\alpha}(M_{n-1}) = \{diag(A_1,\ldots,A_{i-1},I_m,A_{i+1},\ldots,A_{n-1},I_m)\in PSL_{nm}(K) \mid A_j\in SL_m(K), j=1,\ldots,\hat{i},\ldots,n-1.\}$$

hence

$$C_{PSL_{nm}(k)}(\overline{\pi\alpha}(M_1)\ldots\widehat{\overline{\pi\alpha}(M_i)}\ldots\overline{\pi\alpha}(M_{n-1})) = \left\{\begin{bmatrix} C_1 & & & & & \\ & \ddots & & & & \\ & & C_{i-1} & & & \\ & & & A_i & \cdots & B_n \\ & & & \vdots & C_{i+1} & \vdots \\ & & & & \ddots & C_{n-1} \\ & & & B_i & \cdots & A_n \end{bmatrix} \in PSL_{nm}(k) \;\middle|\; A_i,A_n\in GL_m(k),\ B_i,B_n\in Mat_{m\times m}(k),\ C_j\in C(SL_m(k)),\ j=1,\ldots,\hat{i},\ldots n-1.\right\}$$

so that by (4.12.2),

$$C_{PSL_{nm}(k)}(\overline{\pi\alpha}(M_1)\ldots\overline{\pi\alpha}(M_i)\ldots\overline{\pi\alpha}(M_{n-1}))\cap\overline{\pi\alpha}(U_1)(k) = \left\{\begin{bmatrix} I_m & & & \\ & \ddots & & \\ & & I_m & B_i \\ & & & \ddots \\ & & & I_m\end{bmatrix}\in PSL_{nm}(k) \;\middle|\; B_i\in Mat_{m\times m}(k)\right\}$$

From which, the claim follows. So far, we have homomorphisms of abelian groups $\phi_i: D\longrightarrow Mat_{m\times m}(k)$ by $\phi_i(b_i)=B_i$, $i=1,\ldots,n-1$. Similarly, the restriction of $\pi\alpha$ to U_1 induces homomorphisms of abelian groups $\phi_i': D\longrightarrow Mat_{m\times m}(k)$, such that

$$\pi\alpha\begin{pmatrix} 1 & & \\ & \ddots & \\ 0\cdots & c_i & \cdots 1\end{pmatrix} = \begin{bmatrix} I_m & & \\ \vdots & \ddots & \\ 0\cdots & \phi_i'(c_i) & \cdots I_m\end{bmatrix}$$

$\phi_i(1)=Q_i$, $\phi_i(1)=R_i$, $Q_i, R_i\in Mat_{m\times m}(k)$, $1\leq i\leq n-1$. Let $w_i=(I_n+E_{in})(I_n-E_{ni})(I_n+E_{in})\in SL_n(D)$ (cf. (2.1)), then

$$\pi\alpha\, w_i = \begin{bmatrix} I_m & & & \\ & \ddots & & \\ & & I_m+R_iQ_i \cdots & * \\ & & \vdots & \vdots \\ & & * \cdots & I_m+Q_iR_i\end{bmatrix}$$

On the other hand, $\pi\alpha w_i$ is of form $\begin{bmatrix} I_m & & & \\ & \ddots & & \\ & & 0 \cdots & A \\ & & \vdots & \vdots \\ & & B \cdots & 0\end{bmatrix}$ since $w_i M_i w_i^{-1} = M_n$ and $\overline{\alpha}(w_i)\overline{\alpha}(M_i)\overline{\alpha}(w_i)^{-1} = \overline{\alpha}(M_n)$. Therefore $R_i = Q_i^{-1}$. Let $g=diag(Q_1^{-1},\ldots Q_{n-1}^{-1}, I_m)$

and $\beta \in \mathrm{Aut}\, PSL_{nm}(k)$ such that $\beta(x)=gxg^{-1}$, $\forall x \in PSL_{nm}(k)$. Denote $\sigma_\alpha = \beta\pi\alpha$, then σ_α is also a homomorphism with Zariski dense image, hence the preceding lemmas which hold for α & $\pi\alpha$ also hold for σ_α. Moreover, the homomorphisms of abelian groups $\phi_{\alpha,i}(\phi'_{\alpha,i}):D \longrightarrow Mat_{m\times m}(k)$, which are induced by σ_α, satisfy $\phi_{\alpha,i}(1)=I_m$ $(\phi'_{\alpha,i}(1)=I_m$,

so that $\sigma_\alpha(w_i)=\begin{bmatrix} I_m & & & \\ & \ddots & & \\ & & 0 \cdots & I_m \\ & & \vdots \ddots & \vdots \\ & & -I_m \cdots & 0 \end{bmatrix}$. It follows that $\phi_{\alpha,i}=\phi'_{\alpha,i}$, $i=1,\ldots,n-1$.

4.15. Lemma. $\phi_{\alpha,i}=\phi_{\alpha,j}$, $j,i=1,\ldots,n-1$.

Proof. It is enough to show that $\phi_{\alpha,i}=\phi_{\alpha,n-1}$, $1\leq i\leq n-1$. From the identity $I_n+E_{i,n-1}=w_{n-1}(I_n+E_{in})w_{n-1}^{-1}$, we have $\sigma_\alpha(I_n+E_{i,n-1}) =$

$$= \begin{bmatrix} I_m & & & \\ & \ddots & & \\ & & I_m \cdots & I_m \\ & & \ddots & \vdots \\ & & & I_m \\ & & & & I_m \end{bmatrix}, \quad \text{similarly } \sigma_\alpha(I_n-E_{n-1,i})= \begin{bmatrix} I_m & & & \\ & \ddots & & \\ & & I_m & \\ & & \vdots & \ddots \\ & & -I_m \cdots & I_m \\ & & & & I_m \end{bmatrix}$$

It follows that

$$\sigma_\alpha(I_n+E_{i,n-1})(I_n-E_{n-1,i})(I_n+E_{i,n-1})= \begin{bmatrix} I_m & & & \\ & \ddots & & \\ & & 0 \cdots & I_m \\ & & \vdots \ddots & \vdots \\ & & -I_m \cdots & 0 \\ & & & & I_m \end{bmatrix}, \quad \forall\, 1\leq i\leq n-2.$$

Let $b\in D$, $1\leq i\leq n-2$, then

$$I_n+bE_{n-1,n}=(I_n(E_{i,n-1})(I_n+E_{n-1,i})(I_n-E_{i,n-1})(I_n+bE_{in})(I_n+E_{i,n-1})$$
$$(I_n-E_{n-1,i})(I_n+E_{i,n-1})$$

Apply σ_α to the above identity. We get $\phi_{\alpha,n-1}(b)=\phi_{\alpha,i}(b)$ by observing corresponding matrices in $PSL_{nm}(k)$.

Denote $\phi_{\alpha,i}$ by ϕ_α, $1\leq i\leq n-1$.

4.16. Lemma. $\phi_\alpha:D\longrightarrow Mat_{m\times m}(k)$ is a homomorphism of rings. Furthermore, ϕ_α induces an injection of fields $\phi^0:F\longrightarrow k$, such that $\phi_\alpha(td)=\phi^0(t)\phi(d)$, $\forall t\in F$, $\forall d\in D$, where F is the centre of D.

Proof. The proof is similar to (3.9),(3.10),(3.11) & (3.13) by restricting σ_α to Z_{22}. ∎

4.17. Corollary. Let $(a_{ij})\in SL_n(D)$, then $\sigma_\alpha(a_{ij})=[A_{ij}]$, where $A_{ij}=\phi_\alpha(a_{ij})\in Mat_{m\times m}(k)$, $1\leq i,j\leq n$. ∎

4.18. Theorem. Let G be a simple algebraic group defined over k and $\alpha:SL_n(D)\longrightarrow G(k)$ be a homomorphism with Zariski dense image.

Denote by F the centre of D and $M_i, 1 \leq i \leq n$, as in (4.3). Then there is a finite separable extension field k' of k such that

(i) α induces an injection of fields $\phi: F \longrightarrow K'$, in particular, char F=char k.

(ii) α induces a homomorphism of rings $\phi_\alpha: D \longrightarrow \mathrm{Mat}_{m\times m}(k')$. where $m=\mathrm{rank}_{\bar{\alpha}}(M_i)^{o}+1$. Moreover, if take F as a subfield of k', then ϕ_α is a homomorphism of F-algebras.

(iii)* D is a finite dimensional division algebra over F.

(iv) $\sigma_\alpha: SL_n(D) \longrightarrow PSL_{nm}(k')$ defined by

$$\sigma_\alpha \begin{pmatrix} a_{11} \cdots & a_{in} \\ \vdots & \vdots \\ a_{ni} \cdots & a_{nn} \end{pmatrix} = \begin{bmatrix} \phi_\alpha(a_{11}) \cdots & \phi_\sigma(a_{in}) \\ \vdots & \vdots \\ \phi_\sigma(a_{ni}) \cdots & \phi_\sigma(a_{nn}) \end{bmatrix}$$

is a homomorphism with Zariski dense image. Moreover, up to a k'-isomorphism, there exists an unique k'-isogeny $\pi: G \longrightarrow PSL_{nm}(k)$, such that the following diagram is commutative

$$\begin{array}{ccc} & & G(k') \\ & \overset{\alpha}{\nearrow} & \downarrow \pi \\ SL_n(D) & \underset{\sigma_\alpha}{\searrow} & \\ & & PSL_{nm}(k') \end{array}$$

Furthermore, up to a k'-isomorphism, σ_α is uniquely determined by α. On the other hand, if there is a k'-isogeny $\pi': G \longrightarrow PSL_{nm}(K)$ and a homomorphism $\alpha': SL_n(D) \longrightarrow G(k')$ with Zariski dense image, such that $\pi'\alpha' = \sigma_\alpha$, then $\alpha' = \sigma\alpha$, where σ is a k'-automorphism of G.

(V). If G splits over k, one can take k'=k.

Proof. Since G is defined over k, there exists a finite separable extension k' of k such that G splits over k', so that, without loss of generality, we can suppose G splits over k. (i),(ii), & (iii), as well as first part of (iv) follow from the preceding lemmas and (3.15). Only last part of (iv) remains to prove. If there are $\sigma_\alpha': SL_n(D) \longrightarrow PSL_{nm}(k)$ and k-isogeny $\pi': G \longrightarrow PSL_{nm}(k)$ such that $\sigma_\alpha' = \pi'\alpha$, then there is a k-automorphism Λ of $PSL_{nm}(K)$ such that $\pi' = \Lambda\pi$. Therefore $\sigma_\alpha' = \Lambda\sigma_\alpha$. On the other hand, let $\alpha': SL_n(D) \longrightarrow G(k)$ be a homomorphism with Zariski dense image, $\pi': G \longrightarrow PSL_{nm}(k)$ a k-isogeny and $\pi'\alpha' = \sigma_\alpha = \pi\alpha$, then there is a k-automorphism σ of G such that $\pi' = \pi\sigma$. Since $\sigma\alpha'|_{U_{ij}}$, $1 \leq i,j \leq n$, $i \neq j$, is injective, so is $\pi'\alpha'|_{U_{ij}} = \pi\sigma\alpha'|_{U_{ij}}$. $\pi|_{\alpha(U_{ij})}$ and $\pi|_{\sigma\alpha'(U_{ij})}$ are also injective, hence $\alpha|_{U_{ij}} = \sigma\alpha'|_{U_{ij}}$, which means $\alpha = \sigma\alpha'$ since $SL_n(D)$ is generated by U_{ij}'s $1 \leq i,j \leq n$, $i \neq j$.

---

* Cf. (3.15) (iii).

4.19. Remark. (a). If $\alpha: PSL_n(D) \longrightarrow G(k)$ is a homomorphism with Zariski dense image, we also have results similar to theorems (4.4) (4.5) and (4.18). The proofs of them are same to that of $SL_n(D)$.
(b) It is not difficult to generalize the theorems to the case when G is reductive.
(c) (4.18) (iii) can be deduced by (3.15) & (4.2) directly.

References

[1] A.Borel, Linear Algebraic Groups. notes by H.Bass. New York, W.A.Benjania 1969.

[2] A.Borel, T.A.Springer, Rationality properties of linear algebraic groups II. Tohoku Math. J. 20. 443-497. (1968).

[3] A.Borel, J.Tits, Homomorphismes "abstraits" de groups algébriques simples. Ann. of Math. 97. 499-571 (1973).

[4] A.Borel, J.Tits, Groupes réductifs. Inst. Hautes Etudes Sci. Publ. Math. 27. 55-150 (1960).

[5] A.Borel, J.Tits, Compléments à l'article "Groupes réductifs". Inst. Hautes Etudes Sci. Publ. Math. 41, 253-276 (1972).

[6] N.Bourbaki, Groupes et algébres de Lie. Paris, Hermann ch 4-6. 1968.

[7] C.Chevalley, Séminaire sur la classification des groups de Lie algebriques. Paris, Ecole Norm. Sup 1956-1958.

[8] J.Dieudonné, La géométrie des groupes classiques. 3rd ed. Berlin-Heidellberg-Nes York, Springer 1971.

[9] L.K. Hua, Z.X. Wan, Classical groups, Science and Technomogy press (In Chinese), Schanghai 1963.

[10] D.James, W.Waterhouse, B.Weisfeiler, Abstract homomorphisms of algebraic groups: problems and bibliography. Comm. in Algebra, 9(1), 95-144 (1981).

[11] S.X.Liu, Rings and Algebras. (In Chinese) Science Press. Beijing 1983.

[12] D.Mumford, Introduction to algebraic geometry. Harvard notes.

[13] O.Schreier & B.L.van der Waerden, Die Automorphismen der Projektiven Gruppen. Abh. Math. Sem. Univer. Hamber. 6(1928). 303-322.

[14] R.Steinberg, Abstract homomorphisms of simple algebraic groups. Sem. Bourbaki (1972-1973) Exp 435. Lect. Notes in Math. 383. Berlin-Heidelberg-New York, Springer 1974.

[15] R.Steinberg, Lecture on Chevalley Groups. mimeographed lecture notes. New Haven, Yale Univ. Math. Dept. 1968.

[16] J. Tits, Classification of algebraic semisimple groups. Proceedings of symposia in pure Math. Vol. IX. Amer. Math. Soc. 1966.

[17] B.Weisfeiler, Abstract homomorphisms between subgroups of algebraic groups. Notre Dame University Lecture Notes Series (1982).

A PREHOMOGENEOUS VECTOR SPACE OF CHARACTERISTIC 3

Z. Chen
Department of Mathematics
East China Normal University
Shanghai 200062, PR China

The prehomogeneous vector spaces were originally defined over the complex number field. It is natural to generalize the definition to an algebraically closed field of any characteristic. But there are some differences between characteristic 0 and positive characteristics. In the present article, a new regular prehomogeneous vector space of characteristic 3 is investigated. It has a unique irreducible relative invariant of degree 8. The orbit decomposition is also investigated, but there remain some unsolved problems.

1. Preliminaries

Let K be an algebraically closed field, V a finite-dimensional vector space over K, and G a connected algebraic subgroup of GL(V). A pair (G,V) is called a prehomogeneous vector space (abbrev. PV) if there exists a proper closed subset S (in the sense of Zariski topology) of V such that V-S is a G-orbit. Sometimes we use a triplet (G,ρ,V) to give a PV $(\rho(G),V)$ where G is a connected algebraic group and $\rho: G \to GL(V)$ is a rational representation of G. So the triplets (G_1,ρ_1,V_1) and (G_2,ρ_2,V_2) are equivalent iff the pairs $(\rho_1(G_1),V_1)$ and $(\rho_2(G_2),V_2)$ are isomorphic.

Let (G,V) be a PV, if there exists a nontrivial rational function $P \in K(V)$ such that

$$P(g.x)=\chi(g)P(x) \qquad \forall g\in G,\ x\in V$$

then P is called a relative invariant. $\chi: G \to K^*$ is a character of G, called the character of P.

Example:

Let $G=GL(n)$, $V=\{x\in M(n)/\ {}^tx=x\}$. Assuming $n>1$, there is a representation $\rho: G \to GL(V)$ defined by

$\rho(g).x = gx{}^tg \qquad \forall\, g \in GL(n),\ x \in V.$

Let $P(x)=\det(x)$. Since $S=Z(P)=\{x\in V/\ P(x)=0\}$ is a proper closed subset of V and V-S is a $\rho(G)$-orbit, $(GL(n),\rho,V)$ is a PV. It has relative invariants $P^m(x)$ $(m\in\mathbb{Z})$. The character of P is $\chi(\rho(g))=(\det(g))^2$.

Let (G,ρ,V) be a PV. We denote the dual G-module of V by V^* with the corresponding contragredient representation $\rho^*\colon G\to GL(V^*)$. Taking a dual basis in V^*, let P be a relative invariant of the PV, we can define a morphism

$$(\mathrm{grad}\ P)/P\colon V-S\to V^*$$

$$:\quad x\mapsto \frac{1}{P(x)}\begin{pmatrix}\partial P/\partial x_1(x)\\ \vdots\\ \partial P/\partial x_n(x)\end{pmatrix}.$$

If there exists a relative invariant P such that the morphism (grad P) /P $:V-S\to V^*$ is dominant, then this PV (G,ρ,V) is called a regular PV.

Theorem 1. Let (G,V) be a PV where G is reductive, then the following two conditions are equivalent:

(1) There exists a polynomial $P\in K[V]$ such that $V-Z(P)$ is a G-orbit.

(2) There exists an x in the Zariski open G-orbit O(G) such that G_x is reductive.

Proof. (After [10]) G_x is reductive $\Leftrightarrow G/G_x=O(G)$ is affine $\Leftrightarrow V-O(G)=S$ is an algebraic set of pure codimension 1 $\Leftrightarrow S=\mathbb{Z}(P)$, $P\in K[V]$.

Such a PV is said to be semiregular.

Theorem 2. ([9],[10]) Assume that K is an algebraically closed field of characteristic 0. Let (G,V) be a PV with a reductive algebraic group G, then (G,V) is a regular PV iff it is semiregular.

If char.K=p>0, the theorem 2 is not true. For example, when char.K=3, the triplet $(SL(3)\times GL(2), 2\Lambda_1\times\Lambda_1, V(6)\times V(2))$ is a semiregular PV, but not a regular PV. (See [3])

Hereafter we assume that the prehomogeneous vector spaces are irreducible, that is, V is an irreducible G-module. In that case, the group G must be reductive.

Theorem 3. ([3],[10]) Assume that char.K=p>0 and (G,V) is an irreducible PV. We have the following facts:

(1) If (G,V) is regular, then it is also semiregular.

(2) If (G,V) is semiregular and there exists a relative invariant

polynomial $P\in K[V]$ such that $\deg(P)\not\equiv 0 \pmod p$, then this PV is regular.

The above example is a regular irreducible PV($p>2$). Please refer to [1],[9] for the detail about prehomogeneous vector spaces.

2. Classification of irreducible PV's of characteristic p>2

M. Sato and T. Kimura [9] classified all the irreducible PV's over the complex number field. They introduced an important notion of castling transform and showed that each PV could be obtained by successive castling transforms, starting from a reduced one. It has been shown that there are 29 types of reduced irreducible PV's which are regular. These results can be easily generalized to any algebraically closed field of characteristic 0. Since all these PV's are defined over $\mathbb{Q}$, by the reduction modulo p, we can get corresponding PV's in characteristic p. It turns out that if $p>5$, every regular irreducible PV of characteristic 0 induces a regular irreducible one of characteristic p by the reduction modulo p(See [3]). There is one exception if $p=5$ and 6 exceptions if $p=3$. But not every reduced irreducible PV's of characteristic p can be obtained by the reduction modulo p. In fact, when $p>2$, there are 4 new types(See [3]), that is,

$(GL(n),(1+p^s)\Lambda_1,V(n^2))(s>0,\ n\geq 2)$,

$(GL(n),\Lambda_1+p^s\Lambda_{n-1},V(n^2))(s>0,\ n\geq 3)$,

$(GL(1)\times SL(3),\square\otimes(\Lambda_1+\Lambda_2),V(1)\otimes V(7))(p=3)$,

$(GL(4),\Lambda_1+\Lambda_2,V(16))(p=3)$.

The last one is the most interesting.

3. The construction of $(GL(4),\Lambda_1+\Lambda_2,V(16))(p=3)$

From now on, we assume that char.$K=3$. Let $k=GF(3)$ be the prime field of K. We are going to construct the irreducible representation $\Lambda_1+\Lambda_2$ of GL(4).

Take a 4-dimensional vector space over the complex number field $\mathbb{C}$:

$$V_1(\mathbb{C})=\{{}^t(x_1,x_2,x_3,x_4)/\ x_i\in\mathbb{C},\ i=1,2,3,4\}$$

with a canonical basis

$$f_1={}^t(1,0,0,0)$$
$$f_2={}^t(0,1,0,0)$$
$$f_3={}^t(0,0,1,0)$$
$$f_4={}^t(0,0,0,1).$$

Then the natural representation $\rho_1=\Lambda_1$ of $GL(4,\mathbb{C})$ on $V_1(\mathbb{C})$ is defined as

$$\rho_1(g).v=gv \qquad \forall\, g\in GL(4,\mathbb{C}),\ v\in V_1(\mathbb{C}).$$

Thus $V_1(\mathbb{C})$ is an irreducible $GL(4,\mathbb{C})$-module with the highest weight Λ_1. Put $V_2(\mathbb{C})=\{X\in M(4,\mathbb{C})/{}^tX=-X\}$ and define a representation $\rho_2=\Lambda_2$ of $GL(4,\mathbb{C})$ on $V_2(\mathbb{C})$ as follows:

$$\rho_2(g).X=gX\,{}^tg \qquad \forall\, g\in GL(4,\mathbb{C}),\ X\in V_2(\mathbb{C}).$$

Thus $V_2(\mathbb{C})$ is an irreducible $GL(4,\mathbb{C})$-module with the highest weight Λ_2. Denote

$$e_{ij}=\begin{array}{c}\left(\begin{array}{cccc}0 & & & \\ & 0 & 1 & \\ & -1 & 0 & \\ & & & 0\end{array}\right)\begin{array}{l} \\ i \\ j \\ \\ \end{array} \\ \begin{array}{cccc}\quad & i & j & \quad\end{array}\end{array} \qquad i\neq j,\ i,j=1,2,3,4$$

then $\{e_{ij}/\ 1\leqslant i<j\leqslant 4\}$ form a basis of $V_2(\mathbb{C})$.

The tensor representation $\rho_1\otimes\rho_2$ of $GL(4,\mathbb{C})$ on $V_1(\mathbb{C})\times V_2(\mathbb{C})$ is defined as

$$\rho_1\otimes\rho_2(g)(v\otimes X)=\rho_1(g)v\otimes\rho_2(g)X=gv\otimes gX\,{}^tg$$
$$\forall\, g\in GL(4,\mathbb{C}),\ v\in V_1(\mathbb{C}),\ X\in V_2(\mathbb{C}).$$

As a $GL(4,\mathbb{C})$-module, $V_1(\mathbb{C})\otimes V_2(\mathbb{C})$ has a decomposition into direct summands:

$$V_1(\mathbb{C})\otimes V_2(\mathbb{C})=W_1(\mathbb{C})\oplus W_2(\mathbb{C}).$$

In fact, $W_1(\mathbb{C})$ is the irreducible $GL(4,\mathbb{C})$-module with the highest weight $\Lambda_1+\Lambda_2$, its dimension is 20. Now we take a lattice $GL(4,\mathbb{Z})$ in $GL(4,\mathbb{C})$ and take an admissible $\mathbb{Z}$-form of $W_1(\mathbb{C})$ as follows:

$$\begin{aligned}W_{1,\mathbb{Z}}=\sum_{i\neq j}\mathbb{Z}f_i\otimes e_{ij}&+\mathbb{Z}(f_2\otimes e_{13}+f_3\otimes e_{12})+\mathbb{Z}(f_2\otimes e_{14}+f_4\otimes e_{12})+\mathbb{Z}(f_3\otimes e_{14}+\\&+f_4\otimes e_{13})+\mathbb{Z}(f_3\otimes e_{24}+f_4\otimes e_{23})+\mathbb{Z}(f_1\otimes e_{23}+2f_2\otimes e_{13}+f_3\otimes e_{12})+\\&+\mathbb{Z}(f_1\otimes e_{24}+2f_2\otimes e_{14}+f_4\otimes e_{12})+\mathbb{Z}(f_1\otimes e_{34}+2f_3\otimes e_{14}+f_4\otimes e_{13})+\\&+\mathbb{Z}(f_2\otimes e_{34}+2f_3\otimes e_{24}+f_4\otimes e_{23}).\end{aligned}$$

Let $GL(4,K)=GL(4,\mathbb{Z})\otimes_{\mathbb{Z}}K$ and $W_{1,K}=W_{1,\mathbb{Z}}\otimes_{\mathbb{Z}}K$, then $W_{1,K}$ is a Weyl module of $GL(4)$ with the highest weight $\Lambda_1+\Lambda_2$. There exists a unique maximal $GL(4)$-submodule M in $W_{1,K}$, that is,

$$\begin{aligned}M=(f_1\otimes e_{23}+2f_2\otimes e_{13}+f_3\otimes e_{12})\otimes K&+(f_1\otimes e_{24}+2f_2\otimes e_{14}+f_4\otimes e_{12})\otimes K+(f_1\otimes e_{34}+\\&+2f_3\otimes e_{14}+f_4\otimes e_{13})\otimes K+(f_2\otimes e_{34}+2f_3\otimes e_{24}+f_4\otimes e_{23})\otimes K.\end{aligned}$$

The quotient module $W_{1,K}/M=V$ is an irreducible $GL(4)$-module with the highest weight $\Lambda_1+\Lambda_2$. We denote a basis of V by

$$e_{iij}=(f_i\otimes e_{ij})\otimes 1+M \qquad i\neq j,\ i,j=1,2,3,4$$
$$e_{ijk}=(f_j\otimes e_{ik}+f_k\otimes e_{ij})\otimes 1+M \qquad 1\leqslant i<j<k\leqslant 4.$$

Then $\dim_K V=16$. Let ρ denote the representation of $GL(V)$ on $V(=V(16))$. Let

$$g=(g_{ij})\in GL(4),$$

then

$$\rho(g).e_{iij}=\sum_{p\neq q}(g_{pi}^2 g_{qj}-g_{pi}g_{pj}g_{qi})e_{ppq}+\sum_{p<q<r}(-g_{pi}g_{qi}g_{rj}-g_{pi}g_{qj}g_{ri}-g_{pj}g_{qi}g_{ri})e_{pqr}$$
$$i\neq j,\quad i,j=1,2,3,4,$$

$$\rho(g).e_{ijk}=\sum_{p\neq q}(g_{pi}g_{pj}g_{qk}+g_{pi}g_{pk}g_{qj}+g_{pj}g_{pk}g_{qi})e_{ppq}$$
$$+\sum_{p<q<r}(g_{pi}g_{qj}g_{rk}+g_{pi}g_{qk}g_{rj}+g_{pj}g_{qi}g_{rk}$$
$$+g_{pj}g_{qk}g_{ri}+g_{pk}g_{qi}g_{rj}+g_{pk}g_{qj}g_{ri})e_{pqr},$$
$$1\leq i<j<k\leq 4.$$

This representation ρ: $GL(4)\to GL(V)$ can induce a representation of the Lie algebra $\mathfrak{gl}(4)$ on the same vector space V as follows:

$$d\rho:\ \mathfrak{gl}(4)\to\mathfrak{gl}(V).$$

Let

$$A=(a_{ij})\in\mathfrak{gl}(4),$$

then

$$d\rho(A).e_{iij}=(a_{jj}-a_{ii})e_{iij}-a_{ji}e_{jji}+\sum_{q\neq i,j}a_{qj}e_{iiq}$$
$$+\sum_{p\neq i,j}(-a_{pi})e_{(pij)},\quad i\neq j,\ i,j=1,2,3,4,$$

$$d\rho(A).e_{ijk}=a_{ij}e_{iik}+a_{ji}e_{jjk}+a_{ik}e_{iij}+a_{ki}e_{kkj}+a_{jk}e_{jji}$$
$$+a_{kj}e_{kki}+(a_{ii}+a_{jj}+a_{kk})e_{ijk}+a_{pi}e_{(pjk)}$$
$$+a_{pj}e_{(pik)}+a_{pk}e_{(pij)},\quad 1\leq i<j<k\leq 4.$$

In these formulas, $e_{(pij)}=e_{i_1i_2i_3}$, where (i_1,i_2,i_3) is a rearrangement of (p,i,j) from in ascending order.

4. $(GL(4),\Lambda_1+\Lambda_2,V(16))$ is a regular PV.

Lemma 1. $(GL(4),\Lambda_1+\Lambda_2,V(16))$ is a PV.

Proof. Let

$$x_0=e_{123}+e_{124}+e_{134}+e_{234}\in V,$$

then the stabilizer of $G=GL(4)$ at x_0 is

$$G_{x_0}=\{g\in G/\rho(g).x_0=x_0\}$$
$$=\{E_{1,i_1}+E_{2,i_2}+E_{3,i_3}+E_{4,i_4},\ -E_{i_1,1}-E_{i_1,2}-E_{i_1,3}-E_{i_1,4}$$
$$+E_{i_2,j_2}+E_{i_3,j_3}+E_{i_4,j_4}/(i_1,i_2,i_3,i_4)\text{ is an arrangement}$$
$$\text{of }(1,2,3,4),\ (j_2,j_3,j_4)\text{ is an arrangement of any 3}$$

numbers taken from $\{1,2,3,4\}$.},

where

$$E_{ij}=\begin{pmatrix} & 1 & \\ 0 & & \end{pmatrix}\begin{matrix} i \\ \\ \end{matrix}$$
$$\qquad\quad j$$

Thus G_{x_0} is a finite subgroup of G and

$$\dim \rho(G).x_0=\dim G-\dim G_{x_0}=16=\dim V.$$

Therefore $\rho(G).x_0$ must be a Zariski open subset in the affine space V.

Let $\mathfrak{g}_x$ denote the centralizer of the Lie algebra $\mathfrak{g}=\mathfrak{gl}(4)$ at $x\in V$. Then

$$\mathfrak{g}_x=\{A\in\mathfrak{g}/d\rho(A).x=0\}.$$

If $y=\rho(g).x$, then $\mathfrak{g}_y=\mathrm{Ad}\ g.\mathfrak{g}_x$. Hence $\dim\mathfrak{g}_y=\dim\mathfrak{g}_x$. Let $L(G_x)$ denote the Lie algebra of G_x, then

$$L(G_x)\subseteq\mathfrak{g}_x.$$

So we have

$$\dim G_x=\dim L(G_x)\leq\dim\mathfrak{g}_x.$$

Now we define

$$V_r=\{x\in V/\ \dim d\rho(\mathfrak{g}).x\leq r\}=\{x\in V/\ \dim\mathfrak{g}_x\geq 16-r\},\qquad 0\leq r\leq 16.$$

As $\{E_{i,j}/\ 0\leq i,j\leq 4\}$ is a basis of $\mathfrak{g}$, $\{d\rho(E_{i,j})/\ 0\leq i,j\leq 4\}$ generates $d\rho(\mathfrak{g})$. Construct the 16×16 matrix

$$M_x=(d\rho(E_{1,1}).x,\ldots,d\rho(E_{4,4}).x),$$

then $\dim d\rho(\mathfrak{g}).x\leq r$ iff all the minors of degree r+1 of M_x are null. Hence V_r is closed in V. It is not difficult to see that V_r is stable under the action of G. Since

$$\sum_{i=1}^{4} d\rho(E_{i,i})=0,$$

we have

$$V=V_{16}=V_{15}\supseteq V_{14}\supseteq\ldots\supseteq V_1\supseteq V_0=\{0\}.$$

<u>Theorem 4.</u> $(GL(4),\Lambda_1+\Lambda_2,V(16))$ is a regular PV. The degree of its irreducible relative invariant $P\in k[V]$ is 8, the associated character is $\chi(g)=(\det(g))^6$.

<u>Proof.</u> As G_{x_0} is a reductive group, this PV is semiregular by Theorem 1. There exists a polynomial $P\in K[V]$ such that $V-\rho(G).x_0=Z(P)$. P must be a relative invariant. Since this PV is irreducible, P can be taken to be an irreducible polynomial in K[V] and all the relative invariants of this PV have the form $cP^m(c\in K^*,\ m\in\mathbb{Z})$([1],[9]). Furthermore, since GL(4) is defined over k, P can be taken to be an irreducible polynomial in k[V]([1]).

Since $\dim\mathfrak{g}_{x_0}=1$, $x_0\in V_{15}$, $x_0\notin V_{14}$. Hence $V_{14}\subseteq V-\rho(G).x_0$, that is, $V_{14}\subseteq Z(P)$, where $P\in k[V]$ is the irreducible relative invariant of the PV.

If we take $x_1=e_{112}+e_{334}+e_{442}$, then $\dim G_{x_1}=1$, $\dim \mathfrak{g}_{x_1}=2$. So

$$\dim \overline{\rho(G).x_1}=15.$$

As $x_1\in V_{14}$, $\overline{\rho(G).x_1}\subseteq V_{14}\subseteq Z(P)$. Since $\overline{\rho(G).x_1}$ and $Z(P)$ are all irreducible closed subset, considering the fact that $\dim \overline{\rho(G).x_1}=\dim Z(P)$, we can conclude that

$$\overline{\rho(G).x_1}=V_{14}=Z(P).$$

As an orbit of a connected algebraic group, $\rho(G).x_1$ must be a locally closed subset, so $\rho(G).x_1$ is a relative open subset contained in the irreducible closed subset $Z(P)=V_{14}$. This implies that $\rho(G).x_1$ is the unique 15-dimensional orbit in V.

Now let Q be a nontrivial minor of degree 15 of the matrix M_{x_1}, then

$$Z(P)=V_{14}\subseteq Z(Q).$$

Hence $P|Q$, that is,

$$\deg P\leq \deg Q=15.$$

Let $\deg P=m$ and let χ denote the character associated with P. Then $\chi(g)=(\det(g))^r$. If we take

$$g=t.1=\begin{pmatrix} t & & & \\ & t & & \\ & & t & \\ & & & t \end{pmatrix}\in GL(4), \qquad t\in K^*,$$

then from the relation below

$$P(\rho(g).x)=\chi(g)P(x) \qquad \forall\, g\in G,\ x\in V,$$

we can get

$$P(\rho(g).x)=P(t^3.x)=t^{3m}P(x) \qquad \forall x\in V.$$

Hence

$$t^{3m}=\chi(g)=t^{4r}.$$

So $3m=4r$.

If $g\in G_{x_0}$, then $\chi(g)=(\det(g))^r=1$. But we know that there exists some $g\in G_{x_0}$ such that $\det(g)=-1$. Hence r must be an even integer. Therefore $m=8$, $r=6$.

Corollary. Let $G=SL(4)$, V the irreducible representation space of the highest weight $\Lambda_1+\Lambda_2$, then $K[V]^G=K[P]$, where P is the irreducible relative invariant of the PV $(GL(4),\Lambda_1+\Lambda_2,V(16))$.

It needs a lengthy calculation to express this irreducible relative invariant explicitely.

5. Orbits.

Lemma 2. Let $x_7=e_{112}+e_{331}$, $x_{10}=e_{112}$, then

$$V_9=\rho(G).x_7\cup\rho(G).x_{10}\cup\{0\}.$$

Proof. Assuming that $x\neq 0$, $x\in V_9$ and

$$x=\sum_{i\neq j}x(iij)e_{iij}+\sum_{i<j<k}x(ijk)e_{ijk},$$

if $x(iij)=0$ for all $1\leq i\neq j\leq 4$, then there must exist some $x(ijk)\neq 0$. In this case we can prove that $x\notin V_9$. Hence without loss of generality, we can assume that $x(112)\neq 0$.

Take

$$g_1=\begin{pmatrix}1&&&\\&1&&\\-\frac{x(113)}{x(112)}&&1&\\-\frac{x(114)}{x(112)}&&&1\end{pmatrix}\in GL(4),$$

and put $y=\rho(g_1).x$, then

$$y(112)=x(112),$$
$$y(113)=y(114)=0.$$

There are two cases:

(1) $y(331)$ or $y(441)$ is not equal to 0. We can assume $y(331)\neq 0$. Take

$$g_2=\begin{pmatrix}&1&&\\&&1&\\&&1&\\-\frac{y(334)}{y(331)}&0&-\frac{y(134)}{y(331)}&1\end{pmatrix}\in GL(4),$$

and let $z=\rho(g_2).y$. Then

$$z(112)=y(112),$$
$$z(113)=z(114)=z(334)=z(134)=0,$$
$$z(331)=y(331).$$

As $z\in V_9$, this implies the following equalities:

$$z(441)=z(443)=z(124)=z(442)=z(224)=z(234)=0,$$
$$z(112)z(332)+z(221)z(331)-z(123)^2=0.$$

If we set

$$v=z(112),$$
$$w_2^2=z(331),$$
$$w_1^3=-(z(221)z(123)+z(112)z(223))/w_2,$$
$$u_1^3=\frac{(z(221)z(123)+z(112)z(223))^2}{z(112)^3z(331)}-\frac{z(221)^3}{z(112)^3},$$
$$u_2^3=\frac{z(331)(z(221)z(123)+z(112)z(223))}{z(112)^3}-\frac{z(123)^3}{z(112)^3},$$

then

$$z=\rho(g_3).x_7,$$

where

$$g_3=\begin{pmatrix}1&0&0&0\\u_1&v&w_1&0\\u_2&0&w_2&0\\0&0&0&1\end{pmatrix}\in GL(4).$$

This proves that $x\in\rho(G).x_7$.

(2) Both $y(331)$ and $y(441)$ are equal to zero. Note that $y(113)=y(114)=0$. As $y\in V_9$, this implies that

$$y(134)=y(334)=y(443)=0.$$

Let

$$g_4=\begin{pmatrix}1&&&\\\frac{y(221)}{y(112)}&1&&\\\frac{y(123)}{y(112)}&&1&\\\frac{y(124)}{y(112)}&&&1\end{pmatrix}\in GL(4),$$

$$z=\rho(g_4).y.$$

Then

$$z(113)=z(114)=z(221)=z(331)=z(334)=z(441)$$
$$=z(443)=z(123)=z(124)=z(134)=0.$$

In this case, we have

$$z(442)=z(332)=z(234)=0.$$

If $z(223)=z(224)=0$, then

$$z=z(112)e_{112},$$

so

$$x\in\rho(G).x_{10}.$$

Otherwise, we may assume that $z(223)\neq 0$. Let

$$g_5=\begin{pmatrix}0&1&0&0\\0&0&1&0\\1&0&0&0\\0&0&-\frac{z(224)}{z(223)}&1\end{pmatrix}\in GL(4),$$

then

$$\rho(g_5).z=x_7.$$

Hence

$$x\in\rho(G).x_7.$$

By calculation, we know that $\dim \mathfrak{g}_{x_7}=8$, $\dim G_{x_7}=7$. This implies that $x_7\in V_8\subseteq V_9$ and $\dim \overline{\rho(G).x_7}=9$. Moreover, $\dim \mathfrak{g}_{x_{10}}=\dim G_{x_{10}}=10$, so $x_{10}\in V_6\subseteq V_7$ and $\dim \overline{\rho(G).x_{10}}=6$.

<u>Theorem 5.</u> There exist only one 6-dimensional G-orbit and one 9-di-

mensional G-orbit in V. They are $\rho(G).x_{10}$ and $\rho(G).x_7$ respectively. In addition, when m=7,8 or $1\leq m\leq 5$, there are not any m-dimensional G-orbits.

<u>Proof.</u> We have proved that

$$V_9=V_8=\overline{\rho(G).x_7},$$
$$V_7=V_6=\overline{\rho(G).x_{10}},$$
$$V_5=V_4=\ldots=V_0=\{0\}.$$

Assuming that $m\leq 9$, if x is a point in an m-dimensional orbit, then

$$\dim \mathfrak{g}_x \geq \dim G_x = n-\dim \rho(G).x = n-m \geq n-9,$$

that is,

$$x\in V_9.$$

Hence m=9,6 or 0 by Lemma 2.

By calculation, it turns out that there exist following orbits in V:

(i) a 16-dimensional orbit: $x_0=e_{123}+e_{124}+e_{134}+e_{234}\in V_{15}-V_{14}$;

(ii) a 15-dimensional orbit: $x_1=e_{112}+e_{334}+e_{442}\in V_{14}-V_{13}$;

(iii) at least a 13-dimensional orbit: $x_3=e_{112}+e_{334}+e_{441}\in V_{13}-V_{12}$;

(iv) at least a 12-dimensional orbit: $x_4=e_{112}+e_{334}\in V_{12}-V_{11}$;

(v) at least an 11-dimensional orbit: $x_5=e_{114}+e_{123}\in V_{11}-V_{10}$;

(vi) at least a 10-dimensional orbit: $x_6=e_{123}\in V_{10}-V_9$;

(vii) a 9-dimensional orbit: $x_7=e_{112}+e_{331}\in V_9-V_7$;

(viii) a 6-dimensional orbit: $x_{10}=e_{112}\in V_7-V_5$;

(ix) a 0-dimensional orbit: $V_5=V_4=\ldots=V_0=0$.

<u>Conjecture.</u> All possible dimensions of the GL(4)-orbits are 16,15,14, 12,11,10,9,6,0. For every such dimension, there exists one and only one GL(4)-orbit.

We use the above example to illustrate this conjecture. If $\mathrm{char}.K\neq 2$, then there exist n+1 GL(n)-orbits in $V(\frac{1}{2}n(n+1))$, that is,

$$O_r=\{x\in V/\mathrm{rank}\ x=r\} \qquad r=0,1,\ldots,n.$$

These O_r's are subvarieties in V and

$$\dim O_r=nr-\tfrac{1}{2}r(r-1).$$

REFERENCES

[1] Chen,Z.: On prehomogeneous vector spaces over an algebraically closed field of characteristic p (Chinese), J. of East China Normal Univ., Natural Sci. Edition 2(1983), 11-17.

[2] Chen,Z.: A classification of irreducible prehomogeneous vector spaces over an algebraically closed field of characteristic p (I) (Chinese), Chin. Ann. of Math. 6A(1985), 39-48.

[3] Chen,Z.: A classification of irreducible prehomogeneous vector spaces over an algebraically closed field of characteristic p (II) (Chinese), to appear.

[4] Chen,Z.: A classification of irreducible prehomogeneous vector spaces over an algebraically closed field of characteristic 2 (I) (Chinese), to appear.

[5] Chen,Z.: On the prehomogeneous vector space $(GL(1)\times SL(3),\Box\otimes(\Lambda_1+\Lambda_2), V(1)\otimes V(7))(p=3)$ (Chinese), to appear.

[6] Cline,E., Parshall,B., Scott,L.: Induced modules and affine quotients, Math. Ann. 230(1977), 1-14.

[7] Humphreys,J.E.: Linear Algebraic Groups, Graduate Texts in Math. 21, Springer-Verlag, 1981.

[8] Humphreys,J.E.: Ordinary and Modular Representations of Chevalley Groups, Springer LN 528(1976).

[9] Sato,M. and Kimura,T.: A classification of irreducible prehomogeneous vector spaces and their relative invariants, Nagoya Math. J. 65(1977), 1-155.

[10] Servedio,F.J.: Affine open orbits, reductive isotropy groups and dominant gradient morphisms; a theorem of Mikio Sato, Pacific J. of Math. 72(1977), 537-545.

ON FULL SUBGROUPS OF TWISTED GROUPS

Dong Chongying
Institute of Systems Science
Academia Sinica, Beijing PRC

Introduction

In this article the notations are the same as those in [2].

Let G be a Chevalley group of type $A_{\ell'}$, $(\ell'>2)$, D_ℓ or E_6 over a field K. For a maximal K-torus T of G, let $\Sigma=\Sigma(G,T)$ denote the root system of G with respect to T. Let σ be an automorphism of G derived from field automorphism and graph automorphism. The action of the graph automorphism is as follows:

(1) Type $A_{\ell'}$. The fundamental roots are $\alpha_1,\ldots,\alpha_{\ell'}$, as in the diagram. $\sigma(\alpha_i)=\alpha_{\ell'+1-i}$, $i=1,\ldots,\ell'$.

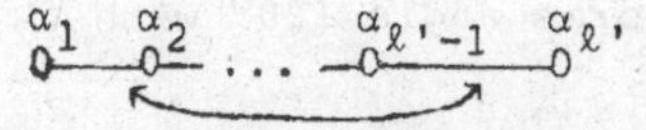

(2) Type D_ℓ. The symmetry is as shown in the diagram.

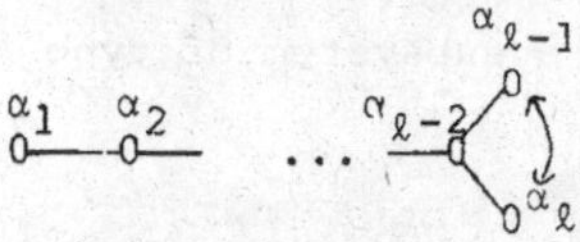

$\sigma(\sigma_i)=\sigma_i$, $i=1,2,\ldots,\ell-2$, $\sigma(\alpha_{\ell-1})=\alpha_\ell$, $\sigma(\alpha_\ell)=\alpha_{\ell-1}$.

(3) Type D_4. The symmetry is as shown in the diagram.

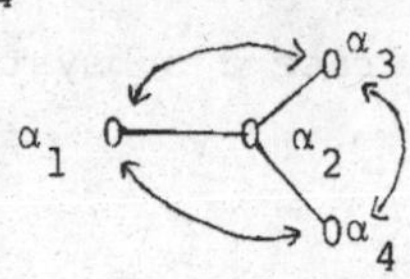

(4) Type E_6. The symmetry is as shown is the diagram.

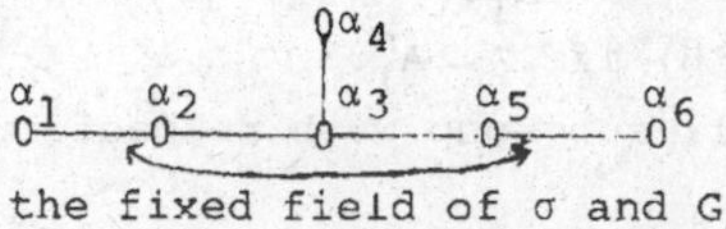

Let K_σ be the fixed field of σ and $G_\sigma=\{g\in G\mid \sigma g=g\}$ be the fixed point set of σ. G_σ is an algebraic group and is called the twisted group of G. Corresponding to the groups considered above, we call G_σ type ${}^2A_{2\ell}$, ${}^2A_{2\ell+1}$ (according to $\ell'=2\ell$, $2\ell+1$), ${}^2D_\ell$, ${}^3D_4$, ${}^2E_6$ respectively. Let the root system of G_σ be $\Sigma_\sigma=\Sigma/\sim$ in which $\alpha\sim\beta$ iff $\alpha(\alpha)=\beta$ or $\alpha(\beta)=\alpha$. Σ_σ is of type B_ℓ, $C_{\ell+1}$, $B_{\ell-1}$, G_2, F_4 respectively. For $\alpha\in\Sigma$ we denote $\bar\alpha$ as the class α belonging to. Then the structure of $\mathfrak{X}_{\bar\alpha,\sigma}$ is as follows:

(1) If $\alpha\sim A_1$, $\mathfrak{X}_{\bar\alpha,\sigma}=\{X_\alpha(t)\mid t\in K_\sigma\}$,

(2) If $\alpha \sim A_1^n$, $\mathfrak{X}_{\bar{\alpha},\sigma}=\{X\sigma X\ldots | X=X_\alpha(t),\ t\in K\}$.

Denote the element of $\mathfrak{X}_{\bar{\alpha},\sigma}$ by $X_{\bar{\alpha}}(t)$.

Definition. Let $t\in K$, we define $T(t)=t+t^\sigma+\ldots+t^{\sigma^{n-1}}$, $N(t)=tt^\sigma\ldots t^{\sigma^{n-1}}$ and call them respectively the trace and norm of t.

Let A be a subset of K. Set $T(A)=\{T(t)|t\in A\}$, $N(a)=\{N(t)|t\in A\}$.

We have the following commutator formulas in G_σ.

1. Type $^3D_4$.

(1) Assume γ, $\beta\in(\Sigma_\sigma)_s$ ($(\Sigma_\sigma)_s$ is the set of short roots of Σ_σ) and γ forms angle 120° with β. Then $[X_\gamma(t),\ X_\beta(u)]=X_{\gamma+\beta}(\pm(ut^\sigma+u^\sigma t))X_{2\gamma+\beta}(\pm(T(ut^\sigma t^{\sigma^2}))X_{\gamma+2\beta}(\pm(tu^\sigma u^{\sigma 2}))$.

(2) Assume γ, $\beta\in(\Sigma_\sigma)_s$ and γ forms angle 60° with β. Then $[X_\gamma(t),\ X_\beta(u)]=X_{\gamma+\beta}(\pm T(tu^\sigma))$.

(3) Assume $\gamma\in(\Sigma_\sigma)_s$ and δ is a long root forming angle 150° with γ. Then

$[X_\gamma(t),\ X_\delta(u)]=X_{\gamma+\delta}(\pm tu)X_{2\gamma+\delta}(\pm utt^\sigma)X_{3\gamma+\delta}(\pm utt^\sigma t^{\sigma^2})X_{3\gamma+2\delta}(\pm 2u^2tt^\sigma t^{\sigma^2})$.

(4) Assume δ_1, $\delta_2\in(\Sigma_\sigma)_\ell$ and δ_1 forms angle 120° with δ_2. Then $[X_{\delta_1}(t),\ X_{\delta_2}(u)]=X_{\delta_1+\delta_2}(\pm tu)$.

2. Type $^2A_{2\ell}$, $^2A_{2\ell+1}$, $^2E_6$.

(1) Assume $\alpha,\beta\in\Sigma_\sigma$. If α,β generate a subsystem of type A_2 and α forms angle 120° with β. Then

$[X_\alpha(t),\ X_\beta(u)]=X_{\alpha+\beta}(\pm tu)$.

(2) Assume $\alpha,\beta\in(\Sigma_\sigma)_s$, α,β generate a subsystem of type B_2 and α forms angle 90° with β. Then

$[X_\alpha(t),\ X_\beta(u)]=X_{\alpha+\beta}(\pm(tu^\sigma+t^\sigma u))$.

(3) Assume $\alpha\in(\Sigma_\sigma)_\ell$, $\beta\in(\Sigma_\sigma)_s$, α,β generate a subsystem of type B_2 and α forms angle 135° with β. Then

$[X_\alpha(t),\ X_\beta(u)]=X_{\alpha+\beta}(\pm tu)X_{\alpha+2\beta}(\pm tuu^\sigma)$.

Let H be a subgroup of G and $\bar{\alpha}\in\Sigma_\sigma$. Set

(1) $R_{\bar{\alpha}}=\{t\in K_\sigma | X_\alpha(t)\in H\}$ if $\bar{\alpha}\sim A_1$.

(2) $R_{\bar{\alpha}}=\{t\in K | X_{\bar{\alpha}}(t)=X\sigma X\ldots\in H\}$ if $\bar{\alpha}\sim A_1^n$.

Definition. H is called full subgroup with respect to G if for any $\bar{\alpha}\in\Sigma_\sigma$ we always have $R_{\bar{\alpha}}\neq\{0\}$.

General definition of full subgroup can be found in [1]. B. Weisfeiler, L.N. Vaserstein have discussed full subgroups of Cheva-groups in [1]. The results of the present paper are modeled on the results of Weisfeiler and Vaserstein. The methods are also similar to theirs. However the situation for $^3D_4$ turns out to be more complicated. Throughout the following we assume H is a full subgroup. The main results of the paper are

1 Type $^3D_4$

Assume

(1) When char(K)=2, $R_\gamma \cap K_\sigma \neq \{0\}$ for every short root γ;

(2) When char(K)=3, there exists a short root γ such that $R_\gamma \cap K_\sigma \neq \{0\}$

Then there exist subrings A, B of K and non-zero a_ε, b_ε in K (for any root ε) such that

(i) If $R_\gamma \subset K_\sigma$ for any short root γ then

(a) $a_\delta B \subset R_\delta \subset b_\delta B$, $R_\delta A^p \subset R_\delta$, $AR_\delta R_{-\delta} \subset A$ for any long root δ.

(b) $a_\gamma A \subset R_\gamma \subset b_\gamma A$, $R_\gamma B \subset R_\gamma$, $B'(R_\gamma R_{-\gamma})^p \subset B$ for any short root γ where if char(K)=2, $B'=B^2$ otherwise $B'=B$

when char(K)=3, p=3 and $A^3 \subset B \subset A$, otherwise B=A p=1.

(c) $A, B \subset K_\sigma$ and a_δ, b_δ, a_γ, $b_\gamma \in K_\sigma$.

(ii) If char(K)$\neq$2,3 and $T(R_\gamma)=\{0\}$ for any short root γ then

(a) $AR_\delta \subset R_\delta$, $AR_\delta R_{-\delta} \subset A$, for any long root δ.

(b) $AR_\gamma \subset R_\gamma$, $6AR_\gamma R_{-\gamma} \subset A$, for any short root γ. Otherwise

(iii)

(a) $a_\delta A \subset R_\delta \subset b_\delta A$, $AR_\delta \subset R_\delta$, $AR_\delta R_{-\delta} \subset A$, $\delta \in (\Sigma_\sigma)_\ell$.

(b) $a_\gamma A \subset T(R_\gamma) \subset b_\gamma A$, $AT(R_\gamma) \subset T(R_\gamma)$, $A^2 T(R_\gamma) T(R_{-\gamma}) \subset A$, $\gamma \in (\Sigma_\sigma)_s$.

(c) $A \subset K_\sigma$, A_δ, b_δ, a_γ, $b_\gamma \in K_\sigma$.

2 Type ${}^2A_{2\ell}$, ${}^2A_{2\ell+1}$, ${}^2D_\ell$, ${}^2E_6$

Assume char(K)$\neq$2. Then there exist a subring A of K_σ and non-zero a_ε, $b_\varepsilon \in K_\sigma$ with $\varepsilon \in \Sigma_\sigma$ such that

(a) $a_\delta A \subset R_\delta \subset b_\delta A$, $AR_\delta \subset R_\delta$, $4A(R_\delta R_{-\delta})^2 \subset A$, $\delta \in (\Sigma_\sigma)_\ell$.

(b) $a_\gamma A \subset N(R_\gamma) \subset b_\gamma A$, $AN(R_\gamma) \subset N(R_\gamma)$, $4AN(R_\gamma)N(R_{-\gamma}) \subset A$, $\gamma \in (\Sigma_\sigma)_s$.

The following result will be cited many times.

Lemma 0 ([1]) Let n, m, N be natural numbers, let non-empty $A, B, R_i \subset K$ and $c_i, d_i \in K$ for i=1,...,N. Assume $0 \neq AB^n \subset A$, $A^m B \subset B$, $0 \neq c_i A \subset R_i \subset d_i A$ (i=1,...,N). Then there is a non-zero b in B such that $d_i A(bB)^n \subset c_i A$ and therefore, $R_i(bB)^n \subset R_i$ for i=1,...,N.

1. G_σ is of type ${}^3D_4$

Proposition 1. If char(k)$\neq$2,3 and there exists a short root γ such that $T(R_\gamma) \neq 0$ then for any short root γ we have $R_\gamma \cap K_\sigma \neq \{0\}$.

Proof: Let δ be a long root forming angle 150° with γ, $t \in R_\gamma$ $T(t) \neq 0$ and $v, u \in R_\delta$. Then

$$H \ni Z_1(t,u) = [X_\gamma(t), X_\delta(u)]$$
$$= X_{\gamma+\delta}(\pm tu) X_{2\gamma+\delta}(\pm utt^\sigma) X_{3\gamma+\delta}(\pm utt^\sigma t^{\sigma^2}) X_{3\gamma+2\delta}(\pm 2u^2 tt^\sigma t^{\sigma^2}),$$

$$H \ni Z_2(t,u) = Z_1(t,u) Z_1(-t,u)^{-1}$$
$$= X_{\gamma+\delta}(\pm 2tu) X_{3\gamma+\delta}(\pm 2utt^\sigma t^{\sigma^2}) X_{3\gamma+2\delta}(\pm 4u^2 tt^\sigma t^{\sigma^2}),$$

$$H \ni Z_3(t,u) = Z_2(t,u) Z_2(-t,u)^{-1}$$
$$= X_{\gamma+\delta}(\pm 4tu) X_{3\gamma+\delta}(\pm 4utt^\sigma t^{\sigma^2}),$$

$H\ni Z_3(t,u)^2Z_3(2t,u)^{-1}=X_{3\gamma+\delta}(\pm 24utt^{\sigma}t^{\sigma^2})$,

$H\ni Z_3(t,6u)X_{3\gamma+\delta}(24utt^{\sigma}t^{\sigma 2})=X_{\gamma+\delta}(\pm 24tu)$.

Thus

$$24tu\in R_{\gamma+\delta} \qquad (1)$$

$H\ni Z_2'(t,v)=Z_1(-t,-v)^{-1}Z_1(t,v)$

$=X_{2\gamma+\delta}(\pm 2vtt^{\sigma})X_{3\gamma+2\delta}(\pm 4u^2tt^{\sigma}t^{\sigma^2})$,

$H\ni Z_2'(t,v)Z_2'(-t,v)=X_{2\gamma+\delta}(\pm 4vtt^{\sigma})$.

Thus

$$4vtt^{\sigma}\in R_{2\gamma+\delta} \qquad (2)$$

In (2) Replacing δ,γ by $-3\gamma-\delta$, $\delta+2\gamma$ and doing t,v by $c_{-3\gamma-\delta}\in R_{-3\gamma-\delta}$, $4vtt^{\sigma}$, we have $16v^2tt^{\sigma}t^{\sigma}t^{\sigma^2}c_{-3\gamma-\delta}\in R_{\gamma+\delta}$. Since $H\ni Z_2'(t,v)Z_2'(t,v)=X_{3\gamma+2\delta}(\pm 8v^2tt^{\sigma}t^{\sigma^2})$, thus $8v^2tt^{\sigma}t^{\sigma 2}\in R_{3\gamma+2\delta}$. However, $R_{3\gamma+2\delta}R_{-3\gamma-\delta}\subset R_{\delta}$ therefore $8v^2tt^{\sigma}t^{\sigma^2}c_{-3\gamma-\delta}\in R_{\delta}$. In (1) setting $u=8v^2tt^{\sigma}t^{\sigma^2}c_{-3\gamma-\delta}$, we have $t_{\gamma+\delta}=24tu$, $t^{\sigma}_{\gamma+\delta}\in R_{\gamma+\delta}$. Clearly, $T(t_{\gamma+\delta})\neq 0$.

Replacing γ,δ by $\gamma+\delta$, $-2\delta-3\gamma$, we have $t_{-2\gamma-\delta}=24c_{-3\gamma-2\delta}t_{\gamma+\delta}$, $t^{\sigma}_{-2\gamma-\delta}$, $t^{\sigma^2}_{-2\delta-\gamma}\in R_{-2\gamma-\delta}$ with $c_{-3\gamma-2\delta}\in R_{-3\gamma-\delta}$. Therefore $T(t_{-2\gamma-\delta})\in R_{-2\gamma-\delta}\cap K_{\sigma}$ and $T(t_{-2\gamma-\delta})\neq 0$. i.e. $R_{-2\gamma-\delta}\cap K_{\sigma}\neq\{0\}$.

Repeating above process, we get at once $R_{\gamma}\cap K_{\sigma}\neq\{0\}$ for any short root γ.

Corollary. If char(K)$\neq 2$ and there exists a short root γ such that $R_{\gamma}\cap(K-K_{\sigma})\neq\phi$ (resp., $R_{\gamma}\cap K_{\sigma}\neq\{0\}$), then for any short γ', $R_{\gamma'}\cap(K-K_{\sigma})\neq\phi$ (resp., $R_{\gamma'}\cap K_{\sigma}\neq\{0\}$).

Proof: From formula (2) appeared in the proof of Proposition 1 we know $4vtt^{\sigma}\in R_{2\gamma+\delta}$ with $t\in R_{\gamma}$, $v\in R_{\delta}$. If $t\in K_{\sigma}$, $4vtt^{\sigma}\in K_{\sigma}$; if $t\in K-K_{\sigma}$, $4vtt^{\sigma}\in K-K_{\sigma}$.

Proposition 2. Assume char(K)$\neq 2,3$ and $T(R_{\beta})=\{0\}$ for any short root β. Then for any short root γ in R_{γ} , let δ be a long root forming angle 150° with γ. we have

(i) $K=K_{\sigma}(t)$ and $t^{\sigma}=\alpha t$ where α is a primitive 3th root of unity.

(ii) R_{γ}, $R_{\gamma+\delta}$, $R_{-2\gamma-\delta}\subset K_{\sigma}t$ and $R_{-\gamma}$, $R_{-\gamma-\delta}$, $R_{2\gamma+\delta}\subset K_{\sigma}t^2$.

Proof: (i) Since $T(t)=0$, by Proposition 1, $T(tt^{\sigma})=0$. Hence $t^{\sigma}=\alpha t$ with a primitive 3th root α of unity in K_{σ} . Clearly $K=K_{\sigma}(t)$.

(ii) Evidently, $R_{\gamma}\subset K_{\sigma}t+K_{\sigma}t^2$. If $v=at+bt^2\in R_{\gamma}$ for some non-zero b in K_{σ}, we will conclude a contrary. By Proposition 1 (2) we know $cvv^{\sigma}\in R_{2\gamma+\delta}$ for some non-zero c in K_{σ}. However $vv^{\sigma}=a^2\alpha t^2+b^2\alpha^2t^4-abt^3$, we can choose $a\neq 0$, therefore $T(cvv^{\sigma})\neq 0$, contrary to the hypothesis. Thus $R_{\gamma}\subset K_{\sigma}t$. Similarly we can prove that (ii) holds for other short roots.

Theorem 1. If $\mathrm{char}(k)\neq 2,3$ and $T(R_{\gamma})=0$ for any short root γ, then there exists a subring A of K such that

(1) $R_{\delta}A\subset R_{\delta}$, $AR_{\delta}R_{-\delta}\subset A$, $\delta\in(\Sigma_{\sigma})_{\ell}$.

(2) $R_{\gamma}A\subset R_{\gamma}$, $6AR_{\gamma}R_{-\gamma}\subset A$, $\gamma\in(\Sigma_{\sigma})_{s}$.

Proof: By Proposition 2, we know that the commutator formulas in type $^3D_4$ is very similar to those in Chevalley group of type G_2 in this case. Hence we can get the theorem by the method used in [1].

In the following discussion, when $\mathrm{char}(K)\neq 2,3$. we always assume that there exists a short root γ such that $T(R_{\gamma})\neq 0$.

Lemma 1 ([1]). There exists a subring A of K_{σ} such that for any long root δ, $R_{\delta}A\subset R_{\delta}$, $AR_{\delta}R_{-\delta}\subset A$.

Lemma 2 If $\mathrm{char}(K)=3$ and $t\in K-K_{\sigma}$, $T(t)=0$, then $T(tt^{\sigma})\neq 0$.

Lemma 3 If $\mathrm{char}(K)=3$ and there exists a short root γ such that $T(t)=0$ (resp., $T(tt^{\sigma})\neq 0$) for some $t\in R_{\gamma}\cap(K-K_{\sigma})$. Then for any short root β, there exists $v\in R_{\beta}\cap(K-K_{\sigma})$ such that $T(v)=0$ (resp., $T(vv^{\sigma})\neq 0$).

Proof: Let β be a short root forming angle 120° with γ. By corollary of Proposition 1, there exists some $u\in R_{\beta}\cap K_{\sigma}$. Since

$$H\ni Z(t,u)=[X_{\gamma}(t),\ X_{\beta}(u)]$$
$$=X_{\gamma+\beta}(\pm(tu^{\sigma}+ut^{\sigma}))X_{2\gamma+\beta}(\pm T(ut^{\sigma}t^{\sigma^2}))X_{\gamma+2\beta}(\pm T(tu^{\sigma}u^{\sigma^2})).$$

$H\ni Z(t,u)Z(-t,-u)=X_{\gamma+\beta}(\pm 2(tu^{\sigma}+t^{\sigma}u)$, $t_{\gamma+\beta}=2(t+t^{\sigma})u\in R_{\gamma+\beta}\cap(K-K_{\sigma})$ and $T(t_{\gamma+\beta})=0$ (resp., $T(t_{\gamma+\beta}t^{\sigma}_{\gamma+\beta})\neq 0$).

Repeating the above process we can get the results.

Lemma 4 If $\mathrm{char}(K)=3$, $R_{\gamma}\cap(K-K_{\sigma})\neq\phi$ for a short root γ, then there exists $u\in R_{\gamma}\cap(K-K_{\sigma})$ such that $T(uu^{\sigma})\neq 0$.

Proof: Take $v\in R_{\gamma}\cap(K-K_{\sigma})$. Suppose $T(vv^{\sigma})=0$. We have known $4vv^{\sigma}t\in R_{2\gamma+\delta}$ where δ is a long root forming angle 150° with γ and $t\in R_{\delta}$. Set $v'=4tvv^{\sigma}$. Then $T(v')=0$ and $v'\in R_{2\gamma+\delta}\cap(K-K_{\sigma})$. By lemma 2 $T(v'v'^{\sigma})\neq 0$, by Lemma 3, we complete the proof.

Lemma 5. There exist $0\neq a_{\gamma},b_{\gamma},d_{\gamma}\in K_{\sigma}$ for γ in $(\Sigma_{\sigma})_s$ such that

(1) $b_{\gamma}T(R_{\gamma})\subset A$;

(2) $a_{\gamma}A\subset R_{\gamma}$, if $\mathrm{char}(K)\neq 3$'

(3) $a_{\gamma}A\subset T(R_{\gamma})$ if $\mathrm{char}(K)=3$ and there exists a short root γ' such that $R_{\gamma}\cap(K-K_{\sigma})\neq\phi$.

(4) $8d_{\gamma}N(R_{\gamma})\subset A$.

Proof: (1) Let δ be a long root forming angle 30° with γ. Then $[X_{\gamma}(t),\ X_{\delta-\gamma}(u)]=X_{\delta}(\pm T(tu^{\sigma}))$. thus $R_{\delta}\supset T(R_{\gamma}R^{\sigma}_{\delta-\gamma})$. Taking $c_{-\delta}\in$

$R_{-\delta}$, $c_{\delta-\gamma}\in R_{\delta-\gamma}\cap K_\sigma$ and $a\in A$, we have $A\supset AR_\delta R_{-\delta}\supset ac_{-\delta}c_{\delta-\gamma}T(R_\gamma)=b_\gamma T(R_\gamma)$ with $b_\gamma=ac_{-\delta}c_{\delta-\gamma}$.

(2) will be proved separately in the following three cases: char(k)$\neq$2; card(A)=2; char(K)=2 and card(A) > 2.

When char(K)$\neq$2, let $t\in R_{3\gamma-2\delta}$ and $u\in R_{\delta-\gamma}$, then $4tuu^\sigma\in R_\gamma$. This implies $R_\gamma\supset 4R_{3\gamma-2\delta}c^2_{\delta-\gamma}\supset 4Ac_{3\gamma-2\delta}c^2_{\delta-\gamma}=a_\gamma A$ where $c_{\delta-\gamma}\in R_{\delta-\gamma}\cap K_\sigma$ and $a_\gamma=4c_{3\gamma-2\delta}c^2{}_{\delta-\gamma}$.

When card(A)=2, then A={0,1} and we have (2) with $a_\gamma=c_\gamma$.

When card(A)>2 and char(K)=2, we pick b$\neq$0,1 in A. For any a in $R_{-\delta}$, d in R_δ and u in R_γ we have

$$H\ni Y_1(a,d)=[[X_\gamma(u),X_{-\delta}(a)],X_\delta(d)]$$
$$=X_\gamma(\pm uad)X_{2\gamma-\delta}(\pm a^2duu^\sigma)X_{3\gamma-2\delta}(\pm a^3duu^\sigma u^{\sigma^2}).$$
$$H\ni Y_2(a,d)=Y_1(ab.d)Y_1(a,db^2)^{-1}$$
$$=X_\gamma(\pm adu(b+b^2))X_{3\gamma-2\delta}(\pm da^3(b^2+b^3)uu^\sigma u^{\sigma^2}).$$
$$H\ni Y_2(ab^3,d)Y_2(ab^2,db^3)Y_2(ab,db^3)Y_2(a,db^6)$$
$$=X_\gamma(adub^4(1+b^4)).$$

Thus $R_\gamma\supset R_\delta R_{-\delta}R_\gamma b^4(1+b^4)\supset c_\delta c_{-\delta}Ac_\gamma b^4(1+b^4)=Aa_\gamma$ where $c_\delta\in R_\delta$, $c_{-\delta}\in R_{-\delta}$, $c_\gamma\in R_\gamma\cap K_\sigma$ and $a_\gamma=c_\delta c_{-\delta}c_\gamma b^4(1+b^4)\neq 0$.

(3) Since $R_\gamma\supset 4R_{3\gamma-2\delta}uu^\sigma$ with $u\in R_{\gamma-\delta}$, $T(R_\gamma)\supset 4R_{3\gamma-2\delta}T(uu^\sigma)\supset 4T(uu^\sigma)c_{3\gamma-2\delta}A=a_\gamma A$ where $c_{3\gamma-2\delta}\in R_{3\gamma-2\delta}$ and $a_\gamma=4c_{3\gamma-2\delta}T(uu^\sigma)\neq 0$.

(4) Let δ be a long root forming angle 150° with γ. From the Proof of Proposition 1, we can clearly see that $8R^2_\delta N(R_\gamma)\subset R_{3\gamma+2\delta}$ where $R^2_\delta=\{t^2\,|\,t\in R_\delta\}$. Thus $A\supset 8c_{-3\gamma-2\delta}ac^2N(R_\gamma)=9d_\gamma N(R_\gamma)$ where $c_{-3\gamma-2\delta}\in R_{-3\gamma-2\delta}$, $a\in A$, $c_\delta\in R_\delta$ and $d_\gamma=c_{-3\gamma-2\delta}ac^2_\delta$.

Lemma 6. If for any short root γ, $R_\gamma\subset K_\sigma$, then $C_\gamma C_\gamma C_\gamma R_\delta\subset R_\delta$ where δ is a long root forming angle 150° with γ and $C_\gamma=12R_\gamma R_{-\gamma}$.

Proof: Let $t,t_1,t_2,t_3\in R_\gamma$, $s,s_1,s_2,s_3\in R_{-\gamma}$, $u\in R_\delta$. then

$$H\ni Z_1(t)=X_{\gamma+\delta}(\pm 4tu)X_{3\gamma+\delta}(\pm 4ut^3).$$
$$H\ni Z_2(t_1,t_2)=Z_1(t_1)Z_1(t_2)Z_1(t_1+t_2)^{-1}$$
$$=X_{\delta+3\gamma}(\pm 4u(t_1+t_2)^3\mp 4ut_1^3\mp 4ut_2^3),$$
$$H\ni Z_3=Z_2(t_1+t_3,t_2)Z_2(-t_1,-t_2)Z_2(-t_3,-t_2)=X_{\delta+3\gamma}(\pm 24ut_1t_2t_3),$$
$$H\ni Z_4(s,u)=[Z_3,X_{-\gamma}(s)],$$
$$H\ni Z_5(s,u)=Z_4(s,u)Z_4(-s,u)^{-1},$$

$H \ni Z_6(s)=Z_5(s,u)Z_5(s,-u)^{-1}$,

$H \ni Z_7(s_1,s_2)=Z_6(-s_1)Z_6(-s_2)Z_6(s_1+s_2)$,

$H \ni Z_7(s_1+s_3,s_2)Z_7(-s_1,-s_2)Z_7(-s_3,-s_2)=X_\delta(\pm 24^2 ut_1t_2t_3s_1s_2s_3)$

Thus $24^2ut_1t_2t_3s_1s_2s_3 \in R_\delta$. This implies $C_\gamma C_\gamma C_\gamma R_\delta \subset R_\delta$

Lemma 7. If char(K)=3 and for any short root γ, $R_\gamma \subset K_\sigma$, then if $\gamma_1,\gamma_2,\gamma_1+\gamma_2 \in (\Sigma_\sigma)_s$, $R_{\gamma_1}R_{\gamma_2} \subset R_{\gamma_1+\gamma_2}$.

Proof: Since $[X_{\gamma_1}(t), X_{\gamma_2}(u)]=X_{\gamma_1+\gamma_2}(\pm 2tu)$, $R_{\gamma_1}R_{\gamma_2} \subset R_{\gamma_1+\gamma_2}$.

Theorem 1. Assume char(K)=3. If for any short root γ, $R_\gamma \subset K_\sigma$, there exist subring A, B of K_σ and non-zero $a_\varepsilon, b_\varepsilon$ with $\varepsilon \in \Sigma_\sigma$ such that

(1) $a_\delta B \subset R_\delta \subset b_\delta B$, $R_\delta A^3 \subset R_\delta$, $AR_\delta R_{-\delta} \subset A$, $\delta \in (\Sigma_\sigma)_\ell$;

(2) $a_\gamma A \subset R_\gamma \subset b_\gamma A$, $R_\gamma B \subset R_\gamma$, $B(R_\gamma R_{-\gamma})^3 \subset B$, $\gamma \in (\Sigma_\sigma)_s$;

(3) $A^3 \subset B \subset A$.

Proof: By Lemma 5(4) and Lemma 7, from [1] we can get the results.

In following discussion if char(K)=3, we always assume $R_\gamma \cap (K-K_\sigma) \neq \phi$ for any short root γ.

Lemma 8. There exist subring A and non-zero a'_γ, b'_γ with $\gamma \in (\Sigma_\sigma)_s$ such that $a'_\gamma A \subset T(R_\gamma) \subset b'_\gamma A$, $AT(R_\gamma) \subset T(R_\gamma)$, $R_\delta A \subset R_\delta$, $AR_\delta R_{-\delta} \subset A$ for δ in $(\Sigma_\sigma)_\ell$.

Proof: By Lemma 5 there exist a'_γ, $b'_\gamma \in K_\sigma$ such that for any short root γ in $(\Sigma_\sigma)_s$, $a'_\gamma A \subset T(R_\gamma) \subset b'_\gamma A$. By Lemma 0 there exists $a \in A$ such that $aAT(R_\gamma) \subset T(R_\gamma)$. Replacing a'_γ, b'_γ, A by $a'a^{-1}$, $b'a^{-1}$, aA we get the results at once.

Theorem 3. There exist subring A of K_σ and non-zero $a_\varepsilon, b_\varepsilon \in K_\sigma$ for ε in Σ_σ such that

(1) $a_\delta A \subset R_\delta \subset b_\delta A$, $AR_\delta \subset R_\delta$, $AR_\delta R_{-\delta} \subset A$, $\delta \in (\Sigma_\sigma)_\ell$;

(2) $a_\gamma A \subset T(R_\gamma) \subset b_\gamma A$, $AT(R_\gamma) \subset T(R_\gamma)$, $A^2T(R_\gamma)T(R_{-\gamma}) \subset A$, $\gamma \in (\Sigma_\sigma)_s$.

Proof: When char(K)≠3, let $\gamma \in (\Sigma_\sigma)_s$ and δ be a long root forming angle 30° with γ. Given $C_\alpha \in R_\alpha \cap K_\sigma$ for any α in Σ_σ, then $9C_\gamma C_{\delta-\gamma} \subset R_\delta$, $9C_{-\gamma}C_{\delta-2\gamma} \subset R_{\delta-3\gamma}$ and $81C_\gamma C_{\delta-\gamma}C_{\delta-2\gamma}C_{-\gamma} \subset R_{2\delta-3\gamma}$. However, $AR_{2\delta-3\gamma}R_{3\gamma-2\delta} \subset A$. This implies $R_{2\delta-3\gamma} \subset Ad_1$ for some $d_1 \in K_\sigma$ with $d_1 \neq 0$. Thus $9C_{\delta-2\gamma}C_{\delta-\gamma}= =b_1d_1$, $81C_{\delta-\gamma}C_{\delta-2\gamma}C_\gamma C_{-\gamma}=b_2d_1$ with $b_1,b_2 \in A$. Hence $C_\gamma C_\gamma = 9^{-1}b_1^{-1}b_2$. By Lemma 7, $3C_{-\gamma}A \subset T(R_{-\gamma})$, $3C_\gamma A \subset T(R_\gamma)$. Thus $9C_\gamma C_{-\gamma}A \subset T(R_\gamma)T(R_{-\gamma}) \subset b_2A \subset A$. Hence $A \supset u_\gamma T(R_\gamma)T(R_{-\gamma})$ with $u_\gamma=b_1b_3 \in A$. Set $u=\Pi u_\gamma$. Then $uT(R_\gamma)T(R_{-\gamma}) \subset A$ for any short root γ. Replacing A by Au, we have $A^2T(R_\gamma)T(R_{-\gamma}) \subset A$, $T(R_\gamma)A \subset T(R_\gamma)$ for any short root γ. We still have $R_\delta A \subset R_\delta$, $AR_\delta R_{-\delta} \subset A$ for any long root δ.

When char(K)=3, firstly we prove that for any short root γ there exist $u_\gamma \in R_\gamma \cap (K-K_\sigma)$, $u_{-\gamma} \in R_{-\gamma} \cap (K-K_\sigma)$ such that $T(u_\gamma)R(u_{-\gamma})=ab^{-1}$ with $a, b \in A$.

Let δ be a long root forming angle 150° with γ . Pick $C_\gamma \in R_\gamma \cap K_\sigma$ $u_{-\delta-\gamma} \in R_{-\delta-\gamma} \cap (K-K_\sigma)$, $C_{\delta+\gamma} \in R_{\delta+\gamma} \cap K_\sigma$, $u_{\delta+2\gamma} \in R_{\delta+2\gamma} \cap (k-K_\sigma)$ such that $T(u_{-\delta-\gamma}) \neq 0$, $T(u_{\delta+2\gamma}) \neq 0$, $C_\gamma \neq 0$. $C_{\delta+\gamma} \neq 0$. Since

$$[X_{\gamma_1}(t), X_{\gamma_2}(u)] = X_{\gamma_1+\gamma_2}(\pm T(ut^\sigma),$$

$$[X_{\delta_1}(t), X_{\delta_2}(u)] = X_{\delta_1+\delta_2}(\pm tu)$$

where $\gamma_1, \gamma_2 \in (\Sigma_\sigma)_s$, $\delta_1, \delta_2 \in (\Sigma_\sigma)_\ell$ and $\gamma_1+\gamma_2$, $\delta_1+\delta_2 \in (\Sigma_\sigma)_\ell$, $C_{\delta+\gamma}T(u_{\delta+2\gamma}) \subset R_{2\delta+3\gamma}$, $C_\gamma T(u_{-\delta-\gamma}) \subset R_{-\delta}$, $C_{\delta+\gamma}C_\gamma T(u_{\gamma+2\delta})T(u_{-\delta-\gamma}) \subset R_{\delta+3\gamma}$. However, $R_{3\gamma+\delta} \subset d_1 A$ for some non-zero $d_1 \in K_\sigma$. Thus $C_{\delta+\gamma}C_\gamma T(u_{\delta+2\gamma})T(u_{-\delta-\gamma})=b_1 d_1$, $C_\gamma T(u_{\delta+2\gamma})=b_2 d_1$ with $0=b_1, b_3 \in A$. Hence $C_{\delta+\gamma}T(u_{-\delta-\gamma})=b_1 b_2^{-1}$. Similarly, we have $C_{-\delta-\gamma}T(u_{\delta+\gamma})=b_3 b_4^{-1}$ with $0 \neq b_3, b_4 \in A$ and $u_{\delta+\gamma} \in R_{\delta+\gamma} \cap (K-K_\sigma)$, $T(u_{\delta+\gamma}) \neq 0$.

Let $t, t' \in R_\gamma$, $u \in R_\delta$. Set $v=4ut$, $Z_1=Z_{\delta+\gamma}(\pm 4tu)X_{\delta+3\gamma}(\pm 4utt^\sigma t^{\sigma^2})$. Then

$$H \ni Z_2(v,t')=[X_\gamma(t'),Z_1]=[X_\gamma(t'),X_{\delta+\gamma}(v)]$$

$$=X_{2\gamma+\delta}(\pm(t'v^\sigma+t'^\sigma v))X_{\delta+3\gamma}(\pm T(vt'^\sigma t'^{\sigma^2}))X_{3\gamma+2\delta}(\pm T(t'v^\sigma v^{\sigma^2})),$$

$$H \ni Z_2(-v,-t')Z_2(v,t')=X_{\delta+2\gamma}(\pm 2(t'v^\sigma+vt'^\sigma)).$$

Therefore $8t't^\sigma u+8t'^\sigma tu \in R_{\delta+2\gamma}$. Take $t' \in R_\gamma \cap K_\sigma$, then $16t'u(t+t^\sigma)$ $=t'u(t+t^\sigma) \in R_{\delta+2\gamma}$. Replacing γ, δ by $\gamma+\delta$, $-3\gamma-2\delta$, we have $u_{-\gamma}=C_{\gamma+\delta}C_{-2\delta-3\gamma}(u_{\delta+\gamma}+u^\sigma_{\gamma+\delta}) \in R_\gamma \cap (K-K_\sigma)$. Similarly, $u_\gamma=C_{-\gamma-\delta}C_{2\delta+3\gamma}(u_{-\gamma-\delta}+u^\sigma_{-\gamma-\delta}) \in R_\gamma \cap (K-K_\sigma)$. Hence $T(u_\gamma)T(u_{-\gamma})=C_{2\delta+3\gamma}C_{-2\delta-3\gamma}b_1 b_3 b_2^{-1} b_4^{-1}$. However, $R_{2\delta+3\gamma}R_{-2\delta-3\gamma}A \subset A$. Thus $C_{2\delta+3\gamma}C_{-2\delta-3\gamma}=b_5 b_6^{-1}$ with b_5, $b_6 \in A$. Therefore $T(u_{-\gamma})T(u_\gamma)=b_1 b_3 b_5 b_2^{-1} b_4^{-1} b_6^{-1}=ab^{-1}$ with $a=b_1 b_3 b_5$, $b=b_2 b_4 b_6 \in A$.

Then we can prove the theorem.

Since $T(u_\gamma)T(u_{-\gamma})A \subset T(R_\gamma)T(R_{-\gamma}) \subset Ad_2$ with $0 \neq d_2 \in K_\sigma$. By Lemma 0 there exists b' in A such that $b'T(R_\gamma)T(R_{-\gamma}) \subset b'd_2 A \subset T(u_\gamma)T(u_{-\gamma})A$. Set $V_\gamma=b'b$. Then $V_\gamma T(R_\gamma)T(R_{-\gamma}) \subset A$. Setting $u=\Pi V_\gamma$, we have $uT(R_\gamma)T(R_{-\gamma}) \subset A$ for any short root γ. Replacing A by uA, we have $T(R_\gamma)A \subset T(R_\gamma)$, $A^2 T(R_\gamma) T(R_{-\gamma}) \subset A$, $\gamma \in (\Sigma_\sigma)_s$. We still have $R_\delta A \subset R_\delta$, $AR_\delta R_{-\delta} \subset A$ for any long root δ.

Corollary. If $char(K) \neq 3$ and $R_\gamma \subset K_\sigma$ for every short root γ, then there exists subring A of K_σ such tha

(1) $AR_\gamma \subset R_\gamma$, $A^2 R_\gamma R_{-\gamma} \subset A$, $\gamma \in (\Sigma_\sigma)_s$;

(2) $AR_\delta \subset R_\delta$, $AR_\delta R_{-\delta} \subset A$, $\delta \in (\Sigma_\sigma)_\ell$.

Proof: Replacing A by 3A, we know this is a direct consequence of Theorem 2.

Theorem 4. Assume $char(K) \neq 2,3$ and $R_\gamma \subset K_\sigma$ for every short root γ.

Then there exists a subring A of K_σ such that $AR_\alpha \subset R_\alpha$, $AR_\alpha R_{-\alpha} \subset A$ for any root α.

Proof: By Lemma 6, the above corollary and 1 we can get the results.

Combining the Theorem 1,2,3,4, we get main results for type $^3D_4$.

2, G_σ is of type $^2A_{2\ell}$ $^2A_{2\ell+1}$, $^2D_\ell$, $^2E_6$.

Let α,β lie in a subsystem Σ' of type A_2 or B_2.

Theorem 1. If Σ' is of type A_2, there exist subring A of K and non-zero a_γ, $b_\gamma \in K$ with γ in Σ' such that $AR_\gamma \subset R_\gamma$, $a_\gamma A \subset R_\gamma \subset b_\gamma A$, $AR_\gamma R_{-\gamma} \subset A$.

Proof: In this case we still have commutator formula $[X_{\gamma_1}(t), X_{\gamma_2}(u)] = X_{\gamma_1+\gamma_2}(tu)$, if $\gamma\ \gamma_2, \gamma_1+\gamma_2 \in \Sigma'$. From [1] we get the Theorem 1.

Remark: when $\alpha,\beta \in (\Sigma_\sigma)_\ell$, $A \subset K_\sigma$, a_γ, $b_\gamma \in K_\sigma$.

Corollary. If $\alpha,\beta \in (\Sigma_\sigma)_s$, there exists subring A of K_σ such that $AN(R_\gamma) \subset N(R_\gamma)$, $AN(R_\gamma)N(R_{-\gamma}) \subset A$.

Proof: Replacing A by the subring generated by N(A), we can get the corollary.

In following discussion, we assume that Σ' is of type B_2. Furthermore, we may suppose $\alpha \in (\Sigma_\sigma)_\ell$, $\beta \in (\Sigma_\sigma)_s$ and α forms angle 135° with β.

Definition. A pair (γ,δ) of roots is called admissible, if $\gamma \in (\Sigma_\sigma)_s$, $\delta \in (\Sigma_\sigma)_\ell$ and $\delta-\gamma \in (\Sigma_\sigma)_s$.

Set $R_{\gamma\ \delta} = \{(t,u) \in K \oplus K \mid X_\gamma(t)X_\delta(u) \in H\}$. Let $R'_{\gamma,\delta}$ (resp., $R''_{\gamma,\delta}$) be the projection of $R_{\gamma,\delta}$ on the first factor (resp., second). Clearly, $R'_{\gamma,\delta} \supset R_\gamma$, $R''_{\gamma,\delta} \supset R_\delta$.

Lemma 1. Let (γ,δ) be a admissible pair of roots, $a \in R_{2\gamma-\delta}$, $b \in R_{\delta-2\gamma}$, $(c,d) \in R_{\gamma,\delta}$, $t, t_1, t_2 \in R_{\gamma-\delta}$. Then

(1) $(abc,\ ab^2cc^\sigma) \in R_{\gamma,\delta}$;

(2): $d(t_1t_2^\sigma + t_1^\sigma t_2) \in R_{2\gamma-\delta}$.

Proof: Set $\varepsilon = \delta-2\gamma$,

(1) $H \ni [X_{-\varepsilon}(a), [X_\varepsilon(b), X_\gamma(c)X_\delta(d)]] = X_\gamma(\pm abc)X_\delta(\pm ab^2cc^\sigma)$.

Thus $R_{\gamma,\delta} \ni (abc, ab^2cc^\sigma)$.

(2) $H \ni Y(t) = [X_{\gamma-\delta}(t), X_\gamma(c)X_\delta(d)] = X_\gamma(\pm t^\sigma d)X_{2\gamma-\delta}(\pm(tc+t^\sigma c^\sigma) \pm tt^\sigma d)$,

$Y(t_1+t_2)Y(-t_1)Y(-t_2) + X_{2\gamma-\delta}(\pm(t_1t_2^\sigma + t_2t_1^\sigma)d)$. So $R_{2\gamma-\delta} \ni d(t_1^\sigma t_2 + t_1t_2^\sigma)$

Lemma 2. In the notations of Lemma 1, then

(1) $R_\gamma \supset 2R_\varepsilon R_{-\varepsilon} R'_{\gamma,\delta}$, $R_\delta \supset 8R_\varepsilon R_{-\varepsilon} R_\varepsilon N(R_\gamma)$;

(2) $R_\delta \supset 2R_\delta N(R_{\gamma-\delta})N(R_{\delta-\gamma})$.

Proof: (1) Let a,b,c,d be as in Lemma 1 and $c' \in R_\gamma$, $b' \in R_\varepsilon$.

By Lemma 1 $R_{\gamma,\delta} \ni Z(c)=(abc, ab^2cc^{\sigma}) \in K \oplus K$. Since $X_\delta(K)$ and $X_\gamma(K)$ commutate, $R_{\gamma,\delta}$ is a subgroup of $K \oplus K$. Thus $R_{\gamma,\delta} \ni Z(c)-Z(-c)=(2abc,0)$, i.e. $R_\gamma \supset 2R_{-\varepsilon}R_\varepsilon R'_{\gamma,\delta}$. Similarly, $R_\delta \ni 2ab^2(cc'^{\sigma}+c^{\sigma}c')$. Set $c=c'$, $d=0$. Then $R_\delta \supset 4ab^2N(c')$. Therefore

$R_\delta \ni 4a(b+b')^2N(c')-4aN(c')b^2-4aN(c')b'^2=8abb'N(c')$, i.e.

$R_\delta \ni 8R_\varepsilon R_\varepsilon R_{-\varepsilon}N(R_\gamma)$.

(2) By Lemma 1 (2) $R_{2\gamma-\delta} \supset 2N(R_{\gamma-\delta})R''_{\gamma\ \delta}$. Replacing (γ,δ) by $(\gamma,2\gamma-\delta)$ we have $R_\delta \supset 2N(R_{\delta-\gamma})R''_{\gamma\ ,2\gamma-\delta} \supset 2N(R_{\delta-\gamma})R_{2\delta-\gamma}$. Thus $R_\delta \supset 4N(R_{\delta-\gamma})N(R_{\gamma-\delta})R_\delta$.

Theorem 2. There exist subring A and non-zero a_α, $b_\alpha \in K_\sigma$ with $\alpha \in \Sigma'_\sigma$ such that

(1) $a_{\delta'}A \subset R_{\delta'} \subset b_{\delta'}A$, $AR_{\delta'} \subset R_{\delta'}$, $4A(R_{\delta'}R_{-\delta'})^2 \subset A$ for long root δ' in Σ'.

(2) $a_{\gamma'}A \subset N(R_{\gamma'}) \subset b_{\gamma'}A$, $AN(R_{\gamma'}) \subset N(R_{\gamma'})$, $4N(R_{\gamma'})N(R_{-\gamma'})A \subset A$ for any short root γ' in Σ'_σ .

Proof: For every root $\phi \in \Sigma'$ we pick a non-zero $c_\phi \in R_\phi$. By Lemma 2, $R_\delta \supset c_{\varepsilon,\delta}R_\varepsilon$, $R_\delta \supset c_{\delta-\gamma,\delta}N(R_{\delta-\gamma})$ where $c_{\varepsilon,\delta}$, $c_{\delta-\gamma,\delta} \in K_\sigma$ and $c_{\varepsilon,\delta}$, $c_{\delta-\gamma,\delta} \neq 0$. Changing δ and ε, we have for any $\delta \in \Sigma'_\ell$ there exists non-zero $c_{\delta',\delta}$ such that $R_\delta \supset c_{\delta',\delta}R_{\delta'}$,

Let $u \in R_\gamma$, $t \in R_{\delta-2\gamma}$. Then $H \ni [X_\gamma(t), X_{\delta-2\gamma}(u)]$ $=X_{\delta-\gamma}(\pm tu)X_\delta(\pm tuu^\sigma)$. Thus $X_{\delta-\gamma}(\pm 2ut)X_\delta(4tuu^\sigma)$, $X_{\delta-\gamma}(\pm 2tu)X_\delta(\pm 2uu^\sigma t)$ $\in H$. This implies $X_\delta(\pm 2tuu\ ) \in H$ and $X_{\delta-\gamma}(\pm 2ut) \in H$. Hence $R_{\delta-\gamma} \supset 2R_{\delta-2\gamma}R_\gamma$. Forthermore $N(R_{\delta-\gamma}) \supset c_{\delta-\gamma,\gamma}N(R_\gamma)$ with $0 \neq c_{\delta-\gamma,\gamma} \in K_\sigma$. Changing γ and δ we know that for any $\gamma,\gamma' \in \Sigma'_s$ and $\delta' \in \Sigma'_\ell$. there exist $0 \neq c_{\gamma,\gamma'} \in K_\sigma$, $0 \neq c_{\delta',\gamma'} \in K_\sigma$, such that $N(R_\gamma) \supset c_{\gamma,\gamma'}N(R_{\gamma'})$, $R_{\delta'} \supset N(R_{\gamma'})c_{\delta',\gamma'}$.

Let A be a subring generated by $2R_\alpha R_{-\alpha}$. Then $A \supset b_\alpha^{-1} R_\alpha$ with $b^{-1}=2c_{-\alpha}$. However, $R_\alpha \supset c_{\alpha,\delta'}R_{\delta'}$ for any long root δ' in Σ'. Thus $R_{\delta'} \subset b_{\delta'}A$. By Lemma 2(1) $R_{\delta'} \supset Aa_{\delta'}$. Therefore $a_{\delta'}A \subset R_{\delta'} \subset Ab_{\delta'}$. By Lemma 0 there exists $a \in A$ such that $aAR_{\delta'} \subset R_{\delta'}$ for every $\delta' \in \Sigma'_\ell$. By Lemma 2 (1) $R_\beta \supset AR_\beta$, Replacing A by the subring generated by $(aA)^2=\{a^2b^2 \mid b \in A\}$ we have $N(R_\beta) \supset AN(R_\beta)$. Still we have $AR_{\delta'} \subset R_{\delta'}$, $a_{\delta'}A \subset R_{\delta'} \subset b_{\delta'}A$ for any long root δ' in Σ'. Hence $a_{\gamma'}A \subset N(R_{\gamma'}) \subset b_{\gamma'}A$ where $\gamma' \in \Sigma'_s$, $0 \neq a_{\gamma'}, b_{\gamma'} \in K_\sigma$. By Lemma 0 we have

(1) $a_{\delta'}A \subset R_{\delta'} \subset b_{\delta'}A$, $R_{\delta'}A \subset R_{\delta'}$ with $\delta' \in \Sigma'_\ell$;

(2) $a_{\gamma'}A \subset N(R_{\gamma'}) \subset b_{\gamma'}A$, $N(R_{\gamma'})A \subset N(R_{\gamma'})$ for $\gamma' \in \Sigma'_s$, where A is a subring of K_σ.

Set $A_1 = \prod_{\gamma \in \Sigma'_s} N(R_\gamma)$, $(R_\delta R_{-\delta})^2=\{a^2b^2 \mid a \in R_\delta,\ b \in R_{-\delta}\}$. By Lemma 2(1) $4A_1(R_\delta R_{-\delta})^2 \subset A_1$ for any long root δ in Σ'. By (2) there exists c such that $cA_1 \subset A$. Replacing A by the subring generated by cA_1A we have

$4A(R_\delta R_{-\delta})^2 \subset A$ for any long root δ in Σ'. Changing appropriately $a_{\delta'}$, $b_{\delta'}$, $a_{\gamma'}$, $b_{\gamma'}$, still we have (1) and (2).

Set $A_2 = \prod_{\alpha\in\Sigma_\ell} R_\alpha$. By Lemma 2 (2) $4A_2 N(R_\gamma)N(R_{-\gamma}) \subset A_2$ for any short root γ in Σ'. Clearly, there exists c_1 such that $c_1 A_2 \subset A$. Replacing A by the subring generated by $c_1 A_2 A$ we can pick $a_\delta, b_\delta, a_\gamma, b_\gamma$ such that

(1) $a_\delta A \subset R_\delta \subset b_\delta A$, $R_\delta A \subset R_\delta$, $4A(R_\delta R_{-\delta})^2 \subset A$ for $\delta \in \Sigma'_\ell$;

(2) $a_\gamma A \subset N(R_\gamma) \subset b_\gamma A$, $N(R_\gamma)A \subset N(R_\gamma)$, $4AN(R_\gamma)N(R_{-\gamma}) \subset A$ for $\gamma \in \Sigma'_s$.

Lemma 3. Let $\phi, \psi \in \Sigma_\sigma$ and have same length. Then $R_\phi \supset c_{\phi,\psi} R_\psi$ for some non-zero $c_{\phi,\psi}$.

Proof: From Theorem 1 and Theorem 2 we know that when ϕ, ψ lie in a subsystem of type A_2 or B_2, Lemma 3 holds.

Generally. there exist $\gamma_1, \ldots, \gamma_k \in \Sigma_\sigma$ such that $\gamma_1 = \phi$, $\gamma_k = \psi$ and γ_i, γ_{i+1} generate a subsystem of type A_2 or B_2. Therefore the Proof is completed.

In the following we will complete the proof.

Now we pick $\alpha \in (\Sigma_\sigma)_\ell$, $\beta \in (\Sigma_\sigma)_s$ and α, β generate a subsystem of typw B_2. By Theorem 2 and Lemma 3. there exist subring A of K_σ and non-zero $a_\delta, b_\delta, a_\gamma, b_\gamma \in K_\sigma$ for $\delta \in (\Sigma_\sigma)_\ell$ and $\gamma \in (\Sigma_\sigma)_s$ such that $a_\delta A \subset R_\delta \subset b_\delta A$, $a_\gamma A \subset N(R_\gamma) \subset b_\gamma A$. By Lemma 0 there exists $a \in A$ such that $aAR_\delta \subset R_\delta$, $aAN(R_\gamma) \subset N(R_\gamma)$. Replacing A by aA and changing $a_\delta, b_\delta, a_\gamma, b_\gamma$, we have

(1) $a_\delta A \subset R_\delta \subset b_\delta A$, $AR_\delta \subset R_\delta$ for $\delta \in (\Sigma_\sigma)_\ell$;

(2) $a_\gamma A \subset N(R_\gamma) \subset b_\gamma A$, $AN(R_\gamma) \subset N(R_\gamma)$ for $\gamma \in (\Sigma_\sigma)_s$.

For any short root γ it can be thought as a short root in a subsystem A_2 or B_2. Thus there exists a subring A_γ of K_σ such that $N(R_\gamma)A_\gamma \subset N(R_\gamma)$, $4N(R_\gamma)N(R_{-\gamma})A_\gamma \subset A_\gamma$. By (2) there exist u_γ, v_γ such that $u_\gamma A \subset A_\gamma \subset v_\gamma A$ (when $\gamma=\beta$, $A_\gamma = A$). Set $A_s = \prod A_\gamma$. Then $4A_s N(R_\gamma)N(R_{-\gamma}) \subset A_s$ for any γ in $(\Sigma_\sigma)_s$. Clearly, there exists c such that $cA_s \subset A$. Replacing A by the subring generated by $cA_s A$, we have $4AN(R_\gamma)N(R_{-\gamma}) \subset A$ for any short root γ in Σ_σ. Still we have (1), (2).

Similarly we perform the procedure to long roots. Then there exists a subring A of K_σ such that $4A(R_\delta R_{-\delta})^2 \subset A$. Still we have (1), (2) and $4AN(R_\gamma)N(R_{-\gamma}) \subset A$. Thus we complete the proof.

Problems remain to be solved.

(1) If char(K)=2 and G_σ is of type ${}^3D_4$ there is no results for general case.

(2) If char(K)=2, there is no results for G_σ being other cases.

(3) If char(K)=3, G_σ is of type ${}^3D_4$ and $R_\gamma \cap K_\sigma = \{0\}$ for any short root γ, there is no results.

The author hopes to express his sincere thanks to Prof.B.Weisfeiler who suggested the problem when he was visiting Beijing in 1983. The author is also indebted to Prof. Zhexian Wan for valuable discussion.

References

1 L.N. Vaserstein, B. Weisfeiler, On full subgroups of Chevalley groups. No. 83005, Department of Mathematics research report. The Pennsylvania State University, 1983.

2 R. Steinberg, Lectures on Chevalley groups. Yale University. 1967.

3 Nathan, Jacobson. Basic Algebra 1. W.H. Freeman and Company, San Francisco 1974.

A CHARACTERIZATION OF THE 2-LOCAL SUBGROUPS OF RUDVALIS GROUP AND HIGMAN-SIMS GROUP

Huang Jianhua

Department of Mathematics

University of Zhengzhow, PR China

In /3/ Goldschmidt considered groups G, such that:

i) $G = \langle M_1, M_2 \rangle$, where M_1 and M_2 are finite subgroups of G,

ii) no non-trivial normal subgroup of G is contained in $M_1 \cap M_2$,

iii) $|M_i/M_1 \cap M_2| = 3$ for $i = 1, 2$.

He gave the exact structure of all possible pairs of subgroups M_1 and M_2. In his proof he used a graph theoretical approach:

Any group G with properties i) and ii) operates as an edge-transitive group of automorphisms on a graph Γ whose vertex set is $\{M_1x \mid x \in G\} \cup \{M_2y \mid y \in G\}$. Two vertices are adjacent if and only if they have non-empty intersection. G operates on Γ by right multiplication.

The method will be used in this paper. At first, we make the following hypothesis.

Hypothesis A. Let G be a group such that:

i) $G = \langle M_1, M_2 \rangle$, where M_1 and M_2 are finite subgroups of G,

ii) no non-trivial normal subgroup of G is contained in $M_1 \cap M_2$,

iii) $M_1/O_2(M_1) \cong L_3(2)$, $M_2/O_2(M_2) \cong S_5$ and $M_1 \cap M_2/O_2(M_i) \cong S_4$,

iv) $C_{M_i}(O_2(M_i)) \leqslant O_2(M_i)$, $i = 1,2$.

For the statement of the theorem, we give a definition.

Definition. Let (G, M_1, M_2) be a triple which satisfies Hypothesis A. Let $Q_i = O_2(M_i)$, $Z_i = \Omega_1(Z(Q_i))$ and $V_i = [Q_i, M_i]$.

(I) The pair (M_1, M_2) is called parabolic of type HS, if the following hold:

a) $V_1 \cong C_4 \times C_4 \times C_4$, and Z_1 and V_1/Z_1 are isomorphic natural modules for M_1/Q_1,

b) $V_2 \cong C_4 \text{ Y } Ex(2^5)$, and $V_2/Z(V_2)$ is a natural module for M_2/Q_2.

This paper was the author's Ph. D. dissertation under the advisorship of B. Stellmacher at the University of Bielefeld, Spring, 1983.

c) We have either $V_i = Q_i$ for $i = 1,2$ or $Q_2 \cong E\mathbb{X}(2^7)$ and $Z_1 = \Phi(Q_1) = Z(Q_1)$.

(II) The pair (M_1, M_2) is called parabolic of type Ru if the following hold:

a) Q_1/Z_1 is an irreducible module of order 2^8 for M_1/Q_1 and Z is a natural module for M_1/Q_1,

b) In M_2 there exists a normal series $Z_2 \leq B \leq A \leq Q_2 \leq M_2$ such that

i) Q_2/A and B/Z_2 are natural modules for M_2/Q_2.

ii) A/B is a trivial module of order 4 for $O^2(M_2/Q_2)$ and Z_2 is a trivial module of order 2 for $O^2(M_2/Q_2)$.

iii) Q_2/B and B are elementary abelian, $Q_2' = B$ and $[Q_2, B] = Z_2$.

Here HS denotes Higman-Sims group and Ru denotes Rudvalis group.

Now we give an explanation of the terms " type HS " or " type Ru ". . Assume that $G = Ru$, HS or Aut(HS) and $S \in Syl_2(G)$. Then there exist two maximal 2-local subgroups M_1 and M_2 in G, which contain S, and (M_1, M_2) is parabolic of type HS or Ru.

Then we can state:

Theorem. Assume Hypothesis A. Then (M_1, M_2) is parabolic of type HS or Ru.

As we have mentioned above, we investigate the " coset graph ". In our investigation, a certain parameter b, which was discovered by B. Stellmacher, plays a central role. (The definition of b is in chapter 2). The main part of the proof is to determine b, which is 2 or 3 in this paper. With this information, we may without difficulty show that (M_1, M_2) is of type HS, if $B = 2$; and (M_1, M_2) is of type Ru, if $b = 3$.

1. Some preliminary results about GF(2)-Modules

(1.0) Hypothesis. Let G be a finite group, and V a faithful GF(2)-module of finite dimension for G.

In the following we list some well known properties of modules for $L_3(2)$ and S_5 without proof.

Definition. An involution $t \in G$ induces a transvection in V, if $|V/C_V(t)| = 2$. A subgroup U of G induces transvections in V, if each element from U induces a transvection in V.

V is a natural module for G, if $G \cong L_3(2)$ and $|V| = 2^3$.

Let $G = A_5$, $S \in Syl_2(G)$, then V is called a natural module for G if $|V| = 2^4$ and $[V, S, S] = 1$; and V is called an orthogonal module for G, if $|V| = 2^4$ and $[V, S, S] = 1$.

Let $G \cong S_5$. V is called a natural module (orthogonal module) for G, if V is a natural module (orthogonal module) for G'.

(1.1) Let $G = L_3(2)$ and V be irreducible. Then V is a natural module or $|V| = 2^8$.

(1.2) Let $G \cong A_5$ and V be irreducible. Then V is a natural module or an orthogonal module for G.

(1.3) Let $G = L_3(2)$ and V be a natural module for G. Then:

(a) Each involution from G induces a transvection in V.

(b) All elements of $V^{\#}$ are conjugate under G.

(c) Let $v \in V$ and H be a hyperplane of V. Then $C_G(V) = N_G(H) \cong S_4$.

(d) $|C_V(S)| = 2$ for $S \in Syl_2(G)$.

(1.4) Let $G \cong S_5$, $S \in Syl_2(G')$ and V be an orthogonal module for G.

(a) Each element of order 3 in G does not operate fix-point-freely on V.

(b) $|[V, S]| = 2^3$, $|[V, S, S]| = 2$ and $[V, S, S] = C_V(S)$.

(c) The involutions of $G \setminus G'$ induce transvections in V.

(d) For each involution $t \in G'$, we have $|V/C_V(t)| = 4$.

(1.5) Let $G \cong S_5$ and V be a natural module for G. Then:

(a) The elements of order 3 in G operate fix-point-freely on V.

(b) For each involution $t \in G$, we have $|V/C_V(t)| = 4$.

(c) All elements of $V^{\#}$ are conjugate under G.

(1.6) Let $G \cong L_3(2)$. Then G may be generated by three involutions.

(1.7) Let $G \cong L_3(2)$ or S_5. If there exists an element in G, which induces a transvection in V, then $V/C_V(G)$ is a natural or orthogonal module for G.

(1.8) Let $G \cong A_5$ and $S \in Syl_2(G)$. If $|V/C_V(S)| = 4$, then $V/C_V(G)$ is a natural module for G.

(1.9) Let $G \cong L_3(2)$, $S \in Syl_2(G)$ and M be a subgroup of G such that:

(a) $V = \langle C_V(S)^g \mid g \in G \rangle$.

(b) $C_V(G) = 1$ and $|V| \leq 2^6$.

(c) $M \cong S_4$ and $|C_V(M)| \geq 4$.

Then V is a direct product of two natural modules for G.

(1.10) Let $G = L_3(2)$ and M_1 and M_2 be two subgroups of G, which are isomorphic to S_4. For $v \in V$ and $E = \langle v^m \mid m \in M_2 \rangle$, the following hold :

(a) $v \in C_V(M_1)$,

(b) $|E| = 4$. Then $\langle v^g \mid g \in G \rangle$ is a natural module for G.

(1.11) Let X be a finite group with the following properties:

(a) $X/O_2(X) \cong A_5$ or $L_3(2)$,

(b) $O_2(X)/Z(X)$ is a natural module for $X/O_2(X)$,

(c) $Z(X)$ is elementary abelian.

Then $O_2(X)$ is elementary abelian.

2. The coset graph

(2.0) Hypothesis. Let G be a group and M_1 and M_2 be finite subgroups of G such that:

(a) $G = \langle M_1, M_2 \rangle$,

(b) no non-trivial normal subgroup of G is contained in $M_1 \cap M_2$.

Let $\Gamma = \Gamma(G, M_1, M_2)$ be the set of right cosets of G with respect to M_1 and M_2. We say two cosets are adjacent, if they are different and have non-empty intersection. Then Γ is a graph, which is called the coset graph of G with respect to M_1 and M_2, and G operates on Γ by right multiplication.

For $\delta \in \Gamma$, we define $\Delta(\delta) = \{\lambda \mid \lambda \sim \delta\}$. An arc of length n is an ordered n-tuple $(\alpha_0, \alpha_1, \ldots, \alpha_n)$ with $\alpha_i \in \Delta(\alpha_{i+1})$, $0 \leq i \leq n-1$, and $\alpha_i \neq \alpha_{i+1}$ for $0 \leq i \leq n-1$. Γ is connected if every pair of points is joined by an arc.

Two vertices δ, λ are conjugate under G, if there exists $g \in G$ with $\delta^g = \lambda$. The notation $d(\delta, \lambda)$ denote the distance of δ and λ.

In the following we list some elementary properties of Γ.

(2.1)(a) Γ is bipartite and connected.

(b) $G \leq \mathrm{Aut}(\Gamma)$.

(c) G operates edge- but not vertex-transitive on Γ.

(d) For $\delta \in \Gamma$, G_δ is transitive on $\Delta(\delta)$.

(e) For $\delta \in \Gamma$, G_δ is conjugate to M_1 or M_2 in G.

(f) The stabilizer of an edge $G_\alpha \cap G_\beta$, $\alpha \in \Delta(\beta)$, is conjugate to $M_1 \cap M_2$.

(2.2) Let α and β be two adjacent vertices in Γ. Suppose that there exists a subgroup $U \leq G_\alpha \cap G_\beta$ such that $N_G(U)_\delta$ is transitive on $\Delta(\delta)$ for $\delta = \alpha, \beta$. Then $U = 1$.

Proof. See /3/, (2.2).

From now on we assume that (G, M_1, M_2) fulfils Hypothesis A. We fix the following notations:

Notations. For $\delta \in \Gamma$: $Q_\delta = O_2(G_\delta)$; $A_\delta = O^2(G_\delta)Q_\delta$;

$Z_\delta = \langle \Omega_1(Z(T)) \mid T \in Syl_2(G_\delta) \rangle$; $V_\delta = \langle Z_\lambda \mid \lambda \in \Delta(\delta) \rangle Z_\delta$;

$D = \langle x \in G_\delta \mid x \in G_\lambda$ for all λ with $d(\delta,\lambda) \leq 2 \rangle$;

$b = \min \{ d(\delta,\lambda) \mid Z_\delta \nleq Q_\lambda \}$; $b = \min \{ b_\delta \mid \delta \in \Gamma \}$.

By (2.1) there are two vertices α and β such that $\{G_\alpha, G_\beta\} = \{M_1, M_2\}$ with $b = b_\alpha$ or $b = b_\beta$. We choose notation so that $b = b_\alpha$. Then there exists an α' with $d(\alpha,\alpha') = b$ and $Z_\alpha \nleq Q_{\alpha'}$, and by (2.1) we may choose β so that $d(\beta,\alpha') = b-1$. Let $\gamma = (\alpha, \beta, \ldots, \alpha')$ be an arc of length b from α to α'. We denote the vertices in γ by: $\gamma = (\alpha, \alpha+1, \ldots, \alpha+b)$ or $\gamma = (\alpha'-b, \ldots, \alpha')$, i.e. $\alpha = \alpha'-b$, $\alpha' = \alpha+b$ and $\beta = \alpha+1 = \alpha'-(b-1)$.

(2.3) Let $\delta \in \Gamma$. Then:

(a) $Q_\delta = \bigcap_{\lambda \in \Delta(\delta)} (G_\delta \cap G_\lambda)$.

(b) $G_\delta / Q_\delta \cong S_5$ or $L_3(2)$.

(c) $Z_\delta \leq Q_\delta$, i.e. $b \geq 1$.

(d) $C_{G_\delta}(Z_\delta) = Q_\delta$ or $Z_\delta = \Omega_1(Z(G_\delta))$.

Proof. Directly by (2.1) and the structure of M_1 and M_2.

(2.4) Let δ, λ be adjacent in Γ, and let $T \in Syl_2(G_\delta \cap G_\lambda)$ and d be an element of order 3 from $G_\delta \cap G_\lambda$. Then:

(a) $G_\lambda \cap G_\delta / Q_\delta \cong S_4$.

(b) $Q_\lambda Q_\delta = O_2(G_\lambda \cap G_\delta)$.

(c) $C_{Q_\lambda Q_\delta}(d) \leq Q_\lambda \cap Q_\delta$.

(d) If $Z_\delta \leq Z(G_\delta)$, then $Z(G_\lambda) = 1$.

(e) If $R \leq Z(Q_\delta) \cap C_G(O^2(G_\lambda))$, then $R \leq Z(Q_\lambda)$.

Proof. (a)-(c). By (2.1) and the structure of M_1, M_2 and $M_1 \cap M_2$.

(d) It is clear that $\Omega_1(Z(G_\lambda)) \leq \Omega_1(Z(T)) \leq Z_\delta \leq Z(G_\delta)$. Then (2.2) implies $\Omega_1(Z(G_\lambda)) = 1$.

(e) By (c), $C_{Q_\lambda}(d) \leq Q_\delta$, so $Q_\lambda = (Q_\lambda \cap O^2(G_\lambda))C_{Q_\lambda}(d) \leq (Q_\lambda \cap O^2(G_\lambda))Q_\delta \leq C_G(R)$. Then $R \leq Z(Q_\lambda)$.

(2.5)(a) $Z_\alpha \leq O_2(G_{\alpha'} \cap G_{\alpha'-1})$ and $Z_{\alpha'} \leq O_2(G_\alpha \cap G_\beta)$.

(b) $C_{G_\alpha}(Z_\alpha) = Q_\alpha$.

(c) $C_{G_\beta}(V_\beta) \leq Q_\beta$.

(d) Suppose that $[Z_\alpha, Z_{\alpha'}] = 1$, then $b \equiv 1 (2)$.

Proof. The definition of b and (2.4)(a) yield (a).

Suppose that $C_{G_\alpha}(Z_\alpha) \neq Q_\alpha$ and $T \in Syl_2(G_\alpha \cap G_\beta)$. By (2.3)(d) $Z_\alpha = \Omega_1(Z(G_\alpha))$, in particular $Z_\alpha = \Omega_1(Z(T))$. Then $Z_\alpha \leq Z_\beta$ and so $Z_\beta \not\leq Q_{\alpha'}$. And it follows that $b_\beta \leq b-1$, which contradicts the minimality of b. Thus (b) holds.

(d) is an obvious consequence of (2.1).

(2.6)(a) $[Z_\alpha, Z_{\alpha'}, Z_{\alpha'}] = [Z_{\alpha'}, Z_{\alpha'}, Z_\alpha] = 1$.

(b) If $[Z_\alpha, Z_{\alpha'}] = 1$ and $b > 1$, then V_β and $V_{\alpha'}$ are elementary abelian, and $[V_\beta, V_{\alpha'}, V_{\alpha'}] = 1 = [V_{\alpha'}, V_\beta, V_\beta]$.

Proof. (a) follows directly from (2.5)(a).

(b) is a consequence of (2.5)(d).

3. The case $[Z_\alpha, Z_{\alpha'}] \neq 1$

(3.0) Hypothesis. We assume Hypothesis A and $[Z_\alpha, Z_{\alpha'}] \neq 1$.

Under (3.0) we prove:

Theorem 1. (G_α, G_β) is parabolic of type HS.

We divide the proof in some lemmas.

(3.1) Suppose that $[V_\alpha, Z_{\alpha'}] = [Z_\alpha, Z_{\alpha'}]$, then $Z_\beta = \Omega_1(Z(G_\beta))$.

Proof. By assumption $[V_\alpha, Z_{\alpha'}] \leq Z_\alpha$, so $[\langle Z_{\alpha'}^{G_\alpha}\rangle, V_\alpha] \leq Z_\alpha$. This implies that $Z_\alpha Z_\beta$ is normalized by $\langle Z_{\alpha'}^{G_\alpha}\rangle(G_\alpha \cap G_\beta) = G_\alpha$.

If $Z_\beta = \Omega_1(Z(G_\beta))$. By (2.3)(d) $C_{G_\beta}(Z_\beta) = Q_\beta$ and so $C_{Q_\alpha}(Z_\alpha Z_\beta) = Q_\alpha \cap Q_\beta \trianglelefteq G_\alpha$. Let d be an element of order 3 from $G_\alpha \cap G_\beta$. Then $d \in \langle Q_\beta^{G_\alpha}\rangle$ and $[Q_\alpha, Q_\beta] \leq Q_\alpha \cap Q_\beta$ implies that $[Q_\alpha, d] \leq Q_\alpha \cap Q_\beta$, i.e. $Q_\alpha = C_{Q_\alpha}(d)(Q_\alpha \cap Q_\beta)$. By (2.4) we get $Q_\alpha = Q_\alpha \cap Q_\beta$, contrary to (2.4)(b).

(3.2) α is conjugate to α', i.e. $b \equiv 0(2)$.

Proof. Assume, by way of contradiction, that α is not conjugate to α'.

(1) We have either $Z(G_\alpha) = 1$ and $|Z_\alpha/Z_\alpha \cap Q_{\alpha'}| = 4$ or $Z(G_{\alpha'}) = 1$ and $|Z_{\alpha'}/Z_{\alpha'} \cap Q_\alpha| = 4$.

Set $\{\delta, \delta'\} = \{\alpha, \alpha'\}$. By (2.5)(a) $|Z_\delta/Z_\delta \cap Q_{\delta'}| \leq 4$, i.e. $|Z_\delta/C_{Z_\delta}(Z_{\delta'})| \leq 4$. The notations δ and δ' may be so selected, that $G_\delta/Q_\delta \cong L_3(2)$ and $G_{\delta'}/Q_{\delta'} \cong S_5$.

By (1.4)(d) and (1.5)(b), we get:

(*) $|Z_{\delta'}/Z_{\delta'} \cap Q_\delta| = 4$ and $|Z_\delta/Z_\delta \cap Q_{\delta'}| = 2$ or 4.

Let $\lambda \in \Delta(\delta)$ and d be an element of order 3 from $G_\delta \cap G_\lambda$.

Now we assume that $|Z_\delta/Z_\delta \cap Q_{\delta'}| = 4$. By (1.8) $Z_{\delta'}/\Omega_1(Z(A_{\delta'}))$ is a natural module for $G_{\delta'}/Q_{\delta'}$, and $Z_\lambda/\Omega_1(Z(A_\lambda))$ for G_λ/Q_λ. It follows from (1.5)(a) that $C_{Z_\lambda}(d) \leq \Omega_1(Z(A_\lambda))$, and then $\Omega_1(Z(G_\delta)) \leq C_{Z_\lambda}(d)$ and (2.2) imply that $Z(G_\delta) = 1$.

If $|Z_\delta/Z_\delta \cap Q_{\delta'}| = 2$. Then by (1.7) $Z_\delta/\Omega_1(Z(G_\delta))$ is a natural module for G_δ/Q_δ, and $(*)$ implies that $C_{Z_\delta \cap Z_\lambda}(d) = \Omega_1(Z(G_\delta))$, then as above $\Omega_1(Z(G_\lambda)) = 1$, and so $\Omega_1(Z(G_{\delta'})) = 1$.

By the symmetry of α and α', and (1), we may select the notations, such that $Z(G_\alpha) = 1$ and $|Z_\alpha/Z_\alpha \cap Q_{\alpha'}| = 4$.

(2) $b > 1$.

Suppose that $b = 1$. Let d be an element of order 3 from $G_\alpha \cap G_\beta$. Then it is clear that $|Z_\alpha/Z_\alpha \cap Q_\beta| = |Z_\beta/Z_\beta \cap Q_\alpha| = 4$. In particular, as in (1), it follows that $Z(G_\beta) = 1$. We choose notation so that $G_\alpha/Q_\alpha \cong S_5$. By the assumption $b = 1$, we can easily show that $Q_\alpha = [Z_\beta, d] \times C_{Q_\alpha \cap Q_\beta}(d)$ and $Q_\beta = [Z_\beta, d] \times C_{Q_\alpha \cap Q_\beta}(d)$. Thus $\Phi(Q_\alpha) = \Phi(Q_\beta)$, and (2.2) yields that $\Phi(Q_\alpha) = \Phi(Q_\beta) = 1$, i.e. Q_α and Q_β are elementary abelian. Now we observe that Q_α is a natural module for G_α/Q_α, i.e. $|Q_\alpha| = |Q_\beta| = 2^4$ and $C_{Q_\alpha Q_\beta}(d) = 1$, which contradicts (1.1) and the operation of d on Q_β.

(3) $Z_\lambda \leq Q_{\alpha'-1}$ for all $\lambda \in \Delta(\alpha)$.

Suppose that there exists a $\lambda \in \Delta(\alpha)$ with $Z_\lambda \not\leq Q_{\alpha'-1}$. By the minimality of b, we have $Z_\lambda \leq Q_{\alpha'-1}Q_{\alpha'}$. Set $R = [Z_\alpha, Z_{\alpha'}]$. Then $R \leq Z_\alpha$ and so $[R, Z_\lambda] = 1$, i.e. R is normalized by $\langle Z_\lambda, G_{\alpha'-1} \cap G_{\alpha'} \rangle$.

If $Z_\lambda \leq G_{\alpha'} \cap G_{\alpha'-1}$, then $[Z_\lambda, Z_\alpha] = 1$ and $O_2(G_{\alpha'-1} \cap G_{\alpha'}) = Q_{\alpha'}Z_\alpha = Q_{\alpha'}Q_{\alpha'-1}$ imply that $Z_\lambda \leq Q_{\alpha'}Q_{\alpha'-1}$. So that $Z_\lambda \leq Q_{\alpha'-1}Q_{\alpha'-2} \cap Q_{\alpha'-1}Q_{\alpha'} = Q_{\alpha'-1}$, a contradiction.

Now we assume that $Z_\lambda \not\leq G_{\alpha'-1} \cap G_{\alpha'}$. Then $\langle Z_\lambda, G_{\alpha'-1} \cap G_{\alpha'} \rangle = G_{\alpha'-1}$, and R is normalized by $G_{\alpha'-1}$ and centralized by Z_λ. These together mean that $[R, O^2(G_{\alpha'-1})] = 1$ and so $Z(G_{\alpha'-1}) = 1$, which contradicts that $Z(G_\alpha) = 1$ and $\alpha' - 1$ is conjugate to α.

By (3) $V_\alpha \leq Q_{\alpha'-1}$ and so $V_\alpha \leq Z_\alpha Q_{\alpha'}$. Then $[V_\alpha, Z_{\alpha'}] = [Z_\alpha, Z_{\alpha'}]$, and it follows from (3.1) that $Z_\beta = \Omega_1(Z(G_\beta))$, which contradicts that $Z_{\alpha'} \neq \Omega_1(Z(G_{\alpha'}))$ and β is conjugate to α'.

(3.3) $G_\alpha/Q_\alpha = L_3(2)$.

Proof. Assume that $G_\alpha/Q_\alpha = S_5$. Then by (3.2) $G_{\alpha'}/Q_{\alpha'} = S_5$, and so by (1.4)(d) and (1.5)(b), we have $|Z_\alpha/Z_\alpha \cap Q_{\alpha'}| = |Z_{\alpha'}/Z_{\alpha'} \cap Q_\alpha| = 4$.

By (1.8), $Z_\alpha/\Omega_1(Z(A_\alpha))$ is a natural module for $G_{\alpha'}/Q_{\alpha'}$, and by (1.5)(a) and (2.2), we have $Z(G_\beta) = 1$. Then it follows from (3.2) that $Z(G_{\alpha'}) = 1$. It is clear that $[Z_\alpha, Z_{\alpha'}]$ is centralized by V_α and is normalized by $\langle V_\alpha, G_{\alpha'} \cap G_{\alpha'-1}\rangle$. Hence $V_\alpha \leq G_{\alpha'} \cap G_{\alpha'-1}$, i.e. $V_\alpha \leq Z_\alpha Q_{\alpha'}$ and so $[V_\alpha, Z_{\alpha'}] = [Z_\alpha, Z_{\alpha'}]$, a contradiction to (3.1) and $Z(G_\beta) = 1$.

(3.4) If $b = 2$, Then $|Z_\alpha / Z_\alpha \cap Q_{\alpha'}| = 4$.

Proof. By (3.3), $G_\beta / Q_\beta \cong S_5$. And so $3 \mid |G_\alpha \cap G_\beta \cap G_{\alpha'}|$, and the assertion follows from the operation of an element of order 3 of the intersection $G_\alpha \cap G_\beta \cap G_{\alpha'}$.

(3.5) $Z_\beta = \Omega_1(Z(G_\beta))$.

Proof. Assume first that there exists an $\alpha - 1 \in \Delta(\alpha)$ with $Z_{\alpha-1} \not\leq Q_{\alpha'-1}$. Then for $\alpha - 1$ and $\alpha' - 1$ Hypothesis (3.0) holds, and a contradiction follows from (3.3). So we have $V_\alpha \leq Q_{\alpha'-1}$.

Assume by way of contradiction that $Z_\beta \neq \Omega_1(Z(G_\beta))$. Let $R = [V_\alpha, Z_{\alpha'}]$. Then by (3.1) $[Z_\alpha, Z_{\alpha'}] = R$, i.e. $|Z_\alpha / Z_\alpha \cap Q_{\alpha'}| = 2$ and $V_\alpha Q_{\alpha'} = Q_{\alpha'} Q_{\alpha'-1}$. In particular, by (3.4) $b \geq 4$, and so $[V_\lambda, V_\alpha] = 1$ for all $\lambda \in \Delta(\alpha)$. With the same argument as above, we have $|Z_{\alpha'} / Z_{\alpha'} \cap Q_\alpha| = 2$, and so Z_α induces transvection in $Z_{\alpha'}$. By (1.7) $R = [Q_{\alpha'-1}, Z_{\alpha'}] \leq Z_{\alpha'} \cap Z_{\alpha'-1}$ and

(*) $|R| = 4$ and $R \not\leq Z(O^2(G_{\alpha'} \cap G_{\alpha'-1}))$.

If there exists a $\lambda \in \Delta(\alpha)$ with $V_\lambda \not\leq Q_{\alpha'-2}$. Then R is centralized by V_λ and $Q_{\alpha'-1}$ and is normalized but not centralized by $G_{\alpha'} \cap G_{\alpha'-1} \cap G_{\alpha'-2}$. Now we investigate $L = \langle V_\lambda, G_{\alpha'-2} \cap G_{\alpha'-1} \cap G_{\alpha'}\rangle$.

If $V_\lambda \leq G_{\alpha'-1}$, then $[V_\lambda, R] = 1$ implies that $[O^2(L), R] = 1$, i.e. R is centralized by $O^2(L) Q_{\alpha'-1} Q_{\alpha'}$, a contradiction to (*).

If $V_\lambda \not\leq G_{\alpha'-1}$, then $L Q_{\alpha'-2} = G_{\alpha'-2}$, and R is centralized by $O^2(G_{\alpha'-2})$, in particular by $O^2(G_{\alpha'} \cap G_{\alpha'-1} \cap G_{\alpha'-2}) Q_{\alpha'-1} Q_{\alpha'}$, a contradiction to (*).

Now we assume that $V_\lambda \leq Q_{\alpha'-2}$ for all $\lambda \in \Delta(\alpha)$. Then R is normalized by $\langle V_\lambda, G_{\alpha'} \cap G_{\alpha'-1}\rangle$ and is centralized by V_λ. If $V_\lambda \not\leq G_{\alpha'} \cap G_{\alpha'-1}$, then $\langle V_\lambda, G_{\alpha'} \cap G_{\alpha'-1}\rangle = G_{\alpha'-1}$ and $[R, O^2(G_{\alpha'-1})] = 1$, a contradiction to (*). Then $V_\lambda \leq G_{\alpha'-1} \cap G_{\alpha'}$, and it follows from $[V_\lambda, V_\alpha] = 1$ that $V_\lambda \leq V_\alpha Q_\alpha$ and $[V_\lambda, Z_\lambda] \leq R$ for all $\lambda \in \Delta(\alpha)$. The structure of $L_3(2)$ and $|Z_{\alpha'} / Z_{\alpha'} \cap Q_\alpha| = 2$ imply that there exists a $\lambda \in \Delta(\alpha)$ with $Z_{\alpha'} \leq G_\lambda$ and $Z_{\alpha'} \not\leq Q_\lambda Q_\alpha$.

If $|Z_\lambda / Z_\lambda \cap Q_{\alpha'}| = 4$, then $[Z_\lambda, Z_{\alpha'}] = [Z_{\alpha'}, V_\lambda]$, so by (3.1) $Z_\alpha = \Omega_1(Z(G_\alpha))$, a contradiction. Now we assume that $|Z_\lambda / Z_\lambda \cap Q_{\alpha'}| = 2$. Then $Z_{\alpha'}$ induces transvections in Z_λ, and by (1.7), $Z_\lambda / \Omega_1(Z(G_\lambda))$ is an orthogonal module for G_λ / Q_λ. In particular, $Z_{\alpha'-1} Z_{\alpha'} \leq Z(G_\lambda \cap G_\alpha)$ and so $Z_{\alpha'-1} \cap Z_{\alpha'} \leq Z(G_{\alpha'-1} \cap G_{\alpha'})$, a contradiction to (*).

(3.6) Z_α is a natural module for G_α / Q_α. In particular, we have

$|Z_\beta| = 2$, $[Q_\beta, Z_\alpha] = Z_\beta$ and $Z_\alpha = [Z_\alpha, d]Z_\beta$ for an element of order 3 of $G_\alpha \cap G_\beta$.

Proof. By (3.5), we simply need to show that Z_α is a natural module. We select $\lambda \in \Delta(\alpha)$ so that $Z_\lambda \not\leq Q_{\alpha'}$. It follows from the structure of $L_3(2)$ that $|Z_{\alpha'}/Z_{\alpha'} \cap G_\lambda| = 2$. On the other hand, by (3.5) we have $[Z_{\alpha'} \cap G_\lambda, Z_\lambda] = 1$, i.e. Z_λ induces transvections in $Z_{\alpha'}$. Then by (3.2), there exists transvections in $G_{\alpha'}$, and then (3.5) and (2.2) imply $Z(G_\alpha) = 1$. Now the assertion follows from (1.7).

(3.7) $b = 2$.

Proof. Suppose the contrary. We select $\lambda \in \Delta(\alpha)$ so that $Z_{\alpha'} \not\leq G_\lambda$. Then $\langle G_\lambda \cap G_\alpha, Z_{\alpha'} \rangle = G_\alpha$ and so by (3.5) and (2.2) $Z_\lambda \not\leq Q_{\alpha'}$.

(1) $b \geq 8$.

By (3.2), $b \equiv 0(2)$. If $b = 4$. Then (3.6) implies $[Z_\alpha, Z_{\alpha+4}] = Z_\beta = Z_{\alpha+3}$, and so $Z_\beta \trianglelefteq \langle G_\beta \cap G_{\alpha+2}, G_{\alpha+2} \cap G_{\alpha+3} \rangle = G_{\alpha+2}$, a contradiction.

If $b = 6$, then $[Z_\alpha, Z_{\alpha+6}] = Z_\beta = Z_{\alpha+5}$. Let d be an element of order 3 from $G_{\alpha+2} \cap G_{\alpha+3} \cap G_{\alpha+4}$. By (3.6) we get $Z_{\alpha+2} = \langle Z_\beta^{\langle d \rangle} \rangle Z_{\alpha+3} = \langle Z_{\alpha+5}^d \rangle \cdot Z_{\alpha+3} = Z_{\alpha+4}$. But then $Z_{\alpha+2} \trianglelefteq \langle G_{\alpha+2} G_{\alpha+3}, G_{\alpha+3} G_{\alpha+4} \rangle$, a contradiction.

(2) $V_\lambda \leq Q_{\alpha'-2}$ and $V_\lambda \not\leq Q_{\alpha'-1}$.

If $V_\lambda \not\leq Q_{\alpha-2}$, then there is $\alpha-2$ with $Z_{\alpha-2} \not\leq Q_{\alpha'-2}$. It follows by (1) and (3.6) $[Z_{\alpha-2}, Z_{\alpha'-2}] = Z_\lambda \leq Z_{\alpha'-2} \leq Q_{\alpha'}$, a contradiction.

Assume that $V_\lambda \leq Q_{\alpha'-1}$, then $[V_\lambda, Z_{\alpha'}] = Z_{\alpha'-1} = Z_\beta \leq Z_\alpha \leq V_\lambda$, i.e. $V_\lambda \trianglelefteq \langle G_\alpha \cap G_\lambda, Z_{\alpha'} \rangle = G_\alpha$, a contradiction to (2.2).

(3) Set $U = O^2(A_\beta)$. Then $V_\beta/C_{V_\beta}(U)$ is a natural module for G_β/Q_β.

Set $V = V_\beta/C_{V_\beta}(U)$ and $\widetilde{Z}_\alpha = Z_\alpha C_{V_\beta}(U)/C_{V_\beta}(U)$. By (3.3), (3.5) and (3.6), we have $G_\beta/Q_\beta \cong S_5$, $Z_\beta \leq C_{V_\beta}(U)$ and $V_\beta/Z_\beta \leq Z(Q_\beta/Z_\beta)$. Therefore V is a module for G_β/Q_β. By (1.2) there exists $V_o \leq V$, and V_o is a natural module or an orthogonal module for G_β/Q_β. If $V_o \cap \widetilde{Z}_\alpha = 1$, then by (3.6) and (1.8), $\widetilde{Z}_\alpha \leq V_o = V$, and so V is a natural Module.

We assume $V_o \cap \widetilde{Z}_\alpha = 1$, then we have the following:

(*) $[V, U] \not\leq V_o$ and $|V| \geq 2^8$.

Notice that then there exists $Z_\lambda \leq V_1 \lneq V_\lambda$ with $V_1 \cap Z_\alpha = Z_\lambda$, and $V_1/C_{V_1}(O^2(A_\lambda))$ is a natural or orthogonal module for G_λ/Q_λ.

It follows from the structure of $|V_{\alpha'-1}/V_{\alpha'-1} \cap Q_\lambda| \leq 2^3$. In addition $[V_{\alpha'-1} \cap Q_\lambda, V_\lambda] \leq V_{\alpha'-1} \cap Z_\lambda$, and from $V_{\alpha'-1} \leq Q_{\alpha'}$ and $|Z_\lambda| = 2$ we get $[V_{\alpha'-1} \cap Q_\lambda, V_\lambda] = 1$. On the other hand, A_5 may be generated by three involutions or by two 2-Sylow subgroups, It follows from these facts and (2): $|V_{\alpha'-1}/C_{V_{\alpha'-1}}(O^2(A_{\alpha'-1}))| \leq 2^6$ and $|V_\lambda/V_\lambda \cap Q_{\alpha'-1}| = 4$, or $|V_{\alpha'-1}/C_{V_{\alpha'-1}}(O^2(A_{\alpha'-1}))| \leq 2^9$ and $|V_\lambda/V_\lambda \cap Q_{\alpha'-1}| = 2$.

The first case contradicts $(*)$, for $\alpha'-1$ is conjugate to β. So it holds that $|V_{\alpha'-1}/V_{\alpha'-1} \cap Q_\lambda| = 2^3$ and $|V_\lambda/V_\lambda \cap Q_{\alpha'-1}| = 2$. In particular, by (1.4)(d) and (1.5)(b) $[V_1 \cap Q_{\alpha'-1}, V_{\alpha'-1} \cap G_\lambda] = Z_{\alpha'-1} = Z_\beta \leq V_1$, contrary to $Z_\alpha \cap V_1 = Z_\lambda$.

(4) V_β/Z_β is a natural module for G_β/Q_β.

Since λ is conjugate to β, it is enough to show $C_{V_\lambda}(O^2(A_\lambda)) = Z_\lambda$. Set $H = C_{V_\lambda}(O^2(A_\lambda))$. By (3.6) $[H, A_\lambda] = [H, Q_\lambda] = Z_\lambda$. Let Y be a subgroup of H, then $YZ_\alpha \trianglelefteq G_\alpha$, and so $[YZ_\alpha, Q_\alpha] \trianglelefteq G_\alpha$. On the other hand, $[YZ_\alpha, Q_\alpha] \leq Z_\lambda$. So it follows that $Y \leq \Omega_1(Z(Q_\alpha))$, and Q_λ induces a transvection in YZ_α. Then by (1.7) we get $Y \leq Z(G_\lambda \cap G_\alpha)$, i.e. $Y \leq Z_\lambda$.

If $H \leq Q_{\alpha'-1}$, then $\langle G_\lambda \cap G_\alpha, Z_{\alpha'}\rangle = G_\alpha$ normalizes HZ_α, so $H = Z_\lambda$.

If $H \not\leq Q_{\alpha'-1}$, then by (3) H induces no transvection in $V_{\alpha'-1}$. Therefore $[V_{\alpha'-1} \cap G_\lambda, H] \neq 1$, and as above G_α normalizes $[V_{\alpha'-1} \cap G_\lambda, H]\cdot Z_\alpha$. So $[V_{\alpha'-1} \cap G_\lambda, H] = Z_\lambda \leq Q_{\alpha'}$, contrary to the selection of Z_λ.

(5) $[V_\lambda, V_{\alpha'-1}] = Z_{\alpha'-2}$.

Proof. The assertion follows directly from (3.3), (2) and (4).

We now define: $W = \langle Z_\delta \mid d(\delta,\lambda) \leq 3\rangle$, $V = \langle Z_\delta \mid d(\delta,\alpha) \leq 2\rangle$.

By (1) W and V are both elementary abelian.

(6) $W \not\leq Q_{\alpha'-2}$.

If $W \leq Q_{\alpha'-2}$, then by (4) and (5) $[W, V_{\alpha'-1}] = [V_\lambda, V_{\alpha'-1}] = Z_{\alpha'-2} \leq V \leq W$, so $W \trianglelefteq \langle G_\lambda \cap G_\alpha, Z_{\alpha'}\rangle = G_\alpha$, contrary to (2.2).

For $\delta \in \Gamma$ with $d(\delta,\lambda) \leq 3$, then $Z_{\alpha'-2} \leq V \leq W$ and so $[Z_\delta, Z_{\alpha'-2}] = 1$. Now it follows from (3.6) and (4) $Z_\delta \leq Q_{\alpha'-2}$, i.e. $W \leq Q_{\alpha'-2}$, contrary to (6). Thus we prove (3.7).

The proof of Theorem 1.

We use the result of (3.2) - (3.7):

$b = 2$, $G_\alpha/Q_\alpha = L_3(2)$ and $G_\beta/Q_\beta = S_5$, Z_α is a natural module for G_α/Q_α, $|Z_\beta| = 2$, $|Z_\alpha/Z_\alpha \cap Q_{\alpha'}| = |Z_{\alpha'}/Z_{\alpha'} \cap Q_\alpha| = 4$.

We fix d for the element of order 3 from $G_\alpha \cap G_\beta \cap G_{\alpha'}$.

(1) $\Phi(Q_\alpha) \leq D_\alpha$ and $\Phi(D_\alpha) = 1$.

It follows from the structure of $G_\alpha \cap G_\beta$ that $\Phi(Q_\alpha) \leq D_\alpha$ and $\Phi(D_\alpha) \leq Q_\lambda$ for all $\lambda \in \Gamma$ with $d(\alpha,\lambda) = 2$. Therefore $[\Phi(D_\alpha), \langle Z_{\alpha'}^{G_\alpha}\rangle] = 1$, which means $\Phi(D_\alpha) = 1$, otherwise we should have $\Phi(D_\alpha) \cap Z(G_\alpha) \neq 1$ and Z_α a non-natural module.

(2) $Z_{\alpha'}$ induces transvections in D_α.

By (3.6) $[Q_\beta, Z_\alpha] = Z_\beta$ and $[Q_\beta, Z_{\alpha'}] = Z_\beta$. So by $D_\alpha \leq Q_\beta$ we conclude that $|[D_\alpha, Z_{\alpha'}]| = 2$.

(3) Q_α/D_α and Z_α are isomorphic natural modules for G_α/Q_α.

Set $V = Q_\alpha/D_\alpha$. Then $[Q_\alpha \cap Q_\beta, Z_{\alpha'}] = Z_\beta$ and by the operation of d, we have (*) $|V/C_V(Z_{\alpha'})| = |Z_{\alpha'}/Z_{\alpha'} \cap Q_\alpha| = 4$.

Since $|Z_\alpha/C_{Z_\alpha}(Z_{\alpha'})| = 4$, then V and Z_α are isomorphic, if V is a natural module. By (1.6) G_α/Q_α is generated by three conjugates of $Z_{\alpha'}Q_\alpha/Q_\alpha$. It follows by (*) that $|V/C_V(G_\alpha)| \leq 2^6$. On the other hand, by (2.4)(d) the preimage of $C_V(G_\alpha)$ is contained in D_α, i.e. $|V| \leq 2^6$. By (2.1), there exists a submodule $V_o \leq V$ with $|V_o| = 2^3$, and for V_o, we have $|V_o/C_{V_o}(Z_{\alpha'})| = 4$. And so $V = V_oC_V(Z_{\alpha'})$, and the elements from $Z^\#$ induce transvections in V. Then it follows from (2.7) that V is a natural module.

(4) $|D_\alpha| \leq 2^4$ and $\Phi(Q_\alpha) \leq Z_\alpha$.

By (1.6), we may select $g, h \in G$ with $U = \langle Z_{\alpha'}, Z_{\alpha'}^h, Z_{\alpha'}^g \rangle$, such that $UQ_\alpha = G_\alpha$. So by (2), $D_\alpha = C_{D_\alpha}(U) \times Z_\alpha$.

Set $D_o = C_D(U)$, $x \in (V_\beta \cap Q_\alpha) \setminus D_\alpha$ and s an element of order 7 in U. Then by (1) and (3), $C_{D_o}(x)$ is centralized by $D_\alpha[\langle x \rangle, \langle s \rangle] = Q_\alpha$, so $C_{D_o}(x) = 1$. On the other hand, $[D_o, x] \leq Z_\beta$ and $|Z_\beta| = 2$, i.e. $|D_o| \leq 2$ and $|D_\alpha| \leq 2^4$. Then $|Q_\alpha/Z_\alpha| \leq 2^4$, and it follows from (1.11) that $\Phi(Q_\alpha) \leq Z_\alpha$.

From now on, we use the following notations:

$D_o = C_D(d)$, $\tilde{U} = O^2(A_\beta)$ and $Z = C_{V_\beta}(\tilde{U})$.

(5)(a) $V_\beta = [Q_\beta, G_\beta]$ and $V_\beta \cong Ex(2^5)\, Y\, C_4$.

(b) $V_\beta/Z(V_\beta)$ is a natural module for G_β/Q_β.

(c) It holds either $D_o = 1$ and $V_\beta = Q_\beta$, or $D_o = 1$ and $Q_\beta = Ex(2^7)$.

By (3) and (4) $|Q_\beta| = 2^6$ or 2^7. Then it follows from (1.2) and the operation of d on $Z_{\alpha'}$, that V_β/Z is a natural module for G_β/Q_β. In particular $Z = Z(V_\beta)$. Moreover, by (1.11) Z is not elementary abelian, otherwise V_β is also elementary abelian.

Likewise, by (3) $Q_\beta = V_\beta D_o$. In particular, $Z \cong C_4$ and $V_\beta \cong C_4\, Y\, Ex(2^5)$.

Now assume that $D_o = 1$. As we have observed from the proof of (4), no element $y \in D_o$ is centralized by an element from $Q_\alpha \setminus D_\alpha$. Therefore Z is inversed by Y, and $Q_\beta \cong Ex(2^7)$.

(6) $Q_\alpha = [Q_\alpha, G_\alpha]D_o$, and $[Q_\alpha, G_\alpha] \cong C_4 \times C_4 \times C_4$.

Let $g \in G_\alpha \setminus G_\beta$. Then $[Z, Z^g] \leq Z \cap Z^g = 1$. So $\langle Z^{G_\alpha} \rangle$ is elementary abelian, and (3) and (4) imply that $\langle Z^{G_\alpha} \rangle = 2^6$. Then $\langle Z^{G_\alpha} \rangle$ is the only abelian subgroup of this order in Q_α, and this establishes (6).

The assertion (3), (5) and (6) show that (G_α, G_β) is parabolic of type HS. Thus we establish Theorem 1.

4. The case $[Z_\alpha, Z_{\alpha'}] = 1$.

(4.0) Hypothesis. In this chapter we assume Hypothesis A and $[Z_\alpha, Z_{\alpha'}] = 1$. The notations are as in chapter 2.

In this chapter we prove the following:

Theorem. (G_α, G_β) is parabolic of type Ru.

We divide the proof in some lemmas and we shall use the following additional notations:

It follows from (2.3) and (2.5) that $Z_\alpha = \langle Z_\beta^{G_\alpha}\rangle$. Then there exists $\alpha-1 \in \Delta(\alpha)$ with $Z_{\alpha-1} \not\leq Q_{\alpha'}$. We select a $z \in Z_{\alpha-1} \setminus Q_{\alpha'}$ and define for $g \in G$, $\delta = (\alpha-1)^g$ and $\lambda \in \Delta(\delta)$:

$\widetilde{Z}_{\alpha-1} = \langle z\rangle$, $\widetilde{Z}_\delta = \widetilde{Z}^g_{\alpha-1}$, $\widetilde{Z}_\lambda = \langle \widetilde{Z}_\delta^{G_\lambda}\rangle$, $\widetilde{V}_\delta = \langle \widetilde{Z}_\lambda^{G_\delta}\rangle$.

These definitions are independent of the selection of g, since $(\alpha-1)^g = (\alpha-1)^h$ implies that $gh^{-1} \in G_{\alpha-1} \leq C_G(z)$.

We shall in (4.3) show that $Z_{\alpha-1} = \widetilde{Z}_{\alpha-1}$. For $g \in G$, $\mu = \alpha^g$ and $\rho \in \Delta(\mu)$, we define:

$$z_\mu^{(1)} = [\widetilde{Z}_\mu, Q_\rho], \quad V_\rho^{(1)} = \langle \widetilde{Z}_\mu^{(1)G_\rho}\rangle.$$

And for simplification we write $Z_1 = \widetilde{Z}_\alpha^{(1)}$, $V_1 = V_\beta^{(1)}$ and $V_{1'} = \widetilde{V}_{\alpha'}^{(1)}$.

(4.1) $G_\alpha/Q_\alpha = L_3(2)$ and $G_\beta/Q_\beta = S_5$.

Assume, by way of contradiction, that $G_\alpha/Q_\alpha \cong S_5$, and by (2.5) (d) $G_{\alpha'}/Q_{\alpha'} \cong L_3(2)$.

(1) $\widetilde{Z}_\alpha$ is an orthogonal module for G_α/Q_α. In particular, $[Z_1, Q_\beta] = \widetilde{Z}_\beta$.

By (2.4)(d) $Z(G_\alpha) = 1$. Then the assertion follows by $|\widetilde{Z}_\beta^{G_\alpha}| = 5$, (1.2) and (1.5)(a).

(2) $b > 1$, and V_β is elementary abelian.

Assume $b = 1$. By the operation of $G_\alpha \cap G_\beta$, we get $|\widetilde{Z}_\alpha/\widetilde{Z}_\alpha \cap Q_\beta| = 4$. On the other hand, $Z_1 = [\widetilde{Z}_\alpha, Q_\beta] \leq Q_\beta$ and so $|\widetilde{Z}_\alpha/\widetilde{Z}_\alpha \cap Q_\beta| \leq |\widetilde{Z}_\alpha/Z_1| = 2$. Thus $b > 1$. Then by (2.6) V_β is elementary abelian.

(3) Let U be a subgroup of G_β with $UQ_\beta = G_\beta$. Then

(a) $C_{\widetilde{Z}_\delta}(U) = \widetilde{Z}_\beta$ for all $\delta \in \Delta(\beta)$,

(b) $|\widetilde{V}_\beta/C_{\widetilde{V}_\beta}(U)| > 2^3$.

Assume that (a) holds. Then by (1) $|\widetilde{Z}_\delta/C_{\widetilde{Z}_\delta}(U)| = 2^3$. On the other hand, by (2.2) $[\tilde{V}_\beta, O^2(U)] \neq 1$, i.e. $\tilde{V}_\beta \neq \tilde{Z}_\beta C_{\tilde{V}_\beta}(U)$, which implies (b). So we need only to show (a).

Set $X = C_{\tilde{Z}_\delta}(U)$. Since U operates on $\Delta(\beta)$ transitively, $3 \mid |U \cap G_\delta|$. Then by (1) $|X| \leq 4$, and it follows from $|X| = 2$ that $X = \tilde{Z}_\beta$.

Assume $|X| = 4$. Then by (1) $Q_\delta = C_{Q_\delta Q_\beta}(X)$, i.e. $O_2(C_{G_\beta}(X)) = Q_\delta \cap Q_\beta$ and $[Q_\beta, O^2(U)] \leq Q_\delta \cap Q_\beta$. But then by (2.4)(c) $Q_\beta \leq Q_\delta$, a contradiction.

(4) No involution from G_β induces a transvection in V_β.

Assume that there exists $t \in G_\beta \setminus Q_\beta$ with $|V_\beta / C_{\tilde{V}_\beta}(t)| = 2$. Then by (1.6) there exist $g, g' \in G_\beta$, such that $G_\beta = UQ_\beta$ for $U = \langle t, t^g, t^{g'} \rangle$. But then $|\tilde{V}_\beta / C_{\tilde{V}_\beta}(U)| \leq 2^3$, contrary to (3).

For involutions in Q_β, the assertion follows from (1) and (1.4)(d).

(5) Let $\lambda \in \Delta(\alpha')$. Then $C_{Z_\lambda}(V_\beta \cap G_\lambda) \leq C_{G_\beta}(V_\beta)$.

The assertion follows from the structure of $L_3(2)$ and (4).

(6) $C_{G_\beta}(V_1) \leq Q_\delta$ for all $\delta \in \Delta(\beta)$.

This is an immediate consequence of (1).

(7) $\tilde{V}_{\alpha'} \not\leq Q_\beta$.

Assume $\tilde{V}_{\alpha'} \leq Q_\beta$. If $[\tilde{V}_{\alpha'}, V_1] = 1$, so $V_\beta \leq Q_{\alpha'}$, a contradiction.

Assume now $[\tilde{V}_{\alpha'}, V_1] = 1$. Then $|[V_{\alpha'}, V_1]| = 2$. But then V_1 induces transvections in $\tilde{V}_{\alpha'}$, a contradiction follows from (4).

(8) $|\tilde{V}_\beta / \tilde{V}_\beta \cap Q_{\alpha'}| = 4$.

Suppose $|\tilde{V}_\beta / \tilde{V}_\beta \cap Q_{\alpha'}| = 2$. Then there exists $\lambda \in \Delta(\alpha')$ with $\tilde{V}_\beta \leq G_\lambda$ and $V_\beta \not\leq Q_{\alpha'} Q_\lambda$. Let $x \in \tilde{V}_\beta \setminus Q_{\alpha'} Q_\lambda$. Then (1) and (1.4)(c) imply $|\tilde{Z}_\lambda / C_{\tilde{Z}_\lambda}(x)| = 2$. By (4), $[C_{\tilde{Z}_\lambda}(x), \tilde{V}_\beta] = 1$, so $|\tilde{V}_\beta / \tilde{V}_\beta \cap Q_\lambda| = 2$. But then by (4), $[\tilde{Z}_\lambda, \tilde{V}_\beta] = 1$, a contradiction.

(9) $|\tilde{V}_\beta / \tilde{V}_\beta \cap Q_\beta| = 4$.

By (7), there exists $\lambda \in \Delta(\alpha')$ with $\tilde{Z}_\lambda \not\leq Q_\beta$. So the Hypothesis holds for (λ, β), so we can apply (8) to $\tilde{V}_{\alpha'} / \tilde{V}_{\alpha'} \cap Q_\beta$.

(10) $V_1 / \tilde{Z}_\beta$ and $V_{1'} / \tilde{Z}_{\alpha'}$ are natural modules for G_β / Q_β and $G_{\alpha'} / Q_{\alpha'}$ respectively.

We need only to show that the assertion holds for α'.

Let $\lambda \in \Delta(\alpha') \setminus \{\alpha'-1\}$. Then (8) implies that $|V_\beta / V_\beta \cap G_\lambda| = 2$ and $\tilde{V}_\beta \cap G_\lambda \not\leq Q_{\alpha'} Q_\lambda$. So by (1) and (1.4) there exists $x \in \tilde{V}_\beta \setminus Q_{\alpha'} Q_\lambda$ with $[\tilde{Z}_\lambda, x] \leq \tilde{Z}_\lambda^{(1)} \leq V_{1'}$, $[\tilde{Z}_\lambda, x] \neq \tilde{Z}_{\alpha'}$, $|[\tilde{Z}_\lambda, x]| = 2$.

$[\widetilde{Z}_\lambda, x] \leq \widetilde{V}_\beta \cap \widetilde{V}_{\alpha'}$ and $b>1$ imply that $[\widetilde{Z}_\lambda, x, \widetilde{V}_\beta] = 1$. So $C_{G_{\alpha'}}([\widetilde{Z}_\lambda, x])Q_{\alpha'}/Q_{\alpha'} = S_4$. The assertion follows from (1.10).

(11)(a) $V_1 \leq Q_{\alpha'}$ and $V_{1'} \leq Q_\beta$,

(b) $[V_1, V_{1'}] = \widetilde{Z}_\beta = \widetilde{Z}_{\alpha'}$,

(c) $[V_1, \widetilde{Z}_\lambda] \neq 1$ for all $\lambda \in \Delta(\alpha') \setminus \{\alpha'-1\}$ and $[V_{1'}, \widetilde{Z}_\delta] = 1$ for all $\delta \in \Delta(\beta) \setminus \{\alpha+2\}$.

Assume $[V_1, \widetilde{Z}_\lambda] = 1$ for a $\lambda \in \Delta(\alpha') \setminus \{\alpha'-1\}$. Then by (6) $[\widetilde{V}_\beta, Z_\lambda] = 1$, and this together with (8), imply that $Z_\lambda \trianglelefteq \langle \widetilde{V}_\beta, G_{\alpha'} \cap G_\lambda \rangle = G_{\alpha'}$, contrary to (2.2). By a symmetry argument, we can show that the assertion holds for $\delta \in \Delta(\beta) \setminus \{\alpha+2\}$. Thus (c) holds.

Assume $V_1 \not\leq Q_{\alpha'}$. Then by (10) $|V_1/\widetilde{Z}^{(1)}_{\alpha+2}| = 2$ and $V_1 \cap Q_{\alpha'} = \widetilde{Z}^{(1)}_{\alpha+2} = C_{V_1}(\widetilde{V}_{\alpha'})$. In particular $|V_1/V_1 \cap Q_{\alpha'}| = 2$, and there exists $\lambda \in \Delta(\alpha')$ with $V_1 \leq G_\lambda$ and $V_1 \not\leq Q_{\alpha'} Q_\lambda$. But then by (1) and (1.4) $|C_{\widetilde{Z}_\lambda}(V_1)| = 2^3$, so by (6) $\widetilde{Z}_\lambda = C_{\widetilde{Z}_\lambda}(\widetilde{V}_\beta)\widetilde{Z}^{(1)}_\lambda$. Thus $\widetilde{Z}_\lambda V_1 \trianglelefteq \langle V_\beta, G_{\alpha'} \cap G_\lambda \rangle = G_{\alpha'}$, i.e. $\widetilde{V}_{\alpha'} = C_{\widetilde{Z}_\lambda}(\widetilde{V}_\beta)V_1$, which means $|\widetilde{V}_{\alpha'}/C_{\widetilde{V}_{\alpha'}}(\widetilde{V}_\beta)| = 2$, contrary to (4).

The same argument as above shows that $V_{1'} \leq Q_\beta$. So (a) holds.

Then (a) and (c) imply (b).

(12) $\widetilde{V}_\delta \leq Q_{\alpha'-1}$ for all $\delta \in \Delta(\alpha)$.

Assume $\widetilde{V}_\delta \not\leq Q_{\alpha'-1}$ for a $\delta \in \Delta(\alpha)$. By (2) $\widetilde{V}_\delta \leq G_\beta$. Then $[\widetilde{V}_\delta, \widetilde{Z}_\beta] = 1$ and by (11) $[\widetilde{V}_\delta, \widetilde{Z}_{\alpha'}] = 1$. Hence $\widetilde{V}_\delta \leq G_{\alpha'-2}$ by the minimality of b.

Since $\widetilde{Z}_{\alpha'}$ is centralized by $\langle \widetilde{V}_\delta \cap Q_{\alpha'-2}, G_{\alpha'-1} \cap G_\alpha \rangle$, so that by (2.2) and the structure of S_5, we have (*) $\widetilde{V}_\delta \cap Q_{\alpha'-2} \leq Q_{\alpha'-1}$.

So $\widetilde{V}_\delta \not\leq Q_{\alpha'-2}$, i.e. there exists $\alpha-2 \in \Delta(\delta)$ with $\widetilde{Z}_{\alpha-2} \not\leq Q_{\alpha'-2}$, and Hypothesis (4.0) applies to $\alpha-2$ and $\alpha'-2$. In particular, we may apply (11) to $V^{(1)}_\delta$ and get $V^{(1)}_\delta \leq Q_{\alpha'-2}$ and $[\widetilde{V}^{(1)}_\delta, \widetilde{Z}_{\alpha'-1}] \neq 1$, contrary to (*).

Let $W_\alpha = \langle \widetilde{V}_\delta \mid \delta \in \Delta(\alpha) \rangle$. By (12) $W_\alpha \leq Q_{\alpha'-1}$ and by (8) $W_\alpha \leq V_\beta Q_{\alpha'}$, so (11)(a) and (c) imply $|\widetilde{V}_\beta \cap G_\lambda / \widetilde{V}_\beta \cap Q_\lambda| = 4$ for all $\lambda \in \Delta(\alpha') \setminus \{\alpha'-1\}$.

On the other hand, $W_\alpha \leq Q_{\alpha+2}$, and $G_\alpha \cap G_\beta$ operates transitively on $\Delta(\beta) \setminus \{\alpha\}$. So that $[W_\alpha, \widetilde{V}_\beta] = 1$. In particular $[\widetilde{V}_\beta \cap G_\lambda, W_\alpha \cap G_\lambda] = 1$, and so $W_\alpha \cap G_\lambda \leq (\widetilde{V}_\beta \cap G_\lambda)Q_\lambda$, which means $[W_\alpha \cap Q_{\alpha'}, \widetilde{Z}_\lambda] \leq [\widetilde{V}_\beta \cap G_\lambda, \widetilde{Z}_\lambda]$ for all $\lambda \in \Delta(\alpha')$. But then $[W_\alpha, V_{\alpha'}] = [\widetilde{V}_\beta, \widetilde{V}_{\alpha'}] \leq \widetilde{V}_\beta \leq W_\alpha$, and W_α is normalized by $\langle G_\alpha \cap G_\beta, V_{\alpha'} \rangle = G_\beta$, contrary to (2.2).

(4.2) $\widetilde{Z}_\alpha$ is a natural module for G_α/Q_α, and $[\widetilde{V}_\beta, Q_\beta] = \widetilde{Z}_\beta$.

Proof. It follows from the definition of $\widetilde{Z}_\alpha$ and $|G_\alpha/G_\alpha \cap G_\beta| = 7$ and (2.4)(d), that $|\widetilde{Z}_\alpha| \leq 2^6$. So by (1.1) there exists $Z \leq \widetilde{Z}_\alpha$, and Z is a natural module for G_α/Q_α. If $\widetilde{Z}_\beta \leq Z$, then $Z = \widetilde{Z}_\alpha$. If $\widetilde{Z}_\beta \not\leq Z$, then $|\widetilde{Z}_\beta| \geq 4$, and $\widetilde{Z}_\alpha$ is a natural module from (1.9).

By $C_{\tilde{Z}_\alpha}(Q_\beta) = \tilde{Z}_\beta$ we get $[\tilde{Z}_\alpha, Q_\beta] = \tilde{Z}_\beta$ and so $[\tilde{V}_\beta, Q_\beta] = \tilde{Z}_\beta$.

(4.3) $b = 3$ and $\tilde{V}_{\alpha'} \not\leq Q_\beta$.

Proof. We prove the assertion in some lemmas

(1) $b > 1$.

Supposing $b = 1$. The operation of $G_\alpha \cap G_\beta$ on $\tilde{Z}_\alpha$ implies that $|\tilde{Z}_\alpha/\tilde{Z}_\alpha \cap Q_\beta| = 4$, and by (4.2) $\tilde{Z}_\alpha \cap Q_\beta = \tilde{Z}_\beta$. But then $[Q_\beta, \tilde{Z}_\alpha, \tilde{Z}_\alpha] = 1$, contrary to $C_{G_\beta}(Q_\beta)$.

(2) $\tilde{V}_{\alpha'} \not\leq Q_\beta$.

Since $\tilde{V}_\beta \not\leq Q_{\alpha'}$, we have $[\tilde{V}_\beta, \tilde{V}_{\alpha'}] \neq 1$. If $\tilde{V}_{\alpha'} \leq Q_\beta$, then $[\tilde{V}_{\alpha'}, \tilde{V}_\beta] = Z_\beta$, i.e. V_β induces transvections in $\tilde{V}_{\alpha'}$, contrary to (1.4)(d).

(3) Set $V = \tilde{V}_{\alpha'}/\tilde{Z}_{\alpha'}$ and assume that $|\tilde{V}_{\alpha'}/\tilde{V}_{\alpha'} \cap Q_\beta| = 2$. Then $V/C_V(A_{\alpha'})$ is a natural module for $G_{\alpha'}/Q_{\alpha'}$. In particular, $\tilde{V}_{\alpha'} \cap Q_\beta \not\leq Q_\alpha$.

By (4.2), V is a module for $G_{\alpha'}/Q_{\beta'}$. Since $\tilde{Z}_\alpha$ induces no transvection in $\tilde{V}_{\alpha'}$, then $\tilde{V}_{\alpha'} \cap Q_\beta \not\leq Q_\alpha$ and $[V_{\alpha'} \cap Q_\beta, \tilde{V}_\beta] = \tilde{Z}_\beta$. So that

(*) $|V_{\alpha'}/C_{V_{\alpha'}}(x)| \leq 4$ for $x \in V_\beta \setminus Q_{\alpha'}$.

We choose $g, h \in G_{\alpha'}$, so that for $W = \langle x, x^g, x^h \rangle$, we have $WQ_{\alpha'} = A_{\alpha'}$. It follows by (*) that $|V/C_V(A_{\alpha'})| \leq 2^6$. By (1.2) there exists a submodule V_o of V with $C_V(A_{\alpha'}) \leq V_o$ and $|V_o/C_V(A_{\alpha'})| = 2^4$ and $[V, A_{\alpha'}] \leq V_o$. Since $\alpha'-1$ is conjugate to α, we can apply (4.2) to $G_{\alpha'-1}$, then the operation of $G_{\alpha'-1} \cap G_{\alpha'}$ on $Z_{\alpha'-1}$ together with (1.4) and (1.5) imply $V_o = V$ and so the assertion follows.

In the folllwing, we assume that $b > 3$. By (2) there exists $\lambda \in \Delta(\alpha')$ with $\tilde{Z}_\lambda \not\leq G_\alpha$. In particular, Hypothesis is fulfilled by α' and β. We indicate this as the symmetry in α and β.

(4) Assume that $|\tilde{V}_\beta/\tilde{V}_\beta \cap Q_{\alpha'}| = 4$. Then $\tilde{V}_\beta \cap Q_{\alpha'} \leq C_{Q_{\alpha'}}(\tilde{V}_{\alpha'})$.

Supposing $\tilde{V}_\beta \cap Q_{\alpha'} \not\leq C_{Q_{\alpha'}}(\tilde{V}_{\alpha'})$. Then by (4.2) $\tilde{Z}_{\alpha'} = [\tilde{V}_\beta \cap Q_{\alpha'}, \tilde{V}_{\alpha'}] \leq \tilde{V}_\beta \cap \tilde{V}_{\alpha'}$. And by $b > 3$, $[\tilde{Z}_{\alpha'}, \tilde{V}_\delta] = 1$ for all $\delta \in \Delta(\alpha)$.

At first we assume that $\tilde{V}_\delta \not\leq Q_{\alpha'-2}$ and $Y = \langle \tilde{V}_\delta, Q_{\alpha'-1} \rangle$. Then $[Y, \tilde{Z}_{\alpha'}] = 1$ and $YQ_{\alpha'-2} = A_{\alpha'-2}$. In particular, $3 \mid |Y \cap G_{\alpha'-1}|$, and $\tilde{Z}_{\alpha'}$ is centralized by $G_{\alpha'-1}$ contrary to (2.2).

Now supposing $\tilde{V}_\delta \leq Q_{\alpha'-2}$. Then $\tilde{Z}_{\alpha'} \leq Z(\langle \tilde{V}_\delta, G_{\alpha'} \cap G_{\alpha'-2} \rangle)$, and (2.2) implies $\tilde{V}_\delta \leq G_{\alpha'} \cap G_{\alpha'-1}$, so $\tilde{V}_\delta \leq Q_{\alpha'-2} \cap G_{\alpha'} \leq Q_{\alpha'-1}$. Set $W_\alpha = \langle \tilde{V}_\delta \mid \delta \in \Delta(\alpha) \rangle$, Then $W_\alpha \leq Q_{\alpha'-1}$ and so $W_\alpha \leq \tilde{V}_\beta Q_{\alpha'}$. It implies that $[W_\alpha, \tilde{V}_{\alpha'}] = [\tilde{V}_\beta, \tilde{V}_{\alpha'}] \leq \tilde{V}_\beta \leq W_\alpha$. We conclude by the selection of α that $W_\alpha \trianglelefteq \langle G_\alpha \cap G_\beta, \tilde{V}_{\alpha'} \rangle = G_\beta$, contrary to (2.2).

(5) Supposing $|\tilde{V}_\beta/\tilde{V}_\beta \cap Q_{\alpha'}| = 4$. Then $|\tilde{V}_{\alpha'}/\tilde{V}_{\alpha'} \cap Q_\beta| = 2$.

Assume that $|\tilde{V}_{\alpha'}/\hat{V}_{\alpha'}\cap Q_{\beta}| = 4$. By the symmetry in α' and β, we have $\tilde{V}_{\alpha'}\cap Q_{\beta} \le C_{Q_{\beta}}(\tilde{V}_{\beta})$.

Set $W = \langle\tilde{V}_{\beta}, \tilde{V}_{\beta}^{g}\rangle$, $g\in G_{\alpha'}\setminus G_{\alpha'-1}$, and $U = \langle\tilde{V}_{\alpha'}, V_{\alpha'}^{h}\rangle$, $h\in G_{\beta}\setminus G_{\alpha+2}$. Then $UQ_{\beta} = A_{\beta}$ and $WQ_{\beta} = A_{\beta}$, and by (1.8) $\tilde{V}_{\beta}/C_{\tilde{V}_{\beta}}(U)$ and $\tilde{V}_{\alpha'}/C_{\tilde{V}_{\alpha'}}(W)$ are natural modules for $U/U\cap Q_{\beta}$ and $W/W\cap Q_{\alpha'}$ respectively.

Set $R = [\tilde{V}_{\beta}, \tilde{V}_{\alpha'}]$. Then $|R/C_R(W)| = 4$ and $R = (R\cap Z_{\alpha'-1})C_R(W)$.

Let $\delta\in\Gamma$ with $d(\delta, \beta) = 2$ and $\delta \ne \alpha+2$. If $V_{\delta}\not\le Q_{\alpha'-2}$, so $\langle\tilde{V}_{\delta}, Q_{\alpha'-1}\rangle Q_{\alpha'-2} = A_{\alpha'-2}$ and $[R\cap Z_{\alpha'-1}, \langle V_{\delta}, Q_{\alpha'-1}\rangle] = 1$. It is impossible, since by (4.2) there is no element of order 3 in $\langle\tilde{V}_{\delta}, Q_{\alpha'-1}\rangle\cap G_{\alpha'-1}$ centralizing a subgroup of order 4 in $Z_{\alpha'-1}$.

Suppose now $\tilde{V}_{\delta}\le Q_{\alpha'-2}$ and $V_{\delta}\not\le Q_{\alpha'-1}$, and set $H = \langle Q_{\alpha}, V_{\alpha'}\rangle$. By (4.2), $[\tilde{V}_{\delta}, \tilde{Z}_{\alpha'-1}] = Z_{\alpha'-2}$, i.e. $\tilde{Z}_{\alpha'-1}$ induces transvections in $\tilde{V}_{\delta}$. Then (1.4)(d) and (1.5)(b) yield $\tilde{Z}_{\alpha'-1}\le Q_{\delta}$, so $[\tilde{V}_{\delta}, \tilde{Z}_{\alpha'-1}] = \tilde{Z}_{\alpha'-2} = \tilde{Z}_{\delta}$. In particular, $[\tilde{V}_{\alpha'}, \tilde{Z}_{\delta}] = 1$ and so $[H, \tilde{Z}_{\delta}] = 1$. On the other hand, $HQ_{\beta} = A_{\beta}$. Let $\tilde{\alpha}\in\Delta(\delta)\cap\Delta(\beta)$. Then $\langle G_{\delta}\cap G_{\tilde{\alpha}}, H\cap G_{\tilde{\alpha}}\rangle = G_{\tilde{\alpha}}$, so $\tilde{Z}_{\delta}$ is a normal subgroup of $G_{\tilde{\alpha}}$, a contradiction to (2.2).

Set $W_{\beta} = \langle\tilde{V}_{\delta} \mid d(\delta,\beta) = 2\rangle$. Then by $W_{\beta}\le Q_{\alpha'-1}$, we get $W_{\beta}\le\tilde{V}_{\beta}Q_{\alpha'}$ and $[W_{\beta}, \tilde{V}_{\alpha'}]\le RZ_{\alpha'}$. If $[W_{\beta}, \tilde{V}_{\alpha'}]\not\le\tilde{V}_{\beta}$, then $V_{\alpha'}$ induces transvections in $W_{\beta}/\tilde{V}_{\beta}$, which is impossible.

Assuming now $[W_{\beta}, \tilde{V}_{\alpha'}]\le\tilde{V}_{\beta}$, so for $W_{\alpha} = \langle\tilde{V}_{\delta} \mid \delta\in\Delta(\alpha)\rangle$, we have $[W_{\beta}, \tilde{V}_{\alpha'}] \le \tilde{V}_{\beta}\le W_{\alpha}$ and $W_{\alpha}\trianglelefteq\langle G_{\alpha}\cap G_{\beta}, V_{\alpha'}\rangle = G_{\beta}$, contrary to (2.2).

(6) $|\tilde{V}_{\beta}/\tilde{V}_{\beta}\cap Q_{\alpha'}| = |\tilde{V}_{\alpha'}/\tilde{V}_{\alpha'}\cap Q_{\beta}| = 2$ and $\tilde{Z}_{\beta}\tilde{Z}_{\alpha'}\le[\tilde{V}_{\beta}, \tilde{V}_{\alpha'}]$.

By (3) and the symmetry in α' and β, we need only to show that $|\tilde{V}_{\beta}/\tilde{V}_{\beta}\cap Q_{\alpha'}| = 2$. If it is false, then $|\tilde{V}_{\beta}/\tilde{V}_{\beta}\cap Q_{\alpha'}| = 4$. By (5) $|\tilde{V}_{\alpha'}/\tilde{V}_{\alpha'}\cap Q_{\beta}| = 2$, and by (3) $\tilde{V}_{\alpha'}/C_{\tilde{V}_{\alpha'}}(A_{\alpha'})$ is a natural module for $A_{\alpha'}/Q_{\alpha'}$. So there exists $x\in(\tilde{V}_{\alpha'}\cap Q_{\beta})\setminus C_{\tilde{V}_{\alpha'}}(\tilde{V}_{\beta})$ with $|[\tilde{V}_{\beta}, x]|\ge 4$, a contradiction to $[\tilde{V}_{\beta}, x] = \tilde{Z}_{\beta}$.

(7) $\tilde{V}_{\delta}\le Q_{\alpha'-1}$ for all $\delta\in\Gamma$ with $d(\delta,\beta) = 2$.

By (6) $\tilde{Z}_{\alpha'}\le\tilde{V}_{\beta}$, and by $b>3$ $[\tilde{V}_{\delta}, \tilde{Z}_{\alpha'}] = 1$. Then $Z_{\alpha'}$ is centralized by $\langle\tilde{V}_{\delta}, Q_{\alpha'-1}\rangle$ and $\langle\tilde{V}_{\delta}, G_{\alpha'}\cap G_{\alpha'-1}\rangle$. As in (4), the assertion follows.

(8) Set $W_{\beta} = \langle\tilde{V}_{\delta} \mid d(\delta,\beta) = 2\rangle$. Then $|W_{\beta}/W_{\beta}\cap Q_{\alpha'}| = 4$.

By (7), $W_{\beta}\le Q_{\alpha'-1}Q_{\alpha'}$. Then if $|W_{\beta}/W_{\beta}\cap Q_{\alpha'}| = 2$, it follows that $W_{\beta}\le V_{\beta}Q_{\alpha'}$ and $[W_{\beta}, \tilde{V}_{\alpha'}]\le\tilde{V}_{\beta}$. Now a contradiction follows as in (4).

By (6) and (8), we have the following:

$$|W_{\beta}/\tilde{V}_{\beta}(W_{\beta}\cap Q_{\alpha'})| = 2, \quad [W_{\beta}\cap Q_{\alpha'}, \tilde{V}_{\alpha'}] = \tilde{Z}_{\alpha'} .$$

Set $V = W_{\beta}/\tilde{V}_{\beta}$. Then $|V/C_V(\tilde{V}_{\alpha'})|\le 2$, and by (1.4) and (1.5), we get $[V, V_{\alpha'}] = 1$ and so by (2) $[V, O^2(G_\beta)] = 1$. But then $W_\alpha = \langle \tilde{V}_\delta \mid \delta \in \Delta(\alpha) \rangle$ is normalized by $O^2(G_\beta)\, Q_\beta = A_\beta$ and G_α, contrary to (2.2), thus (4.3) is proved.

The proof of Theorem 2.

We have shown in (4.1) - (4.3), that $b = 3$, $G_\alpha/Q_\alpha \cong L_3(2)$, $G_\beta/Q_\beta \cong S_5$ and $V_{\alpha'} \not\leq Q_\beta$. We divide the proof in some lemmas. Put $\lambda \in \Delta(\alpha')$ with $Z_\lambda \not\leq Q_\beta$.

(1) $\langle Q_\alpha, Z_\lambda \rangle = A_\beta$.

Set $U = \langle Q_\alpha, Z_\lambda \rangle$. Then $UQ_\beta = A_\beta$ and $\langle Q_\alpha^{G_\beta} \rangle \leq U$, so $[Q_\beta, \langle Q_\alpha^{G_\beta} \rangle] \leq Q_\beta \cap U$. It follows from (2.4)(c), that $Q_\beta \leq U$ and $U = A_\beta$.

(2) $C_{Z_\alpha}(Z_\lambda) \leq Z(A_\beta)$.

This is a direct result of (1).

(3) $Z_\beta = Z_\beta$, in particular, Z_α is a natural module for G_α/Q_α.

We need only to show that, $Z_\alpha = \tilde{Z}_\alpha$.

Assume Z is a counterexample. Then $|Z_\beta| \geq 4$, and by (1.9) there exists a normal subgroup N of G_α in Z_α with $N = \tilde{Z}_\alpha \times N_1$, where N_1 is a natural module for G_α/Q_α. Correspondingly, there exists $\tilde{N} = \tilde{Z}_\lambda \times N_2$ in Z. By (2), $|N/C_{\tilde{N}}(Z_\lambda)| = |\tilde{N}/C_{\tilde{N}}(Z_\alpha)| = 2^4$. Then $[N \cap Q_{\alpha'}, N \cap Q_\beta] = 1$. On the other hand $[N \cap Q_{\alpha'}, \tilde{N} \cap Q_\beta] \leq Z_{\alpha'} \cap Z_\beta \leq Z(G_{\alpha+2})$, a contradiction to (2.2).

(4) $V_\beta = Z_\alpha Z_{\alpha+2}$, and V_β/Z_β is a natural module for G_β/Q_β.

Set $V = V_\beta/Z_\beta$. Then $V/C_V(A_\beta)$ is a natural module for G_β/Q_β. We argue that $V_\beta = Z_\alpha Z_{\alpha+2}$, which follows directly by $V_\beta \cap V_{\alpha'} = Z_{\alpha+2}$.

Now we show that $V_\beta \cap V_{\alpha'} = Z_{\alpha+2}$. Put $D = V_\beta \cap V_{\alpha'}$ and $B = O^2(G_\beta)$. On the other hand, $[D, Q_\beta] \leq Z_\beta \leq Z_{\alpha+2} \leq D \leq Z_{\alpha+2} C_{V_\beta}(B)$. So D is normalized by $\langle A_\beta \cap G_{\alpha+2}, A_{\alpha'} \cap G_{\alpha+2} \rangle Q_{\alpha+2} = G_{\alpha+2}$.

Put $x \in Q_\beta \setminus Q_{\alpha+2}$. Then $[D, x] = Z_\beta$, i.e. x induces a transvection in D. Set $C = \langle x, x^g, x^h \rangle$ with $h, g \in G_{\alpha+2}$ and $CQ_{\alpha+2} = G_{\alpha+2}$. Then $D = C_D(C) \times Z_{\alpha+2}$, and it follows from $3 \mid |C \cap B|$ and (1.5)(a) that $C_D(C) \leq C_{V_\beta}(B)$, then by (2.2), $C_D(C) = 1$ and $D = Z_{\alpha+2}$.

(5) Q_β/D_β is a natural module for G_β/Q_β.

By (4), $D_\beta = Q_\alpha \cap Q_{\alpha+2}$ and $|Q_\beta/D_\beta| \leq 2^4$. Then the assertin follows from (1.4), (1.5) and the operation of an element of order 3 from $G_\alpha \cap G_\beta \cap G_{\alpha+2}$.

(6) $[D_\beta, A_\beta] = V_\beta$.

Since $\alpha' = \alpha+3$, then $[D_\beta, V_{\alpha'}] = Z_{\alpha+2} \leq V_\beta$. So $[D_\beta, A_\beta] = V_\beta$.

(7) $\Phi(Q_\alpha) = Z_\alpha$, $\Phi(Q_\beta) = V_\beta$ and $\Phi(D_\beta) \leq Z_\beta$.

By (4) and (6), $D_\beta = D_o V_\beta$ and $Z_\beta = D_o \cap V_\beta$, where $D_o = C_{D_\beta}(d)$ and d is an element of order 3 from $G_\beta \cap G_{\alpha+2}$.

Set $W_{\alpha+2} = \langle V_\delta \mid \delta \in \Delta(\alpha+2)\rangle$. By (4) $[Q_{\alpha+2}, W_{\alpha+2}] = Z_{\alpha+2} = \Phi(W_{\alpha+2})$, and together with (5), we have $Q_{\alpha+2} = D_o W_{\alpha+2}$. So $\Phi(D_o) \trianglelefteq G_\beta$ and $\Phi(D_o)Z_{\alpha+2} \trianglelefteq G_{\alpha+2}$. Put $R = \Phi(D_o)Z_{\alpha+2}$. Then $[R, Q_{\alpha+2}] = [\Phi(D_o), Q_{\alpha+2}] \leq \Phi(D_o)$, i.e. $[\Phi(D_o), Q_{\alpha+2}]$ is a normal subgroup of $C_{\alpha+2}$ in $\Phi(D_o)$. But then by (1) $\Phi(D_o)$ is centralized by $\langle O^2(G_\beta), Q_{\alpha+2}\rangle = A_\beta$.

Take $x \in Q_\beta \setminus Q_{\alpha+2}$. Then $|R/C_R(x)| = 2$, and it follows from (1.7) and (2.4)(d) that $\Phi(D_o) \leq Z_{\alpha+2}$, i.e. $\Phi(D_o) \leq Z_\beta$.

Finally we show that $\Phi(Q_\beta) = V_\beta$. It is clear that $\Phi(Q_\beta) \geq V_\beta$. If Q_β/V_β is not elementary abelian, then there is a contradiction to (1.11) and (5).

(8) $|D_\beta/V_\beta| = 4$, and $Q_{\alpha+2}/Z_{\alpha+2}$ is an irreducible module of order 2^8 for $G_{\alpha+2}/Q_{\alpha+2}$.

Let d be an element of order 3 from $G_\alpha \cap G_\beta \cap G_{\alpha+2}$, and D_o and $W_{\alpha+2}$ be defined as above. Then by (4) and (5), we have:

(*) $|Q_\beta| = 2^8 |D_o/Z_\beta|$, $|Q_{\alpha+2}/Z_{\alpha+2}| = 2^6 |D_o/Z_\beta|$.

Let $\overline{N}_{\alpha+2}$ be a minimal normal subgroup of $G_{\alpha+2}/Z_{\alpha+2}$ in $Q_{\alpha+2}/Z_{\alpha+2}$ and $N_{\alpha+2}$ be its preimage in $Q_{\alpha+2}$. By (1.1), $|\overline{N}_{\alpha+2}| = 2, 2^3$ or 2^8. So we need only to show $\overline{N}_{\alpha+2} = 2^8$.

Assume $|\overline{N}_{\alpha+2}| = 2$. Then $N_{\alpha+2} \leq D_o Z_{\alpha+2} \leq D_\beta$ and $[N_{\alpha+2}, V_\beta] = 1$, so $[N_{\alpha+2}, W_{\alpha+2}] = 1$. On the other hand, $Q_{\alpha+2} = D_o W_{\alpha+2}$, and so $[Q_{\alpha+2}, N_{\alpha+2}] \leq Z_\beta$, which implies that $N_{\alpha+2} \leq Z(Q_{\alpha+2})$. Since α is conjugate to $\alpha+2$, $|N_\alpha/Z_\alpha| = 2$, which means that $N_\alpha \leq N_{\alpha+2}V_\beta$ and by (4) $|N_\alpha \cap N_{\alpha+2}| = 4$. Now we deduce that $N_\alpha \cap N_{\alpha+2} \leq Z(A_\beta)$, thus we have $N_{\alpha+2} = (N_{\alpha+2} \cap Z(A_\beta))Z_{\alpha+2}$ and $[N_{\alpha+2}, Q_\beta] = Z_\beta$, i.e. the elements from $Q_\beta Q_{\alpha+2}$ induce transvections *in* $N_{\alpha+2}$, contrary to (2.4)(d).

Assume $|\overline{N}_{\alpha+2}| = 2^3$. Then by $Q_{\alpha+2} = W_{\alpha+2}D_o$ and (*), $N_{\alpha+2} \cap W_{\alpha+2} = Z_{\alpha+2}$. If $N_{\alpha+2} \not\leq Q_\beta$, then the operation of d implies that $|N_{\alpha+2}/N_{\alpha+2} \cap Q_\beta| = 4$ and $|N_{\alpha+2}/Z_{\alpha+2}| \geq 2^4$, a contradiction.

It is also impossible that $[N_{\alpha+2}, Q_\beta] \leq N_{\alpha+2} \cap V_\beta = Z_{\alpha+2}$, which means that $N_{\alpha+2}/Z_{\alpha+2}$ is a trivial module, a contradiction.

We next argue that $|D_o/Z_\beta| = 4$.

Put $\overline{Q}_{\alpha+2} = Q_{\alpha+2}/Z_{\alpha+2}$ and $x \in (Q_\alpha \cap Q_\beta) \setminus Q_{\alpha+2}$. Then $[\overline{D}_o, x] \leq Z_\alpha$, i.e. $|\overline{D}_o/C_{\overline{D}_o}(x)| = 4$, and $C_{\overline{D}_o}(x)$ is centralized by $\langle d, x\rangle Q_{\alpha+2}/W_{\alpha+2} \cong A_4$. Assume that $C_{\overline{D}_o}(x) \neq 1$. Then ther exists $1 \neq \overline{y} \in C_{\overline{D}_o}(x)$ with

$|\bar{Y}^{G_{\alpha+2}}| = 7$, then $|\langle\bar{Y}^{G_{\alpha+2}}\rangle| \leq 2^7$. A contradiction to the fact that each minimal module for $G_{\alpha+2}/Z_{\alpha+2}$ in $Q_{\alpha+2}$ is of order 2^8.

Since α is conjugate to $\alpha+2$, so all the assertions for $G_{\alpha+2}$ also hold for G_α.

Set $Q_1 = Q_\alpha$, $Z_1 = Z_\alpha$, $Q_2 = Q_\beta$, $Z_2 = Z_\beta$, $A = D_\beta$, and $B = V_\beta$. Then it follows from (4.1) and (1) - (8), that (G_α, G_β) is parabolic of type Ru.

References

1. A. Chermak, Index-p systems in finite groups, part I, preprint.
2. A.L. Delgado, On edge-transitive automorphism groups of graphs, Ph. D. thesis, University of California, Berkeley.
3. D.M. Goldschidt, Automorphisms of trivalent graphs, Ann. of Math., 111(1980), 377-406.
4. B. Stellmacher, On graphs with edge-transitive automorphism groups, Ill. J. Math., 28(1984), 211-266.
5. G. Stroth, Graphs with subconstituents containing Sz(q) and $L_2(r)$, J. Algebra 80(1983), 186-215.
6. G. Stroth, Kantentransitive Graphen und Gruppen vom Rang 2, J. Algebra, to appear.

ON 2-CONNECTEDNESS OF GEOMETRIES

Li Huiling

Department of Mathematics
Lanzhow University
Lanzhow, Gansu
People's Republic of China

In [4] Tits defined a kind of geometry and proved that, with some exceptions, every such geometry is the image of a building under a certain morphism and the latter is the universal 2-cover of the former. After that many such geometries were found by Kantor and other mathematicians. For some of those whose universal 2-covers are buildings people have determined their universal 2-covers. But for only few geometries whose universal 2-covers are not buildings people can determine their universal 2-covers. Generally speaking, it is hard to determine the universal 2-cover of a given geometry. Here we suggest a method with which we can settle this problem under some special conditions. Up to now this method has settled the problems of determining the universal 2-covers for two infinite classes of such geometries.

In the following we will call such geometries GABs (Geometries that are almost buildings).

1. Preliminaries

A geometry over a finite set I is a system $\Gamma=(V,*,\tau)$ consisting of a set V, whose elements are called vertices, a binary symmetric relation * and a map τ from V to I, such that if $x,y\in V$, then $x*y$ and $x^\tau=y^\tau$ imply $x=y$. We say Γ is of rank n if $|I|=n$. If $x*y$, we say that x is incident to y. x^τ is called the type of x. A flag of Γ is a set of vertices in which every two vertices are incident. Two flags are incident if their union is also a flag. The type of flag X is $\{x^\tau, x\in X\}$. We say the geometry Γ is connected if for every two vertices x,y, there is a sequence

$$x_1, x_2, \cdots, x_d,$$

such that $x_1=x$ and $x_d=y$ and two consecutive terms are incident.

Let X be a flag, and let Y be the set of all vertices in V-X incident to X. Writing the restrictions of * and τ to Y as $*|_Y$ and $\tau|_Y$, $(Y,*|_Y,\tau|_Y)$ is a geometry and is called the residue of X. So this residue is of rank $|I-(X)^\tau|$ and of type $I-X^\tau$. If for a geometry $(V,*,\tau)$ there are integers $n_{i,j}$ for $i,j\in I$, $i\neq j$, such that every residue of type $\{i,j\}$, $i,j\in I$, is a generalized n_{ij}-gon, then we call $(V,*,\tau)$ a GAB. It is easy to see that buildings are GABs. We define the diagram of a GAB to be a graph whose nodes are the elements of I, and distinct nodes i and j are joined by $n_{ij}-2$ edges if $n_{ij}\in\{2,3,4\}$, by $\frac{1}{2}n_{ij}$ edges if $n_{ij}=6$ or 8, by an edge labeled n_{ij} in all other cases.

Let $(V',*',\tau')$ and $(V,*,\tau)$ be two GABs over a set I. If φ is a type-preserving map from $(V',*',\tau')$ to $(V,*,\tau)$ such that for $x',y'\in V'$, $x'*y'$ implies $x'^{\varphi}*y'^{\varphi}$, we say φ is a morphism. A morphism $\varphi:(V',*',\tau')\to(V,*,\tau)$ is called a 2-cover if φ is surjective on flags and φ induces an isomorphism from every residue of rank≤ 2 in $(V',*',\tau')$ to a residue of rank≤ 2 in $(V,*,\tau)$. For this case $(V',*',\tau')$ is also called a 2-cover of $(V,*,\tau)$. Let $(V',*',\tau')$ and $(V,*,\tau)$ and φ be as above and let $(V,*,\tau)$ be connected, if $(V',*',\tau')$ is connected and for every 2-cover $\psi:(V'',*'',\tau'')\to(V,*,\tau)$, there is a morphism α from $(V',*',\tau')$ to $(V'',*'',\tau'')$ such that $\varphi=\alpha\psi$, then we say that $(V',*',\tau')$ is a universal 2-cover of $(V,*,\tau)$. It is easy to see that the universal 2-cover of a GAB is unique up to isomorphism. If the universal 2-cover of a GAB Γ is isomorphic to Γ itself, then we say Γ is 2-connected.

A chamber system $\mathcal{C}=(C,\pi_i,i\in I)$ over I consists of a set C, whose elements are called chambers, and a family $\{\pi_i\}$ of partitions. $|I|$ is called the rank of $\mathcal{C}$. Two chambers x,y are i-adjacent if they are in the same member of the partition π_i. We can define the join of partitions and use π_J to denote the join of π_j, $j\in J$. We say $(C,\pi_i,i\in I)$ is connected if C is the only member of the partition π_I. Now suppose D is a member of partition π_J for $J\subseteq I$. Writing π_j' for the restriction of the partition π_j to D, then $(D,\pi_j',j\in J)$ is a chamber system, which is called a residue of type J and rank $|J|$. If for a chamber system $(C,\pi_i,i\in I)$ there are integers n_{ij}, $i,j\in I$, such that for every subset $\{i,j\}$ of I, every residue of type $\{i,j\}$ is a generalized n_{ij}-gon, we say $(C,\pi_i,i\in I)$ is a SCAB (Chamber system that is almost a building). We define the diagram of a SCAB in a similar way as for GABs.

We can also define morphisms of chamber systems to chamber systems. Let $\mathcal{C}'=(C',\pi_i',i\in I)$ and $\mathcal{C}=(C,\pi_i,i\in I)$ be two SCABs. A morphism $\varphi:\mathcal{C}'\to\mathcal{C}$ is called a 2-cover if φ maps every residue of rank≤ 2 of $\mathcal{C}'$ isomorphic to a residue of rank≤ 2 of $\mathcal{C}$. Similarly we can define the universal 2-cover of a SCAB and 2-connectedness of a SCAB.

Every geometry defines a chamber system. Suppose that $(V,*,\tau)$ is a geometry. Let C be the set of all flags of type I. We say two flags are i-adjacent if their intersection is a flag of type $I-\{i\}$. Then we obtain a chamber system. But not every chamber system can be obtained in this way. A chamber system $\mathcal{C}$ is obtained from a geometry Γ if and only if every set of vertices has nonempty intersection whenever any two vertices of it have nonempty intersection. At that time we say $\mathcal{C}$ is the flag-chamber system of the geometry Γ.

Let Γ be a geometry and $\mathcal{C}$ be its flag-chamber system. Then we can easily prove the following facts: If Γ is a GAB, then $\mathcal{C}$ is a SCAB. If $\varphi:\Gamma'\to\Gamma$ is a morphism (respectively, a 2-cover), and $\mathcal{C}'$ is the flag-chamber system of Γ', then φ induces a morphism (respectively, a 2-cover) from $\mathcal{C}'$ to $\mathcal{C}$. Finally, if Γ is a connected GAB and $\mathcal{C}$ is 2-connected, then Γ is 2-connected.

In this paper we will consider GABs and SCABs defined by groups. Let G be a group and let $\{P_i, i\in I\}$ be a family of subgroups of G. For $J\subseteq I$, set $P_J = \bigcap \{P_j, j\in J\}$. Suppose that

(i) $\quad P_J = \langle P_{J\cup\{i\}}, i\in I-J\rangle$

Then we consider the disjoint union $\Delta = \coprod G/P_i$ of the set G/P_i of right cosets of P_i in G. Define $\tau(P_i\cdot x)=i$ and $P_i x * P_j y$ if and only if $P_i x \cap P_j y = \emptyset$ Then $(\Delta,*,\tau)$ is a geometry. We often require that

(ii) $\quad$ G is transitive on maximal flags.

Then every flag of type J has the form $P_J\cdot g$, $g\in G$, and every residue of type $\{i,j\}$ is isomorphic to the geometry defined by the group P_J and its subgroups $P_{J\cup\{i\}}$ and $P_{J\cup\{j\}}$, where $J=I-\{i,j\}$. For this case in the corresponding flag-chamber system, the chambers are just the cosets $P_I g$, $g\in G$ and two chambers $P_I\cdot g_1$ and $P_I\cdot g_2$ are i-adjacent if $P_{I-\{i\}}g_1 = P_{I-\{i\}}g_2$.

In the following the geometry defined by G and $\{P_i, i\in I\}$ will be denoted by $\Delta(G; P_i, i\in I)$.

2. A principle

Lemma Let C be a set and let $\pi_1, \pi_2, \cdots, \pi_n, \pi_n'$ be n+1 partitions. Suppose that

i) $\mathcal{C}=(C,\pi_1,\cdots,\pi_n)$ is a 2-connected SCAB and $\mathcal{C}'=(C,\pi_1,\cdots,\pi_{n-1},\pi_n')$ is a connected SCAB;

ii) The partitions $\pi_{n-1}\cup\pi_n$ and $\pi_{n-1}\cup\pi_n'$ are identical;

iii) For every i, $1\le i\le n-2$, every residue of type $\{i,n-1,n'\}$ in $\mathcal{C}'$ is 2-connected.

Then $\mathcal{C}'=(C,\pi_1,\cdots,\pi_{n-1},\pi_n')$ is 2-connected.

Proof Suppose $\hat{\mathcal{C}}'=(\hat{C},\hat{\pi}_1,\cdots,\hat{\pi}_{n-1},\hat{\pi}_n)$ is a 2-cover of $\mathcal{C}'=(C,\pi_1,\cdots,\pi_{n-1},\pi_n')$ and φ is the corresponding morphism. Then φ induces an isomorphism from every residue of rank≤ 2 of $\hat{\mathcal{C}}'$ to a residue of $\mathcal{C}'$.

In $\hat{\mathcal{C}}'$ we define a partition $\hat{\pi}_n$ by means of the above isomorphisms:

Two chambers $\hat{c}$ and $\hat{d}$ in $\hat{\mathcal{C}}'$ are n-adjacent if they are in the same member of partition $\hat{\pi}_{n-1} \cup \hat{\pi}_n'$ and their image $c=\hat{c}^{\varphi}$ and $d=\hat{d}^{\varphi}$ are n-adjacent in $\mathcal{C}$. Then φ is a morphism from $\hat{\mathcal{C}}=(\hat{C}, \hat{\pi}_1, \cdots, \hat{\pi}_{n-1}, \hat{\pi}_n)$ to $\mathcal{C}$.

We want to show that φ is a 2-cover from $\hat{\mathcal{C}}$ to $\mathcal{C}$. To do this it suffices to show that φ induces an isomorphism from every residue of type J with $|J| \leq 2$ in $\hat{\mathcal{C}}$ to a residue of $\mathcal{C}$. This is obvious if J= j or J={i,j} where $i,j < n$. Because of our definition of π_n this is also right if J={n} or J={n-1,n}. Now suppose J={i,n} where $i < n-1$. By the definition of π_n every member of partition $\hat{\pi}_i \cup \hat{\pi}_n$ in $\hat{\mathcal{C}}$ is contained in a member of partition $\hat{\pi}_i \cup \hat{\pi}_{n-1} \cup \hat{\pi}_n'$ in $\hat{\mathcal{C}}'$. Now we fix a residue $\hat{\mathcal{X}}$ of type J and and take the residue $\hat{\mathcal{D}}$ of type {i,n-1,n'} containing it. By iii) φ induces an isomorphism from $\hat{\mathcal{D}}$ to a residue $\mathcal{D}$ of type {i, n-1,n'} in $\mathcal{C}'$. So two chambers $\hat{c}, \hat{d}$ in $\hat{\mathcal{D}}$ are i-adjacent (respectively:(n-1)-adjacent, n-adjacent) if and only if their images $c=\hat{c}^{\varphi}$ and $d=\hat{d}^{\varphi}$ are i-adjacent (respectively:(n-1)-adjacent,n-adjacent) in $\mathcal{D}$. Now suppose two chambers $\hat{c}$ and $\hat{d}$ in $\hat{\mathcal{C}}$ are n-adjacent then by definition $c=\hat{c}^{\varphi}$ and $d=\hat{d}^{\varphi}$ are n-adjacent. Conversely, if $c=\hat{c}^{\varphi}$ and $d=\hat{d}^{\varphi}$ are n-adjacent, then c and d are in the same residue of type {n-1,n'} and as φ induces an isomorphism from $\hat{\mathcal{D}}$ to $\mathcal{D}$, $\hat{c}$ and $\hat{d}$ are in the same residue of type {n-1,n'} of $\hat{\mathcal{D}}$, and so c,d are n-adjacent. Thus φ induces an isomorphism from $(\hat{D}, \hat{\pi}_i, \hat{\pi}_{n-1}, \hat{\pi}_n)$ to $(D, \pi_i, \pi_{n-1}, \pi_n)$, where $\hat{D}$ (respectively,D) is the set of chambers in $\hat{\mathcal{D}}$ (respectively, in $\mathcal{D}$). So φ maps $\hat{\mathcal{X}}$ isomorphically to a residue $\mathcal{X}$ of type J in $\mathcal{C}$. Hence φ is a 2-cover from $\hat{\mathcal{C}}$ to $\mathcal{C}$. By i) φ is an isomorphism from $\hat{\mathcal{C}}$ to $\mathcal{C}$. It is easy to see that φ is also an isomorphism from $\hat{\mathcal{C}}'$ to $\mathcal{C}'$. So $\mathcal{C}'$ is 2-connected. This completes our proof.

3. An example

Let Q be the field of rational numbers and Q^8 be a 8-dimensional vector space. Let f(x,y)=(x,y) be the usual dot product. Take a standard basis

$$\{u_i, \ i=1,2,\cdots,8\}.$$

Set

$$v_{2i-1} = u_{2i-1} + u_{2i} \quad \text{and} \quad v_{2i} = u_{2i-1} - u_{2i}$$

where i=1,2,3,4. Then $\{v_i\}$ form a second orthogonal basis.

Consider the following four root systems of type E_8:

$$\Phi^+ = \{\pm u_i \pm u_j, \ i \neq j\} \cup \{\tfrac{1}{2}\sum \epsilon_i u_i, \ \epsilon_i = \pm 1, \ \prod \epsilon_i = 1\}$$

$$\Phi^- = \{\pm u_i \pm u_j, \ i \neq j\} \cup \{\tfrac{1}{2}\sum \epsilon_i u_i, \ \epsilon_i = \pm 1, \ \prod \epsilon_i = -1\}$$

$$\Phi^{\#} = \{\pm v_i \pm v_j, \ i \neq j\} \cup \{\tfrac{1}{2}\sum \epsilon_i v_i, \ \epsilon_i = \pm 1, \ \prod \epsilon_i = 1\}$$

$$\Phi^{b} = \{ \pm v_i + v_j, \ i \neq j \} \cup \{ \tfrac{1}{2} \textstyle\sum \epsilon_i v_i \ , \ \epsilon_i = \pm 1, \ \prod \epsilon_i = -1 \}$$

Set $A=\{+,-,\#,b\}$. For $\alpha \in A$, let W^{α} be the commutator subgroup of the corresponding Weyl group. Then $W^{\alpha} = 2\cdot\Omega^{+}(8,2)$.

Four groups W^{α}, $\alpha \in A$, share a Sylow 2-subgroup S of order 2 . We define a group

$$C = \langle S, \ (u_1, u_2, u_3), \ (u_5, u_6, u_7), \ (v_1, v_2, v_3), \ (v_5, v_6, v_7) \rangle.$$

Let $G_8 = \langle C, \ W^{\alpha}, \alpha \in A \rangle$. Then $G_8 \leq \Omega = GO(Q,f)$, where $GO(Q,f)$ is the group of all $g \in GL(8,Q)$ such that $(u^g, v^g) = c_g(u,v)$ for some $c_g \in Q^*$ and all u, v in Q^8. Also $G \leq GL(8, Z[\tfrac{1}{2}])$.

Now define $\Delta_8 = \Delta(G \ ; W^{\alpha}, C, \alpha \in A)$. Kantor [1] has proved that Δ_8 is a GAB with diagram

Set $H=C_{\Omega}(r_{7-8})'$, $K=C_{\Omega}(\langle r_{7-8}, r_{7+8} \rangle)'$, $G_7 = \langle W^{+} \cap H, W^{-} \cap H \rangle$, $G_6 = \langle W^{+} \cap K, \ W^{\#} \cap K \rangle$, where r_{7-8} , r_{7+8} are the reflections defined by $u_7 - u_8$ and $u_7 + u_8$. Define $\Delta_7 = \Delta(G_7; \ C \cap H, \ W^{\alpha} \cap H, \ \alpha \in \{+,-,\#\})$ and $\Delta_6 = \Delta\,(G_6 ; \ W^{+} \cap K, \ C \cap K, \ W^{\#} \cap K)$, then Δ_7 and Δ_6 are two GABs with diagrams

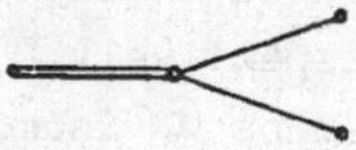

Now define a GAB $\Delta_{7'} = \ (G_7; \ G_7 \cap C, \ W^{+} \cap G_7, \ W^{\#} \cap G_7, \ G_6^{*})$, where $G_6^{*} = \langle G_6 , \ r_{7+8} \rangle$. Then $\Delta_{7'}$ has diagram

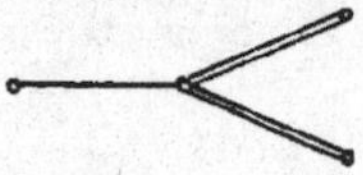

Kantor proved that GABs Δ_8, Δ_7 and Δ_6 are 2-connected and conjectured that $\Delta_{7'}$ is 2-connected.

Applying the lemma in section 2 to Δ_7 and $\Delta_{7'}$ we obtained the following result:

<u>Theorem</u> $\Delta_{7'}$ <u>is 2-connected</u>.

<u>Proof</u> Consider the chamber system $\mathcal{C}$ and $\mathcal{C}'$ corresponding to Δ_7 and $\Delta_{7'}$. Since Δ_7 and $\Delta_{7'}$ are both associated with G_7 and

$$G_6^{*} \cap W^{+} \cap W^{\#} \cap C = W^{+} \cap W^{-} \cap W^{\#} \cap C \cap H,$$

the set of chambers of $\mathcal{C}$ and $\mathcal{C}'$ are identical. Also, the partitions in $\mathcal{C}$ are defined by groups

$$X_1 = W^{+} \cap W^{-} \cap W^{\#} \cap G_7$$

$$X_2 = W^+ \cap W^- \cap C \cap G_7,$$
$$X_3 = W^+ \cap W^\# \cap C \cap G_7,$$
$$X_4 = W^- \cap W^\# \cap C \cap G_7$$

and the partitions in $\mathcal{C}'$ are defined by groups

$$Y_1 = W^+ \cap W^\# \cap G_6^* \cap G_7,$$
$$Y_2 = W^+ \cap C \cap G_6^* \cap G_7,$$
$$Y_3 = W^+ \cap W^\# \cap C \cap G_7,$$
$$Y_4 = W^\# \cap C \cap G_6^* \cap G_7.$$

Since $W^+ \cap G_6^* \cap G_7 = W^+ \cap W^- \cap G_7$, we know that $X_1 = Y_1$ and $X_2 = Y_2$. Also, $X_3 = Y_3$. So we see that $\mathcal{C}$ is defined by the partitions $\pi_1, \pi_2, \pi_3, \pi_4$ and $\mathcal{C}'$ by π_1, π_2, π_3, π_4', where π_i is defined by X_i, i=1,2,3,4 and π_4' is defined by Y_4. Also, $\langle X_3, X_4 \rangle = \langle X_3, Y_4 \rangle = W^\# \cap C \cap G_7$, so the join of π_3 and π_4 is just that of π_3 and π_4'. Moreover, every residue of type $\{1,3,4'\}$ is isomorphic to the flag-chamber system corresponding to GAB $\Delta(W^\# \cap H;\ W^\# \cap G_6^* \cap H,\ W^\# \cap C \cap H,\ W^+ \cap W^\# \cap H)$ with diagram

It is obvious that this GAB is isomorphic to the GAB $\Delta(A_7;\ A_6,\ PSL(2,7),\ (A_3 \times A_4) \cdot 2)$ with the same diagram, whose 2-connectedness has been proved by Ronan [3]. So every residue of type $\{1,3,4'\}$ is 2-connected. Now consider the residues of type $\{2,3,4'\}$. Each of them consists of 27 chambers and has rank 3, in which every rank 1 residue consists of 3 chambers and every residue of rank 2 is a generalized 2-gon. So they are 2-connected. Thus $\mathcal{C}$ and $\mathcal{C}'$ satisfy the assumption of the lemma in section 2. Hence $\mathcal{C}'$ is 2-connected and so is Δ_7'.

<u>References</u>

1. W. M. Kantor, Some exceptional 2-adic buildings, J.Algebra 92 (1985), pp. 208-223.

2. W. M. Kantor, Finite simple groups via p-adic groups, pp. 175-181 in Proc. Rutgers Groups Theory Year, 1983-1984 (Eds. M.Aschbacher et al.), Cambridge University Press, Cambridge 1984.

3. M. A. Ronan, (unpublished).

4. J. Tits, A local approach to buildings. pp. 519 - 547 in "The Geometric Vein, The Coxeter Festschift" Springer, New York-Heidelberg - Berlin 1981.

ON THE NUMBER OF P-BLOCKS WITH GIVEN DEFECT GROUPS AND SOME APPLICATIONS OF THE P-POWER HOMOMORPHISM

Shi Shengming
Department of Mathematics
Peking University
Beijing, China

For a p-subgroup D of a finite group G, how many p-blocks does G have with D as a defect group? This is an important problem in modular representation theory. R. Brauer and K. A. Fowler [1], Y. Tsushima [2], T. Wada [3], K. Iizuka and A. Watanable [4], and W. Willems [5] discussed this problem. G. R. Robinson [6] gave a precise formula for the number in the case $D \triangleleft G$. Many results in [1], [2], [3] are easier corollaries of his formula. In this paper we obtain a generalized formula and many corollaries without the restriction $D \triangleleft G$. The key to doing this is the homomorphism in Lemma 1. Making use of the argument in [6], we calculate the homomorphism image and generalize the formula of Robinson. In addition, we give some applications of the homomorphism.

We need to fix our notation. G is a finite group, p is a prime, P is a fixed S_p-subgroup of G. D is a p-subgroup of G and $D < P$ (including $D = 1_G$). $D \lesssim D_1 (D \lesssim\neq D_1)$ means that D is conjugate to a subgroup (a proper subgroup) of D_1. $C_1, C_2, \cdots, C_r$ are the full set of conjugate classes of G, we also denote by the same symbol the sum of the elements in C_i, and, in brief, called the class sum.

We take $(K, R, \bar{R})$ as a splitting modular system of G. For example, we may take K a finite extension of the p-adic field Q_p such that the polynomial $x^{|G|} - 1$ splits into linear factors over K. Let R be the algebraic closure of Z_p (the p-adic integer ring) in K. π is the unique maximal ideal of R. Then $\bar{R} = R/\pi R$ is a finite field with characteristic p, and $x^{|G|} - 1$ also splits into linear factors over $\bar{R}$. Thus both K and $\bar{R}$ are splitting fields of G, all the subgroups and the quotient groups of G. For $r \in R$, we denote by r^* the congruence class of r in $\bar{R}$.

Let $\chi^1, \chi^2, \cdots, \chi^r$ be the full set of the irreducible characters of KG, $\omega^1, \omega^2, \cdots \omega^r$ be the central characters of KG corresponding to $\chi^1, \chi^2, \cdots, \chi^r$. χ_i^ℓ are the values of χ^ℓ for the elements of C_i. Let $C_{i'}$ be the conjugate class of the inverses of the elements of C_i, and h_i or $|C_i|$ the number of the elements of C_i. Let $e_1, e_2, \cdots, e_m$ be the full set of primitive idempotents of $Z(\bar{R}G)$. It is well known that $e_i e_j = \delta_{ij} e_j$, for $1 \le i, j \le m$ and $e_1 + e_2 + \cdots + e_m = 1$. $\bar{R}G$ has m blocks; we denote them by $B_1, B_2, \cdots, B_m$. Two central characters ω^i and ω^j of KG belong to one block of $\bar{R}G$ if and only if $(\omega_k^i)^* = (\omega_k^j)^*$, for $k = 1, 2, \cdots, r$. Let $\bar{\omega}^i$ be the functions of $Z(\bar{R}G)$, $\bar{\omega}_k^i = (\omega_k^i)^*$. They are central characters of $\bar{R}G$, and let $\bar{\omega}^1, \cdots, \bar{\omega}^m$ be the central characters corresponding to $B_1, \cdots, B_m$ (after a rearrangement of the order of $\bar{\omega}^1, \bar{\omega}^2, \cdots, \bar{\omega}^r$). Then

$$\bar{\omega}^i(e_j) = \delta_{ij}, \quad \text{for } 1 \le i, j \le m.$$

Let J(Z) be the radical of $Z(\bar{R}G)$. It is an ideal of $Z(\bar{R}G)$ consisting of all the nilpotent elements. Let $V_D (V_D^\circ)$ be the ideal of $Z(\bar{R}G)$ generated by all the class sums of G with defect groups $\lesssim D$ ($\lesssim\neq D$). Let V_D' be the $\bar{R}$-subspace of $Z(\bar{R}G)$ generated by all the class sums of G with defect groups $\not\gtrsim D$.

Lemma 1. There exist sufficiently large positive integers b such that the mapping from $Z(\bar{R}G) \to Z(\bar{R}G)$,

$$v \to v^{p^b}, \quad v \in Z(\bar{R}G),$$

is a $\bar{R}$-algebra homomorphism with the following properties,

1) For every $r \in \bar{R}$, $r^{p^b} = r$;

2) For any ideal V of $Z(\bar{R}G)$, let $e_1, e_2, \cdots, e_s$

be the full set of the primitive idempotents in V and $V(p^b)$ be the homomorphism image of V, then $V(p^b) = \bar{R}e_1 + \cdots + \bar{R}\, e_s$ and so s=

the dimension of $V(p^b)$ (in brief, $d(V(p^b))$.

Proof. Since $Z(\overline{R}G)$ is commutative and $\overline{R}$ is of characteristic p, for $v_1, v_2 \in Z(\overline{R}G)$ and any positive integer b,

$$(v_1 + v_2)^{p^b} = v_1^{p^b} + v_2^{p^b} ,$$

$$(v_1 v_2)^{p^b} = v_1^{p^b} v_2^{p^b} .$$

Let $|\overline{R}| = p^{b_1}$, then for any $r \in \overline{R}$, $r^{p^{b_1}} = r$.

Thus

$$(rv_1)^{p^{b_1}} = r^{p^{b_1}} v_1^{p^{b_1}} = r\, v_1^{p^{b_1}} .$$

This shows that the mapping corresponding to b_1 is a $\overline{R}$-algebra homomorphism.

From $e_1 + e_2 + \cdots + e_m = 1$, $V = Ve_1 + \cdots + Ve_m$.

Each Ve_i is an ideal of $Z(\overline{R}G)$. The primitive idempotents in Ve_i is, of course, in V. For $i > s$, assume that some e_j, $1 \le j \le s$, is in Ve_i. Then there exists $v \in V$, $ve_i = e_j$, Since $ve_i \cap ve_j = (0)$, for $i \ne j$, it is impossible. So for $i > s$, no idempotent is in ve_i and it is nilpotent, namely, $ve_i \subseteq V \cap J(Z)$. For $i = 1, 2, \cdots, s$, $Ve_i \subseteq Z(\overline{R}G)e_i$. Because $\overline{R}$ is a splitting field of G,

$Z(\overline{R}G) = \overline{R}e_1 + \cdots + \overline{R}e_m + J(Z)$. Then $Ve_i \subseteq Z(\overline{R}G)e_i \subseteq \overline{R}e_i + V \cap J(Z)$ and $V = Ve_1 + \cdots + Ve_m \subseteq \overline{R}e_1 + \cdots + \overline{R}e_s + \overline{R}e_{s'} + V \cap J(Z)$

Since $V \cap J(Z)$ is nilpotent, for a sufficient large positive integer b_2, $\{V \cap J(Z)\}^{p^{b_2}} = (0)$, thus $V(p^{b_2}) = \overline{R}e_1 + \cdots + \overline{R}e_{s'}$.

Choosing $b = kb_1 > b_2$, where k is a positive integer, the mapping $v \to v^{p^b}$, $v \in Z(\overline{R}G)$ meets the requirement of this lemma.

Remark. It is obvious that we can choose a common b such that for all the subgroups and the quotient groups of G, Lemma 1 holds. From now on, we fix the integer b.

When V is an ideal generated as a $\bar{R}$-vector space by some class sums of G, let $C_1, C_2, \cdots, C_t$ be the full set of class sums in V, then $V = \bar{R}C_1 + \cdots + \bar{R}C_t$ and $V(p^b) = \bar{R}C_1^{p^b} + \cdots + \bar{R}C_t^{p^b}$. We have

Corollary. Suppose $V = \bar{R}C_1 + \cdots + \bar{R}C_t$ is an ideal of $Z(\bar{R}G)$, $e_1, e_2, \cdots, e_s$ are the full set of primitive idempotents in V, then

$$V(p^b) = \bar{R}C_1^{p^b} + \cdots + \bar{R}C_t^{p^b} = \bar{R}e_1 + \cdots + \bar{R}e_s$$

and

$$s = \text{the rank of the vector set } \{C_1^{p^b}, \cdots, C_t^{p^b}\} \qquad (1)$$

Lemma 2. For each class sum C of G, we have

$$C^{p^b} = \sum_{\ell=1}^{m} \bar{\omega}^{\ell}(C) e_\ell ,$$

If C is in an ideal V of $Z(\bar{R}G)$ and let $e_1, \cdots, e_s$ be the full set of primitive idempotents in V.

Then
$$C^{p^b} = \sum_{\ell=1}^{s} \bar{\omega}^{\ell}(C) e_\ell , \qquad (2)$$

namely, $\bar{\omega}^{\ell}(C) = 0$, for all such ℓ that $e_\ell \notin V$.

Proof. Lemma 1 asserts that

$$C^{p^b} = \sum_{t=1}^{m} r_t e_t ,$$

so $\bar{\omega}^{\ell}(C^{p^b}) = \sum_{t=1}^{m} r_t \bar{\omega}^{\ell}(e_t) = \sum_{t=1}^{m} r_t \delta_{\ell t} = r_\ell$.

And the left side is $(\bar{\omega}^{\ell}(C))^{p^b} = \bar{\omega}^{\ell}(C)$, consequently $\bar{\omega}^{\ell}(C) = r_{\ell}$,

namely,
$$C^{p^b} = \sum_{\ell=1}^{m} \bar{\omega}^{\ell}(C)e_{\ell} .$$

When $C \in V$, it follows from Lemma 1 that $C^{p^b} = \sum_{\ell=1}^{s} \bar{\omega}^{\ell}(C)e_{\ell}$.

The proof is complete.

Corollary. $$C^{p^b}e_{\ell} = \bar{\omega}^{\ell}(C)e_{\ell} \tag{3}$$

Because of the form of (2), by using the method in [6], we can derive an explicit expression for C^{p^b} by the class sums of G and show that $d(V(p^b))$ is equal to the rank of some matrix. The latter result is a generalization of results in [6].

As in [6], let

$$\Omega_{ij} = \{(a,b) \mid a \in C_i,\ b \in C_j,\ b^{-1}a \in P\} , \tag{4}$$

then

$$|\Omega_{ij}| = \frac{h_j}{[G:P]} \sum_{\ell=1}^{r} \omega_i^{\ell} x_{j'}^{\ell} (x^{\ell}, 1)_P ,$$

where $(x^{\ell}, 1)_P$ is the inner product of class functions $x^{\ell}|_P$ and 1_P on P, it is an integer.

Let p^{d_j} be the order of the defect groups of C_j and $p^a \| |G|$ then $p^{a-d_j} \| h_j$, so $p^{a-d_j} \mid |\Omega_{ij}|$.

We define
$$S_{ij} = \left(\frac{|\Omega_{ij}|}{p^{a-d_j}}\right)^* = \frac{(h_j/p^{a-d_j})^*}{[G:P]^*} \sum_{\ell=1}^{r} (\omega_i^{\ell})^* (x_{j'}^{\ell})^* (x^{\ell}, 1)_P^*$$

$$= \frac{(h_i/p^{a-d_j})^*}{[G:P]^*} \sum_{\ell=1}^{m} \bar{\omega}_i^{\ell} \sum_{x \in B_{\ell}} (x_{j'})^* (x\ , 1)_P^* , \tag{5}$$

Theorem 3. Let $V = \bar{R}C_1 + \cdots + \bar{R}C_t$ be an ideal of $Z(\bar{R}G)$ and $C_1, \cdots, C_{t_1}$ $(t_1 \le t)$ be the full set of p'-class sums in V. Let the matrix $S_V = (S_{ij})_{1 \le i,\, j \le t_1}$, then

$$d(V(p^b)) = \text{rank } (S_V) \tag{6}$$

Proof. From (1), $d(V(p^b)) = \text{rank } \{C_1^{p^b}, \cdots, C_t^{p^b}\}$ and from (2),

$$C_i^{p^b} = \sum_{\ell=1}^{m} \bar{\omega}_i^{\ell} \, e_\ell \ , \ (i=1,2,\cdots, t).$$

For each idempotent e_ℓ of $Z(\bar{R}G)$, we can lift it to an idempotent of $Z(RG)$; it is a linear combination of p'-class sums of G and the coefficient of C_j in the combination is equal to

$$\frac{1}{|G|} \sum_{X \in B_\ell} x(1) x_{j'} \ ,$$

The block orthogonality relations show that $\sum_{X \in B_\ell} x(u) x_j = 0$ for $u \in P \setminus \{1\}$. It follows that

$$\frac{1}{|G|} \sum_{x \in B_\ell} \sum_{u \in P} x(u) x_{j'} = \frac{1}{[G:P]} \sum_{x \in B_\ell} (x,1)_P x_{j'} \ .$$

V is an ideal of $Z(\bar{R}G)$, $C_i^{p^b} \in V = \bar{R}C_1 + \cdots + \bar{R}C_t$.

And e_ℓ is a $\bar{R}$-linear combination of p'-class sums of G, so is $C_i^{p^b}$. Hence $C_i^{p^b} \in \bar{R}C_1 + \cdots + \bar{R}C_{t_1}$. Consequently,

$$C_i^{p^b} = \frac{1}{[G:P]^*} \sum_{j=1}^{t_1} \sum_{\ell=1}^{m} \bar{\omega}_i^{\ell} \sum_{x \in B_\ell} (x,1)_P^* \, x_j^* \cdot C_j$$

$$= \sum_{j=1}^{t_1} S_{ij}/(h_j/p^{a-d_j})^* \, C_j \qquad (7)$$

From (2) again, $C_i^{p^b} = \sum_{\ell=1}^{m} \bar{\omega}_i^{\ell} e_\ell$, so $C_i^{p^b} = (C_i^{p^b})^{p^b}$.

(7) shows that $C_i^{p^b}$ $(i=1,\cdots,t)$ are linear combinations of $C_1^{p^b},\cdots, C_{t_1}^{p^b}$; we get $\text{rank}\ \{C_1^{p^b},\cdots, C_t^{p^b}\} = \text{rank}\ \{C_1^{p^b},\cdots, C_{t_1}^{p^b}\}$.

Since $\{C_1,\cdots, C_{t_1}\}$ is a linear independent vector set, it follows from (7) that

$$\text{rank}\ \{C_1^{p^b},\cdots, C_{t_1}^{p^b}\} = \text{rank}\ (S_{ij}/(h_j/p^{a-d_j})^*)$$

$$=\text{rank}\ (S_{ij}) = \text{rank}\ (S_V),\quad 1 \le i,\ j \le t_1 .$$

The proof is complete.

Corollary 1. Let V_D be the ideal of $Z(\overline{R}G)$ generated by the full set of class sums of G with defect groups $\lesssim D$, then the number of the blocks of $\overline{R}G$ with defect groups $\lesssim D$ is equal to rank (S_{V_D}).

Proof. The number of the primitive idempotents in V_D is just the number of the blocks of $\overline{R}G$ with defect groups $\lesssim D$. The result follows from Lemma 1 and Theorem 3.

Corollary 2. Let V_D° be the ideal of $Z(\overline{R}G)$ generated by the full set of class sums with defect groups $\lesssim\neq D$. Then the number of the blocks of $\overline{R}G$ with defect groups $\lesssim\neq D$ is equal to rank $(S_{V_D^\circ})$.

Proof. The number of the primitive idempotents in V_D° is just the number of the blocks of $\overline{R}G$ with defect groups $\lesssim\neq D$. The result follows from Lemma 1 and Theorem 3 again.

From corollaries 1 and 2, we have immediately

Corollary 3. The number of the blocks of $\overline{R}G$ with the defect group D is equal to rank (S_{V_D})−rank $(S_{V_D^\circ})$.

Corollary 4. Let $C_1, \cdots, C_{t_1}$ be the full set of p'-class sums in V_D and $C_1, \cdots, C_{t_2}$ $(t_2 \le t_1)$ be the full set of p'-class sums with the defect group D. Let

$$S_{V_D}^\circ = (S_{ij})_{1 \le i,\, j \le t_2}$$

Then

$$S_{V_D} = \begin{pmatrix} S_{V_D}^\circ & * \\ 0 & S_{V_D^\circ} \end{pmatrix} \tag{8}$$

In particular, in the case $D \triangleleft G$, the number of the blocks of $\overline{R}G$ with the defect group D is equal to rank $(S_{V_D}^\circ)$. (G.R. Robinson)

Proof. The defect groups of $C_{t_2+1}, \cdots, C_{t_1} \underset{\sim}{<} \ne D$, so C_i are in V_D°, for $i=t_2+1, \cdots, t_1$. Since V_D° is an ideal and

$$C_i^{p^b} = \sum_{j=1}^{t_1} S_{ij}/(h_j/p^{a-d_j})^* C_j ,$$

$S_{ij} = 0$ for $t_2+1 \le i \le t_1$, $1 \le j \le t_2$. so (8) holds.

In the case $D \triangleleft G$, if we can prove that (*) in (8) is the zero matrix, then the result follows from corollary 3 and (8).

Since the defect group of each block of $\overline{R}G$ contains D, the defect group of the block corresponding to any primitive idempotent in V_D is D. Let $e_1, \cdots, e_s$ be the full set of primitive idempotents in V_D.

From (2), $C_i^{p^b} = \sum_{\ell=1}^{s} \bar{\omega}_i^\ell e_\ell$, for $1 \le i \le t_2$. e_i, $i=1, \cdots, s$, are of the defect group D and $D \triangleleft G$, so they are linear combinations of the p'-class sums with the defect group D, that is, linear combinations of $C_1, \cdots, C_{t_2}$. Consequentley $C_i^{p^b}$ $(i=1,\cdots, t_2)$ are linear combinations of $C_1,\cdots, C_{t_2}$ thus $S_{ij}=0$, for $i=1,\cdots, t_2$, $t_2+1 \leq j \leq t_1$, namely, (*) = the zero matrix. The proof is complete.

Now denote S matrices of G by $S(G)$ and the ones of $N=N_G(D)$ by $S(N)$. As an application of the First Main Theorem on blocks, we proved

Lemma 4. $\quad \mathrm{rank}\,(S(G)^{o}_{V_D}) = \mathrm{rank}(S(N)^{o}_{V_D})$,

the the number of the blocks of $\overline{R}G$ with the defect group D is equal to rank $(S(G)^{o}_{V_D})$.

Proof. Let σ be the Brauer homomorphism with respect to D, $Z(\overline{R}G) \to Z(\overline{R}N)$. Let $C_1,\cdots, C_{t_2},\cdots, C_{t_1}$ be the full set of p'-class sums of G with defect groups $\lesssim D$ and $C_1, C_2,\cdots, C_{t_2}$ the ones with the defect group D. It is well known that $\sigma(C_1), \sigma(C_2), \cdots \sigma(C_{t_2})$ are the full set of p'-class sums of $N=N_G(D)$ with the defect group D and $\sigma(C_j)=0$, for $t_2<j\leq t_1$.

Assume that $\quad C_i^{p^b} = \sum_{j=1}^{t_1} M_{ij}\, C_j, \quad 1\leq i\leq t_2$.

Since $\quad \sigma(C_i^{p^b}) = \sigma(C_i)^{p^b}, \quad \sigma(C_i)^{p^b} = \sigma\,(\sum_{j=1}^{t_1} M_{ij}\, C_j) = \sum_{j=1}^{t_1} M_{ij}\, \sigma(C_j)$

$$= \sum_{j=1}^{t_2} M_{ij}\, \sigma(C_j).$$

Denote (hj/p^{a-d_j}) by $(h_j)p'$ and the one corresponding to $\sigma(C_j)$ by $(h_{\sigma(C_j)})_{p'}$. It follows from (7) that $M_{ij} = S(G)_{ij}/(h_j)^*_{p'}$. But

$\sigma(C_i)^{p^b} = \sum_{j=1}^{t_2} M_{ij}\, \sigma(C_j)$ implies that

$$M_{ij} = S(N)_{ij}/(h_{\sigma(C_j)})^*_{p'} \quad .$$

Hence

$$S(G)_{ij} = S(N)_{ij}(h_j)^*_{p'}/(h_{\sigma(C_j)})^*_{p'} \;, \quad 1 \le i,j \le t_2 \;,$$

and it follows that

$$\text{rank } (S(G)^0_{V_D}) = \text{rank } (S(N)^0_{V_D}) \;.$$

From the Brauer's first main theorem and corollary 4 of Theorem 3, we get

$$\text{rank } (S(G)^0_{V_D}) = \text{rank } (S(N)^0_{V_D}) = \text{the number of the blocks of } \overline{R}N$$

with the defect group D = the number of blocks of $\overline{R}G$ with the defect group D.

The proof is complete.

Corollary 5. D is a defect group of a block of $\overline{R}G$ if and only if $(S^{\circ}_{V_D}) \neq 0$

Now we will give another explicit formula of S_{ij}, it relates closely to the structure of G. The idea is due to [6].

$$\Omega_{ij} = \{(x_1, x_2) \mid x_1 \in C_i,\ x_2 \in C_j,\ x_2^{-1}x_1 \in P\} \;.$$

And $x_2^{-1} x_1 \in P$ if and only if x_1, x_2 belong to one coset gP of P in G. If gP contains a elements in C_i and b elements in C_j, then it contains ab pairs of Ω_{ij}. For any $x \in P$, $xgP = xgPx^{-1}$ also contains ab pairs of Ω_{ij}. So the contributions to $|\Omega_{ij}|$ from PgP is $[P: gPg^{-1} \cap P]$ab, since PgP contains $[P: gPg^{-1} \cap P]$ left cosets of P in G.

Theorem 5. Let V be V_D or V°_D and $C_1, \cdots, C_{t_1}$ be the p'-class sums in V. Suppose that the decomposition of G into the union of the P-P double cosets is

$$G = \bigcup_{i=1}^{k_1} Pg_iP \;.$$

Let, whenever possible, $g_1, \cdots, g_k (k \le k_1)$ be the full set of such g_ℓ, $1 \le \ell \le k_1$, that satisfy

(1) $P \cap g_\ell P g_\ell^{-1} \lesssim D$;

(2) $g_\ell P \cap C_i \ne \phi$, for some i, $1 \le i \le t_1$.

If it is not possible to choose any g_ℓ which satisfies conditions (1), (2), then $S_V = 0$. Otherwise we define $A = (a^*_{ij})_{t_1 \times k}$, $B = (b^*_{ij})_{t_1 \times k}$

where $a_{ij} = |C_i \cap g_j P|$ and $b_{ij} = a_{ij}[P : g_j P g_j^{-1} \cap P]/P^{a-d_i}$,

then $S_{ij} = \sum_{\ell=1}^{k} a^*_{i\ell} b^*_{j\ell}$, namely, $S_V = AB^T$.

Proof. $\frac{1}{p^{a-d_j}} |\Omega_{ij}| = \sum_{\ell=1}^{k_1} a_{i\ell}\, a_{j\ell}[P : P \cap g_\ell P g_\ell^{-1}]/p^{a-d_j}$

$$= \sum_{\ell=1}^{k_1} a_{i\ell} b_{j\ell} .$$

We show that $b^*_{j\ell}$ is well-defined, that is, $b_{j\ell}$ is an integer. Note that the group $P \cap g_\ell P g_\ell^{-1}$ permutes the elements of $C_i \cap g_\ell P$ under conjugations. For $y \in C_j$, the length t of its orbit under $P \cap g_\ell P g_\ell^{-1}$ is

$$t = \frac{|P \cap g_\ell P g_\ell^{-1}|}{|P \cap g_\ell P g_\ell^{-1} \cap C_G(y)|}. \tag{9}$$

Since the order of the defect group D_j of C_j is p^{dj} and $P \cap g_\ell P g_\ell^{-1} \cap C_G(y)$ is a p-subgroup of G which is conjugate to a subgroup of D_j, then

$$\frac{[P : P \cap g_\ell P g_\ell^{-1}]t}{p^{a-d_j}} = \frac{|P|}{|P \cap g_\ell P g_\ell^{-1}|} \cdot \frac{p^{dj}|P \cap g_\ell P g_\ell^{-1}|}{|P||P \cap g_\ell P g_\ell^{-1} \cap C_G(y)|} \tag{10}$$

$$= \frac{p^{d_j}}{|P \cap g_\ell P g_\ell^{-1} \cap C_G(y)|} = \text{an integer.}$$

And $a_{j\ell}$ is a sum of some t's, so

$$b_{j\ell} = \frac{[P : P \cap g_\ell P g_\ell^{-1}]}{p^{a-d_j}} \; a_{j\ell} = \text{an integer.}$$

Now we claim that for $\ell > K$, $a^*_{i\ell} = 0$, namely, $S_{ij} = \sum_{\ell=1}^{k} a^*_{i\ell} b^*_{j\ell}$.

From (9), $t \equiv 0$ or $1 (\mathrm{mod}\, p)$ and $t \equiv 1$ if and only if there exists a S_p-subgroup of $C_G(y)$ which contains $P \cap g_\ell P g_\ell^{-1}$ or equivalently, $P \cap g_\ell P g_\ell^{-1} \lesssim D_j$. If $a^*_{j\ell} \neq 0$, then $a_{j\ell} \not\equiv 0 \ (\mathrm{mod}\, p)$ and we have at least one $t \not\equiv 0 \ (\mathrm{mod}\, p)$. Thus $P \cap g_\ell P g_\ell^{-1} \lesssim D_j$. This implies that if $P \cap g_\ell P g_\ell^{-1} \not\lesssim D_j$, then $a^*_{j\ell} = 0$. In particular, since $D_j \lesssim D$, so if $P \cap g_\ell P g_\ell^{-1} \not\lesssim D$, then $a^*_{j\ell} = 0$, $j = 1, \cdots, t_1$.

Clearly the condition (2) is necessary, as in the case $C_i \cap g_\ell P = \phi$, $a_{i\ell} = 0$, for all i.

In summary, if no g_ℓ satisfies conditions (1) and (2), then all $a^*_{i\ell} = 0$ and $S_{ij} = \sum_{\ell=1}^{K_1} a^*_{i\ell} b^*_{j\ell} = 0$, for all i, j, namely, $S_v = 0$.

If $k \neq 0$, then for $\ell > K$, $a^*_{j\ell} = 0$ and $S_{ij} = \sum_{\ell=1}^{k_1} a^*_{i\ell} b^*_{j\ell} = \sum_{i=1}^{k} a^*_{i\ell} b^*_{j\ell}$, namely, $S_v = AB^T$. The proof is complete.

Remark. From the proof of Theorem 5, if $a^*_{i\ell} \neq 0$, then $P \cap g_\ell P g_\ell^{-1} \lesssim D_i$. From (10), if $b^*_{j\ell} \neq 0$, then $b_{j\ell} \not\equiv 0 \ (\mathrm{mod}\ p)$ and there exists some $y \in C_j \cap g_\ell P$ such that some S_p-subgroup of $C_G(y)$ is contained in $P \cap g_\ell P g_\ell^{-1}$, then $D_j \lesssim P \cap g_\ell P g_\ell^{-1}$. So if $S_{ij} \neq 0$, then $a^*_{i\ell} \neq 0$ for at least one ℓ. Thus $D_j \lesssim P \cap g_\ell P g_\ell^{-1} \lesssim D_i$.

From Theorem 5 and corollaries 1 and 2 of Theorem 3, we can calculate the number of the blocks of $\overline{R}G$ with defect groups $\lesssim D$ or $\lneqq D$.

Corollary 1 (G.R. Robinson). Suppose $D \triangleleft G$. Let $C_1, \cdots, C_{t_2}$ be the p'class sums of G with the defect group D and $g_1, \cdots, g_k$ be the full set of representatives of the P-P double cosets in G which satisfy

(1) g_j is p-regular.

(2) D is a S_p-subgroup of $C_G(g_j)$.

(3) $P \cap g_j P g_j^{-1} = D$.

If no such g_j exists, then $S^0_{V_D} = 0$. If $k \neq 0$, let

$A = (a^*_{ij})_{t_2 \times k}$, where $a_{ij} = |C_i \cap g_j C_P(D)|$, then $S^0_{V_D} = AA^T$.

Proof. From Theorem 5, $S_{ij} = \sum_{\ell=1}^{k} a^*_{i\ell} b^*_{j\ell}$, where $a^*_{i\ell}, b^*_{j\ell}, g_\ell$ are as in Theorem 5. $S^o_{V_D} = (S_{ij})_{1 \le i, j \le t_2}$. For $1 \le i \le t_2$, C_i is of the defect group D. Since $D \triangleleft G$ and $D < P$, $D < P \cap g_\ell P g_\ell^{-1}$, for $1 \le \ell \le k$. Then the condition (1) in Theorem 5 implies that $D = g_\ell P g_\ell^{-1} \cap P$, for $1 \le \ell \le K$. If for some ℓ, $1 \le \ell \le K$, $g_\ell P \cap C_i = \phi$, for all i, $1 \le i \le t_2$, then all $a_{i\ell} = 0$, for $1 \le i \le t_2$. So we can omit all terms with such ℓ's in $S_{ij} = \sum_{\ell=1}^{k} a^*_{i\ell} b^*_{j\ell}$. For remaiming ℓ's, $g_\ell P \cap C_i \neq \phi$, for some i, $1 \le i \le t_2$. Thus we can choose g_ℓ in some C_i, $1 \le i \le t_2$, for any such ℓ. For all these g_ℓ the condition (3) still holds, and so do conditions (1) and (2). In brief, we still denote the remaining g_ℓ's by $g_1, g_2, \cdots, g_k$. If no such g_ℓ exists, then all $a^*_{i\ell}$ 0, so $S^o_{V_D} = 0$.

If $k \neq 0$, then $S_{ij} = \sum_{\ell=1}^{k} a^*_{i\ell} b^*_{j\ell}$ for $1 \le i, j \le t_2$.

Now we claim that for $1 \le \ell \le k$, $1 \le i, j \le t_2$, $b_{j\ell} = a_{j\ell}$ and the $a_{i\ell}$'s in Theorem 5 and in this corollary are the same.

Since for all i, $1 \le i \le t_2$, C_i is of defect group D, so $p^{d_i} = |D|$.

$$b_{i\ell} = \frac{[P : P \cap g_\ell P g_\ell^{-1}]}{p^{a-d_i}} a_{i\ell} = \frac{[P:D]}{p^{a-d_i}} a_{i\ell} = a_{i\ell}.$$

Note that if $y \in g_\ell P \cap C_i$, then both y and g_ℓ are in $C_G(D)$, so $g_\ell^{-1} y \in C_G(D)$. But $y \in g_\ell P$, this implies that $g_\ell^{-1} y \in P$, then $g_\ell^{-1} y \in C_P(D)$, namely, $y \in g_\ell C_P(D)$. Consequently $a_{i\ell} = |g_\ell C_P(D) \cap C_i|$. The result is proved.

In the remaining of this paper, we give some applications of Theorem 5.

Lemma 6. Suppose that the p-subgroup D of G is a S_p-subgroup of $DC_G(D)$. Let C be a p'-class sum with a defect group containing D. Then

$$C^{p^b} \equiv C \pmod{V_D'}$$

where V_D' is the $\bar{R}$-subspace of $Z(\bar{R}G)$ generated by all the class sums of G with defect groups $\not\supseteq D$.

Proof. At first we observe the subgroup $DC_G(D)$. Let $\hat{C}_1, \cdots, \hat{C}_u$ be the full set of p'-classes of $DC_G(D)$. We show that $\hat{C}_i^{p^b} = \hat{C}_i$ (i=1, ..., u).

Since the p'-elements of $DC_G(D)$ are all in $C_G(D)$, $\hat{C}_i \subseteq C_G(D)$, (i=1,···, u). We calculate S_{ij}, for $1 \le i, j \le u$, by using Theorem 5. For any coset $g_\ell D$ which satisfies the conditions (1) and (2), $g_\ell \in C_G(D)$, so it is the only p'-element in $g_\ell D$. It implies that there exist $|\hat{C}_i|$ cosets $g_\ell D$ such that $g_\ell \in \hat{C}_i$. At this moment $a_{i\ell} = 1$ and $a_{j\ell} = 0$ for $j \ne i$. Since every $\hat{C}_i$ is of defect group D, so $p^{a-d_i} = 1$.

Then $b_j = \dfrac{[D:D \cap g_\ell D g_\ell^{-1}]}{p^{a-d_j}} a_{j\ell} = \dfrac{[D:D]}{1} a_{j\ell} = a_{j\ell}$.

Consequently,

$$S_{ij} = \sum_{g_\ell \in \hat{C}_i} a_{i\ell}^* b_{j\ell}^* = \sum_{g_\ell \in \hat{C}_i} a_{i\ell}^* a_{j\ell}^* = \sum_{g_\ell \in \hat{C}_i} a_{i\ell}^* \delta_{ij}$$

$$= \sum_{g_\ell \in \hat{C}_i} \delta_{ij} = |\hat{C}_i|^* \delta_{ij} .$$

Note that $(h_j/p^{a-d_j})^* = h_j^* = |\hat{C}_j|^* \ne 0$, so

$$\hat{C}_i^{p^b} = \sum_{j=1}^{u} S_{ij}/(h_j/p^{a-d_j})^* \hat{C}_j = \sum \frac{|\hat{C}_i|^* \delta_{ij}}{|\hat{C}_i|^*} \hat{C}_j = \hat{C}_i \quad , \tag{11}$$

Let C be a p'-class sum of G with a defect group containing D. Let σ be the Brauer homomorphism from $Z(\overline{R}G)$ to $Z(\overline{R}N)$ with respect to D. Then $\sigma(C)\neq 0$ and $\sigma(C)\subseteq C_G(D)$.

It follows that

$$\sigma(C) = \sum_{i\in I} \hat{C}_i ,$$

Where I is a subset of $\{1,2,\cdots, u\}$.

Then

$$\sigma(C)^{p^b} = (\sum_{i\in I} \hat{C}_i)^{p^b} = \sum_{i\in I} \hat{C}_i^{p^b} = \sum_{i\in I} \hat{C}_i = \sigma(C), \tag{12}$$

consequently

$$0 = \sigma(C)^{p^b} - \sigma(C) = \sigma(C^{p^b} - c)$$

and

$$C^{p^b} - C \in \ker \sigma = V_D' .$$

Thus

$$C^{p^b} \equiv C (\mathrm{mod}\ V_D')$$

Theorem 7. Suppose that the p-subgroup D of G is a S_p-subgroup of $DC_G(D)$. Then the number of the blocks of $\overline{R}G$ with a defect group D_1 containing D is equal to the number of the p'-classes of G with the defect group D_1 . Therefore the number of the blocks of $\overline{R}G$ with defect groups containing D equals the number of the p'-classes of G with defect groups containing D.

Proof. Let D_1 be a p-subgroup of G and $D_1 \geq D$. Let $C_1,\cdots,C_s$ be the p'-classes of G with the defect group D_1. We know that $S^{\circ}_{V_{D_1}} = (S_{ij})_{1\leq i,\, j\leq s}$. Since $C_i^{p^b} \equiv C_i (\mathrm{mod}\ V_D')$, for $1\leq i\leq s$, (Lemma 6), $S_{ij}/(h_j/p^{a-d_j})^* = \delta_{ij}$, for $1\leq j\leq s$.

$$S_{ij} = (h_j/p^{a-d_j})^* \delta_{ij} = (h_j/[P:D_1])^* \delta_{ij}$$

and

$$S^{\circ}_{V_{D_1}} = \begin{pmatrix} (h_1/[P:D_1])^* & & 0 \\ & \ddots & \\ 0 & & (h_s/[P:D_1])^* \end{pmatrix},$$

Then from corollary 4, $s = \operatorname{rank}(S^{\circ}_{V_{D_1}})$ = the number of the blocks of $\overline{R}G$ with the defect group D_1. The first part of the theorem is proved.

If D_1 is a defect group of a fixed block of $\overline{R}G$, so are the conjugates of D_1. A similar fact holds for the defect groups of a fixed p'-class of G. Denote the set of the conjugates of D_1 by (D_1). Then the number of the blocks with defect groups containing D is equal to

$\sum_{(D_1),D_1 \geqq D}$ the number of the blocks with (D_1) as defect groups =

$\sum_{(D_1),D_1 \geqq D}$ the number of the p'-classes with (D_1) as defect groups

= the number of the p'-classes with defect groups containing D_1.

Corollary (Brauer). The number of the blocks of $\overline{R}G$ with P as a defect group is equal to the number of the p'-classes of G with P as a defect group.

Proof. Clearly P is a S_p-subgroup of $PC_G(P)$. The result follows from the above theorem.

By using Lemma 6, we can also calculate the idempotent of the principal block of $\overline{R}G$, for the group G in the lemma. We call the idempotent of the principal block the principal idempotent.

Lemma 8. Suppose the p-subgroup D of G is a S_p-subgroup of $DC_G(D)$. Let $\hat{C}_1, \cdots, \hat{C}_n$ be the p'-classes of $DC_G(D)$, then the principal idempotent of $\overline{R}(DC_G(D))$ is equal to $\frac{1}{N_1^*} \sum_{i=1}^{n} \hat{C}_i$, where $N_1 = [DC_G(D):D]$.

Remark. Let f be an primitive idempotent of $Z(\overline{R}G)$ with a central character $\overline{\omega}$. Then f is a principal idempotent if and only if $\overline{\omega}(C) = h_c^*$ for any p'-class sum C of G.

The proof of Lemma 8. Let η be the natural homomorphism from $DC_G(D)$ to $DC_G(D)/D$. It induces a $\bar{R}$-algebra homomorphism β, from $\bar{R}(DC_G(D)) \to \bar{R}(DC_G(D)/D)$. Since D is a S_p-subgroup of $DC_G(D)$, $\bar{R}(DC_G(D)/D)$ is semi-simple. We have that ker β = rad $(\bar{R}DC_G(D))$. [See, for example, the remark in [7], P. 144]. And it is easily seen that there is a one-one correspondence between the p'-elements of $DC_G(D)$ (in fact, $C_G(D)$) and $DC_G(D)/D$ under β. Furthermore, if $\hat{C}_a$ is a p'-class of $DC_G(D)$ to which a belongs, then $\beta\hat{C}_a$ is a p'-class of $DC_G(D)/D$. So $\beta\hat{C}_1, \cdots, \beta\hat{C}_n$ is the full set of p'-classes of $DC_G(D)/D$. Since $DC_G(D)/D$ is a p'-group, it is easy to calculate the principal idempotent of $\bar{R}(DC_G(D)/D)$, that is $\frac{1}{N_1^*} \sum_{i=1}^{n} \beta\hat{C}_i$

Now we claim that $\frac{1}{N_1^*} \sum_{i=1}^{n} \hat{C}_i$ is just the principal idempotent of $\bar{R}(DC_G(D))$. Let I be the subalgebra of $\bar{R}(DC_G(D))$ generated by all the central idempotents of $\bar{R}(DC_G(D))$. From Lemma 1 and (11), $\hat{C}_i \in I$, for all i. So $\frac{1}{N_1^*} \sum_{i=1}^{n} \hat{C}_i \in I$.

Since $\frac{1}{N_1^*} \sum_{i=1}^{n} \beta\hat{C}_i$ is the principal idempotent of $\bar{R}(DC_G(D)/D)$, for any j, $1 \le j \le n$,

$$(\beta\hat{C}_j)\left(\frac{1}{N_1^*} \sum_{i=1}^{n} (\beta\hat{C}_i)\right) = |\beta\hat{C}_j|^* \left(\frac{1}{N_1^*} \sum_{i=1}^{n} \beta\hat{C}_i\right)$$

$$= |\hat{C}_j|^* \left(\frac{1}{N_1^*} \sum_{i=1}^{n} \beta\hat{C}_i\right).$$

Then

$$\beta\left(\hat{C}_j \frac{1}{N_1^*} \sum_{i=1}^{n} \hat{C}_i\right) = |\hat{C}_j|^* \beta\left(\frac{1}{N_1^*} \sum_{i=1}^{n} \hat{C}_i\right)$$

and $$\hat{C}_j\left(\frac{1}{N_1^*}\sum_{i=1}^{n}\hat{C}_i\right)\equiv|\hat{C}_j|^*\left(\frac{1}{N_1^*}\sum_{i=1}^{n}\hat{C}_i\right) \bmod \operatorname{rad} Z(\bar{R}DC_G(D)), \quad (13)$$

Since $\bar{R}$ is splitting, $Z(\bar{R}DC_G(D)) = I + \operatorname{rad} Z(\bar{R}DC_G(D))$.
And both two sides of (13) are in I, so (13) must be an equality, namely

$$\hat{C}_j\left(\frac{1}{N_1^*}\sum_{i=1}^{n}\hat{C}_i\right)=|\hat{C}_j|^*\left(\frac{1}{N_1^*}\sum_{i=1}^{n}\hat{C}_i\right). \quad (14)$$

If $\hat{f}$ is the principal idempotent of $\bar{R}(DC_G(D))$, then from (11) and (3), we have

$$\hat{C}_j\hat{f} = \hat{C}_j^{p^b}\cdot\hat{f} = |\hat{C}_j|^*\hat{f}.$$

Assume that $\hat{f}=\sum_{j=1}^{n} r_j\hat{C}_j$, it follows from $\hat{f}^2=\hat{f}$ that

$\sum_{j=1}^{n} r_j|\hat{C}_j|^* = 1$. Now we have

$$\hat{f}\left(\frac{1}{N_1^*}\sum_{i=1}^{n}\hat{C}_i\right)=\sum_{j=1}^{n} r_j\hat{C}_j\left(\frac{1}{N_1^*}\sum_{i=1}^{n}\hat{C}_i\right)=\frac{1}{N_1^*}\sum_{i=1}^{n}\hat{C}_i.$$

On the other hand $\left(\frac{1}{N_1^*}\sum_{i=1}^{n}\hat{C}_i\right)\hat{f}=\frac{1}{N_1^*}\sum_{i=1}^{n}|\hat{C}_i|^*\hat{f}$.

Since $\sum_{i=1}^{n}|\hat{C}_i| = \sum_{i=1}^{n}|\beta\hat{C}_i| = N_1$, so $\sum_{i=1}^{n}|\hat{C}_i|^* = N_1^*$.

Thus $\frac{1}{N_1^*}\sum_{i=1}^{n}\hat{C}_i\hat{f} = \hat{f}$.

Consequently $\hat{f} = \frac{1}{N_1^*} \sum_{i=1}^{n} \hat{C}_i$. The lemma is proved.

Theorem 9. The p-subgroup D of G is a S_p-subgroup of $DC_G(D)$. Let $C_1, \cdots, C_{t_1}$ be the full set of p'-classes of G with defect groups $\gtrsim D$. Let $N_1 = [DC_G(D):D]$. Then the principal idempotent of $\bar{R}N_G(D)$ is $\frac{1}{N_1^*} \sum_{i=1}^{t_1} \sigma(C_i)$, where σ is the Brauer homomorphism from $Z(\bar{R}G)$ $Z(\bar{R}N_G(D))$ with respect to D, and the principal idempotent of $\bar{R}G$ is $\frac{1}{N_1^*} \sum_{i=1}^{t_1} C_i$, (mod V_D').

Proof. Every p'-class of $N_G(D)$ with defect groups $\geq D$ is contained in $C_G(D)$, so every such p'-class sum is a sum of p'-elements of $C_G(D)$ and the sum of all such p'-class sums of $\bar{R}N_G(D)$ is just the sum of all the p'-elements of $C_G(D)$ and is also the sum of all the p'-class sums of $\bar{R}DC_G(D)$. Let $\hat{f} = \frac{1}{N_1^*} \sum_{i=1}^{n} \hat{C}_i$ defined as in

Lemma 8. Let $\tilde{C}$ be a p'-class of $N_G(D)$ with a defect group $\geq D$. Then $\tilde{C} = \sum_{\text{some } j\text{'s}} \hat{C}_j$, and

$$\tilde{C}\hat{f} = \sum_{\text{some } j\text{'s}} \hat{C}_j\hat{f} = \sum_{\text{some } j\text{'s}} |\hat{C}_j|^* \hat{f} = |\tilde{C}|^* \hat{f} . \qquad (15)$$

From (11), we have

$$\tilde{C}^{p^b} = \sum_{\text{some } j\text{'s}} (\hat{C}_j)^{p^b} = \sum_{\text{some } j\text{'s}} \hat{C}_i = \tilde{C} .$$

$\hat{f}$ is a primitive idempotent of $\bar{R}(DC_G(D))$ and is also a central idempotent of $\bar{R}(N_G(D))$. And it satisfies

$$\tilde{C}^{p^b}\hat{f} = \tilde{C}\hat{f} = |\tilde{C}|^*\hat{f} \quad ,$$

for all the p'-class sums $\tilde{C}$ of $\bar{R}N_G(D)$ with defect groups $\geq D$. From (3), the value of the central character associated with $\hat{f}$ at $\tilde{C}$ is $|\tilde{C}|^*$. This is just the value of the central character associated with the principal idempotent at $\tilde{C}$.

Since $D \triangleleft N_G(D)$, the defect group of $\hat{f}$ contains D. So if $\tilde{C}$ is a p'-class of $N_G(D)$ with a defect group $\not\geq$ D, the value of the central character associated with $\hat{f}$ at $\tilde{C}$ is zero. It is also equal to the value of the one assciated with the principal idempotent. Certainly $\hat{f}$ is just the principal idempotent of $\bar{R}N_G(D)$.

Since $C_1, \cdots, C_{t_1}$ is the full set of p'-classes of G with defect groups $\gtrsim$D, then $\sigma(C_1) \cup \sigma(C_2) \cup \cdots \cup \sigma(C_{t_1})$ is the union of all the p'-classes of $N_G(D)$ with defect groups $\geq D$. So

$$\hat{f} = \frac{1}{N_1^*} \sum_{i=1}^{n} \hat{C}_i = \frac{1}{N_1^*} \sum_{i=1}^{t_1} \sigma(C_i)$$

Now let $e = \frac{1}{N_1^*} \sum_{i=1}^{t_1} C_i^{p^b}$. From Lemma 6,

$$e \equiv \frac{1}{N_1^*} \sum_{i=1}^{t_1} C_i \quad (\text{mod } V_D'). \text{ So } \quad \sigma(e) = \hat{f} .$$

Let $e_1, \cdots, e_m$ be the full set of block idempotents of $\bar{R}G$. From (2)

$$e = \frac{1}{N_1^*} \sum_{i=1}^{t_1} C_i^{p^b} = \ell_1 e_1 + \cdots + \ell_m e_m ,$$

then

$$\sigma(e) = \hat{f} = \ell_1\sigma(e_1) + \cdots + \ell_m\sigma(e_m) .$$

Each $\sigma(e_i)$ is zero or a central idempotent of $\overline{R}N_G(D)$. Even if for some i,j, $\sigma(e_i) \neq 0$ and $\sigma(e_j) \neq 0$, we still have $\sigma(e_i)\sigma(e_j) = 0$. Since $\hat{f}$ is an idempotent, $\ell_i = 0$ or 1 except for $\sigma(e_i) = 0$. And $\hat{f}$ is primitive, so only one ℓ_i is not zero. Thus $\hat{f} = \sigma(e_j)$, for some j. Note that $\ker \sigma = V_D'$ and $\sigma(e) = \sigma(e_j) = \hat{f}$, it follows that

$$e \equiv e_j, \ (\text{mod } V_D').$$

Let $\overline{\omega}^j$ be the central character associated with e_j, from (3),

$$C^{p^b} e_j = \overline{\omega}^j(C) e_j, \qquad \text{for any class of } G.$$

Then $\sigma(C)^{p^b} \sigma(e_j) = \overline{\omega}^j(C)\sigma(e_j)$, namely, $\sigma(C)^{p^b} \hat{f} = \overline{\omega}^j(C)\hat{f}$. Since $\hat{f}$ is the principal idempotent and $\sigma(C)$ is a sum of class sums of $N_G(D)$, then from (3) again, we get

$$\sigma(C)^{p^b} \hat{f} = |\sigma(C)|^* \hat{f} \ \text{ and } \ |\sigma(C)|^* = \overline{\omega}^j(C).$$

Note that $C = (C \cap C_G(D)) \cup (C \setminus C_G(D)) = \sigma(C) \cup (C \setminus C_G(D))$. C and $\sigma(C)$ are invariant under the conjugations of the elements of D, so is $C \setminus C_G(D)$. Since D is a p-subgroup of G and the fixed elements under D are in $C_G(D)$, the length of each orbit of D in $C \setminus C_G(D)$ is a multiple of p. So $|C \setminus C_G(D)| \equiv 0$, (mod p) and $|C| \equiv |\sigma(C)|$ (mod p). Therefore $\overline{\omega}^j(C) = |C|^*$, that is, $\overline{\omega}^j$ is just the central character of the principal idempotent and e_j is the principal idempotent. Consequently,

the principal idempotent of $\overline{R}G$

$$\equiv e = \frac{1}{N_1^*} \sum_{i=1}^{t_1} C_i^{p^b} \equiv \frac{1}{N_i^*} \sum_{i=1}^{t_1} C_i, \ (\text{mod } V_D') .$$

The theorem is proved.

Remark. Using the theorem of R. Brauer on the correspondence between principal blocks, the proof will be much shorter. In fact, we have proved the theorem of Brauer in a special case: Let D be the

Sp-subgroup of $DC_G(D)$. b is a block of $\overline{R}(DC_G(D))$. Then b^G is the principal block of $\overline{R}G$ iff b is the one of $\overline{R}(DC_G(D))$.

Corollary 1. Suppose $D \triangleleft G$ and D is a S_p-subgroup of $DC_G(D)$, then the principal idempotent of $\overline{R}G$ is equal to

$$\frac{1}{N_1^*} \sum_{x \in C_G(D)} x ,$$

where

$$N_1 = [DC_G(D):D] .$$

Corollary 2. Let $C_1 \cdots, C_t$ be the full set of the p'-classes of G with the defect group P and $N_2 = [PC_G(P):P]$, then the principal idempotent of $\overline{R}G$ is equal to

$$\frac{1}{N_2^*} \sum_{i=1}^{t} C_i, \pmod{V_p^0}.$$

Remark. Let N_1 and N_2 be as above. It is easy to prove that $N_1 \equiv N_2 \pmod p$. So no contradiction arises in Theorem 9 and its corollary.

References

[1] R. Brauer and K.A. Fowler: " On groups of even order", Ann. of Math., 62(1955), 565-583.

[2] Y. Tsushima: "On the weakly regular p-blocks with respect to $O_{p'}(G)$". Osaka J. Math. 14(1977, 465-470.

[3] T. Wada: "On the existence of p-blocks with given defect groups", Hokkaido Mathematical Journal, 6(1977), 243-248.

[4] K. Iizuka and A. Watanable: "On the number of blocks of irreducible characters of finite groups with a given defect group", Kumamoto J. Sci. (Math.) 9(1972), 55-61.

[5] W. Willems: "Über die Existenz von Blocken", J. Algebra 53(1978), 402-409.

[6] G. R. Robinson: " The number of blocks with a given defect group", J. Algebra 84(1983), 493-502.
[7] B. H. Puttaswamaiah and J. D. Dixon: "Modular representations of finite groups", Academic Press, New York, London, 1977.

Appendix

" A proof of the first main theorem on blocks——another application of the p-power homemorphism"

The following well known facts will be used,

(i) Each central idempotent of $\bar{R}N_G(D)$ associated with the defect group D is a linear combination of class sums of $N_G(D)$ with the defect group D.

(ii) Let σ be the Brauer homomorphism

$$Z(\bar{R}G) \to Z(\bar{R}N_G(D)),$$

with respect to D.

If C is a class sum of G with the defect group D, then $\sigma(C)$ is a class sum of $N_G(D)$ with the defect group D. Conversly, if $\tilde{C}$ is a class sum of $N_G(D)$ with the defect group D, then $\tilde{C} = \sigma(C)$ for some class sum C of G with the defect group D. So σ defines a one-one correspondence between the set S of all the class sums of G with the defect group D and the set $\tilde{S}$ of all the class sums of $N_G(D)$ with the defect group D.

Theorem (First Main Theorem on Blocks) (R. Brauer). The Brauer homomorphism σ defines a one to one correspondence between the set F of all the blocks of $\bar{R}G$ with the defect group D and the set $\tilde{F}$ of all the blocks of $\bar{R}N_G(D)$ with the defect group D.

Let B be a block of F and $\tilde{B}$ be a block of $\tilde{F}$ corresponding to B. ω and $\tilde{\omega}$ are central characters associated with B and $\tilde{B}$ respectively. Then for any class sum C of G, we have $\omega(C) = \tilde{\omega}(\sigma(C))$.

Proof. Let I(F) be the set of all the primitive idempotents of $Z(\bar{R}G)$ which are associated with blocks in F and $I(\tilde{F})$ be the set associated with blocks in $\tilde{F}$. It is sufficient that we show that σ defines a one to one correspondence between the elements of $I(F)$ and $I(\tilde{F})$

Let f be a primitive idempotent of $I(F)$ and ω be the central character associated with f . Since f is of the defect group D, from the fact (ii), $\sigma(f) \neq 0$. And σ is a homomorphism, so

$$\sigma(f)^2 = \sigma(f^2) = \sigma(f) ,$$

that is, $\sigma(f)$ is a central idempotent of $\overline{R}N_G(D)$. We show that $\sigma(f)$ is primitive. If $\sigma(f)$ is not primitive, then

$$\sigma(f) = \tilde{f}_1 + \cdots + \tilde{f}_s$$

where $\{\tilde{f}_i\}$ is a set of pairwise orthogonal primitive central idempotents of $\overline{R}N_G(D)$. For any class sum C of G, from (3),

$$C^{p^b} f = \omega(C) f .$$

Then $\sigma(C^{p^b} f) = \sigma(\omega(C) f)$ and $\sigma(C)^{p^b} \sigma(f) = \omega(C)\sigma(f)$.

Thus we have

$$\sigma(C)^{p^b}(\tilde{f}_1 + \cdots + \tilde{f}_s) = \omega(C)(\tilde{f}_1 + \cdots + \tilde{f}_s). \qquad \text{(A)}$$

Let $\tilde{\omega}^i$ be the central character associated with $\tilde{f}_i$. From (3) again,

$$\sigma(C)^{p^b} \tilde{f}_i = \tilde{\omega}^i(\sigma(C)) \tilde{f}_i$$

So $\omega(C)(\tilde{f}_1 + \cdots + \tilde{f}_s) = \tilde{\omega}^1(\sigma(C)) \tilde{f}_1 + \cdots + \tilde{\omega}^s(\sigma(C)) \tilde{f}_s$

and

$$\omega(C) = \tilde{\omega}^i(\sigma(C)), \quad i=1,2,\cdots, s . \qquad \text{(B)}$$

For $C \in S$, $\sigma(C)$ is in $\tilde{S}$. The above equations say that for any

$$\tilde{C} \in \tilde{S}, \ \tilde{\omega}^1(\tilde{C}) = \cdots = \tilde{\omega}(\tilde{C}) \ . \tag{C}$$

From the fact (i), $\tilde{f}_1 = \sum_{\tilde{C}\in\tilde{S}} r_c \tilde{C}$. From (C), we have

$$\tilde{\omega}^i(\tilde{f}_1) = \sum_{\tilde{C}\in\tilde{S}} r_c \tilde{\omega}^i(\tilde{C}) = \sum_{\tilde{C}\in\tilde{S}} r_c \tilde{\omega}^1(\tilde{C}) = \tilde{\omega}^1(\tilde{f}_1) = 1$$

$1 \leq i \leq s$.

This contradicts to that $\tilde{\omega}^i(f_1) = 0, \ i \neq 1$.

So $\sigma(f)$ is primitive.

From the facts (i) and (ii), $\sigma(f)$ is of the defect group D, so $\sigma(f) \in I(\tilde{F})$
and the mapping

$$f \to \sigma(f), \quad f \in I(F),$$

maps $I(F)$ into $I(\tilde{F})$.

Since $ff' = 0$, for $f \neq f'$ and f, f' in $I(F)$,

$\sigma(f)\sigma(f') = \sigma(f\,f') = 0.$

Thus the mapping is injective.

Now let $\tilde{f}$ be a primitive idempotent of $I(\tilde{F})$.

From the fact (i), $\tilde{f} = \sum_{\tilde{C}\in\tilde{S}} r_c \tilde{C}$,

Evidently $\tilde{f} = \tilde{f}^{p^b} = \sum_{\tilde{C}\in\tilde{S}} r_c \tilde{C}^{p^b}$.

From the fact (ii), for any $\tilde{C} \in \tilde{S}$, there exists $C \in S$ such that $\sigma(C) = \tilde{C}$. Then

$$\sigma\Big(\sum_{\tilde{C}\in\tilde{S}} r_c \ C^{p^b}\Big) = \sum_{\tilde{C}\in\tilde{S}} r_c \sigma(C)^{p^b} = \sum_{\tilde{C}\in\tilde{S}} r_c \tilde{C}^{p^b} = \tilde{f} \ .$$

Let $f_1, \cdots, f_s$ be all the primitive idempotents in V_D.
From Lemma 1,

$$\sum_{\tilde{C} \in \tilde{S}} r_C C^{p^b} = \sum_{i=1}^{s} \ell_i f_i, \quad \text{for some } \ell_i \in \bar{R},\ i=1,\cdots,s.$$

And we have

$$\sigma\Big(\sum_{i=1}^{s} \ell_i f_i\Big) = \sum_{i=1}^{s} \ell_i \sigma(f_i) = \tilde{f} .$$

By using the same argument as in Theorem 9, we get

$$\sigma(f_i) = \tilde{f} ,$$

for some i. Since $\sigma(f_i) = \tilde{f} \neq 0$, $f_i \notin V_D^{\circ}$.

So f_i is of the defect group D, namely, $f_i \in I(F)$.

The first part of the theorem is proved. The second part of the theorem follows from (B).

THE EQUITYPE AND QUASI-EQUITYPE DECOMPOSITIONS OF ARBITRARY PERMUTATIONS BY PERMUTATIONS OF ORDER 2 OR 3

Wang Efang

Department of Mathematics
Peking University
Beijing, China

The commutator subgroup of a group G is the subgroup generated by all commutators of G. In general, all the commutators of G do not form a subgroup of G. In 1951 O.Ore [3] proved that every element in A_n is a commutator of S_n. And he pointed out that when $n \geq 5$, every element in A_n can be expressed as a commutator of elements in A_n. This claim was proved by N.Ito [4]. Because the commutator subgroup of S_n is A_n, and when $n \geq 5$, the commutator subgroup of A_n is A_n itself. Hence, this result is equivalent to the assertion that the commutator subgroup of S_n (or A_n, $n \geq 5$) is just the set of all commutators of S_n (or A_n).

The problem of expressing an element of A_n as a commutator is equivalent to the problem of expressing an element of A_n as a product of two conjugate elements. In 1972, E.Bertram [5] generalized this problem, and gave a condition for t such that an even permutation can be expressed as a product of two t-cycles, and also a condition for t such that an odd permutation can be expressed as a product of a t-cycle and a (t+1)-cycle.

In a forthcoming paper, the author gives all the possible decompositions of any even permutation as products of two permutations of the same type.

In this paper, we give some conditions and methods about expressing an even permutation as a product of two equitype permutations of order 2 or 3 and about expressing an odd permutation as a product of two quasi-equitype involutions. We also prove that any even permutation can be expressed as a product of two equitype involutions, and except one special case any even permutation can be expressed as a product of two equitype permutations of order 3.

Because of the importance of the permutations of low orders, especially of order 2, in S_n, our results are useful in the study of permutation groups.

1. Notation

The permutations discussed here are all acting on the n-set $\Omega = \{1,2,\dots,n\}$, $n \geq 3$. We denote by S_n and A_n the symmetric and alternating group on Ω respectively. For $a \in \Omega$, $\sigma \in S_n$, a^σ represents the image of a under σ, and define the product $\sigma\tau$ of two permutations σ and τ on Ω by the formula $a^{\sigma\tau} = (a^\sigma)^\tau$.

If the degrees of the cycles in the reduced cyclic form of σ are $h_1, h_2, \dots, h_m$, then σ is called a permutation of type$(h_1,h_2,\dots,h_m)$, or shortly, a $(h_1,h_2,\dots,h_m)$-permutation, and m is called the type length of σ. Obviously, the type and type length of a permutation are uniquely determined except for the order of h_i's. If $h_1,h_2,\dots,h_u$ are all the distinct numbers in $h_1,h_2,\dots,h_m$; and suppose h_i $(1 \leq i \leq u)$ appears t_i times among $h_1,h_2,\dots,h_m$, then the type of σ can be written as$(h_1^{t_1}, h_2^{t_2},\dots, h_u^{t_u})$. If two permutations are of the same type, then we say that they are equitype permutations. If the reduced cyclic forms of two permutations differ by only one cycle, then we say that they are quasi-equitype permutations.

We denote the set of all fixed points of σ by $F(\sigma)$, and the complement of $F(\sigma)$ by $\{\sigma\}$. We also say that σ is generated by $\{\sigma\}$. The degree of σ is denoted by $|\sigma|$.

Assume that σ is expressed as a product of two permutations:

$$\sigma = \sigma_1\sigma_2 \tag{1}$$

If

$$\{\sigma_i\} \subseteq \{\sigma\} , \qquad i = 1, 2,$$

then we call formula (1) an inner decomposition of σ. If σ_1, σ_2 are equitype permutations of type $(k_1,k_2,\ldots,k_v)$, then we call (1) a $(k_1,k_2,\ldots,k_v)$ equitype decomposition, or shortly, an equitype decomposition. If σ_1, σ_2 are quasi-equitype permutations of type $(k,k_1,\ldots,k_v)$ and $(k_1,\ldots,k_v)$ respectively, then we say that (1) is a $(k;k_1,\ldots,k_v)$ quasi-equitype decomposetion, or shortly, a quasi-equitype decomposition.

Suppose σ is decomposed as (1), and τ is a cycle. If

$$\{\tau\} \cap \{\sigma_i\} = \phi , \qquad i = 1, 2,$$

then

$$\sigma = (\sigma_1\tau)(\sigma_2\tau^{-1}). \qquad (2)$$

We say that (2) is obtained from (1) by adding the factor τ. If the original decomposition (1) is (quasi-)equitype, then the decomposition by adding a factor is also (quasi-)equitype. In general, we may add several cyclic factors to a decomposition.

Let

$$\sigma = \sigma_1\sigma_2 \qquad (1)$$

and the reduced cyclic forms of σ_1, σ_2 be

$$\sigma_1 = \sigma_{11}\sigma_{12}\cdots\sigma_{1s},$$

$$\sigma_2 = \sigma_{21}\sigma_{22}\cdots\sigma_{2t}.$$

If $\{\sigma_{11}\}$ and $\{\sigma_{21}\}$ have a common point a_1:

$$\sigma_{11} = (a_1 a_2 \ldots a_p),$$

$$\sigma_{21} = (a_1 b_2 \ldots b_q).$$

Take any $c \notin \{\sigma_1\} \cup \{\sigma_2\}$. Let

$$\sigma_{11}' = (a_1\ a_2\ \dots\ a_p\ \ c),$$
$$\sigma_{21}' = (a_1\ \ c\ \ b_2 \dots b_q),$$
$$\sigma_1' = \sigma_{11}'\ \sigma_{12} \dots \sigma_{1s}\ ,$$
$$\sigma_2' = \sigma_{21}'\ \sigma_{22} \dots \sigma_{2t}.$$

Then

$$\sigma = \sigma_1'\ \sigma_2'. \qquad (3)$$

If (1) is an (quasi-)equitype decomposition and $p = q$, then (3) is also (quasi-)equitype. The expressing (3) is called an outer stretching of (1) by the point c. In general, we can get an outer stretching of a decomposition by several points.

2. The (2^t) equitype or $(2;2^t)$ quasi-equitype decompositions of permutations

Now we show how to express a permutation as a product of two involutions. First we discuss the even permutations.

Lemma 1. (1) Let σ be an even cycle of degree s. If σ is expressed as a product of two involutions $\sigma = \sigma_1\ \sigma_2$, then σ_1 and σ_2 must be of the same type.

(2) A necessary and sufficient condition for an even cycle σ to be expressed as a product of two (2^t)-permutations, is:

(i) in the case of inner decomposition,

$$t = \frac{s - 1}{2}\quad ;$$

(ii) in the general case,

$$\frac{s-1}{2} \leq t \leq [\frac{n-1}{2}] .$$

Proof. Let

$$\sigma = (1\ 2 \ldots s) ,$$

here s is an odd number. We first discuss the case of inner decomposition. If σ is written as the product of two involutions: $\sigma = \sigma_1 \sigma_2$, then because an involution is a product of disjoint transpositions, $|\sigma_i|$ is even. For the case of inner decomposition, $\{\sigma_i\} \subseteq \{\sigma\}$ $i=1,2$. Hence, $F(\sigma_i) \cap \{\sigma\} \neq \phi$. We may suppose

$$1 \in F\{\sigma_1\} \cap \{\sigma\} .$$

Then

$$1^{\sigma_2} = 2, \quad \sigma_2 = (1\ 2)\ldots,$$

$$2^{\sigma_1} = 2^{\sigma_1^{-1}} = 2^{\sigma_2 \sigma^{-1}} = 1^{\sigma^{-1}} = s ,$$

$$\ldots\ldots .$$

Thus, we obtain

$$\begin{aligned} \sigma_1 &= (2\ s)(3\ s-1)\ldots(\tfrac{s+1}{2}-1\ \ \tfrac{s+1}{2}+2)(\tfrac{s+1}{2}\ \ \tfrac{s+1}{2}+1) \\ \sigma_2 &= (1\ 2)(3\ s)\ldots(\tfrac{s+1}{2}-1\ \ \tfrac{s+1}{2}+3)(\tfrac{s+1}{2}\ \ \tfrac{s+1}{2}+2) \end{aligned} \tag{4}$$

So σ_1 and σ_2 are of the same type. And, there is an inner decomposition of σ as a product of two (2^t)-permutations iff $t = \frac{s-1}{2}$.

If $\sigma = \sigma_1 \sigma_2$ is not an inner decomposition, let $a \in F(\sigma) \cap \{\sigma_1\}$:

$$\sigma_1 = (a\ a')\ \ldots .$$

Then

$$\sigma_2 = (a\ a')\ \ldots .$$

Hence a' is also a fixed point of σ . This leads to the fact that, in this case, every decomposition of σ can be got from an inner decomposition by adding several transpositions, which are generated by fixed points of σ . Obviously, this is also an equitype decomposition. Moreover, because $|F(\sigma)| = n-s$, the number of pairs of adding transpositions is at most equal to $[\frac{n-s}{2}]$. Therefore

$$t \leq \frac{s-1}{2} + [\frac{n-s}{2}] = [\frac{n-1}{2}].$$

When σ is an odd cycle, it can not be expressed as a product of two equitype permutations. However, we have the following result.

Lemma 2. Let $\sigma = (1\ 2 \ldots s)$ be an odd cycle, then

(1) If σ is expressed as a product of two involutions $\sigma = \sigma_1 \sigma_2$, where σ_i (i=1,2) is a (2^{t_i})-permutation. Then

$$|t_1 - t_2| = 1.$$

(2) A necessary and sufficient condition that σ has $(2;2^t)$ quasi-equitype decomposition is

(i) inner decomposition: $t = \frac{s}{2} -1$;

(ii) general case: $\frac{s}{2} -1 \leq t \leq [\frac{n}{2}] -1$.

Proof. If $\sigma = \sigma_1 \sigma_2$ is an inner decomposition, and

$$|\sigma_i| = 2^{t_i} , \qquad i=1,2 .$$

Since $(\sigma_1 \sigma_2)^{-1} = \sigma_2 \sigma_1 = \sigma^{-1}$, we may assume $t_1 < t_2$. Then $F(\sigma_1) \cap \{\sigma\} \neq \phi$. Without loss of generality we may set $1 \in F(\sigma_1)$. Then

$$\sigma_2 = (1\ 2) \ldots,$$

$$\sigma_1 = (2\ s) \ldots,$$

$$\ldots\ldots$$

and we get

$$\begin{aligned} \sigma_1 &= (2\ s)(3\ s-1)\ldots(\tfrac{s}{2}\ \ \tfrac{s}{2}+2), \\ \sigma_2 &= (1\ 2)(3\ s)\ldots(\tfrac{s}{2}+1\ \ \tfrac{s}{2}+2\ ). \end{aligned} \qquad (5)$$

Hence $t_1 = \frac{s}{2} -1$, $t_2 = \frac{s}{2}$.

The inequality in the general case may be proved as Lemma 1.

Lemma 3. Let $\sigma = (a_1 \dots a_{h_1})(b_1 \dots b_{h_2})$ be an even permutation of type (h_1, h_2), $2 \le h_1 \le h_2$, $s = h_1 + h_2$. Then, σ can be expressed as a product of two (2^t)-permutations, iff

(1) when $h_1 \ne h_2$

inner decomposition: $t = \frac{s}{2} - 1$;

general case: $\frac{s}{2} - 1 \le t \le [\frac{n}{2}] - 1$.

(2) when $h_1 = h_2$,

inner decomposition: $t = \frac{s}{2} - 1$ or $\frac{s}{2}$;

general case: $\frac{s}{2} - 1 \le t \le [\frac{n}{2}]$.

Proof. Suppose $\sigma = \sigma_1 \sigma_2$ is an inner decomposition, where σ_1 and σ_2 are (2^t)-permutations. We consider the case that σ_1 includes a cyclic factor $(a_i\ b_j)$ first. Without loss of generality, we may assume $\sigma_1 = (a_1\ b_1)\dots$. Then, from $\sigma = \sigma_1 \sigma_2$, we get

$$\begin{aligned} \sigma_1 &= (a_1\ b_1)(a_2\ b_{h_2})\dots(a_{h_1-1}\ b_{h_2-h_1+3})(a_{h_1}\ b_{h_2-h_1+2}) \\ \sigma_2 &= (a_2\ b_1)(a_3\ b_{h_2})\dots(a_{h_1}\ b_{h_2-h_1+3})(a_1\ b_{h_2-h_1+2}) \end{aligned} \tag{6}$$

$$b_1^{\sigma} = b_1^{\sigma_1 \sigma_2} = a_1^{\sigma_2} = b_{h_2-h_1+2}$$

Hence we should have

$$h_2 - h_1 + 2 = 2$$

i.e.

$$h_1 = h_2 .$$

And for this case

$$t = h_1 = s/2 .$$

If all the cyclic factors of $\sigma_i (i=1,2)$ are $(a_i\ a_j)$ or $(b_i\ b_j)$, the decomposition of σ may be obtained from the decompositions of cycles $(a_1\ a_2 \dots a_{h_1})$ and $(b_1\ b_2 \dots b_{h_2})$. When h_1 and h_2 are odd, from Lemma 1, we know that $t = \frac{s}{2} - 1$. When h_1 and h_2 are even, by Lemma 2,

$(a_1\ a_2 \ldots a_{h_1})$ can be expressed as a product of a $(2^{h_1/2-1})$-permutation σ_{11} and a $(2^{h_1/2})$-permutation σ_{12} ; while $(b_1\ b_2 \ldots b_{h_2})$ can be expressed as a product of a $(2^{h_2/2})$-permutation σ_{21} and a $(2^{h_2/2-1})$-permutation σ_{22} . Because permutations σ_{1k} and $\sigma_{2k'}$ $(k,k'=1,2)$ are disjoint, we get

$$\begin{aligned}\sigma &= \sigma_{11}\sigma_{12}\cdot\sigma_{21}\sigma_{22}\\ &= \sigma_{11}\sigma_{21}\cdot\sigma_{12}\sigma_{22}\ .\end{aligned}$$

In this expression, $\sigma_{11}\sigma_{21}$ and $\sigma_{12}\sigma_{22}$ are both $(2^{s/2-1})$-permutations.

When $h_1 \neq h_2$, we must decompose σ_1 and σ_2 respectively. Hence, $t = \frac{s}{2} - 1$.

Thus, the case of inner decomposition is proved.

In the general case, because the decomposition must be got from the inner decomposition by adding several transpositions generated by fixed point of σ , and σ has $n-s$ fixed points, hence, when $h_1 \neq h_2$:

$$\frac{s}{2}-1 \le t \le \frac{s}{2}-1+[\frac{n-s}{2}] = [\frac{n}{2}]-1\ ;$$

while $h_1=h_2$:

$$\frac{s}{2}-1 \le t \le [\frac{n}{2}].$$

From Lemmas 1-3 we get the following theorem.

Theorem 1. Let σ be a $(h_1, h_2, \ldots, h_m)$-permutation, $s=h_1+h_2+\ldots+h_m$. Suppose that after removing all the pairs of equal numbers in $\{h_1, h_2, \ldots, h_m\}$, there are m' distinct numbers. Then a necessary and sufficient condition for that σ has (2^t) equitype decomposition is

$$\frac{s-m}{2} \le t \le [\frac{n-m'}{2}]\ ,$$

i.e. σ can be expressed as a product of two involutions of degree s_1 iff

$$s-m \le s_1 \le 2[\frac{n-m'}{2}].$$

As a corollary of Theorem 1, we have

Theorem 2. Every even permutation of S_n can be expressed as a product of two conjugate involutions.

Because an odd permutation is a product of an even permutation and an odd cycle, from Theorem 1 and Lemma 2, we have the following theorem.

Theorem 3. Let σ be an odd permutation of type $(h_1,h_2,\ldots h_m)$; s and m' are defined as in Theorem 1. Then a necessary and sufficient condition for that σ has a $(2;2^t)$ quasi-equitype decomposition is

$$\frac{s-m-1}{2} \leq t \leq [\frac{n-m'-1}{2}].$$

i.e. σ can be expressed as a product of two involutions of degrees s_1 and s_1+2 respectively, iff

$$s-m-1 \leq s_1 \leq 2[\frac{n-m'-1}{2}].$$

From Lemma 1, we have the following theorem.

Theorem 4. Every even cycle of S_n can be expressed as a product of two (2^t)-permutations, iff

$$t = [\frac{n-1}{2}].$$

If we consider all the even permutations of S_n, then we have the following theorem.

Theorem 5. A necessary and sufficient condition for that every even permutation of S_n can be expressed as a product of two (2^t)-permutations, is:

(1) $n=3,4$: $t=1$;

(2) $n=5,6$: $t=2$;

(3) $n=8$: $t=3$;

(4) $n>6$, $n\neq 8$: t does not exist

Proof. We need only to prove (4). If $n>8$ is even, then the even permutation (1 2...n-1) has equitype decomposition of type $(2^{n/2-1})$ only. Let σ be an even permutation of type (4,3,2). Then σ can be expressed as a product of two (2^t)-permutations iff

$$3 \leq t \leq \frac{n-3}{2} < \frac{n}{2} - 1.$$

Hence, there is no t such that both σ and (1 2...n-1) can be expressed as a product of two (2^t)-permutations.

If $n > 6$ is odd, then there is no t such that both (1 2...n) and (1 2)(3 4 5 6) can be expressed as a product of two (2^t)-permutations.

Theorem 4 and Theorem 5 give also necessary and sufficient conditions for s_1 such that every even permutation or even cycle may have an equitype decomposition by involutions of degree s_1.

3. The (3^t) equitype decompositions of even permutations

In this section, we discuss the problem of decomposition of an even permutation σ as a product of two equitype permutations of order 3. First, we consider the case that σ is an even cycle. The following lemmas give several formulae of decomposition:

Lemma 4. Any even cycle of degree s can be innerly decomposed as a product of two $(3^{[s/3]})$-permutations.

Proof. We give formulae according to different cases of s modula 3.

1. $s = 3k$, where k is odd.

(1) $k = 1$:

$$(1\ 2\ 3)=(1\ 3\ 2)(1\ 3\ 2)$$

(2) $k = 3$:

$$(1\ 2\ldots 9)=[(1\ 4\ 7)(2\ 5\ 8)(3\ 9\ 6)][(1\ 8\ 6)(2\ 9\ 4)(3\ 7\ 5)]$$

(3) $k \geq 5$:

$$\begin{aligned}(1\ 2\ \ldots\ 3k)=[&(1\ \ k+1\ \ 2k+1)(2\ \ k+2\ \ 2k+2)(3\ \ 2k+\tfrac{k+3}{2}\ \ 2k)\\ &(4\ \ 3k\ \ k+3)(5\ \ 2k+\tfrac{k+1}{2}\ \ k+4)(6\ \ 3k-1\ \ k+5)\ldots\\ &(k-1\ \ 3k-\tfrac{k-5}{2}\ \ 2k-2)(k\ \ 2k+3\ \ 2k-1)]\cdot\\ [&(1\ \ 2k+2\ \ k+3)(2\ \ 2k+3\ \ k+1)(3\ \ 2k+1\ \ k+2)\\ &(4\ \ k+4\ \ 2k+\tfrac{k+3}{2})(5\ \ k+5\ \ 3k)(6\ \ k+6\ \ 2k+\tfrac{k+1}{2})\ldots\\ &(k-1\ \ 2k-1\ \ 2k+4)(k\ \ 2k\ \ 2k+\tfrac{k+5}{2})]\end{aligned}$$

2. $s = 3k+1$, where k is even.

(1) $k = 2$

$$(1\ 2\ 3\ 4\ 5\ 6\ 7)=[(1\ 2\ 4)(3\ 6\ 5)][(1\ 5\ 7)(3\ 6\ 4)]$$

$k = 4$

$$(1\ 2\ \ldots\ 13)=[(1\ 2\ 6)(3\ 7\ 10)(4\ 8\ 13)(5\ 11\ 9)]$$
$$\cdot[(1\ 7\ 4)(3\ 11\ 6)(5\ 10\ 8)(9\ 12\ 13)]$$

(3) $k \geqq 6$

$$(1\ 2\ \ldots\ 3k+1)$$
$$\begin{aligned}=[&(1\ \ 2\ \ k+2)(3\ \ k+3\ \ 2k+2)(4\ \ k+4\ \ 3k+1)(5\ \ 2k+\tfrac{k+2}{2}\ \ 2k+1)\\ &(6\ \ 3k-1\ \ k+5)(7\ \ 2k+\tfrac{k}{2}\ \ k+6)(8\ \ 3k-2\ \ k+7)\ldots\\ &(k-1\ \ 2k+4\ \ 2k-2)(k\ \ 2k+\tfrac{k+4}{2}\ \ 2k-1)(k+1\ \ 2k+3\ \ 2k)]\cdot\\ [&(1\ \ k+3\ \ 4)(3\ \ 2k+3\ \ k+2)(5\ \ 2k+2\ \ k+4)(6\ \ k+6\ \ 2k+\tfrac{k+2}{2})\\ &(7\ \ k+7\ \ 3k-1)(8\ \ k+8\ \ 2k+\tfrac{k}{2})\ldots(k-1\ \ 2k-1\ \ 3k-\tfrac{k-6}{2})(k\ \ 2k\ \ 2k+4)\\ &(k+1\ \ 2k+1\ \ 3k-\tfrac{k-4}{2})(k+5\ \ 3k\ \ 3k+1)]\end{aligned}$$

3. $s = 3k+2$, where k is odd.

(1) $k = 1$

$(1\ 2\ 3\ 4\ 5)=(1\ 2\ 3)(1\ 4\ 5)$

(2) $k = 3$

$(1\ 2\ \ldots\ 11)=[(1\ 2\ 3)(4\ 6\ 10)(5\ 8\ 7)][(1\ 4\ 11)(5\ 8\ 6)$
$\cdot(7\ 9\ 10)]$

(3) $k \geq 5$

$$(1\ 2\ \ldots\ 3k+2)$$
$$=[(1\ 2\ 3)(4\ \ k+3\ \ 3k+1)(5\ \ 2k+\tfrac{k+1}{2}\ \ 2k+1)(6\ \ 3k-1\ \ k+4)$$
$$(7\ \ 2k+\tfrac{k-1}{2}\ \ k+5)(8\ \ 3k-2\ \ k+6)\ \ldots\ (k\ \ 2k+3\ \ 2k-2)$$
$$(k+1\ \ 2k+\tfrac{k+3}{2}\ \ 2k-1)(k+2\ \ 2k+2\ \ 2k)][(1\ \ 4\ \ 3k+2)(5\ \ 2k+2\ \ k+3)$$
$$(6\ \ k+5\ \ 2k+\tfrac{k+1}{2})(7\ \ k+6\ \ 3k-1)(8\ \ k+7\ \ 2k+\tfrac{k-1}{2})\ldots(k\ \ 2k-1$$
$$2k+\tfrac{k+5}{2})(k+1\ \ 2k\ \ 2k+3)(k+2\ \ 2k+1\ \ 2k+\tfrac{k+3}{2})(k+4\ \ 3k\ \ 3k+1)].$$

In order to increase the degree of σ_1 in the equitype decomposition we make use of the method of adding 3-cycle factors and outer stretching. Beside this, we will consider the case that $[\frac{s}{3}]+[\frac{n-3}{3}]<[\frac{n}{3}]$. For this purpose, we give the following two lemmas.

Lemma 5. σ is a cycle of degree $s = 3k+2$, k odd. Then

(1) σ can be expressed as a product of two permutations of type $(2,3^k)$.

(2) If $s<n$, σ can be expressed as two (3^{k+1})-permutations by outer stretching one fixed point to the above decomposition.

Proof. (1) The following formulae give the required decompositions.

$(1\ 2\ 3\ 4\ 5)=[(1\ 3)(2\ 5\ 4)][(1\ 4)(2\ 5\ 3)]$

$(1\ 2\ \ldots 11)=[(1\ 6)(2\ 4\ 10)(3\ 9\ 8)(5\ 11\ 7)][(1\ 7)(2\ 11\ 6)$
$(3\ 9\ 4)(5\ 8\ 10)];$

$(1\ 2\ \ldots\ 3k+2)$

$$=[(1\ \tfrac{3k+1}{2}+1)(2\ \tfrac{3k+1}{2}-1\ 3k+1)(3\ \tfrac{3k+1}{2}-3\ 3k-1)(4\ \tfrac{3k+1}{2}-5\ 3k-3)\ldots$$
$$(\tfrac{k+1}{2}\ \tfrac{k+1}{2}+2\ 2k+4)(\tfrac{k+1}{2}+1\ 2k+3\ 2k+2)(\tfrac{k+1}{2}+3\ 2k+5\ 2k+1)\ldots$$
$$(\tfrac{3k+1}{2}-2\ 3k\ \tfrac{3k+1}{2}+3)(\tfrac{3k+1}{2}\ 3k+2\ \tfrac{3k+1}{2}+2)][(1\ \tfrac{3k+1}{2}+2)(2\ 3k+2$$
$$\tfrac{3k+1}{2}+1)(3\ 3k\ \tfrac{3k+1}{2}-1)(4\ 3k-2\ \tfrac{3k+1}{2}-3)\ldots(\tfrac{k+1}{2}\ 2k+5\ \tfrac{k+1}{2}+4)$$
$$(\tfrac{k+1}{2}+1\ 2k+3\ \tfrac{k+1}{2}+2)(\tfrac{k+1}{2}+3\ 2k+2\ 2k+4)\ldots(\tfrac{3k+1}{2}-2\ \tfrac{3k+1}{2}+4\ 3k-1)$$
$$(\tfrac{3k+1}{2}\ \tfrac{3k+1}{2}+3\ 3k+1)] \qquad (k \geq 5)$$

(2) When $s < n$, we write the above formula as $\sigma = \sigma_1\sigma_2$. Let

$$\sigma_1' = (1\ \tfrac{3k+1}{2}+1\ n)\ldots;$$

$$\sigma_2' = (1\ n\ \tfrac{3k+1}{2}+2)\ldots.$$

Then

$$\sigma = \sigma_1'\sigma_2'$$

and σ_1', σ_2' are both (3^{k+1})-permutations.

Lemma 6. When $s = 3k+1$ (k even), $\sigma = (1\ 2\ \ldots\ s)$ can be expressed as a product of two permutations of type $(2, 2, 3^{k-1})$. Hence, if $s \leq n-2$, this expression can be outer stretched to a (3^{k+1}) equitype decomposition.

Proof. We need only to prove that σ can be expressed as a product of two $(2, 2, 3^{k-1})$-permutations $\sigma = \sigma_1\sigma_2$, and the corresponding 2-cycle in σ_1, σ_2 have a common point respectively. We give the decomposition as following:

$k=2$: $(1\ 2\ \ldots\ 7)$

$$=[(1\ 3)(4\ 7)(2\ 6\ 5)][(1\ 4)(5\ 7)(2\ 6\ 3)]$$

$k=4$: $(1\ 2\ \ldots\ 13)$

$$=[(1\ 6)(7\ 13)(2\ 4\ 11)(3\ 10\ 9)(5\ 12\ 8)][(1\ 7)(8\ 13)(2\ 12\ 6)$$
$$(3\ 10\ 4)(5\ 9\ 11)]$$

$k \geq 6$: $(1\ 2\ \ldots\ 3k+1)$

$$=[(1\ 3k/2)(3k/2+1\ 3k+1)(2\ 3k/2-2\ 3k-1)(3\ 3k/2-4\ 3k-3)$$
$$(4\ 3k/2-6\ 3k-5)\ldots(k/2-2\ k/2+6\ 2k+7)(k/2-1\ k/2+4\ 2k+5)$$
$$(k/2\ k/2+2\ 2k+3)(k/2+1\ 2k+2\ 2k+1)(k/2+3\ 2k+4\ 2k)$$
$$(k/2+5\ 2k+6\ 2k-1)\ldots(3k/2-5\ 3k-4\ 3k/2+4)(3k/2-3\ 3k-2\ 3k/2+3)$$
$$(3k/2-1\ 3k\ 3k/2+2)][(1\ 3k/2+1)(3k/2+2\ 3k+1)(2\ 3k\ 3k/2)$$
$$(3\ 3k-2\ 3k/2-2)(4\ 3k-4\ 3k/2-4)\ldots(k/2-2\ 2k+8\ k/2+8)$$
$$(k/2-1\ 2k+6\ k/2+6)(k/2\ 2k+4\ k/2+4)(k/2+1\ 2k+2\ k/2+2)$$
$$(k/2+3\ 2k+1\ 2k+3)(k/2+5\ 2k\ 2k+5)\ldots(3k/2-5\ 3k/2+5\ 3k-5)$$
$$(3k/2-3\ 3k/2+4\ 3k-3)(3k/2-1\ 3k/2+3\ 3k-1)]$$

From Lemmas 4-6, we get the following theorem.

Theorem 6. Every even cycle σ of degree s can be expressed as a product of two (3^t)-permutations iff

$$\frac{s-1}{4} \leq t \leq \frac{n}{3}\ .$$

i.e. iff

$$\frac{3}{4}(s-1) \leq s_1 \leq 3[\frac{n}{3}]$$

σ can be expressed as a product of two permutations of degree s_1 and order 3.

Proof. First we prove the necessity part.

It is obviously, $t \leq [\frac{n}{3}]$. Let $\sigma = \sigma_1\sigma_2$, σ_1、σ_2 are (3^t)-permutations. And suppose $\sigma_1 = \tau_1\tau_2\ldots\tau_t$, where $\tau_1, \tau_2, \ldots, \tau_t$ are disjoint 3-cycles. Then $\{\tau_i\} \cap \{\sigma_2\} \neq \phi$, $(i=1,2,\ldots,t)$, and all τ_j $(1 \leq j \leq t)$ except one, at least have two common points with σ_2. For otherwise, $\sigma_1\sigma_2$ cannot be a cycle. Because

$$|\sigma_1| + |\sigma_2| = 6t,$$

we have

$$6t-(2t-1) \geq 3.$$

i.e.

$$4t \geq s-1.$$

The sufficiency of the condition can be proved from Lemmas 4-6 by use of the method of adding 3-cycle factors and outer stretching.

For the (3^t) equitype decomposition of an even permutation, we need the following lemmas.

Lemma 7. A (2,4)-permutation σ can not be innerly decomposed as a product of two (3^t)-permutations. But it can be innerly decomposed as a product of two (2,3)-permutations. Thus, for $n > 6$, this inner decomposition can be outer stretched to a (3^2) equitype decomposition. Besides, σ can be expressed as a product of a 3-cycle and a (3^2)-permutation.

Proof. Let

$$\sigma = (1\ 2)(3\ 4\ 5\ 6).$$

If σ is innerly decomposed as a product of two (3^h)-permutations: $\sigma = \sigma_1\sigma_2$, then $h = 2$. Therefore

$$\{\tau_1\} = \{\tau_2\} = \{\sigma\},$$

$$F(\tau_i) \cap \{\sigma\} = \phi.$$

We need only consider the following three cases:

$$\tau_1 = (1\ 3\ 5)(2\ 4\ 6);$$

$$\tau_1 = (1\ 3\ 5)(2\ 6\ 4);$$

$$\tau_1 = (1\ 3\ 6)(2\ 5\ 4).$$

But for these τ_1, $\tau_1^{-1}\sigma$ are all not (3^2)-permutation. So σ cannot be decomposed as a product of two (3^2)-permutations.

The other conclusions can be proved by the following formulae:

$$\begin{aligned}&(1\ 2)(3\ 4\ 5\ 6)\\ &=(1\ 3\ 6)[(1\ 3\ 2)(4\ 5\ 6)]\\ &=[(1\ 2\ 3)(4\ 6)][(1\ 4\ 3)(5\ 6)]\\ &=[(1\ 2\ 3)(4\ n\ 6)][(1\ 4\ 3)(5\ 6\ n)] \qquad (n > 6)\end{aligned}$$

Lemma 8. Let σ be an even permutation of type $(h_1,h_2),\{h_1,h_2\}\neq \{2, 4\}$, $s = h_1+ h_2$. Then σ can be expressed as a product of two $(3^{[s/3]})$-permutations.

Proof. We calculate the formulae of decompositions according to the cases for h_1,h_2 modula 3. Here we give a formula for the case $h_1 \equiv h_2 \equiv 2 \pmod 3$ only. The other cases can be discussed in the same way.

Assume $h_1 \equiv h_2 \equiv 2 \pmod 3$.

1. h_1,h_2 are odd. In this case $h_1,h_2 \geq 5$.

(1) $h_1=h_2=5$.

$$(a_1\ a_2\ a_3\ a_4\ a_5)(b_1\ b_2\ b_3\ b_4\ b_5)$$
$$=(a_1\ b_5\ b_1)(a_2\ a_4\ b_2)(a_3\ b_4\ b_3)$$
$$(a_5\ a_1\ b_2)(a_2\ b_3\ b_5)(a_3\ b_4\ a_4)$$
$$=\sigma_1\,\sigma_2$$

where

$$\sigma_1=(a_1\ b_5\ b_1)(a_2\ a_4\ b_2)(a_3\ b_4\ b_3)$$
$$\sigma_2=(a_5\ a_1\ b_2)(a_2\ b_3\ b_5)(a_3\ b_4\ a_4)$$

are both (3^3)-permutations and

$$a_5^{\sigma_1}= a_5\ ;\ b_1^{\sigma_2}= b_1\ .$$

(2) $h_1 = 5$, $h_2 > 5$.

$$\sigma = (a_1\ a_2\ a_3\ a_4\ a_5)(b_1\ b_2\ldots b_{h_2})$$
$$= (a_1\ a_2\ a_3\ a_4\ a_5)(b_1\ b_2\ b_3\ b_4\ b_5)(b_1\ b_6\ldots b_{h_2})$$

where $(b_1\ b_6\ldots b_{h_2})$ is an even cycle of degree h_2-4, it can be innerly decomposed as a product of two $(3^{[(h_2-4)/3]})$-permutations: $\tau_1\,\tau_2$. And because of $h_1-4 \equiv 1 \not\equiv 0 \pmod 3$, we can assume that τ_1 leaves b_1 fixed. Then

$$\sigma = (\sigma_1\,\sigma_2)(\tau_1\,\tau_2)$$
$$= (\sigma_1\,\tau_1)(\sigma_2\,\tau_2)$$

where $\sigma_1\tau_1$, $\sigma_2\tau_2$ are both $(3^{[(s-1)/3]})$-permutations, and $\sigma_1\tau_1$ leaves a_5 fixed.

(3) $h_1 > 5$, $h_2 > 5$.

$$\sigma = (a_1\ a_2 \ldots a_{h_1})(b_1\ b_2 \ldots b_{h_2})$$
$$= (a_5\ a_6 \ldots a_{h_1})(a_1\ a_2 \ldots a_5)(b_1\ b_2 \ldots b_{h_2})$$

where $(a_5\ a_6 \ldots a_{h_1})$ is an even cycle of degree h_1-4, it can be expressed as a product of two $(3^{[(h_1-4)/3]})$-permutations: $\rho_1\rho_2$, and we may assume that ρ_1 leaves a_5 fixed. Then from (2):

$$\sigma = \rho_1\rho_2[(\sigma_1\tau_1)(\sigma_2\tau_2)]$$
$$= (\rho_1\sigma_1\tau_1)(\rho_2\sigma_2\tau_2)$$

where both $\rho_1\sigma_1\tau_1$, $\rho_2\sigma_2\tau_2$ are permutations of type $(3^{[(s-1)/3]})=(3^{[s/3]})$.

2. h_1, h_2 are even.

$$\sigma = (a_1\ a_2 \ldots a_{h_1})(b_1\ b_2 \ldots b_{h_2})$$
$$= (a_2\ a_3 \ldots a_{h_1})(a_1\ a_2)(b_1\ b_2)(b_1\ b_3 \ldots b_{h_2})$$

where $(a_2\ a_3 \ldots a_{h_1})$ is an even cycle of degree h_1-1, it can be innerly decomposed as a product of two $(3^{[(h_1-1)/3]})$-permutations: $\sigma_1\sigma_2$ And we can assume that σ_2 leaves a_2 fixed. By the same reasoning $(b_1\ b_3 \ldots b_{h_2})$ can be expressed as a product of two $(3^{[(h_2-1)/3]})$-permutations: $\tau_1\tau_2$, and $b_1^{\tau_1} = b_1$. Then

$$\sigma = \sigma_1\sigma_2[(a_1\ b_2\ b_1)(a_1\ b_2\ a_2)]\tau_1\tau_2$$
$$= [\sigma_1(a_1\ b_2\ b_1)\tau_1][\sigma_2(a_1\ b_2\ a_2)\tau_2]$$

where both $\sigma_1(a_1\ b_2\ b_1)\tau_1$ and $\sigma_2(a_1\ b_2\ a_2)\tau_2$ are $(3^{[s/3]})$-permutations.

If $h_1=2$, we can remove the factor $(a_2\ a_3 \ldots a_{h_1}) = \sigma_1\sigma_2$ from the above formula, while for $h_2=2$ we can remove the factor $(b_1\ b_3 \ldots b_{h_2}) = \tau_1\tau_2$.

Lemma 9. A (2,4,4,4)-permutation can be innerly decomposed as a product of two (3^4)-permutations.

Proof. $(1\ 2)(3\ 4\ 5\ 6)(7\ 8\ 9\ 10)(11\ 12\ 13\ 14)$

$$=[(1\ 2\ 3)(6\ 4\ 7)(11\ 12\ 8)(5\ 13\ 9)]$$

$$[(1\ 4\ 3)(6\ 8\ 13)(7\ 5\ 10)(9\ 14\ 11)]$$

Lemma 10. Suppose σ is an even permutation of type $(2,4,h)$. Then,

(1) when $h \not\equiv 2 \pmod 3$, σ can be innerly decomposed as a product of two $(3^{[h/3]+1})$-permutations.

(2) when $h \equiv 2 \pmod 3$, σ can be innerly decomposed as a product of two $(3^{[h/3]+2})$-permutations.

Proof. (1) $h \not\equiv 2 \pmod 3$.

$$\sigma = (1\ 2)(3\ 4\ 5\ 6)(a_1\ a_2 \ldots a_h)$$

$$= (1\ 2)(3\ 4\ 5\ 6)(a_1\ a_2\ a_3)(a_1\ a_4 \ldots a_h)$$

where $(a_1\ a_4 \ldots a_h)$ is an even cycle of degree $h-2$, and $h-2 \not\equiv 0 \pmod 3$. Hence it can be innerly decomposed as a product of two $(3^{[(h-1)/3]})$-permutations $\tau_1 \tau_2$, and we can assume $a_1^{\tau_1} = a_1$. Then

$$\sigma = [(1\ 3\ 6)(a_1\ a_2\ a_3)\tau_1][(1\ 3\ 2)(4\ 5\ 6)\tau_2]$$

is a product of two $(3^{[(h-2)/3]+2})$-permutations. Because of $h \not\equiv 2 \pmod 3$,

$$[\frac{h-2}{3}]+2 = [\frac{h}{3}]+1$$

(2) $h \equiv 2 \pmod 3$, $h \geq 5$.

$$\sigma = (1\ 2)(3\ 4\ 5\ 6)(a_1\ a_2 \ldots a_h)$$

$$= (1\ 2)(3\ 4\ 5\ 6)(a_1\ a_2 \ldots a_5)(a_1\ a_6 \ldots a_h)$$

If $h = 5$, we remove the factor $(a_1\ a_6 \ldots a_h)$ from the above expression. Express $(a_1\ a_6 \ldots a_h)$ as a product of two $(3^{[(h-4)/3]})$-permutations $\sigma_1 \sigma_2$, and assume that σ_1 leaves a_1 fixed. Therefore

$$\sigma = [(1\ 2\ 3)(4\ a_3\ 6)(5\ a_5\ a_1)\sigma_1]$$

$$[(1\ 4\ 3)(a_3\ 5\ a_2)(a_4\ a_5\ 6)\sigma_2].$$

This is an inner decomposition of σ. It is an equitype decomposition of type $(3^{[(h-5)/3]+3})$. Because of $h \equiv 2 \pmod 3$,

$$[\frac{h-5}{3}]+3 = [\frac{h}{3}]+2 .$$

The following theorems give the equitype decomposition of an even permutation.

Theorem 7. Let σ be an even permutation of type $(2^m,4)$, $s=2m+4$.

1. If $s=n$, then σ can not possess a (3^t) equitype decomposition, but it possesses a $(3;3^{(m+1)/2})$ quasi-equitype decomposition.

2. If $s<n$, then a necessary and sufficient condition for σ to have a (3^t) equitype decomposition is

(1) When $n \le \frac{m+1}{2}+s$:

$$\frac{m+1}{2}+1 \le t \le \frac{m+1}{2}+[\frac{n-s-1}{2}]$$

(2) When $n > \frac{m+1}{2}+s$:

$$\frac{m+1}{2}+1 \le t \le [\frac{n-1}{3}]$$

For the discussion of the general case, we define a number p as follows. Let σ be a $(h_1, h_2, \ldots, h_m)$-permutation, $s = h_1+h_2+\ldots+h_m$, and $(h_1, h_2, \ldots, h_m) \neq (2^k, 4)$. Let H denote the set $\{h_1,h_2,\ldots,h_m\}$ Let H_1 and H_2 be the subset of H of all odd and even numbers respectively. Suppose that in H_i $(i=1,2)$, after forming pairs of a number residue 1 and a number residue 2 modulas 3, there are m_{i1} numbers residue 1 and m_{i2} numbers residue 2. Obviouly, at least one of m_{i1}, m_{i2} is zero. If $m_{i2} \neq 0$, we form pairs of these m_{i2} numbers of H_i. Put

$$p = m_{01}+ m_{11} + [\frac{m_{02}}{2}] + [\frac{m_{12}}{2}] + 2\,\delta_{m_{02}} + 2\,\delta_{m_{12}}$$

where

$$\delta_a = \begin{cases} 1 & \text{when } a \text{ is odd;} \\ 0 & \text{when } a \text{ is even.} \end{cases}$$

Then when σ is expressed as an (3^t)equitype decomposition $\sigma_1 \sigma_2$, p is the least value of $|\sigma| - |\sigma_1|$. Hence we have the following

Theorem 8. Let σ be an even permutation of type $(h_1, h_2, \ldots h_m) \neq (2, 4)$ and define p as above. Then σ can be innerly decomposed as a product of two (3^t)-permutations iff

$$(s-1)/4 + m'-1 \leq t \leq (s-p)/3 .$$

where m' is the sum of the number of even cyclic factors degree greater than 3 and the number of pairs of odd cyclic factors.

For the general case of decomposition, if we use the method of outer stretching, we must calculate the number of fixed points. Let

$$A = [m_{01}/2] + 2\,\delta_{m_{01}} + [m_{11}/2] + 2\,\delta_{m_{11}} + 2[(m_{02}+1)/2] + 2[(m_{12}+1)/2] .$$

And defined an integral function over [0 ,A]:

$$f(x) = \begin{cases} x_0 , & \text{when } 0 \leq x \leq [\frac{m_{01}}{2}] + [\frac{m_{11}}{2}] + \delta_{m_{02}} + \delta_{m_{12}} = x_0 ; \\ x_0 + [\frac{x-x_0}{2}] , & \text{when } x_0 < x \leq x_0 + 2[\frac{m_{01}+1}{2}] + 2[\frac{m_{02}+1}{2}] = x_1 ; \\ x_1 + [\frac{x-x_1}{2}] = [\frac{x+x_1}{2}] , & \text{when } x_1 < x \leq A ; \\ [\frac{x+x_1}{2}] + [\frac{x-A}{2}] , & \text{when } x > A. \end{cases}$$

Then we have the following theorem.

Theorem 9. Let σ be a $(h_1, h_2, \ldots, h_m)$-permutation, $(h_1, h_2, \ldots, h_m) \neq (2^k , 4)$. Then if t satisfies the following condition, σ can be expressed as a product of two (3^t)-permutations:

Inner decomposition: $\frac{s-1}{4} + m' - 1 \leq t \leq \frac{1}{3}[s-p] = t_0$.

General case: $\frac{s-1}{4} + m' - 1 \leq t \leq t_0 + f(n-s)$

where m' and $f(x)$ are defined as above.

Theorem 10. (1) If $n \not\equiv 2 \pmod 4$ then every permutation in S_n can be expressed as a product of two conjugate permutations of order 3.

(2) If $n = 2m+4$ and m is odd, then, every even permutation of S_n, except the case of type $(2^m,4)$, can be expressed as a product of two equitype permutations of order 3. While an even permutation of type $(2^m,4)$ can be expressed as a product of two quasi-equitype permutations of order 3.

References

1. M. Hall, The Theory of Groups, Macmillan, New York, 1959.
2. H. Wielandt, Finite Permutation Groups, Academic Press, New York, 1964.
3. O. Ore, Some remarks of commutators, Proc. Amer. Math. Soc., 2(1951), 307-314.
4. N. Ito, A theorem on the alternating group $A_n (n \geq 5)$, Math. Japan, 2(1951), 59-60.
5. E. Bertram, Even permutations as a product of two conjugate cycles, J. Comb. Theory, 12(1972), 368-380.
6. G. Boccara, Decompositions d'une permutation d'un ensemble fini product de deux cycles, Discrete Math. 23(1978), 180-205.

ON THE EXTENSIONS OF ABELIAN VARIETIES BY AFFINE GROUP SCHEMES *

XIAOLONG WU
Institute of Mathematics, Academia Sinica
Beijing P.R.China

CONTENTS

* The paper is author's doctoral dissertation written under the supervision of Professors Ellis R.Kolchin and Hyman Bass in Columbia University, 1984. I would like to express my thanks to them.

Introduction

According to Chevalley Theorem, any connected algebraic group has a largest connected linear algebraic subgroup and the corresponding quotient is an abelian variety. Therefore, the study of the structure and classification of connected algebraic groups can be divided into three parts: the study of connected linear algebraic groups, the study of abelian varieties and the study of the extensions of abelian varieties by connected linear algebraic groups. The present paper is devoted to the third part. Several mathematicians have worked on the third part. Let A be an abelian variety. A.Weil showed that $Ext(A,G_m)$ is isomorphic to A*, where G_m is the multiplicative group and A* is the dual abelian variety of A. ("Variétés abéliennes", Colloque d'algèbre et théorie des nombres, Paris, (1949), p125-128). I.Barsotti proved that $Ext(A,G_a)$ is isomorphic to $G_a^{\dim A}$, where G_a is the additive group. ("Structure theorems for group varieties" Annali di Math. 38, (1955) p77-119). M.Rosenlicht proved the same result by means of generalized jacobian varieties. ("Extensions of vector groups by abelian varieties" Amer.J.of Math. 80, (1958), p685-714). J.-P.Serre collected almost all known results in this aspect in his book [10]. In particular, he showed that if H is a connected commutative unipotent algebraic group, then $Ext(A,H)\cong H^1(A,H)$, [10], p187, Théorème 8. It is worth to mention that F.Oort gave a full list of extensions of commutative elementary group schemes [8]. However, the structure of Ext(A,H) for general H is left unknown. This is what we will study in the present paper, provided that the base field is algebraically closed.

In this paper we will treat a generalized version of the problem, i.e. the kernels of the extensions need not be connected nor reduced. This requires very little more effort. In Chapter 1 we will show that two Noether isomorphism theorems hold for the category of group schemes. We will also prove some useful properties of centers of group schemes. In Chapter 2 we will show that if A is an abelian variety and H is an affine group scheme, then there is a canonical bijection between Ext(A,H) and Ext(A,C(H)), where C(H) is the center of H. As a result, Ext(,) is a bifunctor from the category of abelian varieties and the category of affine group schemes and homomorphisms sending centers into centers into the category of abelian groups. Now assume that H is commutative. We will show in Section 5 that $Ext(A,H)\cong Ext(A,H_m)\times Ext(A,H_u)$, where H_m and H_u are the multiplicative factor and the unipotent factor of H respectively. Furthermore, we will show that $Ext(A,H_m)$ is isomorphic to the direct product of $(A^*)^{\dim(H_m)_{red}}$ and a finite group (cf.§6), and that $Ext(A,H_u)$ is isomorphic to $G_a^{\dim A \dim H}$ if the characteristic of the base field is 0, and to $(H_u)^r_{red} \times Hom(N,H_u)$ if the characteristic p of the base field is positive, where r is the p-rank of A and N is certain local-local group subscheme of A (cf. §7). In Section 8 we will show that Ext(A,H) has a canonical algebraic group structure which is compatible with the functorial property of Ext(,). In Chapter 3 we will calculate some special cases of the group $Hom(N,H_u)$ by means of Dieudonné modules. The calculation shows that there is no simple expression for $Hom(N,H_u)$. This last chapter is not closely related to the main trend of this symposium and is therefore omitted.

General Conventions

Everything in this paper is defined over an algebraically closed field k. All schemes are supposed to be of finite type, i.e. each of them can be covered by finite affine schemes $\{SpecB_i\}$, where B_i is a finitely generated k-algebra. We identify the category of algebraic groups with the category of reduced group schemes, which is a full subcategory of the category of group schemes. We denote by G_m and G_a the multiplicative group and the additive group respectively. If V is an n-dimensional k-vector space, we will write GL_n or GL(V) for the general linear group over V. e will stand for the neutral group subschemes of all group schemes. If G is a group scheme, we will denote by C(G) the center of G, by G_{red} the largest reduced group subscheme of G, by G^o the connected component of G containing e and by id_G the identity map of G.

Chapter 1. Preliminaries

§1. A sufficient condition for Noether theorems.

Let C be a category with a universal element e, i.e. for any object X in C there is one and only one element in Mor(X,e) and there is one and only one element in Mor(e,X). We denote by ε_X the only element in Mor(e,X). $\varepsilon_X: e \longrightarrow X$ defines e as a subobject of X. A morphism in C is called trivial if it factors through e. Let $f: X \longrightarrow X'$ be a morphism in C and H,H' be subobjects of X,X' respectively. Then the smallest subobject Y' of X' (if it exists) such that $f|_H$ factors through Y' is called the image of H and will be denoted by f(H). In particular, we call f(X) the image of f and denote it by Imf. If Imf=X', we will call f surjective. The largest subobject Y (if it exists) of X such that f(Y) is contained in H' is called the inverse image of H' and will be denoted by $f^{-1}(H')$. In particular, the inverse image of e is called the kernel of f and will be denoted by Kerf. If Kerf=e, we will call f injective. If the images (resp. inverse images) always exist, we say that C has images (resp. inverse images).

Let $\{X_i\}_{i\in I}$ be a set of subobjects of an object X in C. The intersection of $\{X_i\}_{i\in I}$ is the largest subobject Y (if it exists) of X such that Y is contained in X_i for all $i\in I$. If the intersection exists, we will denote it by $\cap_{i\in I}X_i$. If for any set (resp. finite set) of subobjects of any object in C the intersection exists, we say that C has intersections (resp. finite intersections). The join of $\{X_i\}_{i\in I}$ is the smallest subobject (if it exists) of X such that Y contains all X_i. If the join exists, we will denote it by $\cup_{i\in I}X_i$. If for any set (resp. finite set) of subobjects of any object in C the join exists, we say that C has joins (resp. finite joims).

Now assume that C has inverse images. A morphism $f:X \longrightarrow X'$ in C is called a quotient if for any morphism $g:X \longrightarrow Y$ in C such that g is trivial on Kerf there exists a unique morphism h of X' to Y such that hf=g. When this is so we also call X' a quotient of X by Kerf and call f the canonical projection of X to X'. It is obvious that the quotient of X by a subobject H are uniquely determined upto a unique isomorphism. We will denote the quotient by X/H.

Proposition 1. Let C be a category satisfying the following conditions:

(i) C has images and inverse images,

(ii) if $f:X \longrightarrow X'$ is a surjective morphism in C, then f is a quotient of X by Kerf,

(iii) if $f:X \longrightarrow X'$ is a surjective morphism in C and H' is a subobject of X', then $f(f^{-1}(H'))=H'$,

(iv) if $X \xrightarrow{f} X' \xrightarrow{g} X''$ is a sequence of surjective morphisms in C, then gf is also surjective,

(v) C has finite intersections and finite joins.

Then

(1) if $H_1\subset H_2\subset X$ are objects in C and X/H_1, X/H_2 exist, then H_2/H_1 exists and is naturally a subobject of X/H_1. Furthermore, $(X/H_1)/(H_2/H_1)$ exists and is canonically isomorphic to X/H_2. (The first Noether isomorphism theorem),

(2) if $X=H_1\cup H_2$ and X/H_1 exists, then $H_2/(H_1\cap H_2)$ exists and is

canonically isomorphic to X/H_1. (The second Noether isomorphism theorem).

Proof. (1) Let $p_i:X \longrightarrow X/H_i$ be the projections, $i=1,2$. Then p_2 is trivial on H_1. So there exists a unique morphism $g:X/H_1 \longrightarrow X/H_2$ such that $gp_1=p_2$. Since p_2 is surjective, g is surjective. Let $K=\mathrm{Ker}g$. Then

$$p_2(p_1^{-1}(K))=(gp_1)(p_1^{-1}(K))=g(p_1(p_1^{-1}(K)))=g(K)=e.$$

(To get the second equality note that $(gp_1)(p_1^{-1}(K))$ is obviously contained in $g(p_1(p_1^{-1}(K))$ and the other side inclusion follows from (iv)). Hence $p_1^{-1}(K)$ is contained in $\mathrm{Ker}p_2=H_2$. On the other hand, since $e=p_2(H_2)=(gp_1)(H_2)=g(p_1(H_2))$, so $p_1(H_2)$ is contained in K. Hence H_2 is contained in $p_1^{-1}(K)$ by the definition of inverse images. So $p_1^{-1}(K)=H_2$. Hence by (iii) $K=p_1(p_1^{-1}(K))=p_1(H_2)$. Since $p_1|_{H_2}:H_2 \longrightarrow p_1(H_2)$ is surjective and $\mathrm{Ker}(p_1|_{H_2})=H_1$, we get $p_1(H_2)\cong H_2/H_1$ by (ii). Hence by (ii) again, $(X/H_1)/(H_2/H_1)\cong X/H_2$ since g is surjective with kernel H_2/H_1.

(2) Let $p:X \longrightarrow X/H_1$ be the projection. Let $Y=p(H_2)$. Then both H_1 and H_2 are contained in $p^{-1}(Y)$, so $X=H_1\cup H_2$ is also contained in $p^{-1}(Y)$, so $p^{-1}(Y)=X$. Hence $Y=p(X)$ and $p|_{H_2}$ is surjective. Since $H_1\cap H_2$ is contained in H_1, we have $p|_{H_2}(H_1\cap H_2)=e$. Conversely, if Z is contained in H_2 and $p|_{H_2}(Z)=e$, then $Z\subset H_1$ and hence $Z\subset(H_1\cap H_2)$. So $H_1\cap H_2$ is the kernel of $p|_{H_2}$. Therefore $H_2/(H_1\cap H_2)\cong X/H_1$ by (ii). This completes the proof.

§2. Noether theorems for group schemes.

Now we are going to show that the category of group schemes over k satisfies the conditions in Proposition 1. We omit the definitions of group schemes and their homomorphisms. If f is a homomorphism of group schemes, we denote by f_* the map on the underlying spaces, by f^* the map on the sheaves and by f^*_e the map on the stalks at the neutral points e. A group subscheme H of a group scheme G is a group scheme H and a homomorphism $f:H \longrightarrow G$ such that f^* is surjective and f_* is injective. Note that f^* is surjective if and only if f^*_e is surjective. It is easy to check that a group subscheme is a subobject in the category of group schemes. However, it is not obvious that a subobject in the category is a group subscheme Hence we have to redefine certain terminologies such as images and inverse images.

Let G be a group scheme over k. Let H be a group subscheme of G. Then H determines an ideal sheaf I. We denote by $\underline{a}_H$ the image of I in the stalk of G at e and call it the defining ideal of H in G.

Lemma 1. Let G be a group scheme, μ be its group law, γ be its inverse map and O_G, $O_{G\times G}$ be the stalks of G, G×G at e respectively. Then the map $H \longmapsto \underline{a}_H$ is a bijection from the set of connected group subschemes of G onto the set of ideals $\underline{a}$ in O_G such that $\gamma_e^*(\underline{a})=\underline{a}$ and $\mu_e^*(\underline{a})\subset(\underline{a}\otimes O_G+O_G\otimes\underline{a})O_{G\times G}$.

Proof. [11] p25, Theorem 2.12.

Lemma 2. Any group subscheme of a group scheme G is completely determined by its underlying space and its defining ideal in G.

Proof. Obvious.

Lemma 3. Let $f:G_1 \longrightarrow G_2$ be a homomorphism of group schemes, H_i be a group subscheme of G_i, O_i be the stalk of G_i at e and $\underline{a}_i$ be the defining ideal of H_i in G_i, i=1,2. Then

(1) there exists a smallest group subscheme H_2' of G_2 such that $f|_{H_1}$ decomposes through H_2'. The underlying space of H_2' is $f_*(H_1)$ and the defining ideal of H_2' is $f_e^*(\underline{a}_1)$;

(2) there exists a largest group subscheme H_1' of G_1 such that $f|_{H_1'}$ decomposes through H_2. The underlying space of H_1' is $f_*^{-1}(H_2)$ and the defining ideal of H_1' in G_1 is $f_e^*(\underline{a}_2)O_1$.

Proof. [11] p108 Proposition 8.12, p109 Proposition 8.13 and p111 Proposition 8.14.

We will call H_2' in the lemma the image of H_1 and denote it by $f(H_1)$. We will call H_1' in the lemma the inverse image of H_2 and denote it by $f^{-1}(H_2)$. In particular, $f(G_1)$ is called the image of f and will be denoted by Imf. $f^{-1}(e)$ is called the kernel of f and will be denoted by Kerf. If Imf=G_2, f is called surjective. If Kerf =e, f is called injective.

Lemma 4. Let $f:G_1 \longrightarrow G_2$ be a homomorphism of group schemes. Then

(1) f is surjective if and only if f_* is surjective and f^* is injective;

(2) f is an isomorphism if and only if f is both injective and surjective;

(3) f is injective if and only if f_* is injective and f^* is surjective, and this is so if and only if f is a monomorphism in the category of group schemes over k.

Proof. (1) If f_* is surjective and f^* is injective, then the underlying space of $f(G_1)$ is G_2. Since f^* is injective, f_e^* is also injective. Hence the defining ideal of $f(G_1)$ is $f_e^{*-1}(0)=0$. So $f(G_1) =G_2$, i.e. f is surjective. If f is surjective, then by Lemma 3 $f_*(G_1)=G_2$ and $f_e^{*-1}(0)=0$. Hence f_* is surjective and f^* is injective.

(2) [7] p119.

(3) Let f be injective. Since f factors through Imf, we have an induced homomorphism $f':G_1 \longrightarrow \mathrm{Imf}$. f' is both injective and surjective, so f' is an isomorphism by (2). Hence f_* is injective and f^* is surjective since Imf is a group subscheme of G_2. If f_* is injective and f^* is surjective, then it is easy to check that f is a monomorphism. Finally, assume that f is a monomorphism. If f is not injective, then Kerf is not trivial. Let $g:\mathrm{Kerf} \longrightarrow G_1$ be the inclusion map and $h:\mathrm{Kerf} \longrightarrow G_1$ be the trivial map. Then fg=fh but $g \neq h$. This is a contradiction. So f must be injective. This completes the proof.

Let G and H be group schemes. Then Lemma 4 shows that H is a group subscheme of G if and only if H is a subobject of G in the category of group schemes over k. Hence the words such as "image" and "inverse image" defined by subobjects are the same as those defined by group subschemes.

Lemma 5. If f is a homomorphism of a group scheme G into another group scheme, then the quotient of G by Kerf exists and the projection is flat.

Proof. Gabriel.P. "Généralités sur les groupes algèbrique", Lecture Notes in Math. Vol.151, p287-317.

Proposition 2. The category of group schemes over k satisfies the conditions (i)-(v) in Proposition 1. Hence two Noether theorems hold for this category.

Proof. (i) By Lemma 3.

(iv) By Lemma 4 (1).

(v) [11] p135,p136 and p137.

(ii) and (iii) First we prove (iii) for the special case in which X' is the quotient of X by a group subscheme H and f is the projection. Let H' be a group subscheme of X'. By Lemma 3 the underlying space of $f(f^{-1}(H'))$ is $f_*(f_*^{-1}(H'))=H'$ since f_* is surjective. Let O_X and $O_{X'}$ be the stalks of X and X' at e respectively. Let $\underline{a}'$ be the defining ideal of H' in X'. Then the defining ideal of $f(f^{-1}(H'))$ in X' is $f_e^{*-1}(f_e^*(\underline{a}')O_X)$ by Lemma 3. Hence it suffices to show that $f_e^{*-1}(f_e^*(\underline{a}')O_X)=\underline{a}'$. Since f is flat (i.e. O_X is a flat $f_e^*(O_{X'})$-module) by Lemma 5, so $f_e^{*-1}(f_e^*(\underline{a}')O_X)=\underline{a}'$ if $f_e^*(\underline{m}')O_X \neq O_X$, where $\underline{m}'$ is the maximal ideal of $O_{X'}$, [1] p45 Exercise 16. By the defintion of scheme morphisms, f_e^* sends $\underline{m}'$ into the maximal ideal of O_X. So $f_e^*(\underline{m}')O_X \neq O_X$ and the special case is proven. Now we prove (ii). Let $f:X \longrightarrow X'$ be a surjective homomorphism of group schemes. Then there exists a unique homomorphism $g:X/\mathrm{Ker}f \longrightarrow X'$ such that $g\pi=f$, where $\pi:X \longrightarrow X/\mathrm{Ker}f$ is the projection. g is surjective since f is. If g is not injective, Kerg will be nontrivial. By the special case we just proved there exists a group subscheme K of X, which is strictly larger than Kerf, such that the image of K in X/Kerf is Kerg. But this means that f(K)=e which contradicts the definition of kernels. Hence g is injective. So g is an isomorphism by Lemma 4 (2). So (ii) is true. Now (iii) follows from (ii) and the special case we proved before. The proposition is proven.

§3. Centers of group schemes.

In this section we are going to show that the center of any group scheme exists, and that if a group scheme G is connected or the commutator group subscheme of G is affine, then G/C(G) is affine.

Let G be a group scheme, m be the group law of G and H,K be group subschemes of G. We say that H centralizes K if the following diagram commutes

$$\begin{array}{ccccc} H\times K & \xrightarrow{i} & G\times G & & \\ S\big| & & & \searrow^{m} & G\ , \\ K\times H & \xrightarrow{i} & G\times G & \nearrow_{m} & \end{array}$$

where i is the inclusion map and S is the exchange of factors. The largest H (if it exists) that centralizes K is called the centralizer of K (in G). We say that H is central if it centralizes G. The centralizer of G is called the center of G and will be denoted by C(G).

Proposition 3. Let G be a group scheme. Then C(G) exists.

Proof. First we show that the centralizer of G^o exists. We give here a sketch of the proof (cf. [11], Sections 11 and 13). Define $f:G\times G \longrightarrow G$ to be the composite morphism

$$m(m\times id_G)(inv_G\times id_{G\times G})(S\times id_G)(id_G\times\Delta_G),$$

where m is the group law of G, inv_G is the inverse map of G, S is the exchange of factors and Δ_G is the diagonal map. Let O_G be the stalk of G at e and $\underline{m}$ be the maximal ideal of O_G. Let s be a positive integer. Then f sends the subscheme $\mathrm{Spec}(O_G/\underline{m}^s)\times G$ of $G\times G$ into the subscheme $\mathrm{Spec}(O_G/\underline{m}^s)$ of G. So we get a k-algebra homomorphism f_s^* of $O_G/\underline{m}^s$ into $(O_G/\underline{m}^s)\otimes\Gamma(G)$, where $\Gamma(G)$ is the k-algebra of all global sections of the structure sheaf on G. Let $e_1,\ldots,e_{n_s}$ be a k-vector space basis of $O_G/\underline{m}^s$. Then there exist $a_{ij}\in\Gamma(G)$, $1\le i,j\le n_s$, such that $f_s^*(e_i)=\Sigma_{j=1}^{n_s} e_j\otimes a_{ij}$. Let T_{ij}, $1\le i,j\le n_s$, be the canonical regular functions on $GL(O_G/\underline{m}^s)$ relative to the basis $e_1,\ldots,e_{n_s}$. Define a morphism $Ad_s:G \longrightarrow GL(O_G/\underline{m}^s)$ by $Ad_s^*(T_{ij})=a_{ij}$. It can be checked that Ad_s is a homomorphism of group schemes. A group subscheme H centralizes G^o if and only if it is contained in KerAd_s for all positive integers s. Since KerAd_s, $s=1,2,\ldots$, is a descending series of closed subschemes of G and G is of finite type, so there exists an integer s' such that $\mathrm{KerAd}_s=\mathrm{KerAd}_{s'}$ for all $s\geq s'$. It is clear that $\mathrm{KerAd}_{s'}$ is the center of G^o.

Now we are going to show that the centralizer of G_{red} exists. For any closed point x of G define $g_x:G \longrightarrow G$ to be the composite morphism gj_x, where j_x is the canonical map of G onto $G\times\{x\}$ and

$$g=m(m\times id_G)(m\times id_{G\times G})(id_{G\times G}\times inv_G\times inv_G)(id_G\times S\times id_G)(\Delta_G\times\Delta_G),$$

where m, inv_G, S and Δ_G are defined as in the previous paragraph. $g_x^{-1}(e)$ is a closed subscheme of G. A group subscheme of G centralizes G_{red} if and only if it is contained in $g_x^{-1}(e)$ for all closed point x in G. It is easy to check that $\cap_x g_x^{-1}(e)$, where x runs through all closed points of G, is a group subscheme of G. Hence it is the centralizer of G_{red}.

It is obvious that the intersection of the centralizers of G^o and G_{red} is the center of G. This completes the proof.

Let g be defined as in the above proof. The smallest group subscheme of G containing Img is called the commutator group subscheme of G. Let H be a normal group subscheme of G. (The normal group subschemes are defined by suitable commutative diagrams or equivalently as kernels of homomorphisms from G). Then it is easily seen that G/H is commutative if and only if H contains the commutator group subscheme of G.

Proposition 4. Let G be a group scheme. If G is connected or the commutator group subscheme of G is affine, then G/C(G) is affine.

Proof. Let notations be as in the proof of the previous proposition. If G is connected, then $C(G)=\mathrm{KerAd}_{s'}$. So $G/C(G)\cong\mathrm{ImAd}_{s'}$ is a group subscheme of $GL(O_G/\underline{m}^{s'})$. Hence G/C(G) is affine.

Now assume that the commutator group subscheme K of G is affine. Let D be the smallest group subscheme of G such that G/D is an affine group scheme. This is the same as to say that D is the largest group subscheme of G such that any homomorphism of D into an affine group scheme is trivial. It is easy to check that any morphism of D into an affine scheme is constant. It suffices to show that D is contained in C(G). Since the quotient of G by the centralizer of G^o is affine, so D is contained in the centralizer of G^o. We need to show that D is also contained in the centralizer of G_{red}. This is so if D is contained in $g_x^{-1}(e)$ for all closed points x in G. Let π be the projection of G onto G/K. Since G/K is commutative, πg_x is trivial. So the image of g_x is contained in K. By the definition of D, $g_x|_D$ is a constant map. Since $g_x(e)=e$, D is contained in $g_x^{-1}(e)$ for all closed points x in G. This completes the proof.

Corollary 1. Let G be a group scheme and H be a normal affine group subscheme of G such that G/H is an abelian variety. Then the join $H \cup C(G)=G$ and the intersection $H \cap C(G)=C(H)$.

Proof. $G/(H \cup C(G))$ is both an affine group scheme by the proposition and an abelian variety by the assumption, so is trivial. So the first statement holds. It is obvious that $H \cap C(G)$ is contained in C(H). Conversely, since C(H) centralizes both H and C(G), so it centralizes their join $H \cup C(G)=G$. Hence C(H) is contained in C(G). So $H \cap C(G)=C(H)$. The proof is completed.

Corollary 2. Let assumption be as in Corollary 1. Assume further that H is commutative. Then G is also commutative.

Proof. Since H is commutative, it centralizes both H and C(G). So it is contained in C(G). So C(G) contains $H \cup C(G)=G$, i.e. G is commutative. This completes the proof.

Some mathematicians have defined the algebraic group center of an algebraic group G, which is, in our definition, equal to $C(G)_{red}$. In general, the center of a reduced group scheme is not reduced. We will give a criterion for reduced centers.

If G is a connected algebraic group, we denote by k(G) the field of rational functions over G. If $f:G \longrightarrow G'$ is a surjective homomorphism of connected algebraic groups, then we have an induced embedding $f^*:k(G') \longrightarrow k(G)$ of fields over k. f is called separable if k(G) is a separable extension field of $f^*(k(G'))$. Let $f:G \longrightarrow G'$ be any homomorphism of connected algebraic groups. Then f is called separable if the homomorphism $G \longrightarrow \mathrm{Im}f$ induced by f is separable.

Lemma 6. Let $f:G \longrightarrow G'$ be a homomorphism of connected algebraic groups. Then f is separable if and only if Kerf is reduced.

Proof. We may assume that f is surjective. Since $f|_{(\mathrm{Kerf})_{red}}$ is trivial, we have a unique homomorphism $g:G/(\mathrm{Kerf})_{red} \longrightarrow G'$ making the following diagram commutative

$$\begin{array}{ccc} & \overset{j}{\nearrow} & G/(\mathrm{Kerf})_{red} \\ G & & \downarrow g \\ & \underset{f}{\searrow} & G' \end{array} ,$$

where j is the projection. If f is separable, then g is also separable. But Kerg is local, so g induces a bijection of the underlying spaces of $G/(\mathrm{Kerf})_{red}$ onto that of G'. Hence $k(G/(\mathrm{Kerf})_{red})$ is a purely inseparable extension of $g^*(k(G))$ by Zariski's Main Theorem. Hence $k(G/(\mathrm{Kerf})_{red})=g^*(k(G'))$. Hence g is an isomorphism. So Kerf is reduced. Conversely, if Kerf is reduced, then g is an isomorphism. Since j is separable by [9] p413, Theorem 4, f is also separable. This completes the proof.

Let G be a connected algebraic group. Let O_G be the stalk of G at its neutral point e and $\underline{m}$ be the maximal ideal of O_G. Then $\underline{m}/\underline{m}^2$ has a natural k-vector spaces structure and $\dim_k(\underline{m}/\underline{m}^2)=\dim G$. Denote by $\underline{g}$ the dual space of $\underline{m}/\underline{m}^2$. Then $\underline{g}$ is the Lie algebra of G. Any homomorphism $f:G \longrightarrow G'$ of connected algebraic groups induces a Lie algebra homomophism $df:\underline{g} \longrightarrow \underline{g}'$ which is called the differential of f. Let x be a closed point in G and Int_x be the automorphism of G defined by $\mathrm{Int}_x(y)=xyx^{-1}$ for all closed points y in G. Denote by Ad_x the differential of Int_x. Then the map $\mathrm{Ad}:G \longrightarrow GL(\underline{g})$ defined by $\mathrm{Ad}(x)=\mathrm{Ad}_x$ is a rational homomorphism. Denote by ad the differential of Ad. We know that C(G) is contained in KerAd and that the Lie algebra center $c(\underline{g})$ of $\underline{g}$ is equal to Ker ad. (cf. [11], Sections 11,13 and 15).

Proposition 5. Let G be a connected algebraic group. If $\dim_k c(\underline{g})=\dim C(G)_{red}$, then C(G) is reduced.

Proof. We know that

$$\dim_k c(\underline{g})=\dim_k \mathrm{Kerad} \geq \dim(\mathrm{KerAd})_{red} \geq \dim C(G)_{red}.$$

When $\dim_k c(\underline{g})=\dim C(G)_{red}$, we have $\dim_k \mathrm{Kerad}=\dim(\mathrm{kerAd})_{red}$ and hence Ad is separable. By Lemma 6, KerAd is reduced. Since C(G) is contained in KerAd and $\dim C(G)_{red}=\dim(\mathrm{KerAd})$, we have $C(G)^o=(\mathrm{KerAd})^o$. Hence $C(G)^o$ is reduced. Hence C(G) is reduced. This completes the proof.

The following example was produced by Chevalley to show that the Lie algebra of a noncommutative algebraic group may be commutative. This example shows also that the center of a reduced group scheme may not be reduced and that Ad is not separable in general.

Example. Let G be the algebraic group consisting of all matrices

$$\begin{pmatrix} a & 0 & 0 \\ 0 & a^p & b \\ 0 & 0 & 1 \end{pmatrix} \qquad a,b \text{ in } k,\ a\neq 0,\ \text{char } k=p\neq 0.$$

Define regular functions T_1, T_2 on G by

$$T_1\left(\begin{pmatrix} a & 0 & 0 \\ 0 & a^p & b \\ 0 & 0 & 1 \end{pmatrix}\right)=a-1, \qquad T_2\left(\begin{pmatrix} a & 0 & 0 \\ 0 & a^p & 0 \\ 0 & 0 & 1 \end{pmatrix}\right)=b.$$

Then the coordinate ring of G is $k[T_1,T_2,(T_1+1)^{-1}]$. The comap μ^* of the group law μ is defined by

$$\mu^*(T_1)=T_1\otimes 1+1\otimes T_1+T_1\otimes T_1 \qquad \mu^*(T_2)=T_2\otimes 1+1\otimes T_2+T_1^p\otimes T_2.$$

Let H be a central group subscheme of G, I be the ideal defining H in the coordinate ring k[G] of G and $\overline{T}_1,\overline{T}_2$ be respectively the images of T_1,T_2 in the coordinate ring k[H]=k[G]/I of H. Since H is central, we must have $\overline{T}_2\otimes 1+1\otimes T_2+\overline{T}_1^p\otimes T_2=1\otimes T_2+\overline{T}_2\otimes 1+\overline{T}_2\otimes T_1^p$. So we have $\overline{T}_1^p\otimes T_2=\overline{T}_2\otimes T_1^p$. Let e_i, f_j, i,j in $\mathbb{N}$ be k-vector space bases of k[G],k[H] respectively. Since T_1^p and T_2 are linearly independent over k, we may assume that $e_1=T_1^p$ and $e_2=T_2$. Assume that $\overline{T}_1^p=\Sigma_j a_j f_j$, $\overline{T}_2=\Sigma_j b_j f_j$, a_j, b_j in k. Then we have $\Sigma_j a_j f_j\otimes e_2=\Sigma_j b_j f_j\otimes e_1$. Since $f_j\otimes e_i$, i,j in $\mathbb{N}$, form a k-vector space basis of k[H]$\otimes$k[G], all a_j and b_j are 0. Hence $\overline{T}_1^p=\overline{T}_2=0$. This shows that any central group subscheme of G must be contained in the subscheme $\mathrm{Spec}(k[G]/(T_1^p,T_2))$ of G. On the other hand it is easy to check that $\mathrm{Spec}(k[G]/(T_1^p,T_2))$ is a central group subscheme of G. Hence $C(G)=\mathrm{Spec}(k[G]/(T_1^p,T_2))$. It is easily seen that C(G) is not reduced.

Let

$$x=\begin{pmatrix} a' & 0 & 0 \\ 0 & a'^p & b' \\ 0 & 0 & 1 \end{pmatrix} \quad \text{and} \quad y=\begin{pmatrix} a & 0 & 0 \\ 0 & a^p & b \\ 0 & 0 & 1 \end{pmatrix}.$$

Then

$$xyx^{-1}=\begin{pmatrix} a & 0 & 0 \\ 0 & a^p & -a^pb'+a'^pb+b' \\ 0 & 0 & 1 \end{pmatrix}.$$

So we have $\mathrm{Int}_x^*(T_1)=T_1$ and $\mathrm{Int}_x^*(T_2)=-b'T_1^p+a'^pT_2$. Denote by S_1 and S_2 the image of T_1 and T_2 in $\underline{m}/\underline{m}^2$ respectively. Let S_1' and S_2' be the basis of $\underline{g}$ dual to S_1 and S_2. Then we have $\mathrm{Ad}_x(S_1')=S_1'$ and $\mathrm{Ad}_x(S_2')=a'^pS_2'$. Hence Ad is given by

$$Ad(x)=\begin{pmatrix}1 & 0\\ 0 & a'^p\end{pmatrix}$$

relative to the basis S_1' and S_2'. Therefore, $\mathrm{KerAd}=\mathrm{Spec}(k[G]/(T_1^p))$. So Ad is not separable.

The above calculation shows that $\dim C(G)_{red}=0$, $\dim(\mathrm{KerAd})_{red}=1$ and $\dim_k c(\underline{g})=2$.

Chapter 2. Ext(A,H)

§4. For general affine group scheme H.

In this section we will reduce the extension problem with non-commutative kernels to that with commutative kernels.

An exact sequence of group schemes is a sequence of homomorphisms of group schemes such that the image of each homomorphism is the kernel of the next homomorphism (if it exists). Let A and H be two group schemes. An extension of A by H is a short exact sequence

$$e \longrightarrow H \longrightarrow G \xrightarrow{f} A \longrightarrow e$$

of group schemes. Sometimes we will call G the extension of A by H. When A,H and G are all reduced and f is separable we call G an algebraic group extension of A by H.

Lemma 7. Let A and H be reduced group schemes. Then any extension of A by H is an algebraic group extension.

Proof. Let $e \longrightarrow H \longrightarrow G \xrightarrow{f} A \longrightarrow e$ be an extension of A by H. We need to show that G is reduced and that f is separable. We know that G_{red} is a normal group subscheme of G, so G/G_{red} exists [7] p104. Let $h:G \longrightarrow G/G_{red}$ be the projection. Since H is reduced, H is contained in G_{red}. Hence h is trivial on H. So there exists a unique homomorphism g of A to G/G_{red} such that gf=h. Since h is surjective, g is also surjective. Since A is reduced, $g(A)=G/G_{red}$ is also reduced. Now G/G_{red} is both local and reduced, so is trivial, so $G=G_{red}$. That f is separable follows from Lemma 6.

Let A and H be reduced group schemes. Lemma 7 shows that there is no difference between the extensions of A by H and the algebraic group extensions of A by H.

Let A and H be group schemes. Two extensions G and G' of A by H are called equivalent if there exists an isomorphism f of G onto G' such that the following diagram commutes

$$\begin{array}{ccccccccc} e & \longrightarrow & H & \longrightarrow & G & \longrightarrow & A & \longrightarrow & e \\ & & \downarrow{\scriptstyle id_H} & & \downarrow{\scriptstyle f} & & \downarrow{\scriptstyle id_A} & & \\ e & \longrightarrow & H & \longrightarrow & G' & \longrightarrow & A & \longrightarrow & e. \end{array}$$

We denote by Ext(A,H) the set of all equivalence classes of extensions of A by H and denote by $\underline{G}$ the class represented by an extension G.

Let A be an abelian variety, H be an affine group scheme and $e \longrightarrow H \longrightarrow G \xrightarrow{f} A \longrightarrow e$ be an extension of A by H. Then by Proposition 2 and Corollary 1 to Proposition 4, the sequence $e \longrightarrow C(H) \longrightarrow C(G) \xrightarrow{f|C(G)} A \longrightarrow e$ is exact. Since the centers are characteristic group subschemes, so equivalent extensions of A by H induce equivalent extensions of A by C(H). Hence we get a well defined map

$$t:\mathrm{Ext}(A,H) \longrightarrow \mathrm{Ext}(A,C(H)).$$

Theorem 1. The map t we just defined is a bijection.

Proof. Surjectivity. Let $e \longrightarrow C(H) \longrightarrow G' \xrightarrow{f} A \longrightarrow e$ be an extension of A by C(H). Define $G=(G'\times H)/E$, where E is the image of the composite map

$$C(H) \xrightarrow{\text{inverse}\times id} C(H)\times C(H) \xrightarrow{\text{embedding}} G'\times H.$$

Then G is an extension of A by H, where the map of H into G is induced by the embedding of H onto the second factor of $G'\times H$ and the map of G onto A is induced by f. Define a map $g:G' \longrightarrow G$ to be the map induced by the embedding of G' onto the first factor of $G'\times H$. Then it is easy to check that g is an isomorphism of G' onto the center of G and that $t(\underline{G})=\underline{G}'$. So t is surjective.

Injectivity. Let G_1 and G_2 be extensions of A by H such that $t(\underline{G}_1)=t(\underline{G}_2)$, i.e. there exists an isomorphism f of $C(G_1)$ onto $C(G_2)$ making the following diagram commutative

$$\begin{array}{ccccccccc} e & \longrightarrow & C(H) & \longrightarrow & C(G_1) & \longrightarrow & A & \longrightarrow & e \\ & & \downarrow{\scriptstyle id_{C(H)}} & & \downarrow{\scriptstyle f} & & \downarrow{\scriptstyle id_A} & & \\ e & \longrightarrow & C(H) & \longrightarrow & C(G_2) & \longrightarrow & A & \longrightarrow & e\,. \end{array}$$

We define a homomorphism $g:G_1 \longrightarrow G_2$ by putting $g|_{C(G_1)}=f$ and $g|_H=id_H$. Since $f|_{C(H)}=id_{C(H)}$, g is well defined on $C(G)\cap H=C(H)$. Since $C(G_1)\cup H=G_1$, g is well defined on G_1. Since f and id_H are isomorphisms, g is also an isomorphism and the diagram

$$\begin{array}{ccccccccc} e & \longrightarrow & H & \longrightarrow & G_1 & \longrightarrow & A & \longrightarrow & e \\ & & \downarrow{\scriptstyle id_H} & & \downarrow{\scriptstyle g} & & \downarrow{\scriptstyle id_A} & & \\ e & \longrightarrow & H & \longrightarrow & G_2 & \longrightarrow & A & \longrightarrow & e \end{array}$$

is commutative. Hence G_1 and G_2 are equivalent. This completes the proof.

We will write AG' for the category whose objects are all affine group schemes over k and the morphisms are all homomorphisms sending centers into centers.

Corollary. Ext(,) is a bifunctor from the category of abelian varieties and AG' into the category of abelian groups.

Proof. Let H be an affine group scheme. Then by Corollary 2 to Proposition 4 Ext(A,C(H)) is the same in the category of group schemes as in the category of commutative group schemes for any abelian variety A. Since the category of commutative group schemes is abelian [8] II.6-1, Ext(,C(H)) is a contravariant functor from the category of abelian varieties into the category of abelian groups. By Theorem 1, Ext(,H) is a contravariant functor naturally equivalent to Ext(,C(H)). Let A be an abelian variety. Then Ext(A,) is the composite of the functor C( ) from the category AG' into the category of commutative affine group schemes and the functor Ext(A,) from the category of commutative group schemes into the category of abelian groups. So Ext(,) is a bifunctor. This completes the proof.

Remark. In Section 8 we will show that Ext(,) is in fact a bifunctor from the category of abelian varieties and AG' into the category of commutative algebraic groups.

§5. For commutative H.

From now on we will deal with the extensions of an abelian variety by a commutative affine group scheme. Let G be a commutative group scheme and d be a positive integer. We denote the d-th power map of G by d_G and the kernel of d_G by $G_{(d)}$.

Lemma 8. Let H be a commutative group scheme, d be a positive integer such that d_H is trivial and A be an abelian variety. Then

$$\mathrm{Ext}(A,H)\cong\mathrm{Hom}(A_{(d)},H).$$

Proof. The exact sequence $e \longrightarrow A_{(d)} \longrightarrow A \xrightarrow{d_A} A \longrightarrow e$ induces an exact sequence

$$e=\mathrm{Hom}(A,H) \longrightarrow \mathrm{Hom}(A_{(d)},H) \longrightarrow \mathrm{Ext}(A,H) \xrightarrow{d_A^*} \mathrm{Ext}(A,H),$$

where d_A^* is the map induced by d_A. Hence it suffices to show that d_A^* is trivial. For any extension G of A by H we have a commutative diagram with exact rows

$$\begin{array}{ccccccccc} e & \longrightarrow & H & \longrightarrow & G & \longrightarrow & A & \longrightarrow & e \\ & & \downarrow d_H & & \downarrow g & & \downarrow id_A & & \\ e & \longrightarrow & H & \longrightarrow & G' & \longrightarrow & A & \longrightarrow & e \\ & & \downarrow id_H & & \downarrow f & & \downarrow d_A & & \\ e & \longrightarrow & H & \longrightarrow & G & \longrightarrow & A & \longrightarrow & e, \end{array}$$

where G' is a commutative group scheme, f and g are homomorphisms such that $fg=d_G$. (cf. B.Mitchell "Theory of Categories" (1965) p163 Lemma 1.1). This diagram shows that the endomorphism of Ext(A,H) induced by d_A is equal to that induced by d_H. Since d_H is trivial, it induces a trivial endomorphism. So d_A^* is also trivial. The proof is completed.

Let H be a commutative affine group scheme. H is called multiplicative if any homomorphism of H into the additive group G_a is trivial. (This is one of several equivalent definitions. [3] p475 Théorème 2.2). H is called unipotent if for any nontrivial group subscheme H' of H there exists a nontrivial homomorphism of H' into G_a.

Lemma 9. Let H be a commutative affine group scheme. Then H is isomorphic to a direct product of a multiplicative group scheme and a unipotent group scheme.

Proof. [3] p501 Théorème 1.1 b).

Lemma 10. Let A be an abelian variety and H,H' be commutative group schemes. Then

$$\mathrm{Ext}(A,H\times H')\cong \mathrm{Ext}(A,H)\times \mathrm{Ext}(A,H').$$

Proof. This follows from the elementary theory of abelian categories.

Theorem 2. Let A be an abelian variety, H be a commutative affine group scheme and H_m,H_u be the multiplicative factor and the unipotent factor of H respectively. Then

$$\mathrm{Ext}(A,H)\cong \mathrm{Ext}(A,H_m)\times \mathrm{Ext}(A,H_u).$$

Proof. This is a direct consequence of Lemmas 9 and 10.

Theorem 2 enables us to deal with separately the extensions with multiplicative kernels and those with unipotent kernels.

Now we recall some properties of commutative finite group schemes. A group scheme G is called finite if the underlying scheme of G is an affine scheme Spec(B) and the k-algebra B is a finite dimensional k-vector space. Let G be a commutative finite group scheme and B be the coordinate ring of G. Let $\mathrm{Hom}_k(B,k)$ be the set of all k-linear maps of B into k. Then $\mathrm{Spec}(\mathrm{Hom}_k(B,k))$ has an induced group scheme structure. This group scheme is called the dual group scheme of G and will be denoted by G^*. G^* is also a commutative finite group scheme. G is called reduced-reduced if both G and G^* are reduced; G is called reduced-local if G is reduced and G^* is local; G is called local-reduced if G is local and G^* is reduced; G is called local-local if both G and G^* are local. (G is called local if G_{red} is trivial). All commutative finite group schemes form a category which is the product of the subcategories of reduced-reduced group schemes, of reduced-local group schemes, of local-reduced group schemes and of local-local group schemes. The category is abelian.

Examples. When char k=0, all group schemes are reduced, so all commutative finite group schemes are reduced-reduced.

Let char $k=p\neq 0$. Denote by μ_n the kernel of n-th power map of the multiplicative group G_m, where n is a positive integer. When $(n,p)=1$, μ_n is reduced-reduced and the underlying group of μ_n is isomorphic to $\mathbb{Z}/n\mathbb{Z}$. When $n=p^r$, μ_n is local-reduced. Let α_{p^n} be the kernel of the endomorphism of G_a which sends each closed point with coordinate a to that with coordinate a^{p^n}. Then α_{p^n} is a local-local group scheme. Let ν_{p^n} be the reduced group scheme with underlying group $\mathbb{Z}/p^n\mathbb{Z}$. Then ν_{p^n} is reduced-local and we will identify it with $\mathbb{Z}/p^n\mathbb{Z}$ in the sequel. We know that $(\mu_n)^*\cong\mu_n$, if $(n,p)=1$; $(\mu_{p^n})^*\cong\nu_{p^n}$; $(\nu_{p^n})^*\cong\mu_{p^n}$ and $(\alpha_{p^n})^*\cong\alpha_{p^n}$. The proof of these properties can be found in [8] I.2.

Lemma 11. Let $n\leq m$ be positive integers. Then

$$\mathrm{Hom}(\mu_{p^n},\mu_{p^m})\cong\mathrm{Hom}(\mu_{p^m},\mu_{p^n})\cong\mathbb{Z}/p^n\mathbb{Z}.$$

Proof. "*" induces an antiequivalence between the category of reduced-local group schemes and that of local-reduced group schemes. Hence $\mathrm{Hom}(\mu_{p^n},\mu_{p^m})\cong\mathrm{Hom}(\nu_{p^m},\nu_{p^n})\cong\mathrm{Hom}(\mathbb{Z}/p^m\mathbb{Z},\mathbb{Z}/p^n\mathbb{Z})\cong\mathbb{Z}/p^n\mathbb{Z}$. The other isomorphism can be similarly proven. This completes the proof.

Lemma 12. Let A be an abelian variety and n be a positive integer. Then

(1) the degree of n_A is $n^{2\dim A}$;

(2) when char k=0, the underlying group of $A_{(n)}$ is isomorphic to $(\mathbb{Z}/n\mathbb{Z})^{2\dim A}$;

(3) when char $k=p\neq 0$ and $(n,p)=1$, $A_{(n)}$ is isomorphic to $(\mu_n)^{2\dim A}$;

(4) when char $k=p\neq 0$, there exists an integer r, $0\leq r\leq \dim A$, depending only on A, such that

$$A_{(p^n)}\cong(\mu_{p^n})^r\times(\nu_{p^n})^r\times N,$$

where N is some local-local group scheme. The k-vector space dimension of the stalk of N is $p^{2n(\dim A-r)}$.

Proof. (1),(2) and (3). [7] p64 Proposition.

(4). [7] p147.

The integer r in Lemma 12 is called the p-rank of A.

§6. For multiplicative H.

Lemma 13. Any multiplicative group scheme is of the form

$$G_m^r\times(\prod_i\mu_{n_i}),$$

where r,n_i are positive integers. When char $k=p\neq 0$, n_i can be chosen such that either $(n_i,p)=1$ or n_i is a power of p.

Proof. By [3] p473 Corollaire 1.5, any multiplicative group scheme is the kernel of a homomorphism f of G_m^n into $G_m^{n'}$ for some positive integers n and n'. Since any endomorphism of G_m is a power map there exist integers n_{ij}, $1\leq j\leq n'$, such that f sends point $(1,\ldots,1,a_i,1,\ldots,1)$ to $(a_i^{n_{i1}},\ldots,a_i^{n_{in'}})$. By the theory of integral matrices there exist an $n\times n$ matrix P and an $n'\times n'$ matrix Q over $\mathbb{Z}$ such that P,Q are invertible and

$$P(n_{ij})Q=\begin{pmatrix}M & 0\\ 0 & 0\end{pmatrix},$$

where M is a diagonal $s\times s$ matrix over $\mathbb{Z}$ and s is the rank of (n_{ij}). Let n_i be the (i,i) entry of M. Let P and Q define automorphisms g and h (in the same way as (n_{ij}) defines f) of G_m^n and $G_m^{n'}$ respectively. Then the kernel of hfg is isomorphic to that of f. Hence we may assume that f sends closed point $(1,\ldots,1,a_i,1,\ldots,1)$ to $(1,\ldots,1,a_i^{n_i},1,\ldots,1)$ when $i\leq s$, to e when $i>s$, and that $n_i>0$. So the kernel of f is of the form $G_m^{n-s}\times(\prod_{i=1}^s\mu_{n_i})$. The last statement follows from the fact that $\mu_{np^m}\cong\mu_n\times\mu_{p^m}$ for $(n,p)=1$. This completes the proof.

Let A be an abelian variety, We denote by A* the dual abelian variety of A. The underlying group of A* is the group of all linear equivalence classes of line bundles on A which are algebraically equivalent to 0. If G is a reduced group scheme, we will denote also by G the underlying group of G.

Lemma 14. Let A be an abelian variety. Then
(1) $\mathrm{Ext}(A,G_m)\cong A^*$;
(2) $\mathrm{Ext}(A,G_a)\cong G_a^{\dim A}$.
Proof. (1) [10] p184 Théorème 6.
(2) [10] p186 Théorème 7 and p191 Théorème 10.

Theorem 3. Let A be an abelian variety and H be a multiplicative group scheme. Then
(1) when char k=0, we have

$$\mathrm{Ext}(A,H)\cong (A^*)^{\dim H}\times (H/H^o)^{2\dim A},$$

(2) when char $k=p\neq 0$ and $G_m^d\times(\prod_i \mu_{n_i})\times(\prod_j \mu_{p^{s_j}})$, $(n_i,p)=1$, is a decomposition of H as in Lemma 13, we have

$$\mathrm{Ext}(A,H)\cong (A^*)^d\times\prod_i(\mathbb{Z}/n_i\mathbb{Z})^{2\dim A}\times\prod_j(\mathbb{Z}/p^{s_j}\mathbb{Z})^r,$$

where r is the p-rank of A.

Proof. When char k=0, H^o is divisible. Hence $H\cong H^o\times(H/H^o)$. Since $H^o\cong G_m^{\dim H}$, we get $\mathrm{Ext}(A,H)\cong\mathrm{Ext}(A,G_m)^{\dim H}\times\mathrm{Ext}(A,H/H^o)$ by Lemma 10. By Lemma 14 (1) we get the first factor in the formula. Let d be the order of the underlying group of H/H^o. Then by Lemma 8 we have $\mathrm{Ext}(A,H/H^o)\cong\mathrm{Hom}(A_{(d)},H/H^o)$. By Lemma 12 (2) we have (regarding $\mathbb{Z}/d\mathbb{Z}$ as a reduced group scheme)

$$\mathrm{Hom}(A_{(d)},H/H^o)\cong\mathrm{Hom}(\mathbb{Z}/d\mathbb{Z},H/H^o)^{2\dim A}\cong(H/H^o)^{2\dim A}.$$

This gives the second factor in the formula and (1) is proven.

When char $k=p\neq 0$, we have, by the decomposition of H,

$$\mathrm{Ext}(A,H)\cong\mathrm{Ext}(A,G_m)^d\times\prod_i\mathrm{Ext}(A,\mu_{n_i})\times\prod_j\mathrm{Ext}(A,\mu_{p^{s_j}}).$$

By Lemma 14 (1) we get the first factor in the formula. For the second factor we have $\mathrm{Ext}(A,\mu_{n_i})\cong\mathrm{Hom}(A_{(n_i)},\mu_{n_i})$ by Lemma 8. Since $A_{(n_i)}$ is isomorphic to $(\mu_{n_i})^{2\dim A}$ by Lemma 12 (3), we get

$$\mathrm{Hom}(A_{(n_i)},\mu_{n_i})\cong\mathrm{Hom}(\mu_{n_i},\mu_{n_i})^{2\dim A}.$$

Since μ_{n_i} is reduced and its underlying group is isomorphic to $\mathbb{Z}/n_i\mathbb{Z}$, so $\mathrm{Hom}(\mu_{n_i},\mu_{n_i})\cong\mathrm{Hom}(\mathbb{Z}/n_i\mathbb{Z},\mathbb{Z}/n_i\mathbb{Z})\cong\mathbb{Z}/n_i\mathbb{Z}$ and we get the second factor in the formula. For the last factor we have, by Lemma 8 and Lemma 12 (4),

$$\mathrm{Ext}(A,\mu_{p^{s_j}})\cong\mathrm{Hom}(\nu_{p^{s_j}},\mu_{p^{s_j}})^r\times\mathrm{Hom}(\mu_{p^{s_j}},\mu_{p^{s_j}})^r\times\mathrm{Hom}(N,\mu_{p^{s_j}}),$$

where N is a local-local group scheme. The first and the third factors on the right are trivial and the second factor is isomorphic to $(\mathbb{Z}/p^{s_j}\mathbb{Z})^r$ by Lemma 11. This completes the proof of (2) and hence completes the proof of the theorem.

§7. For commutative unipotent H.

Lemma 15. Let char $k=p\neq 0$. Let H be a commutative unipotent group scheme over k. Then there exists a positive integer d such that $(p^d)_H$ is trivial.

Proof. We construct a descending sequence $H_0\supset H_1\supset H_2\supset\ldots$ of group subschemes of H as follows: let $H_0=H$; if H_i is nontrivial, then by the definition of unipotent group schemes there exists a nontrivial homomorphism f of H_i into G_a, define H_{i+1} to be the kernel of f; if $H_i=e$, define $H_{i+1}=e$. We know that any group subscheme of H is a closed subscheme of H. Since H is of finite type, the descending sequence H_i is stable, i.e. there exists a positive integer d such that $H_d=H_{d+1}=\ldots$. If H_d is not trivial, H_{d+1} will be strictly smaller than H_d. This is a contradiction. So $H_d=e$. Since p_{G_a} is trivial, the image of p_{H_i} is contained in H_{i+1}. Hence $(p^d)_H$ is contained in $H_d=e$. This completes the proof.

Let H be a commutative unipotent group scheme. We call the smallest integer d such that $(p^d)_H$ is trivial the vanishing index of H, where p=char k.

Theorem 4. Let A be an abelian variety and H be a commutative unipotent group scheme. Then

(1) when char $k=0$, we have $\mathrm{Ext}(A,H)\cong H^{\dim A}$;

(2) when char $k=p\neq 0$, let r be the p-rank of A, d be the vanishing index of H, d' be any integer larger than or equal to d, N be the local-local factor of $A_{(p^{d'})}$. Then we have $\mathrm{Ext}(A,H)\cong(H_{red})^r\times\mathrm{Hom}(N,H^o)$.

Proof. (1) When char $k=0$, any unipotent group scheme is connected since there do not exist nontrivial finite unipotent group schemes. By [10] p172 Corollaire any commutative unipotent group scheme is a direct product of group schemes G_a. Hence by Lemma 10 and 14 we have

$$\mathrm{Ext}(A,H)\cong\mathrm{Ext}(A,G_a^{\dim H})\cong\mathrm{Ext}(A,G_a)^{\dim H}\cong G_a^{\dim A\dim H}\cong H^{\dim A}.$$

This proves (1).

When char $k=p\neq 0$, since $d'\geq d$, $(p^{d'})_H$ is trivial. Hence $\mathrm{Ext}(A,H)\cong \mathrm{Hom}(A_{(p^{d'})},H)$ by Lemma 8. We have by Lemma 12 (4) that

$$\mathrm{Hom}(A_{(p^{d'})},H)\cong\mathrm{Hom}(\nu_{p^{d'}},H)^r\times\mathrm{Hom}(\mu_{p^{d'}},H)^r\times\mathrm{Hom}(N,H).$$

Since $\nu_{p^{d'}}$ is reduced, $\mathrm{Hom}(\nu_{p^{d'}},H)$ is isomorphic to $\mathrm{Hom}(\nu_{p^{d'}},H_{red})$. Since the underlying group of $\nu_{p^{d'}}$ is a cyclic group of order $p^{d'}$

and $(p^{d'})_{H_{red}}$ is trivial, $\mathrm{Hom}(\nu_{p^{d'}}, H_{red})$ is isomorphic to H_{red}. This gives the first factor in the formula. The second factor in the above decomposition is trivial since $\mu_{p^{d'}}$ is multiplicative. Finally since N is connected, the last factor in the above decomposition is isomorphic to $\mathrm{Hom}(H,H^o)$. This gives the second factor in the formula. The proof is completed.

Let char $k=p\neq 0$. Let A be an abelian variety. Then A is called ordinary if the p-rank of A is equal to the dimension of A.

Corollary. Let A be an ordinary abelian variety and H be a commutative unipotent group scheme with char $k=p\neq 0$. Then

$$\mathrm{Ext}(A,H)\cong (H_{red})^{\dim A}.$$

Proof. This is a direct consequence of Theorem 4 (2) and Lemma 12 (4).

§8. Algebraic structure.

In the corollary to Theorem1 we showed that Ext(,) is a bifunctor from the category of abelian varieties and AG' into the category of abelian groups. Now we are going to show that Ext(,) is in fact a bifunctor from the category of abelian varieties and AG' into the category of commutative algebraic groups.

First we define an algebraic group structure on Ext(A,H), where A is an abelian variety and H is an affine group scheme. By Theorem 1 we may assume that H is commutative. By Theorem 2 we may treat separately the multiplicative H and unipotent H.

Assume that H is a multiplicative group scheme and write L for $(H^o)_{red}$. Then by Theorem 3, Ext(A,H) is a direct product of Ext(A,L) and a finite group. Since any finite group can be given a unique algebraic group structure, we need to define the algebraic group structure on only the factor Ext(A,L). Since $\mathrm{Ext}(A,G_m)$ is canonically isomorphic to A*, it inherits an algebraic group structure from A*. We know that L is isomorphic to a power of G_m. Each isomorphism of L onto such a power induces an isomorphism of Ext(A,L) onto a power of $\mathrm{Ext}(A,G_m)$, and hence induces on Ext(A,L) an algebraic group structure. It is easy to check that different isomorphisms of L onto the power of G_m induce the same algebraic group structure on Ext(A,L). Hence we get a well defined algebraic group structure on Ext(A,H).

Assume that H is a commutative unipotent group scheme. We know that $\mathrm{Ext}(A,G_a)$ is canonically isomorphic to the tangent space of A* at its neutral point e [10] p196, so $\mathrm{Ext}(A,G_a)$ has an induced algebraic group structure. When char k=0, H is isomorphic to a power of G_a, so arguing as in the previous paragraph, we can define an algebraic group structure on Ext(A,H). When char $k=p\neq0$, let d be a positive integer such that $p^d{}_H$ is trivial and let L and N be the reduced-local and local-local factors of $A_{(p^d)}$ respectively. Then by Lemma 8 and the proof of Theorem 4, there is a canonical isomorphism of Ext(A,H) onto Hom(L,H)×Hom(N,H). Hence it suffices to define algebraic structure on Hom(L,H) and on Hom(N,H) and to show that the induced algebraic group structure on Ext(A,H) does not depend on the choice of d (as long as $p^d{}_H$ is trivial). We know that $\mathrm{Hom}(\mathbb{Z}/p^d\mathbb{Z},H)$ is canonically isomorphic to H_{red}. Hence $\mathrm{Hom}(\mathbb{Z}/p^d\mathbb{Z},H)$ has an algebraic group structure induced from that of H_{red} and so Hom(L,H) has an induced algebraic group structure via. any isomorphism of L onto a power of $\mathbb{Z}/p^d\mathbb{Z}$. To define the algebraic group structure on Hom(N,H) we need the following Lemma.

Lemma 16. Let N and N' be commutative local group schemes. Let k[N] and k[N'] be the coordinate rings of N and N' respectively, i.e. N=Spec(k[N]) and N'=Spec(k[N']). Let $e_1,\ldots,e_n$ and $e'_1,\ldots,e'_{n'}$ be k-vector space bases of k[N] and k[N'] respectively. Then each homomorphism f of N to N' corresponds to a matrix $q(f)=(b_{ii'})_{n\times n'}$ defined by $f^*(e'_{i'})=\Sigma^n_{i=1}b_{ii'}e_i$. q is an injection of Hom(N,N') onto a Zariski closed subset of $k^{nn'}$ and so q induces an algebraic set structure on Hom(N,N'). This structure makes Hom(N,N') an algebraic group. Furthermore, this structure does not depend on the choices of bases of k[N] nor on that of k[N'].

Proof. It is obvious that q is injective. We know that $e_i\otimes e_j$, $1\le i,j\le n$, and $e'_{i'}\otimes e'_{j'}$, $1\le i',j'\le n'$, form bases of k[N×N] and k[N"×N"] respectively. Let C and C' be the matrices, relative to these bases, corresponding to the group laws of N and N' respectively. Let D and D' be the matrices, relative to these bases, corresponding to the diagonal maps of N into N×N and of N' into N'×N' respectively. Then an n×n' matrix $B=(b_{ii'})$ is in the image of q if and only if $BD'=D(B\otimes B)$, $B1'=1$ and $CB=(B\otimes B)C'$, where 1 and 1' are the vectors corresponding to the units in k[N] and k[N'] respectively and $B\otimes B$ is an $n^2\times n'^2$ matrix with $b_{ii'}b_{jj'}$ as ((i,j),(i',j')) entry. Since these equations are all polynomial equations in $b_{ii'}$, the image of q is Zariski closed. Hence q induces on Hom(N,N') an algebraic set structure. If f and f' in Hom(N,N') correspond to matrices B and B' respectively, then ff' corresponds to the matrix $D(B\otimes B)C'$. Hence the group law in Hom(N,N') is rational. Similarly we can show that the inverse map of Hom(N,N') is also rational. So q induces on Hom(N,N') an algebraic group structure. Let E_1, E_2 be bases of k[N] and E_1', E_2' be bases of k[N']. Let q_1 and q_2 be the maps defined as q of Hom(N,N') to $k^{nn'}$ relative to bases E_1, E_1' and E_2, E_2' respectively. Then there exist invertible matrices P and Q such that $E_2=E_1P$ and $E_2'=E_1'Q$. If B_1 and B_2 correspond to $q_1(f)$ and $q_2(f)$ for some f in Hom(N,N'), then we have $B_2=P^{-1}B_1Q$. Hence $q_2q_1^{-1}$ is a birational isomorphism of $\mathrm{Im}q_1$ onto $\mathrm{Im}q_2$. This shows that the algebraic group structure induced by q on Hom(N,N') does not depend on the choices of bases of k[N] and k[N']. The proof is completed.

Let G be a group scheme, O_G be its stalk at e, $\underline{m}$ be the maximal ideal of O_G and $\underline{m}^{(s)}=O_G[x^{p^s}\,;\ x\in\underline{m}]$, where s is a positive integer. Then $\underline{m}^{(s)}$ is an ideal of O_G and $\mathrm{Spec}(O_G/\underline{m}^{(s)})$ is canonically a group subscheme of G. [11] p18. We will denote this group subscheme by $G^{(s)}$.

Now we return to Hom(N,H) we discussed before Lemma 16. Since N is local, $N^{(s)}$ is equal to N for some positive integer s. Hence the image of any homomorphism of N into H is contained in $H^{(s)}$. So we get a canonical isomorphism of Hom(N,H) onto $\mathrm{Hom}(N,H^{(s)})$. $\mathrm{Hom}(N,H^{(s)})$ has an algebraic group structure defined in Lemma 16. So Hom(N,H) inherits an algebraic group structure. It is easy to check that this structure does not depend on the choices of s as long as $N^{(s)}=N$. Hence we get a canonical algebraic group structure on $\mathrm{Hom}(A_{(p^d)},H)$. This induces on Ext(A,H) an algebraic group structure. We need to show that this structure does not depend on the choices of d. Let d' be another positive integer such that $p^{d'}{}_H$ is trivial. We may assume that d>d'. So we have a commutative diagram with exact rows

$$\begin{array}{ccccccccc}
e & \longrightarrow & A_{(p^d)} & \longrightarrow & A & \xrightarrow{p^d{}_A} & A & \longrightarrow & e \\
 & & \downarrow{\scriptstyle p^{d-d'}} & & \downarrow{\scriptstyle p^{d-d'}} & & \downarrow{\scriptstyle id_A} & & \\
e & \longrightarrow & A_{(p^{d'})} & \longrightarrow & A & \xrightarrow{p^{d'}{}_A} & A & \longrightarrow & e
\end{array}$$

This diagram induces a commutative diagram

$$\begin{array}{ccc}
\mathrm{Hom}(A_{(p^d)},H) & \longrightarrow & \mathrm{Ext}(A,H) \\
\downarrow & & \downarrow{\scriptstyle id} \\
\mathrm{Hom}(A_{(p^{d'})},H) & \longrightarrow & \mathrm{Ext}(A,H),
\end{array}$$

in which all four arrows are isomorphisms of abstract groups. It is easy to check that the left isomorphism is birational relative to their algebraic group structures we just defined. Hence the algebraic group structures induced by the horizontal isomorphisms on

Ext(A,H) are the same. Hence we get a canonical algebraic group structure on Ext(A,H).

Theorem 5. Ext(,) is a bifunctor from the category of abelian varieties and the category AG' into the category of commutative algebraic groups, where for any abelian variety and any affine group scheme H the algebraic group structure on Ext(A,H) is that we just defined

Proof. By Theorem 1 and its corollary we need to show only that

(1) If A is an abelian variety, then Ext(A,) sends homomorphisms of commutative affine group schemes to <u>rational</u> homomorphisms.

(2) If H is a commutative affine group scheme, then Ext(,H) sends homomorphisms of abelian varieties to <u>rational</u> homomorphisms.

To show (1), let $h:H \longrightarrow H'$ be a homomorphism of commutative affine group schemes. By Theorem 2 we may discuss the multiplicative case and the unipotent case separately.

Assume that H and H' are multiplicative group schemes. Let $L=(H^0)_{red}$ and $L'=(H'^0)_{red}$. Then h sends L into L'. We know that Ext(A,H) is the direct product of Ext(A,L) and a finite algebraic group and that any homomorphism of a finite algebraic group into an algebraic group is rational. So it suffices to show that $h_*|_{Ext(A,L)}$ is rational. The commutative diagram

$$\begin{array}{ccc} L & \xrightarrow{h|L} & L' \\ i\downarrow & & \downarrow i' \\ H & \xrightarrow{h} & H' \end{array},$$

where i and i' are inclusions, induces a commutative diagram

$$\begin{array}{ccc} Ext(A,L) & \xrightarrow{h_*|Ext(A,L)} & Ext(A,L') \\ i_*\downarrow & & \downarrow i'_* \\ Ext(A,H) & \xrightarrow{h_*} & Ext(A,H'). \end{array}$$

It is obvious that i_* and i'_* are rational. So we need to show only that the upper homomorphism in the diagram is rational, i.e. we may assume that $H\overset{f}{\cong}G_m^d$ and $H'\overset{f'}{\cong}G_m^{d'}$ for some positive integers d and d'. By suitable choice of f and f' we may assume that the homomorphism h', induced by h, of G_m^d to $G_m^{d'}$ is defined by $h'((x_1,\dots,x_d))=(x_1^{n_1},\dots,x_s^{n_s},1,\dots,1)$ for some integers $s\geq 0$, $n_1,\dots,n_s$. The commutative diagram

$$\begin{array}{ccc} H & \xrightarrow{h} & H' \\ f\downarrow & & \downarrow f' \\ G_m^d & \longrightarrow & G_m^{d'} \end{array}$$

induces a commutative diagram

$$\begin{array}{ccc} Ext(A,H) & \xrightarrow{h_*} & Ext(A,H') \\ f_* \downarrow & & \downarrow f'_* \\ Ext(A,G_m^d) & \xrightarrow{h'_*} & Ext(A,G_m^{d'}). \end{array}$$

By the definition of the algebraic group structure on Ext(A,H) and on Ext(A,H'), f_* and f'_* are birational isomorphisms. So it suffices to show that h'_* is rational. Let $Ext(A,G_m^d) \cong A^{*d}$ and $Ext(A,G_m^{d'}) \cong A^{*d'}$ be the canonical isomorphisms. Then h'_* sends $(x_1,\ldots,x_d) \in A^{*d}$ to $(x_1^{n_1},\ldots,x_s^{n_s},e,\ldots,e) \in A^{*d'}$. This is obviously rational. So the multiplicative case is proven.

Assume now that H and H' are unipotent. Let $h:H \longrightarrow H'$ be a homomorphism. When char k=0, H and H' are the additive groups of some k-vector spaces and h is a k-linear map between them. It is easy to check that Ext(A,H) and Ext(A,H') have canonical k-vector space structures and h induces a k-linear map h_* of Ext(A,H) into Ext(A,H'). So h_* is rational. When char $k=p \neq 0$, let d be a positive integer such that $p^d{}_H$ and $p^d{}_{H'}$ are both trivial. Let L and N be the reduced-local and the local-local factors of $A_{(p^d)}$ respectively. Then it suffices to show that the induced maps $h_{*L}:Hom(L,H) \longrightarrow Hom(L,H')$ and $h_{*N}:Hom(N,H) \longrightarrow Hom(N,H')$ are rational, where the four Hom groups are equipped with the algebraic group structures we defined before the theorem. Since L is isomorphic to a power of $\mathbb{Z}/p^d\mathbb{Z}$, h_{*L} is a direct product of some homomorphisms $Hom(\mathbb{Z}/p^d\mathbb{Z},H) \longrightarrow Hom(\mathbb{Z}/p^d\mathbb{Z},H')$ induced by h. We know that $Hom(\mathbb{Z}/p^d\mathbb{Z},H) \cong Hom(\mathbb{Z}/p^d\mathbb{Z},H_{red})$, $Hom(\mathbb{Z}/p^d\mathbb{Z},H') \cong Hom(\mathbb{Z}/p^d\mathbb{Z},H'_{red})$ and that the following diagram is commutative

$$\begin{array}{ccc} Hom(\mathbb{Z}/p^d\mathbb{Z},H_{red}) & \longrightarrow & H_{red} \\ \downarrow & & \downarrow h \\ Hom(\mathbb{Z}/p^d\mathbb{Z},H'_{red}) & \longrightarrow & H'_{red}. \end{array}$$

The horizontal maps in the diagram are birational isomorphisms by definitions. Since h is rational, the left vertical map in the diagram is also rational. So h_{*L} is rational. Choose a positive integer s such that $N^{(s)}=N$. Then we have $Hom(N,H)=Hom(N,H^{(s)})$ and $Hom(N,H')=Hom(N,H'^{(s)})$. Fix bases of k[N], $k[H^{(s)}]$ and $k[H'^{(s)}]$. Since h sends $H^{(s)}$ into $H'^{(s)}$, we have a corresponding matrix P relative to the fixed bases of $k[H^{(s)}]$ and $k[H'^{(s)}]$. If $f \in Hom(N,H^{(s)})$ corresponds to the matrix B relative to the fixed bases of k[N] and $k[H^{(s)}]$, then the matrix corresponding to $h_{*N}(f)$ is BP. So h_{*N} is rational by the definition of algebraic group structures on Hom(N,H) and on Hom(N,H'). So h_* is rational. This proves (1).

To show (2), let $g:A \longrightarrow A'$ be a homomorphism of abelian varieties. If $H=H_1\times H_2$, then $g^*:\mathrm{Ext}(A',H) \longrightarrow \mathrm{Ext}(A,H)$ is the direct product of induced homomorphisms $g_i^*:\mathrm{Ext}(A',H_i) \longrightarrow \mathrm{Ext}(A,H_i)$, $i=1,2$. So we may devide the proof into four cases:

(i) $H=G_m$;

(ii) H is a finite multiplicative group scheme;

(iii) $H=G_a$, char $k=0$;

(iv) H is a commutative unipotent group scheme, char $k=p\neq 0$.

Case (i). g induces a homomorphism of A'^* into A^* by the duality of abelian varieties. g also induces a homomorphism of A'^* into A^* via. the canonical isomorphisms of $\mathrm{Ext}(A,G_m)$ onto A^* and of $\mathrm{Ext}(A',G_m)$ onto A'^*. By the proof of the formula $\mathrm{Ext}(A,G_m)\cong A^*$, [10] p184, the two induced homomorphisms coincide. Since the former is rational, so is the latter.

Case (ii). In this case $\mathrm{Ext}(A',H)$ is a finite algebraic group, so g^* is rational.

Case (iii). g induces a homomorphism of A'^* into A^*. This homomorphism induces a k-linear map of the tengent space of A'^* at e into that of A^* at e, which is obviously rational. Hence $g^*:\mathrm{Ext}(A',H) \longrightarrow \mathrm{Ext}(A,H)$ is rational by the definition of the algebraic group structure on $\mathrm{Ext}(A',H)$ and on $\mathrm{Ext}(A,H)$.

Case (iv). Let d be an integer such that $p^d{}_H$ is trivial. Then g sends $A_{(p^d)}$ into $A'_{(p^d)}$ and it suffices to show that the induced map $g^*:\mathrm{Hom}(A'_{(p^d)},H) \longrightarrow \mathrm{Hom}(A_{(p^d)},H)$ is rational. Let L and L' be the reduced-local factors of $A_{(p^d)}$ and $A'_{(p^d)}$ respectively. Let N and N' be the local-local factors of $A_{(p^d)}$ and $A'_{(p^d)}$ respectively. Then it suffices to show that the induced homomorphisms $g_L^*:\mathrm{Hom}(L',H) \longrightarrow \mathrm{Hom}(L,H)$ and $g_N^*:\mathrm{Hom}(N',H) \longrightarrow \mathrm{Hom}(N,H)$ are rational. Let r and r' be the p-ranks of A and A' respectively. We can choose isomorphisms $f:L \longrightarrow (\mathbb{Z}/p^d\mathbb{Z})^r$ and $f':L' \longrightarrow (\mathbb{Z}/p^d\mathbb{Z})^{r'}$ such that there exist integers $s\geq 0$ and $n_1,\ldots,n_s$ and the induced homomorphism g' of $(\mathbb{Z}/p^d\mathbb{Z})^r$ to $(\mathbb{Z}/p^d\mathbb{Z})^{r'}$ sends $(x_1,\ldots,x_r)$ to $(n_1x_1,\ldots,n_sx_s,0,\ldots,0)$. The commutative diagram

$$\begin{array}{ccc} L & \xrightarrow{f} & (\mathbb{Z}/p^d\mathbb{Z})^r \\ {\scriptstyle g|_L}\downarrow & & \downarrow{\scriptstyle g'} \\ L' & \xrightarrow{f'} & (\mathbb{Z}/p^d\mathbb{Z})^{r'} \end{array}$$

induces a commutative diagram

$$\begin{array}{ccc} \mathrm{Hom}(L,H) & \xleftarrow{f^*} & \mathrm{Hom}(Z/p^dZ,H)^r \\ {\scriptstyle g_L^*}\uparrow & & \uparrow{\scriptstyle g'^*} \\ \mathrm{Hom}(L',H) & \xleftarrow{f'^*} & \mathrm{Hom}(Z/p^dZ,H)^{r'}. \end{array}$$

The horizontal maps are birational isomorphisms by the definition of the algebraic group structures on Hom(L,H) and Hom(L',H). g'* sends $(x_1,\dots,x_{r'})$ to $(x_1^{n_1},\dots,x_s^{n_s},e,\dots,e)$. Hence g'* is rational. So g_L^* is rational. To show that g_N^* is rational, let s be a positive integer such that $N^{(s)}=N$ and $N'^{(s)}=N'$. Then the image of any homomorphism in Hom(N,H) or in Hom(N',H) is contained in $H^{(s)}$. Fix bases of k[N], k[N'] and $k[H^{(s)}]$. Denote by P the matrix corresponding to $g|_N:N \longrightarrow N'$ relative to the bases. Then if $f\in \mathrm{Hom}(N',H^{(s)})$ corresponds to matrix B, $g_N^*(f)$ will correspond to matrix PB. This shows that g_N^* is rational, so is g*. This completes the proof.

References

1 M.F.Atiyah and L.G.Macdonald. "Introduction to Commutative Algebra" (1969)

2 I.Barsotti. "Analytical method for abelian varieties in positive characteristic" Colloque sur la théorie des groupes algèbriques. Bruxelles (1962) p77-86.

3 M.Demazure and P.Gabriel. "Groupes Algèbriques" Tomb I. (1970).

4 M.Hazewinkel. "Formal Groups and Applications" (1978).

5 S.Lang. "Algebra" (1965).

6 Yu.I.Manin. "The theory of commutative formal groups over fields of finite characteristics" Russian Math. Surveys. 18 (1963) p1-83.

7 D.Mumford. "Abelian Varieties" (1970).

8 F.Oort. "Commutative Group Schemes" Springer Lecture Notes 15. (1966).

9 M.Rosenlicht. "Some basic theorems on algebraic groups" Amer. J.of Math. 78 (1956) p401-443.

10 J.-P.Serre. "Groupes Algèbriques et Corps de Classes" (1975).

11 Yamagihara. "Theory of Hopf Algebras Attached to Group Schemes" Springer Lecture Notes 614. (1977).

ON THE FIRST CARTAN INVARIANT OF THE GROUPS $SL(3,p^n)$ AND $SU(3,p^n)$

Ye Jia-chen
Department of Applied Mathematics
Tongji University
Shanghai 200092, People's Republic of China

The purpose of this paper is to obtain a formula for the first Cartan invariant of the groups $SL(3,p^n)$ and $SU(3,p^n)$ (For convenience, we denote by $SU(3,p^n)$ the group $SU(3,p^{2n})$).

THEOREM Let $\Gamma_n = SL(3,p^n)$ (resp. $\Gamma_n' = SU(3,p^n)$) and $R(n,\theta)$ (resp. $R'(n,\theta)$) be the projective indecomposable module of $K\Gamma_n$ (resp. $K\Gamma_n'$) associated with the trivial irreducible module $M(n,\theta)$ (resp. $M'(n,\theta)$) for the prime $p \geqslant 7$. Then

$$c^{(n)} = \left[R(n,\theta) : M(n,\theta)\right]_{K\Gamma_n} = \left[R'(n,\theta) : M'(n,\theta)\right]_{K\Gamma_n'}$$
$$= a^n + b^n + 6^n - 2 \cdot 8^n ,$$

where a, b are two roots of $x^2 - 18x + 48 = 0$.

COROLLARY

$$\lim_{n\to\infty} \frac{c^{(n)}}{(9+\sqrt{33})^n} = 1.$$

We will prove the Theorem in Section 3, 4, 5.

The above result is inspired by Humphreys' [5] and Cheng's [1]. The author welcomes this opportunity to express his deepest gratitude to Professor Cao Xihua for his guidance and encouragement throughout this work. The author also wishes to give his thanks to Mr. Dong Bainian for his valuable helps.

1. NOTATION

The following notation will be used throughout this paper: n is a natural number, $G = SL(3,K)$, K is the algebraic closure of the finite field of order $p \geqslant 7$, $\Gamma_n = SL(3,p^n)$, $\Gamma_n' = SU(3,p^n)$, $X = \{(r,s) \mid r,s \in Z\}$ is the weight lattice of G, $X^+ = \{(r,s) \mid r,s \in Z^+\}$ is the set of dominant weights, $X_n = \{(r,s) \in X^+ \mid 0 \leq r,s < p^n\}$ is the index set of distinct irreducible $K\Gamma_n$ - (resp. $K\Gamma_n'$ - or $\underline{\underline{u}}_n$-) modules, U_K is the hyperalgebra of G, $\underline{\underline{u}}_n$ is the p^{8n} - dimensional subalgebra of U_K. Set $\theta = (0,0)$, $\rho = (p-1, p-1)$, $\eta = (p-1,0)$, $\xi = (p-1,1)$. For $\lambda \in X^+$ $M(\lambda)$ is an irreducible G-module with highest weight λ and $V(\lambda)$ is a Weyl module with highest weight λ. For $\lambda \in X_n$ $M(n,\lambda)$ is the irreducible $K\Gamma_n$- (or $\underline{\underline{u}}_n$-) module, $R(n,\lambda)$ (or $Q(n,\lambda)$) is the projective indecomposable $K\Gamma_n$- (or $\underline{\underline{u}}_n$-) module having $M(n,\lambda)$ as its top (and bottom) composition factor, and $R'(n,\lambda)$ is the projective indecomposable $K\Gamma'_n$-module associated with the irreducible $K\Gamma'_n$-module $M'(n,\lambda)$. St_n is the Steinberg module. Let $[V:M]_G$ (resp. $[V:M]_{K\Gamma_n}$, $[V:M]_{K\Gamma'_n}$) denote the multiplicity M occurs as a G- (resp. $K\Gamma_n$-, $K\Gamma'_n$-) composition factor of V. Let "$\overset{}{\underset{G}{\longleftrightarrow}}$" (resp. "$\underset{K\Gamma_n}{\longleftrightarrow}$" "$\underset{K\Gamma'_n}{\longleftrightarrow}$") denote that both sides of the sign have the same G- (resp. $K\Gamma_n$-, $K\Gamma'_n$-) composition factors. Let $[Q(n,\lambda):R(n,\mu)]_{K\Gamma_n}$ (resp. $[Q(n,\lambda):R'(n,\mu)]_{K\Gamma'_n}$) denote the number of times $R(n,\mu)$ (resp. $R'(n,\mu)$) occurs as a $K\Gamma_n$- (resp. $K\Gamma'_n$-) summand of $Q(n,\lambda)$, where $\lambda, \mu \in X_n$. W is the Weyl group of G, w_o is the longest element of W. For $\lambda = (r,s) \in X^+$ define $\lambda^* = -w_o(\lambda) = (s,r)$, then $M(\lambda^*)$ is the dual module of $M(\lambda)$.

2. THE PROJECTIVE INDECOMPOSABLE MODULE $R(n,\theta)$ OF $K\Gamma_n$

In this section we will decompose the projective indecomposable $\underline{\underline{u}}_n$-module $Q(n,\theta)$ into a direct sum of some projective indecomposable $K\Gamma_n$-modules $R(n,\mu)$, and compute the dimension of $R(n,\theta)$. We assume $p \geqslant 3$.

By [2] Formula (4.4), let $\lambda = \lambda_o + p\lambda_1 + \dots + p^{n-1}\lambda_{n-1}$ and $\mu = \mu_o + p\mu_1 + \dots + p^{n-1}\mu_{n-1}$ be p-adic expressions of λ and μ, we have

2.1
$$[Q(n,\lambda):R(n,\mu)]_{K\Gamma_n} = \sum_{\substack{\nu_i \in X^+ \\ \nu_n = \nu_o}} \prod_{i=o}^{n-1} [M(\mu_i) \otimes M(\nu_i) : M(\lambda_i + p\nu_{i+1})]_G$$

It is easy to see that we can only take $\nu_i = (o,o)$; $(1,o)$, $(o,1)$ or $(1,1)$, $o \leqslant i \leqslant n-1$

2.2 LEMMA. Let λ, μ be as above, $\lambda_i = (r_i, s_i) \in X_1$, $m_i = [M(\mu_i) \otimes M(\nu_i) : M(\lambda_i + p\nu_{i+1})]_G$, $o \leqslant i \leqslant n-1$, $\nu_n = \nu_o$.

Then we have

(a) If $\nu_i = (o,o)$, then $\mu_i = \lambda_i$, $\nu_{i+1} = \nu_i$, $m_i = 1$, $o \leqslant i \leqslant n-1$, $\nu_n = \nu_o$;

(b) If $\nu_i = (1,o)$, $\lambda_i = (o,s_i)$, then $\mu_i = (p-1, s_i)$, $\nu_{i+1} = \nu_i$, $m_i = 1$; or $\lambda_i = (r_i, o)$. $o \leqslant r_i \leqslant p-2$, then $\mu_i = (r_i+1, p-1)$, $\nu_{i+1} = \nu_i^*$,
$$m_i = \begin{cases} 2 & \text{if } r_i = p-2 \\ 1 & \text{if } r_i < p-2 \end{cases} \qquad o \leqslant i \leqslant n-1, \quad \nu_n = \nu_o;$$

(c) If $\nu_i = (o,1)$, $\lambda_i = (r_i, o)$, then $\mu_i = (r_i, p-1)$, $\nu_{i+1} = \nu_i$, $m_i = 1$; or $\lambda_i = (o, s_i)$, $o \leqslant s_i \leqslant p-2$, then $\mu_i = (p-1, s_i+1)$, $\nu_{i+1} = \nu_i^*$,
$$m_i = \begin{cases} 2 & \text{if } s_i = p-2 \\ 1 & \text{if } s_i < p-2 \end{cases} \qquad o \leqslant i \leqslant n-1, \quad \nu_n = \nu_o;$$

(d) If $\nu_i = (1,1)$; then $\lambda_i = (o,o)$, $\mu_i = (p-1, p-1)$, $\nu_{i+1} = \nu_i$, $m_i = 1$; $o \leqslant i \leqslant n-1$, $\nu_n = \nu_o$.

PROOF It is well known that if $(a,b) \in X^+$, then

$$V(a,b) \otimes M(0,0) \xleftrightarrow[G]{} V(a,b)$$

$$V(a,b) \otimes M(1,0) \xleftrightarrow[G]{} V(a+1,b) \oplus V(a,b-1) \oplus V(a-1,b+1)$$

$$V(a,b) \otimes M(0,1) \xleftrightarrow[G]{} V(a,b+1) \oplus V(a-1,b) \oplus V(a+1,b-1)$$

$$\begin{aligned} V(a,b) \otimes M(1,1) \xleftrightarrow[G]{} 2V(a,b) &\oplus V(a+1,b+1) \\ &\oplus V(a-1,b-1) \oplus V(a+2,b-1) \\ &\oplus V(a+1,b-2) \oplus V(a-1,b+2) \\ &\oplus V(a-2,b+1). \end{aligned}$$

Set $V(-1,b) = V(a,-1) = 0$, $V(-2,b) = -V(0,b-1)$, $V(a,-2) = -V(a-1,0)$.
By the generic decomposition pattern of Weyl module for the group $G=SL(3,k)$ (cf.[7]), the Lemma readily follows.
Now the following is obvious.

2.3 PROPOSITION Let $\mu = \mu_0 + p\mu_1 + \dots + p^{n-1}\mu_{n-1}$ be the p-adic expression of μ, then $[Q(n,\theta) : R(n,\mu)]_{K\Gamma_n} \neq 0$ if and only if (μ_i, μ_{i+1}) is of the following types:
(1) (θ,θ); (2) (ρ,ρ); (3) (η,η); (4) (η,ξ^*); (5) (ξ,ξ^*);
(6) (ξ,η); (7) (η^*,η^*); (8) (η^*,ξ); (9) (ξ^*,ξ); (10) (ξ^*,η^*);
where $0 \leq i \leq n-1$, $\mu_n = \mu_0$.
Moreover, we have $[Q(n,\theta) : R(n,\mu)]_{K\Gamma_n} = 1$ for such a weight μ.
Let $\mathcal{M}$ be the set of $\mu = \mu_0 + p\mu_1 + \dots + p^{n-1}\mu_{n-1} \in X_n$, where the type of (μ_i, μ_{i+1}) is one of (3) - (10) in (2.3), $0 \leq i \leq n-1$, $\mu_n = \mu_0$.
We have

2.4 $Q(n,\theta) \xleftrightarrow[K\Gamma_n]{} R(n,\theta) \oplus St_n \oplus \sum_{\mu \in \mathcal{M}} R(n,\mu)$.
Again by (2.2), we can readily obtain the following

2.5 PROPOSITION let $\zeta = \eta + p\eta + \dots + p^{n-1}\eta = (p^n - 1, 0)$, $\zeta^* = \eta^* + p\eta^* + \dots + p^{n-1}\eta^* = (0, p^n - 1)$. Then we have $Q(n,\zeta) \xleftrightarrow[K\Gamma_n]{} R(n,\zeta) \oplus St_n$, $Q(n,\zeta^*) \xleftrightarrow[K\Gamma_n]{} R(n,\zeta^*) \oplus St_n$, and $Q(n,\mu) \xleftrightarrow[K\Gamma_n]{} R(n,\mu)$ for all $\mu \in \mathcal{M} \setminus \{\zeta,\zeta^*\}$.
Moreover, we have $[R(n,\mu) : M(n,\theta)]_{K\Gamma_n} = [Q(n,\mu) : M(n,\theta)]_{K\Gamma_n}$ for all $\mu \in \mathcal{M}$.
Now we want to compute the dimension of $R(n,\theta)$.
Notice the following facts:

$$\dim Q(1,\eta) = \dim Q(1,\eta^*) = \dim Q(1,\xi) = \dim Q(1,\xi^*) = 3p^3.$$

$$\dim Q(1,\theta) = 12p^3 \text{ and } \dim St_1 = p^3.$$

It is well known that each $Q(n,\lambda)$ with $\lambda \in X_n$ admits a unique G-module structure, and there is a G-module isomorphism:

2.6 $Q(n,\lambda) \simeq Q(1,\lambda_o) \otimes Q(1,\lambda_1)^{(p)} \otimes \dots \otimes Q(1,\lambda_{n-1})^{(p^{n-1})}$

for $\lambda = \lambda_o + p\lambda_1 + \dots + p^{n-1}\lambda_{n-1}$ with $\lambda_i \in X_1$.

Hence we have

$$\dim Q(n,\lambda) = \dim Q(1,\lambda_o)\ \dim Q(1,\lambda_1) \dots \dim Q(1,\lambda_{n-1}),$$

and $\dim Q(n,\mu) = 3^n p^{3n}$ for each $\mu \in \mathcal{M}$.

$$\dim Q(n,\theta) = 12^n p^{3n}, \quad \dim St_n = p^{3n}.$$

We have to compute $\sum_{\mu\in\mathcal{M}} \dim Q(n,\mu)$, it can be reduced to a problem in graph theory. We construct a digraph G_1 with vertices η, η^*, ξ, ξ^* and edges $\eta\eta$, $\eta\xi^*$, $\xi\eta$, $\xi\xi^*$, $\eta^*\eta^*$, $\eta^*\xi$, $\xi^*\xi$, $\xi^*\eta^*$ (each has 3 parallel edges). Therefore the adjacency matrix of the digraph G_1 is

$$Y_1 = \begin{bmatrix} 3 & 0 & 0 & 3 \\ 3 & 0 & 0 & 3 \\ 0 & 3 & 3 & 0 \\ 0 & 3 & 3 & 0 \end{bmatrix}$$

Since $\sum_{\mu\in\mathcal{M}} \dim Q(n,\mu)/p^{3n}$ is equal to the number of all closed walks of length n in the digraph G_1, by [6] Theoren 16.8 we have

2.7 $\sum_{\mu\in\mathcal{M}} \dim Q(n,\mu) = tr(Y_1^n)\,p^{3n}$, where $tr(Y_1^n)$ is the sum of the n-th powers of the eigenvalues of Y_1.

The eigenvalues of Y_1 are 0 , 0 , 0 and 6. Hence $\sum_{\mu\in\mathcal{M}} \dim Q(n,\mu) = 6^n p^{3n}$. Thus we get

2.8 $$\dim R(n,\theta) = (12^n - 6^n + 1)\,p^{3n}.$$

By (2.4) and (2.5), we have

2.9 <u>THEOREM</u> The first Cartan invariant of Γ_n is equal to

$$c^{(n)} = \left[R(n,\theta) : M(n,\theta)\right]_{K\Gamma_n}$$
$$= \left[Q(n,\theta) : M(n,\theta)\right]_{K\Gamma_n} - \sum_{\mu\in\mathcal{M}} \left[Q(n,\mu) : M(n,\theta)\right]_{K\Gamma_n}.$$

3. <u>COMPUTATIONS OF</u> $\left[Q(n,\theta) : M(n,\theta)\right]_{K\Gamma_n}$

In this section we will compute $\left[Q(n,\theta) : M(n,\theta)\right]_{K\Gamma_n}$.

By [9] § 2, to get the $K\Gamma_n$-composition factors of $Q(n,\lambda)$ $(\lambda \in X_n)$ as a $K\Gamma_n$-module, it is enough to decompose each of all possible $M = M_o \otimes M_1^{(p)} \otimes \dots \otimes M_{n-1}^{(p^{n-1})}$, where M_i ranges over all G-composition factors of $Q(1,\lambda_i)$, $\lambda = \lambda_o + p\lambda_1 + \dots + p^{n-1}\lambda_{n-1}$ $(\lambda_i \in X_1)$. Using the notation of [9], if $M_i = M_{io} \otimes M_{i1}^{(p)}$ is a G-composition factor of $Q(1,\lambda_i)$, then we have

$$3.1\quad M \underset{G}{\simeq} M_{oo} \otimes (M_{1o} \otimes M_{o1})^{(p)} \otimes \dots \otimes (M_{n-1o} \otimes M_{n-21})^{(p^{n-1})} \otimes M_{n-11}^{(p^n)}$$

$$\underset{K\Gamma_n}{\simeq} (M_{oo} \otimes M_{n-11}) \otimes (M_{1o} \otimes M_{o1})^{(p)} \otimes \dots \otimes (M_{n-1o} \otimes M_{n-21})^{(p^{n-1})}$$

It is possible to decompose M into $K\Gamma_n$-composition factors by finite steps of decomposing tensor of two irreducible G-modules into G composition factors. In particular, we have

$$3.2\quad Q(n,\theta) = Q(1,\theta) \otimes Q(1,\theta)^{(p)} \otimes \dots \otimes Q(1,\theta)^{(p^{n-1})} .$$

To compute $[Q(n,\theta) : M(n,\theta)]_{K\Gamma_n}$, it is enough to decompose each of all possible $M = M_o \otimes M_1 \otimes \dots \otimes M_{n-1}^{(p^{n-1})}$, where M_i ranges over some G-composition factors of $Q(1,\theta)$ such that M has $M(n,\theta)$ as its $K\Gamma_n$-composition factors, and then compute various $[M : M(n,\theta)]_{K\Gamma_n}$.

By [9] § 1.5 Theorem 10, we have

$$3.3\quad Q(1,\theta) \underset{G}{\longleftrightarrow} A \oplus B \oplus C \oplus D \oplus E \oplus E^* \oplus 4M(p-3,o) \otimes M(o,1)^{(p)} \oplus 4M(o,p-3) \otimes M(1,o)^{(p)}$$
$$\oplus\, 2M(p-3,o) \otimes M(2,o)^{(p)} \oplus 2M(o,p-3) \otimes M(o,2)^{(p)},$$

where $A = 8M(o,o)$, $B = 4M(p-2,p-2)$, $C - 2M(o,o) \otimes M(1,1)^{(p)}$, $D = M(p-2,p-2) \otimes M(1\ 1)^{(p)}$, $E = 4M(p-2,1) \otimes M(o,1)^{(p)}$, $E^* = 4M(1,p-2) \otimes M(1,o)^{(p)}$.

3.4 <u>LEMMA</u> For $\lambda, \mu \in X^+$ we have $M(\lambda^*) \otimes M(\mu^*) \underset{G}{\longleftrightarrow} (M(\lambda) \otimes M(\mu))^*$

<u>PROOF</u> This is obvious.

3.5 <u>LEMMA</u>

(1) $M(p-2,p-2) \otimes M(1,1) \underset{G}{\longleftrightarrow} 4M(p-2,p-2) \oplus M(p-1,p-1) \oplus M(p-3,p-3) \oplus M(p-1,p-4)$
$\oplus M(p-4,p-1) \oplus M(o,p-3) \otimes M(1,o)^{(p)} \oplus M(p-3,o) \otimes M(o,1)^{(p)} \oplus 2M(o,o)$.

$M(p-2,p-2) \otimes M(1,o) \underset{G}{\longleftrightarrow} M(p-1,p-2) \oplus M(p-2,p-3) \oplus M(p-3,p-1)$.

$M(p-2,p-2) \otimes M(o,1) \underset{G}{\longleftrightarrow} M(p-2,p-1) \oplus M(p-3,p-2) \oplus M(p-1,p-3)$.

$M(p-2,p-2) \otimes M(2,o) \underset{G}{\longleftrightarrow} M(p-1,p-3) \oplus M(p-2,p-4) \oplus M(o,p-2) \otimes M(1,o)^{(p)} \oplus M(o,1)$
$\oplus M(p-4,o) \otimes M(o,1)^{(p)} \oplus 2M(p-2,p-1) \oplus 2M(p-3,p-2)$.

$M(p-2,p-2) \otimes M(o,2) \underset{G}{\longleftrightarrow} M(p-3,p-1) \oplus M(p-4,p-2) \oplus M(p-2,o) \otimes M(o,1)^{(p)} \oplus M(1,o)$
$\oplus M(o,p-4) \otimes M(1,o)^{(p)} \oplus 2M(p-1,p-2) \oplus 2M(p-2,p-3)$;

(2) $M(p-2,1) \otimes M(1,1) \underset{G}{\longleftrightarrow} M(p-1,2) \oplus 2M(p-2,1) \oplus M(p-3,o) \oplus M(p-3,3) \oplus$
$\oplus M(o,o) \otimes M(1,o)^{(p)}$.

$M(p-2,1) \otimes M(1,o) \underset{G}{\longleftrightarrow} M(p-1,1) \oplus M(p-3,2)$.

$M(p-2,1) \otimes M(o,1) \xleftrightarrow[G]{} M(p-2,2) \oplus M(p-1,o)$

$M(p-2,1) \otimes M(2,o) \xleftrightarrow[G]{} M(o,1) \otimes M(1,o)^{(p)} \oplus 2M(p-2,2) \oplus M(p-4,3) \oplus M(p-4,o)$

$M(p-2,1) \otimes M(o,2) \xleftrightarrow[G]{} M(p-2,3) \oplus M(p-1,1) \oplus M(p-2,o)$

(3) $M(p-3,o) \otimes M(1,1) \xleftrightarrow[G]{} M(p-2,1) \oplus 2M(p-3,o) \oplus M(p-4,2) \oplus M(p-5,1)$

$M(p-3,o) \otimes M(1,o) \xleftrightarrow[G]{} M(p-2,o) \oplus M(p-4,1)$

$M(p-3,o) \otimes M(o,1) \xleftrightarrow[G]{} M(p-3,1) \oplus M(p-4,o)$

$M(p-3,o) \otimes M(2,o) \xleftrightarrow[G]{} M(p-1,o) \oplus M(p-3,1) \oplus M(p-5,2)$

$M(p-3,o) \otimes M(o,2) \xleftrightarrow[G]{} M(p-3,2) \oplus 2M(p-4,1) \oplus M(p-5,o)$

For $M(1,p-2) \otimes M(\nu)$ and $M(o,p-3) \otimes M(\nu)$, where $\nu = (1,1), (1,o), (o,1), (2,o), (o,2)$, the decomposition are symmetric from Lemma (3.4).

<u>PROOF</u>: Since $V(p-2,p-2) \xleftrightarrow[G]{} M(p-2,p-2) \oplus M(o,o)$,

$V(p-2,1) \xleftrightarrow[G]{} M(p-2,1) \oplus M(p-3,o)$,

$V(p-3,o) \xleftrightarrow[G]{} M(p-3,o)$,

$V(1,p-2) \xleftrightarrow[G]{} M(1,p-2) + M(o,p-3)$,

$V(o,p-3) \xleftrightarrow[G]{} M(o,p-3)$ and $V(\nu) \xleftrightarrow[G]{} M(\nu)$ for $\nu = (1,1), (1,o), (o,1),$ $(2,o)$ or $(o,2)$, the results follow from decomposing tensor of two Weyl modules into G-composition factors.

Consider $M = M_o \otimes M_1^{(p)} \otimes \dots \otimes M_{n-1}^{(p^{n-1})}$, where $M_i = M_{io} \otimes M_{i1}^{(p)}$ is a G-composition factor of $Q(1,\theta)$, if $M_i^{(p^i)} \otimes M_{i+1}^{(p^{i+1})} = M_{ic}^{(p^i)} \otimes (M_{i+1,o} \otimes M_{i1})^{(p^{i+1})} \otimes M_{i+11}^{(p^{i+2})}$

$\xleftrightarrow[G]{} M_{i,o}^{(p^i)} \otimes m_{i+1} M(o,o)^{(p^{i+1})} \otimes M_{i+1}^{(p^{i+2})} \oplus$ other terms, then we call that M_{i+1} contributes the multiplicity m_{i+1} to $[M : M(n,\theta)]_{K\Gamma_n}$, $o \leq i \leq n-1$. It is clear that $m_{i+1} = 2, 4$ or 8. Now we can readily get the following.

3.6 <u>PROPOSITION</u> Suppose $p \geq 7$ and M is as above, then $[M : M(n,\theta)]_{K\Gamma_n} \neq 0$ if and only if the following conditions are satisfied:

(a) $M_i^{(p^i)} \otimes M_{i+1}^{(p^{i+1})}$ is of the following types:

(1) AA, BA, CB, DB, EA, E*A, EB, E*B with $m_{i+1} = 8$,

(2) AC, BC, CD, DD, EC, E*C, ED, E*D with $m_{i+1} = 2$.

(3) CE, DE, CE*, DE*, EE, EE*, E*E, E*E* with $m_{i+1} = 4$, where

$0 \leq i \leq n-1$, $M_n^{(p^n)} = M_o$.

(b) If $M_{i+1}, \ldots, M_{i+k}$, $1 \leq k \leq n-1$, are taken E or E*, then we must take M_i = C or D,

$-1 \leq i \leq n-2$, $M_{-1}^{(p^{-1})} = M_{n-1}^{(p^{n-1})}$, $M_{i+j}^{(p^{i+j})} = M_{i+j-n}^{(p^{i+j-n})}$, $1 \leq j \leq k$, $i + j \geq n$.

<u>PROOF</u> Notice that $M(1,o) \otimes M(o,1) \underset{G}{\longleftrightarrow} M(1,1) \oplus M(o,o)$, by (3.5) we have

$$AA \underset{G}{\longleftrightarrow} 8M(o,o)^{(p^i)} \otimes 8M(o,o)^{(p^{i+1})} \otimes M(o,o)^{(p^{i+2})}.$$

$$BA \underset{G}{\longleftrightarrow} 4M(p-2,p-2)^{(p^i)} \otimes 8M(o,o)^{(p^{i+1})} \otimes M(o,o)^{(p^{i+2})}.$$

$$CB \underset{G}{\longleftrightarrow} 2M(o,o)^{(p^i)} \otimes 8M(o,o)^{(p^{i+1})} \otimes M(o,o)^{(p^{i+2})} \oplus \ldots$$

$$DB \underset{G}{\longleftrightarrow} M(p-2,p-2)^{(p^i)} \otimes 8M(o,o)^{(p^{i+1})} \otimes M(o,o)^{(p^{i+2})} \oplus \ldots$$

$$AC \underset{G}{\longleftrightarrow} 8M(o,o)^{(p^i)} \otimes 2M(o,o)^{(p^{i+1})} \otimes M(1,1)^{(p^{i+2})} \oplus \ldots$$

$$BC \underset{G}{\longleftrightarrow} 4M(p-2,p-2)^{(p^i)} \otimes 2M(o,o)^{(p^{i+1})} \otimes M(1,1)^{(p^{i+2})} \oplus \ldots$$

$$CD \underset{G}{\longleftrightarrow} 2M(o,o)^{(p^i)} \otimes 2M(o,o)^{(p^{i+1})} \otimes M(1,1)^{(p^{i+2})} \oplus \ldots$$

$$DD \underset{G}{\longleftrightarrow} M(p-2,p-2)^{(p^i)} \otimes 2M(o,o)^{(p^{i+1})} \otimes M(1,1)^{(p^{i+2})} \oplus \ldots$$

$$CE \underset{G}{\longleftrightarrow} 2M(o,o)^{(p^i)} \otimes 4M(o,o)^{(p^{i+1})} \otimes (M(o,o) \oplus M(1,1))^{(p^{i+2})} \oplus \ldots$$

$$CE^* \underset{G}{\longleftrightarrow} 2M(o,o)^{(p^i)} \otimes 4M(o,o)^{(p^{i+1})} \otimes (M(o,o) \oplus M(1,1))^{(p^{i+2})} \oplus \ldots$$

$$DE \underset{G}{\longleftrightarrow} M(p-2,p-2)^{(p^i)} \otimes 4M(o,o)^{(p^{i+1})} \otimes (M(o,o) \oplus M(1,1))^{(p^{i+2})} \oplus \ldots$$

$$DE^* \underset{G}{\longleftrightarrow} M(p-2,p-2)^{(p^i)} \otimes 4M(o,o)^{(p^{i+1})} \otimes (M(o,o) \oplus M(1,1))^{(p^{i+2})} \oplus \ldots$$

By analysing carefully what G-composition factor of $Q(1,\theta)$ may occur as a factor of M, we deduce the proposition.

We can now construct a digraph G_2 with vertices A, B, C, D, E, E* and edges AA, BA, CB, DB, EB, EA, E*A, E*B (each has 8 parallel edges); AC, BC, CD, DD, EC, ED, E*C, E*D (each has 2 parallel edges) CE, CE*, DE, DE*, EE, EE*, E*E, E*E* (each has 4 parallel edges). Then we obtain the adjacency matrix of the digraph G_2

$$Y_2 = \begin{bmatrix} 8 & 0 & 2 & 0 & 0 & 0 \\ 8 & 0 & 2 & 0 & 0 & 0 \\ 0 & 8 & 0 & 2 & 4 & 4 \\ 0 & 8 & 0 & 2 & 4 & 4 \\ 8 & 8 & 2 & 2 & 4 & 4 \\ 8 & 8 & 2 & 2 & 4 & 4 \end{bmatrix}$$

By (3.6 (b), there is a subgraph G_3 in the digraph G_2 with vertices E, E* and edges EE, EE*, E*E, E*E* (each has 4 parallel edges). The adjacency matrix of the digraph G_3 is

$$Y_3 = \begin{bmatrix} 4 & 4 \\ 4 & 4 \end{bmatrix}$$

Since there is a one-one correspondence between the set of all $K\Gamma_n$-composition factors of $Q(n,\theta)$ which is isomorphic to $M(n,\theta)$ and the set of all closed walks of length n in the digraph G_2 but one in the digraph G_3, as the same reason in Section 2, we have

3.7 $[Q(n,\theta) : M(n,\theta)]_{K\Gamma n} = \mathrm{tr}(Y_2^n) - \mathrm{tr}(Y_3^n)$, where $\mathrm{tr}(Y_2^n)$ (resp. $\mathrm{tr}(Y_3^n)$) is the sum of the n-th powers of the eigenvalues of Y_2(resp. Y_3).

The eigenvalues of Y_2 are 0, 0, 0, 0, a, b, where a, b are two roots of $x^2 - 18x + 48=0$ and the eigenvalues of Y_3 are o, 8. Thus we have

3.8 $[Q(n,\theta) : M(n,\theta)]_{K\Gamma n} = a^n + b^n - 8^n$.

REMARK If p = 5, n>1, other four G-composition factors of $Q(1,\theta)$ will be involved in computations of $[Q(n,\theta) : M(n,\theta)]_{K\Gamma n}$, this makes computations of $[Q(n,\theta) : M(n,\theta)]_{K\Gamma n}$ to be more complicated. Here we do not consider this case.

4. COMPUTATIONS OF $\sum_{\mu\in\mathcal{M}} [Q(n,\mu) : M(n,\theta)]_{K\Gamma n}$

In this section we will compute $\sum_{\mu\in\mathcal{M}} [Q(n,\mu) : M(n,\theta)]_{K\Gamma n}$.

Let $\mu = \mu_o + p\mu_1 + \dots + p^{n-1}\mu_{n-1}$ be an arbitrarily fixed element of $\mathcal{M}$ throughout this section, by (2.6) we have a G-module isomorphism.

4.1 $Q(n,\mu) \simeq Q(1,\mu_o) \otimes Q(1,\mu_1)^{(p)} \otimes \dots \otimes Q(1,\mu_{n-1})^{(p^{n-1})}$.

Again by [9] § 1.5 Theorem 10, we have

4.2 $Q(1,\eta) \xleftrightarrow[G]{} F \oplus H \oplus 2M(o,p-2) \otimes M(o,1)^{(p)}$

$Q(1,\eta^*) \xleftrightarrow[G]{} F^* \oplus H^* \oplus 2M(p-2,o) \otimes M(1,o)^{(p)}$

$Q(1,\xi) \xleftrightarrow[G]{} L \oplus N \oplus 2M(1,p-3) \otimes M(o,1)^{(p)}$

$Q(1,\xi^*) \xleftrightarrow[G]{} L^* \oplus N^* \oplus 2M(p-3,1) \otimes M(1,o)^{(p)}$

where $F = 3M(p-1, o)$, $F^* = 3M(o,p-1)$,

$H = M(p-2,p-1) \otimes M(1,o)^{(p)}$, $H^* = M(p-1,p-2) \otimes M(o,1)^{(p)}$,

$L = 3M(p-1,1)$, $L^* = 3M(1,p-1)$,

$N = M(p-3,p-1) \otimes M(1,o)^{(p)}$, $N^* = M(p-1,p-3) \otimes M(o,1)^{(p)}$.

As the same discussion in Section 3, we obtain the following

4.3 <u>PROPOSITION</u> Suppose $p \geqslant 5$ and $M = M_o \otimes M_1^{(p)} \otimes \ldots \otimes M_{n-1}^{(p^{n-1})}$, where $M_i = M_{io} \otimes M_{i1}^{(p)}$ is a G-composition factor of $Q(1, \mu_i)$, $0 \leqslant i \leqslant n-1$. Then $[M : M(n, \theta)]_{K\Gamma_n} \neq 0$ if and only if the following conditions are satisfied.

(a) If (μ_i, μ_{i+1}) is of type (η, η). Then $M_i^{(p^i)} \otimes M_{i+1}^{(p^{i+1})}$ is of types HH, FH with m_{i+1} and HF, FF with $m_{i+1} = 3$;

(b) If (μ_i, μ_{i+1}) is of type (η, ξ^*). Then $M_i^{(p^i)} \otimes M_{i+1}^{(p^{i+1})}$ is of types HN*, FN* with m_{i+1} and HL*, FL* with $m_{i+1} = 3$;

(c) If (μ_i, μ_{i+1}) is of the type (ξ, η). Then $M_i^{(p^i)} \otimes M_{i+1}^{(p^{i+1})}$ is of types NH, LH with m_{i+1} and NF, LF with $m_{i+1} = 3$;

(d) If (μ_i, μ_{i+1}) is of the type (ξ, ξ^*). Then $M_i^{(p^i)} \otimes M_{i+1}^{(p^{i+1})}$ is of types NN*, LN* with $m_{i+1} = 1$ and NL*, LL* with $m_{i+1} = 3$;

(e) If (μ_i, μ_{i+1}) is of the type (η^*, η^*). Then $M_i^{(p^i)} \otimes M_{i+1}^{(p^{i+1})}$ is of types H*H*, F*H* with $m_{i+1} = 1$ and H*F*, F*F* with $m_{i+1} = 3$;

(f) If (μ_i, μ_{i+1}) is of the type (η^*, ξ). Then $M_i^{(p^i)} \otimes M_{i+1}^{(p^{i+1})}$ is of types H*N, F*N with m_{i+1} and H*L, F*L with $m_{i+1} = 3$;

(g) If (μ_i, μ_{i+1}) is of the type (ξ^*, η^*). Then $M_i^{(p^i)} \otimes M_{i+1}^{(p^{i+1})}$ is of types N*H*, L*H* with $m_{i+1} = 1$ and N*F*, L*F* with $m_{i+1} = 3$;

(h) If (μ_i, μ_{i+1}) is of the type (ξ^*, ξ). Then $M_i^{(p^i)} \otimes M_{i+1}^{(p^{i+1})}$ is of types N*N, L*N with $m_{i+1} = 1$ and N*L, L*L with $m_{i+1} = 3$;

(i) M_o, M_1, ... , M_{n-1} can not be simultaneously taken only F, F*, L, L*, where $o \leqslant i \leqslant n-1$, $M_n^{(p^n)} = M_o$, $m_n = m_o$.

We then construct a digraph G_4 with vertices F, F*, H, H*, L, L*, N, N*, and edges HH, FH, HN*, FN*, NH, LH, NN*, LN*, H*H*, F*H*, H*N, F*N, N*H*, L*H*, N*N, L*N; and HF, FF, HL*, FL*, NF, LF, NL*, LL*, H*F*, F*F*, H*L, F*L, N*F*, L*F*, N*L, L*L (each has 3 parallel edges). Therefore the adjacency matrix of the digraph G_4 is

$$Y_4 = \begin{bmatrix} 3 & 0 & 1 & 0 & 0 & 3 & 0 & 1 \\ 0 & 3 & 0 & 1 & 3 & 0 & 1 & 0 \\ 3 & 0 & 1 & 0 & 0 & 3 & 0 & 1 \\ 0 & 3 & 0 & 1 & 3 & 0 & 1 & 0 \\ 3 & 0 & 1 & 0 & 0 & 3 & 0 & 1 \\ 0 & 3 & 0 & 1 & 3 & 0 & 1 & 0 \\ 3 & 0 & 1 & 0 & 0 & 3 & 0 & 1 \\ 0 & 3 & 0 & 1 & 3 & 0 & 1 & 0 \end{bmatrix}$$

By (4.3(i)), there is a subgraph G_5 in the digraph G_4 with vertices F, F^*, L, L^* and edges FF, F^*F^*, FL^*, F^*L, LF, LL^*, L^*F^*, L^*L (each has 3 parallel edges). The adjacency matrix of the degraph G_5 is

$$Y_5 = \begin{bmatrix} 3 & 0 & 0 & 3 \\ 0 & 3 & 3 & 0 \\ 3 & 0 & 0 & 3 \\ 0 & 3 & 3 & 0 \end{bmatrix}$$

As the same reason in section 3, we have

4.4 $\sum_{\mu\in\mathcal{M}} [Q(n,\mu) : M(n,\theta)]_{K\Gamma_n} = tr(Y_4^n) - tr(Y_5^n)$, where $tr(Y_4^n)$ (resp. $tr(Y_5^n)$) is the sum of the n-th powers of the eigenvalues of the eigenvalues of Y_4 (resp. Y_5).

The eigenvalues of Y_4 are 0, 0, 0, 0, 0, 0, 0, 8 and the eigenvalues of Y_5 are 0, 0, 0, 6. Thus we have

4.5 $\sum_{\mu\in\mathcal{M}} [Q(n,\mu) : M(n,\theta)]_{K\Gamma_n} = 8^n - 6^n.$

Combining (2.8), (3.8) and (4.5), we have proved

4.6 <u>THEOREM</u> $c^{(n)} = [R(n,\theta) : M(n,\theta)]_{K\Gamma_n}$

$= [Q(n,\theta) : M(n,\theta)]_{K\Gamma_n} - \sum_{\mu\in\mathcal{M}} [Q(n,\mu) : M(n,\theta)]_{K\Gamma_n}$

$= a^n + b^n + 6^n - 2\cdot 8^n$.

where a, b are two roots of $x^2 - 18x + 48 = 0$.

5. <u>COMPUTATIONS OF THE FIRST CARTAN INVARIANT OF THE GROUP</u> $SU(3,p^n)$

In this section we will compute the first Cartan invariant of the group $SU(3,p^n)$. From [8] § 1.3 we have isomorphisms of $K\Gamma_n'$-modules:

5.1 $M(\lambda)^{(p^n)} \simeq M(\lambda^*)$ for any $\lambda \in X^+$

and $M(\lambda + p^n\nu) \simeq M'(n,\lambda) \otimes M(\nu^*)$ for any $\lambda \in X_n$, $\nu \in X^+$.

Let $\lambda = \lambda_0 + p\lambda_1 + \ldots + p^{n-1}\lambda_{n-1}$ and $\mu = \mu_0 + p\mu_1 + \ldots + p^{n-1}\mu_{n-1}$ be the p-adic expressions of λ and μ.

Following Jantzen's formula (cf. [8] § 2.10) we can modify (2.1) as follows:

5.2 $$[Q(n,\lambda) : R(n,\mu)]_{K\Gamma_n'}$$

$$= \sum_{\substack{\nu_i \in X^+ \\ \nu_n = \nu_o^*}} \prod_{i=o}^{n-1} [M(\mu_i) \otimes M(\nu_i) : M(\lambda_i + p\nu_{i+1})]_G$$

In order to compute $[Q(n,\theta) : R'(n,\mu)]_{K\Gamma_n'}$ we have to replace $\mu_n = \mu_o$ by $\mu_n = \mu_o^*$ in (2.3) as $\nu_n = \nu_o^*$, the set $\mathcal{M}$ by the set $\mathcal{M}'$, and all closed walks of length n in the digraph G_1 by all walks of length n from η to η^*. η^* to η, ξ to ξ^*, ξ^* to ξ in the digraph G_1.

Notice that $\dim Q(n,\mu) = 3^n p^{3n}$ for all $\mu \in \mathcal{M}'$, we get $\sum_{\mu\in\mathcal{M}'} \dim Q(n,\mu) = 6^n p^{3n}$.

Similarly, we have the following

5.3 $$Q(n.\theta) \xleftrightarrow[K\Gamma_n]{} R'(n,\theta) \oplus St_n \oplus \sum_{\mu\in\mathcal{M}'} R'(n,\mu)$$

$$Q(n,\mu) \xleftrightarrow[K\Gamma_n]{} R'(n,\mu) \quad \text{for all } \mu \in \mathcal{M}'.$$

5.4 $\dim R'(n,\theta) = (12^n - 6^n - 1)\, p^{3n}$, where $p \geq 3$.

Moreover, (3.1) can be modified as follows:

5.5 $$M \underset{G}{\simeq} M_{oo} \otimes (M_{1o} \otimes M_{o1})^{(p)} \otimes \dots \otimes (M_{n-1o} \otimes M_{n-2\,1})^{(p^{n-1})} \otimes M_{n-1\,1}^{(p^n)}$$

$$\underset{K\Gamma_n'}{\simeq} (M_{oo} \otimes M_{n-1 1}^*) \otimes (M_{1o} \otimes M_{o1})^{(p)} \otimes \dots \otimes (M_{n-1o} \otimes M_{n-2\,1})^{(p^{n-1})}$$

We then have

$$[M : M'(n,\theta)]_{K\Gamma n}$$

$$= [M_{oo} \otimes M_{n-11}^* : M'(1,\theta)]_{K\Gamma_1'} [M_{1o} \otimes M_{o1} : M'(1,\theta)]_{K\Gamma_1'} \cdots [M_{n-1o} \otimes M_{n-2\,1} : M'(1,\theta)]_{K\Gamma_1'}$$

$$= [M_{oo}^* \otimes M_{n-1\,1} : M'(1,\theta)]_{K\Gamma_1'} [M_{1o} \otimes M_{o1} : M'(1,\theta)]_{K\Gamma_1'} \cdots [M_{n-1o} \otimes M_{n-2\,1} : M'(1,\theta)]_{K\Gamma_1'}.$$

In order to compute $[Q(n,\theta) : M'(n,\theta)]_{K\Gamma_n'}$ and $\sum_{\mu\in\mathcal{M}'} [Q(n,\mu) : M'(n,\theta)]_{K\Gamma_n'}$ for all $\mu \in \mathcal{M}'$, we have to replace $M_n^{(p^n)} = M_o$ by $M_n^{(p^n)} = M_o^*$ in (3.6) and (4.3), and all closed walks of length n in the digraphes G_2, G_3, G_4, G_5, respectively, by all walks of length n from V to V^* in the digraphes G_2, G_3, G_4, G_5, respectively, where V ranges over all vertices of the corresponding digraph.

Notice the following facts:

(1) $A^* = A$, $B^* = B$, $C^* = C$, $D^* = D$;

(2) Let $Y_2^n = (A_{ij})_{1 \leq i,j \leq 6}$. Then $A_{55} = A_{56}$, $A_{65} = A_{66}$ as the 5-th column and the

5-th row are the same as the 6-th column and the 6-th row, respectively. Thus we have $tr(Y_2^n) = A_{11} + A_{22} + A_{33} + A_{44} + A_{56} + A_{65}$.

(3)

$$Y_1^n = 2^{n-2} \cdot 3^n \begin{bmatrix} 1 & 1 & 1 & 1 \\ 1 & 1 & 1 & 1 \\ 1 & 1 & 1 & 1 \\ 1 & 1 & 1 & 1 \end{bmatrix}$$

$$Y_3^n = 2^{3n-1} \begin{bmatrix} 1 & 1 \\ 1 & 1 \end{bmatrix}$$

$$Y_4^n = 2^{3n-4} \begin{bmatrix} 3 & 3 & 1 & 1 & 3 & 3 & 1 & 1 \\ 3 & 3 & 1 & 1 & 3 & 3 & 1 & 1 \\ 3 & 3 & 1 & 1 & 3 & 3 & 1 & 1 \\ 3 & 3 & 1 & 1 & 3 & 3 & 1 & 1 \\ 3 & 3 & 1 & 1 & 3 & 3 & 1 & 1 \\ 3 & 3 & 1 & 1 & 3 & 3 & 1 & 1 \\ 3 & 3 & 1 & 1 & 3 & 3 & 1 & 1 \\ 3 & 3 & 1 & 1 & 3 & 3 & 1 & 1 \end{bmatrix}$$

$$Y_5^n = 3^n \cdot 2^{n-1} \begin{bmatrix} 1 & 1 & 1 & 1 \\ 1 & 1 & 1 & 1 \\ 1 & 1 & 1 & 1 \\ 1 & 1 & 1 & 1 \end{bmatrix}$$

We can now get

5.6 $$[Q(n,\theta) : M'(n,\theta)]_{K\Gamma'_n} = [Q(n,\theta) : M(n,\theta)]_{K\Gamma_n}$$

and $$\sum_{\mu\in\mathcal{M}'} [Q(n,\mu) : M'(n,\theta)]_{K\Gamma'_n} = \sum_{\mu\in\mathcal{M}} [Q(n,\mu) : M(n,\theta)]_{K\Gamma_n}.$$

Combining (5.3), (5.6) and (4.6) we complete the proof of the main theorem. Finally, we make some remarks to end this paper. If $p \geq 5$ and $n = 1$, then $C^{(1)} = 8$ as in [3] § 2 Theorem. There are 8 distinct ordinary characters involved in the projective indecomposable $K\Gamma_1$-module $R(1,\theta)$. Moreover, $R(1,\theta)$ involves 16 distinct $K\Gamma_1$-composition factors and 47 $K\Gamma_1$-composition factors in all.

REFERENCES

1 Cheng, Y., On the First Cartan Invariants in Characteristic 2 of the Group $SL_3(2^m)$ and $SU_3(2^m)$, J. Algebra 82 (1983), 194-244.

2 Chastkofsky, L., Projective Characters for Finite Chevalley Groups, J. Algebra 69 (1981), 347-357.

3 Humphreys, J.E., Modular Representations of Finite Groups of Lie Type, in "Finite Simple Groups", Academic Press, London (1980), 259-290.

4 Humphreys, J.E., Ordinary and Modular Characters of SL(3,p), J. Algebra, 72 (1981), 8-16.

5 Humphreys, J.E., Cartan Invariants (preprint).

6 Harary, F., Graph Theory, Addison-Wesley (1969).

7 Jantzen, J.C., Über Decompositionsverhalten gewisser modularer Darstellungen halbeinfacher Gruppen und ihrer Lie-Algebren, J. Algebra 49 (1977), 441-469.

8 Jantzen, J.C., Zur Reduction modulo p der Charaktere von Deligne und Lusztig, J. Algebra 70 (1981), 452-474.

9 Ye Jia-chen, The Cartan Invariants of $SL(3,p^n)$, J. Math. Research and Exposition Vol. 2, No.4 (1982), 9-19.

APPENDIX

List of other Chinese participants and titles of their lectures

Cao Chongguang (Heilongjiang University, Heilongjiang)
On the normal subgroups of the symplectic groups over a ring with one in its stable range.

Cao Xihua (East China Normal University, Shanghai)

Chen Chengdong (Tongji University, Shanghai)
On the structure of the submodules of the PIM's for SL(2,p) and their invariants.

Chen Zhonghu (Xiangtan University, Hunan)
On the construction of simple groups over real and p-adic fields.

Chen Zhongmu (The Southwest Teachers College, Sichuan)
Outer-Σ groups of finite order.

Du Jie (East China Normal University, Shanghai)
Tensor products of certain Weyl modules and computations of Ext^1 for groups of type A_ℓ.

Duan Xuefu (Tuan Hsio-Fu) (Peking University, Beijing)
Some recent works on finite group theory by my colleagues and graduate students.

Fan Yun (Wuhan University, Hubei)
On groups of odd order and rank 2.

Hou Zixin, Zhang Zhixue (Nankai University, Tianjing, Hebei University, Hebei)
Structures of Weyl groups of real semi-simple Lie algebras.

Jing Naihuan (Wuhan Teachers College, Hubei)
The determination of order n of which every group is always a $\mathcal{Y}$-group.

Li Jiongsheng (China University of Science and Technology, Anhui)
On coverings of a finite group by abelian subgroups.

Li Shang-zhi, Zha Jian-guo (China Univ. of Science and Technology, Anhui)
Certain classes of maximal subgroups in classical groups.
(see: Li Shang-zhi, Maximal subgroups containing root subgroups in finite classical groups, Kexue Tongbao, 28 (1983), 257-260 (Chinese Edition), 29 (1984) 14-18 (English Edition).
and: Li Shang-zhi, Maximal subgroups in $P\Omega(n,F,Q)$ containing root subgroups, Scientia Sinica (Series A), 1985, 3: 193-205 (Chinese Edition), 28 (1985), 48-60 (English Edition).

Li Shirong, Li Shiyu (Guangxi University, Guangxi)
Finite groups in which every non-maximal subgroup is 3-closed.

Li Zunxian (Shanxi Normal University, Shanxi)
Amalgams of several groups and their cohomology.

Liu Changkun (Tongji University, Shanghai)
The irreducible representations of affine hyperalgebra and its subalgebra U_n.

Meng Daoji (Nankai University, Tianjing)
On complete Lie algebras.

Qiu Sen (East China Normal University, Shanghai)
Ext* for the Lie algebras of simple algebraic groups.

Shen Guangyu (East China Normal University, Shanghai)
Graded modules of graded Lie algebras.

Shi He (Institute of System Science, Academia Sinica, Beijing)
On the characters of symmetric groups.

Shi Wujie (The Southwest Teachers College, Sichuan)
A new characterization of some finite simple groups.

Tang Shouwen, Hong Jiawei (Beijing Computer College, Beijing)
On decision problems of groups, rings and fields.

Tang Xiangpu, Lin Zhongzhu (Harbin Shipbuilding College, Heilongjiang)
The isomorphisms of linear groups over $X_0 - \Phi$-surjective rings.

Wan Zhexian (Institute of System Science, Academia Sinica, Beijing)
Number of cycles of short length in de Bruijn-Good graph.

Wang Jianpan (East China Normal University, Shanghai)
The inverse image of an induced sheaf.

Wang Luqun (Heilongjiang University, Heilongjiang)
Isomorphism of two dimensional linear groups over ϕ-surjective rings.

Wang Yangxian, Wang Chunsen (Hebei Teachers College, Hebei)
Generators and related length problem of pseudo-orthogonal groups.

Xu Mingyao (Peking University, Beijing)
On the solvability of finite groups with at most two conjugate classes of maximal subgroups.

Xu Yichao (Institute of Mathematics, Academia Sinica, Beijing)
Classification of homogeneous Kählerian manifolds acted by reducible Lie groups.

Yan Zhida (Nankai University, Tianjing)

You Hong (Northeast Normal University, Jilin)
Generation theorems of general linear groups over a kind of non-commutative rings.

Zeng Kencheng (Graduate School of China University of Science and Technology, Beijing)

Zhai Qibin (Zhengzhou Institute of Engineering and Technology, Henan)
On a class of finite solvable complete groups.

Zhang Guangxiang (The Southwest Teachers College, Sichuan)

On two theorems of Thompson.

Zhang Haiquan (Northeast Normal University, Jilin)

Normal subgroups of symplectic groups over ϕ-surjective rings.

Zhang Jiping (Peking University, Beijing)

Influence of S-quasinormal condition on almost minimal subgroups of finite groups.

Zhang Laiwu (Peking University, Beijing)

The outer structure of formation and its applications.

Zhang Yongzheng (Northeast Normal University, Jilin)

Automorphisms of two-dimensional linear groups over ϕ-surjective rings.

Zhang Yuanda (Wuhan University, Hubei)

Zheng Yanlu (Wuhan University, Hubei)

Π-properties of Π-solvable groups.

Zou Yiming (Nakai University, Tianjing)

Class 1 representations and minimal immersions of rank one symmetric space.